ADVANCES IN PROTEIN CHEMISTRY

Volume 68

Cell Surface Receptors

ADVANCES IN PROTEIN CHEMISTRY

EDITED BY

FREDERIC M. RICHARDS
Department of Molecular Biophysics
and Biochemistry, Yale University
New Haven, Connecticut

DAVID S. EISENBERG
Department of Chemistry and Biochemistry
Center for Genomics and Proteomics
University of California, Los Angeles
Los Angeles, California

JOHN KURIYAN
Department of Molecular and Cellular Biology
University of California, Berkeley
Berkeley, California

VOLUME 68

Cell Surface Receptors

EDITED BY

K. Christopher Garcia
Stanford University School of Medicine
Stanford, California

ELSEVIER
ACADEMIC
PRESS

AMSTERDAM • BOSTON • HEIDELBERG • LONDON
NEW YORK • OXFORD • PARIS • SAN DIEGO
SAN FRANCISCO • SINGAPORE • SYDNEY • TOKYO

Elsevier Academic Press
525 B Street, Suite 1900, San Diego, California 92101-4495, USA
84 Theobald's Road, London WC1X 8RR, UK

This book is printed on acid-free paper. ♾

For all information on all Academic Press publications
visit our Web site at www.academicpress.com

ISBN: 0-12-034268-5

Printed and bound by CPI Group (UK) Ltd, Croydon, CR0 4YY

Transferred to Digital Print 2011

CONTENTS

Shared Cytokine Signaling Receptors: Structural Insights from the gp130 System

Martin J. Boulanger and K. Christopher Garcia

The Structural Basis for Biological Signaling, Regulation, and Specificity in the Growth Hormone–Prolactin System of Hormones and Receptors

Anthony A. Kossiakoff

Structural Analysis of IL-10 and Type I Interferon Family Members and Their Complexes with Receptor

Mark R. Walter

Assembly of Post-Receptor Singnaling Complexes for the Tumor Necrosis Factor Receptor Superfamily

Hao Wu

NKG2D and Related Immunoreceptors

Roland K. Strong and Benjamin J. McFarland

Ionotropic Glutamate Receptor Recognition and Activation

Robert E. Oswald

Chemokine-Receptor Interactions: GPCRs, Glycosaminoglycans and Viral Chemokine Binding Proteins

ELAINE K. LAU, SAMANTHA ALLEN, ANDRO R. HSU, AND TRACY M. HANDEL

Chemotaxis Receptors and Signaling

AARON F. MILLER AND JOSEPH J. FALKE

PREFACE

Receptors are the gateway through which our cells sense and respond to their changing environments. Cell surface receptors have the unique capacity to both engage exracellular ligands, as well as relay this information through the plasma membrane to intracellular components, and beyond. Some receptors, such as growth factor and cytokine receptors, contain only a single-transmembrane segment, connecting modular extracellular and intracellular domains that can be studied as soluble fragments retaining full ligand-binding activity. Other receptors, such as G-protein coupled receptors are multi-pass transmembrane proteins requiring a lipid bilayer for functional reconstitution. Others systems, exemplified here by the glutamate receptor, are combinations of autonomous ligand binding extracellular domains connected to multiple transmembrane helix polypeptide cores. Regardless of structural classification, the fundamental question persists, how is ligand engagement outside the cell coupled to receptor activation within the cell? This question remains elusive even for the most intensively studied systems, and remains one of the most challenging problems in biochemistry. Structural biology has made great strides in understanding this question through the determination of structures of receptors and their complexes with ligands. Ultimately a clear elucidation of a detailed mechanistic picture of receptor activation will require a combination of static biophysical methods such as x-ray crystallography, with dynamic methods such as NMR and single-molecule spectroscopy, which together can be dovetailed with functional studies.

While the structural database now has many examples of protein-protein complexes which have taught us about the first principles of molecular recognition, receptors remain somewhat enigmatic in that they are protein machines, rather than simple binding proteins. The chemistry of receptor-ligand recognition is part of an overall process by which ligand orients, or otherwise perturbs the structure of a receptor in such a way that the intracellular adaptor proteins initiate signaling cascades. The basis by which this phenomenon occurs appears far more complex than we originally thought based on simple models of receptor homodimerization. Not so long ago there was a common assumption that simply bringing receptors together is all one needs for activation. More recently, it is being appreciated that subtle orientational differences in the extracellular domains of dimerized receptors can translate into very significant downstream signaling differences. Thus, the transmembrane helices may not simply be loose tethers between the extracellular and intracellular domains. Rather, in a number of receptors systems such as Erythropoetin, the connecting peptides and transmembrane helices appear capable of transmitting torque though the lipid bilayer to effect orientational strain in the intracellular parts of the receptors. It is a fact that most cell surface receptors are oligomerized in some fashion by their ligands, but the geometric and conformational details of this clustering are critical to the appropriate signals being transduced. Examples exist now of ligands inducing large-scale conformational changes in preformed receptors dimers, as well as receptor activation through disruption of dimerization. Most recently, GPCR are now being shown to require some form of ligand-induced dimerization, in concert with conformational change of the helices, for activation. Given the technical challenges inherent in studying receptors at a biophysical level, it is likely we have currently seen only a small fraction of the universe of potential mechanisms for receptor activation.

This issue of Advances in Protein Chemistry presents detailed chapters on several important receptor systems, with an emphasis on relating the structural basis of extracellular ligand recognition to the activation of intracellular signaling events. Some chapters focus on structural aspects of ligand recognition, while others are more focused on mechanistic questions. However in all cases there is an attempt to paint a structural portrait for how ligand engagement may activate the receptor. We have emphasized systems for which a significant amount of structural information is known on either the extracellular or intracellular regions of the receptors. In most cases the chemistry of ligand recognition relates in subtle, yet still unclear ways to receptor activation. The receptors discussed in this edition range from type-I receptors for growth factors hormones, and immunoregulatory ligands (Leahy, Nikolov, Garcia, Kossiakoff, Walter, Wu, Springer, Strong) to multi-pass integral membrane proteins (Oswald, Falke, Handel). For several of these systems structures are known for both extracellular domains and intracellular domains, allowing models to be constructed for the entire receptor. Clearly, these all-encompassing models are an important future direction for the field of receptor structural biology.

I wish to thank all of the authors who contributed to this edition, and to the editorial staff and senior editors for making this a pleasant experience for all involved.

K. Christopher Garcia
Stanford University School of Medicine

STRUCTURE AND FUNCTION OF THE EPIDERMAL GROWTH FACTOR (EGF/ErbB) FAMILY OF RECEPTORS

By DANIEL J. LEAHY

Department of Biophysics & Biophysical Chemistry and HHMI,
Johns Hopkins University School of Medicine,
Baltimore, Maryland 21205

I. Introduction

Epidermal Growth Factor (EGF) was among the first growth factors discovered, and study of EGF and its receptors has established many paradigms for growth factor–mediated signaling (Carpenter, 1987; Cohen, 1986, 1987; Schlessinger, 2000; Yarden and Sliwkowski, 2001). Initially isolated from the mouse submaxillary gland based on its ability to stimulate premature eye opening and tooth eruption in neonatal mice, EGF is a 53 amino-acid polypeptide derived by proteolysis from a larger precursor (Carpenter and Cohen, 1979; Cohen, 1986). The ability of EGF to stimulate growth and differentiation of epidermal and mesodermal tissues led both to its name and to keen interest in its mode of action.

Following its isolation, EGF was shown to bind with high affinity to a specific receptor in the cell membrane and stimulate rapid activation of a protein kinase activity (Carpenter *et al.*, 1975, 1978, 1979; Das *et al.*, 1977; Hock *et al.*, 1979; Wrann *et al.*, 1979). Purification and characterization of the EGF receptor (EGFR) showed it to be a ~170 kD molecular weight integral membrane glycoprotein (Cohen *et al.*, 1982). The ligand-inducible kinase activity co-purified with EGFR, suggesting a physical linkage between the ligand binding and kinase activities, which was later verified by molecular cloning (Cohen *et al.*, 1980; Ullrich *et al.*, 1984). Early on, the EGFR kinase activity was shown to result in phosphorylation of tyrosine residues, the first such demonstration for any receptor (Ushiro and Cohen, 1980).

ADVANCES IN PROTEIN CHEMISTRY, Vol. 68

Molecular cloning of EGFR revealed it to be a 1186 amino acid protein with a 621 amino acid extracellular region followed by a single membrane-spanning region and a cytoplasmic tyrosine kinase (Ullrich *et al.*, 1984). Despite nonhomologous ligand binding regions, this overall architecture is shared by many other receptors, including those for insulin, PDGF, FGF, and VEGF, and these receptors are now collectively known as *receptor tyrosine kinases* (RTKs) (Schlessinger, 2000). The extracellular ligand-binding region of EGFR is made up of four subdomains arranged as a tandem repeat of two types of domains (Fig. 1). The first and third domains are homologous to one another and have been designated domains I and III or L1 and L2, respectively; the second and fourth domains are also homologous to one another and have been designated domains II and IV or CR1 and CR2, respectively (Bajaj *et al.*, 1987; Lax *et al.*, 1988; Ward *et al.*, 1995). The CR in this case is short for "cysteine-rich" and reflects the fact that nearly 50 conserved cysteines are found in these two domains. For simplicity, this review will employ the I, II, III, and IV domain nomenclature. Studies with mutant and chimeric EGF receptors and receptor fragments demonstrated that ligand binding is mediated primarily by domain III with some contribution from regions on domain I (Kohda *et al.*, 1993; Lax *et al.*, 1989). Curiously, the presence of domain IV was shown to be slightly inhibitory to ligand binding (Elleman *et al.*, 2001).

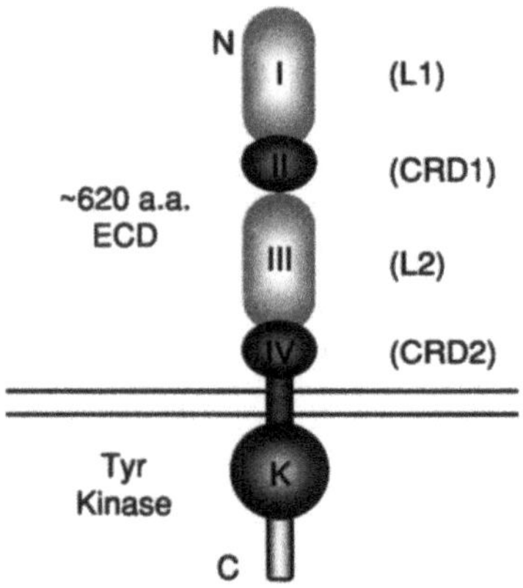

FIG. 1. Domain architecture of ErbB receptors. ErbB receptor extracellular regions are composed of four subdomains arranged as a tandem repeat of two types of domains. Two domain nomenclatures have been proposed (Bajaj *et al.*, 1987; Lax *et al.*, 1988; Ward *et al.*, 1995). The domains in order from the N-terminus are referred to as domain I (L1), II (CR1), III (L2), and IV (CR2). Domains I and III are homologous; domains III and IV are homologous. The extracellular region is followed by a single membrane-spanning region, a cytoplasmic tyrosine kinase, and variable length tail that harbors several phosphorylation sites.

Once the topology and functional organization of EGFR became apparent, questions about receptor activation focused on how extracellular ligand binding activates the intracellular kinase. Early studies with fluorescent-labeled EGF showed aggregation of EGFR on the cell surface in response to ligand binding (Schechter *et al.*, 1979), implicating receptor cross-linking as an activation mechanism. EGFR is endocytosed following ligand binding, however, and it was difficult to distinguish between aggregation as the trigger of signaling as opposed to a downstream response to receptor activation (Haigler *et al.*, 1979). A key piece of the puzzle emerged with the observation of ligand-induced dimers of EGFR (Yarden and Schlessinger, 1987), which was the first indication that dimerization might play a role in signaling for any receptor (Heldin, 1995). Ligand-induced dimerization—more broadly induction of a specific oligomeric conformation by ligand binding—is now accepted as the signaling trigger for all RTKs (Heldin, 1995; Schlessinger, 2000), as memorably illustrated by the crystal structure of the complex of human growth hormone with two of its receptors (de Vos *et al.*, 1992).

Prior to molecular cloning, amino-acid sequence data from EGFR revealed that the ErbB oncogene of the avian erythroblastosis virus encodes a truncated form of EGFR (Downward *et al.*, 1984). This truncated form is missing most of the extracellular region but includes a constitutively active kinase region that is responsible for unregulated growth of infected cells (Frykberg *et al.*, 1983; Yamamoto *et al.*, 1983). Demonstration that an oncogene encoded an activated form of a growth factor receptor provided exciting insight into the origins of cancer and presaged discovery of the involvement of EGFR and related receptors in the genesis and severity of many human cancers (Blume-Jensen and Hunter, 2001; Holbro *et al.*, 2003; Tang and Lippman, 1998). The nature of the ErbB oncogene also indicated that the extracellular region not only mediates ligand-dependent activation but also contributes to maintaining the kinase in an inactive state in the absence of ligand.

II. The EGF and EGFR Families

Both EGF and EGFR are archetypes of protein families that have undergone duplication and diversification throughout animal evolution (Muller and Schacke, 1996; Stein and Staros, 2000). *C. elegans* utilizes a single homolog of both EGFR (Let-23) and EGF (Lin-3), *Drosophila* utilizes a single EGFR (DER) and four EGF homologs (Vein, Spitz, Gurken, and Argos), and humans utilize four EGFR and at least 11 EGF homologs (Stein and Staros, 2000). The four human EGFR homologs are known as both the HER (HER1, HER2, HER3, and HER4 for Human EGF Receptor) and ErbB (ErbB1, ErbB2, ErbB3, and ErbB4) families. EGFR is still commonly used to indicate HER1/ErbB1, and HER2 (ErbB2) is the cellular form of the *neu* oncogene product (Bargmann *et al.*, 1986b). This review will refer to the family of receptors as ErbB receptors and the individual receptors as EGFR, HER2, HER3, or HER4. The soluble extracellular regions of these receptors will be referred to as sEGFR, sHER2, sHER3, and sHER4.

EGF-related ligands include EGF, transforming growth factor-α (TGFα), heparin-binding EGF-like growth factor (HB-EGF), amphiregulin, betacellulin, and several isoforms of heregulin/neuregulin (Yarden and Sliwkowski, 2001). These ligands, which are derived by proteolysis from divergent precursors, are typically ~55 amino acids in length (Stein and Staros, 2000) and are characterized by a conserved pattern of 3 disulfide bonds and a loop-rich structure. EGF-related ligands may be categorized into the neuregulins, which primarily activate HER3 and/or HER4, and all others, which primarily activate EGFR (Fig. 2). Some cross-reaction between these ligand and receptor classes does occur, however. The sequences of EGF family ligands can be quite diverse; EGF and TGFα both bind and activate EGFR but share only 40% sequence identity.

The four human EGFR homologs share 40–45% sequence identity (Stein and Staros, 2000) but have become functionally specialized (Carpenter, 1987, 2003; Citri *et al.*, 2003; Olayioye *et al.*, 2000; Yarden and Sliwkowski, 2001). Each ErbB receptor has a ~220 amino-acid region following the kinase region that harbors multiple phosphorylation sites that become modified when the receptor is activated. Phosphorylation of the different sites recruits binding of different activators and initiates a characteristic pattern of downstream signaling events (Yarden and Sliwkowski, 2001). Expression patterns for each ErbB receptor and ligand are different and reflect involvement of each receptor in mediating growth and differentiation of diverse cell and tissue types (Olayioye *et al.*, 2000). Knockout of any of the ErbB receptors is embryonic lethal in the mouse with overlapping defects. Affected tissues include brain, skin, lung, and gastrointestinal tract (ErbB1) (Miettinen *et al.*, 1995; Sibilia and Wagner, 1995; Threadgill *et al.*, 1995), heart and brain (ErbB2 and ErbB3) (Lee *et al.*, 1995; Morris *et al.*, 1999; Riethmacher *et al.*, 1997), and heart (ErbB4) (Gassmann *et al.*, 1995).

Much of the phenotypic overlap observed in ErbB knockout mice appears to arise from the importance of heteromeric receptor combinations in mediating growth and development of specific tissues. In particular, HER2 and HER3 are unable to signal by themselves and must pair with another ErbB receptor to generate a signal. Despite much effort, no high-affinity

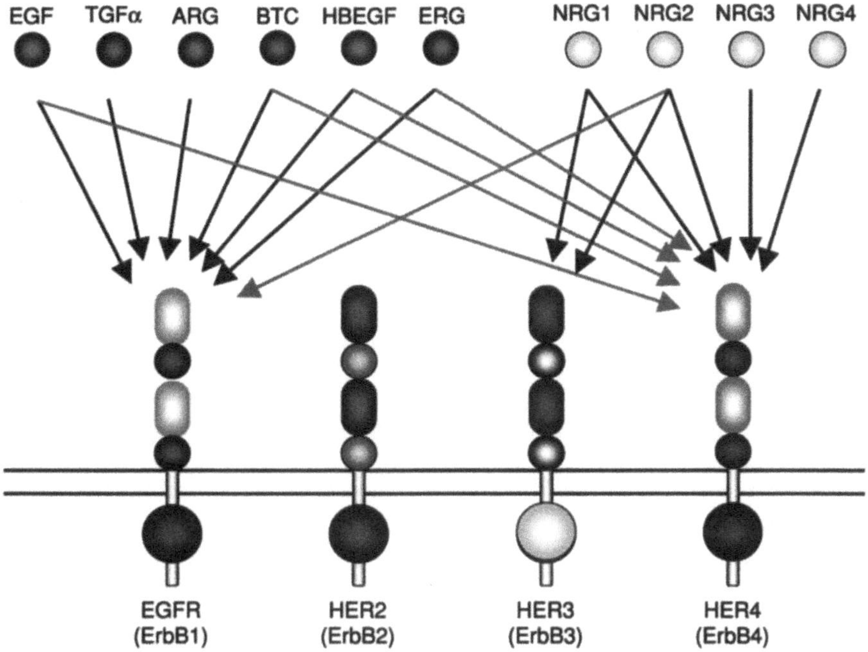

FIG. 2. Cognate ligand/receptor pairs. EGF-like ligands that activate ErbB receptors include Transforming Growth Factor α (TGFα), Amphiregulin (Arg), Betacellulin (Btc), Heparin-binding EGF (HBEGF), Epiregulin (Erg), and Neuregulin (Nrg, also known as Heregulin) (Yarden and Sliwkowski, 2001). Ligand-receptor interactions that induce strong responses are indicated by black arrows. Interactions that induce moderate responses are indicated by gray arrows (Klapper *et al.*, 2000).

ligand for HER2 has been identified, and all HER2-mediated signaling appears secondary to activation of other ErbB receptors (Klapper *et al.*, 1999). Indeed, all other ErbB receptors are able to form heteromeric signaling complexes with HER2 when co-expressed on the same cell. HER3, on the other hand, lacks an active kinase domain owing to mutation of several residues required for kinase activity (Guy *et al.*, 1994). Without an active kinase, HER3 must pair with and activate another ErbB receptor to generate a signal. The preferred partner of HER3—and indeed of EGFR and HER4—appears to be HER2, and the HER2/HER3 pair generates the strongest proliferative signal in many assays (Citri *et al.*, 2003). Because of the absence of a HER2 ligand and the inability of HER3 to activate an intrinsic kinase activity in response to ligand, these receptors have been referred to as the "deaf and the dumb" of the ErbB receptors (Citri *et al.*, 2003).

HER4 is unique among ErbB receptors in having been shown to undergo multiple proteolytic cleavages followed by translocation of its cytoplasmic region to the cell nucleus (Carpenter, 2003). The physiological significance of this processing is as yet poorly understood but raises interesting new possibilities for the generation and regulation of HER4-mediated signals.

III. Association of ErbB Receptors with Human Disease

EGFR became the first cell-surface receptor linked to cancer when Cohen and colleagues demonstrated downregulation of EGFR following transformation of cultured cells with specific oncogenic viruses (Todaro *et al.*, 1976). The significance of this association was not immediately clear, but discovery that the ErbB oncogene encoded an activated form of the EGFR kinase established a clear link between inappropriate EGFR activity and cancer (Downward *et al.*, 1984). In the 20 years since this discovery, dysregulated forms of ErbB receptors have been found to contribute to the genesis and severity of many human cancers (Blume-Jensen and Hunter, 2001; Holbro *et al.*, 2003; Tang and Lippman, 1998). ErbB dysregulation has been shown to occur through mutation (Humphrey *et al.*, 1990), overexpression of the receptor (Arteaga, 2002), or as a secondary effect downstream of inappropriate ligand expression (Sizeland and Burgess, 1992).

HER2 presents a particularly instructive example of growth factor receptor involvement in cancer. Molecular cloning of HER2 was nearly contemporaneous with its identification as the normal counterpart of the *neu* oncogene (Bargmann *et al.*, 1986b; Yamamoto *et al.*, 1986). The oncogene *neu* is responsible for the appearance of mutagen-induced neuroblastomas in rats and encodes a mutant form of the HER2 receptor, which frequently contains a single valine to glutamate change in the membrane-spanning region (Bargmann *et al.*, 1986a). This mutation appears to cause association and activation of the mutant receptors, which in turn leads to uncontrolled cell growth (Bargmann and Weinberg, 1988a,b).

Although the *neu* mutation has not been implicated in any human cancers, HER2 is overexpressed in 20–25% of human breast cancers, and this overexpression is correlated with more aggressive tumors and a significantly reduced survival time (Slamon *et al.*, 1987). This connection stimulated development of strategies targeting HER2 in HER2-overexpressing breast cancers, and a monoclonal antibody directed against HER2, Herceptin, has been approved for treatment of these cancers (Slamon *et al.*, 2001). Treatment with Herceptin has in some cases led to long-term remission, but the mean improvement in life expectancy is currently 4–5 months (Slamon *et al.*, 2001). Earlier treatment with Herceptin (which has few side effects), combination of Herceptin with other therapies, or a better understanding of mechanisms of Herceptin resistance hold the promise of improved treatment strategies and outcomes, however.

IV. Structure of Individual ErbB Receptor Domains

Despite intense interest in ErbB receptors, high-resolution structural information has been slow in coming, owing largely to difficulties expressing and crystallizing these cysteine-rich glycoproteins. Fortunately, this situation has been remedied in the last year with publication of high-resolution crystal structures of active and inactive forms of the extracellular region of EGFR (sEGFR) (Ferguson *et al.*, 2003; Garrett *et al.*, 2002; Ogiso *et al.*, 2002), the extracellular region of HER2 (sHER2) both alone and complexed with the Herceptin Fab (Cho *et al.*, 2003; Garrett *et al.*, 2003), an inactive form of the extracellular region of HER3 (sHER3) (Cho and Leahy, 2002), and the EGFR kinase domain both alone and complexed with an inhibitor (Stamos *et al.*, 2002). These structures have yielded many unexpected but satisfying insights into the activation and regulation of ErbB receptors. At the same time, many new questions have been raised, and our appetite has been whetted for an even deeper molecular understanding of ErbB receptor function.

All structures of ErbB extracellular domains confirm the expected structural homology of the 'L' domains (domains I and III) to one another and to corresponding domains in type I insulin-like growth factor receptor (IGF1R). The structure of the first three domains of IGF1R, which are homologous to the first three domains of ErbB receptors, was determined in 1998 (Garrett *et al.*, 1998). Domains I and III share a beta-helical structure often referred to as a "solenoid" because it consists of 4–5 complete beta-helical turns containing a core of 22–24 amino acids per turn that stack to form an oblate cylinder (Fig. 3). The first beta-helical turn of domain I in the sEGFR/ligand complexes contains a 3-turn alpha helix that is adjacent to the bound ligand. The homologous region is disordered in the unliganded sHER3 structure, suggesting that interaction with ligand results in ordering of this helix in ErbB receptors.

A curious feature of domains I and III is a set of conserved asparagines—one per turn—with side chains that form hydrogen bonds to both main-chain atoms and the side chain atoms of the conserved asparagines of the next and preceding turns. These asparagines mediate a bend in the

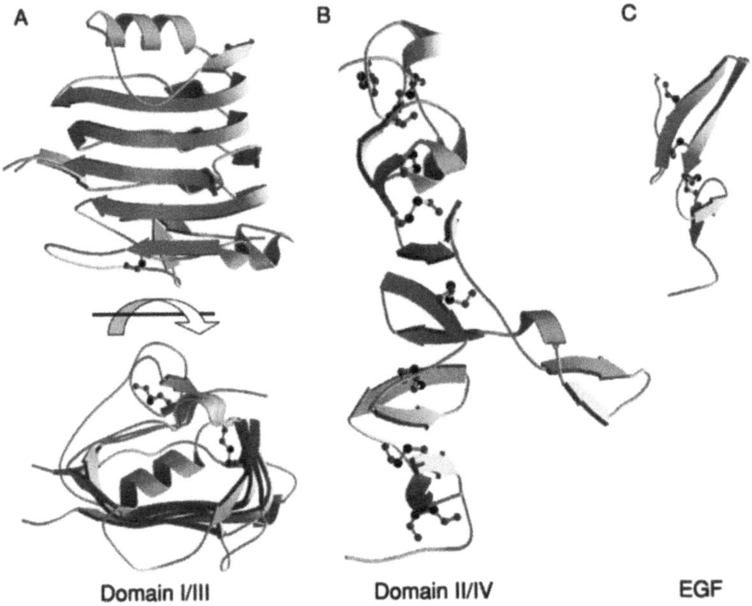

FIG. 3. Ribbon diagrams of ErbB receptor and ligand domains. (A) Orthogonal views of domain I from HER2 (Cho *et al.*, 2003). (B) Domain II from HER2 (Cho *et al.*, 2003). (C) EGF (Ogiso *et al.*, 2002). All ribbon diagrams in this and other figures were made using MOLSCRIPT (Kraulis, 1991).

main-chain within each turn, and the burial of their side-chains in the domain core appears to be accommodated by the inter-turn hydrogen bonds between the asparagine side chains of successive turns. These interactions result in an aligned "line" of buried asparagine side chains running along the beta-helical axis.

The cysteine-rich domains II and IV of ErbB receptors are homologous both to one another and to the cysteine-rich domain II of IGF1R (Burgess *et al.*, 2003; Garrett *et al.*, 1998). These domains can be characterized as a catenation of small disulfide bonded modules of ~15–20 amino acids that contain either a single disulfide bond that constrains a "bow-like" loop (C1 module) or two disulfide bonds in a Cys1-Cys3 and Cys2-Cys4 knot-like structure (C2 module). The pattern of modules in ErbB receptor domains II and IV are C2-C2-C2-C1-C1-C1-C1-C1 and C2-C1-C1-C2-C1-C1-C2, respectively (Ferguson *et al.*, 2003; Garrett *et al.*, 1998). Given the homology of the domain I/II and III/IV pairs, it seems likely that ErbB receptors arose as a duplication of an original domain I/II-like pair, suggesting that C1 and C2 modules are able to interconvert through mutation.

V. Structure of Entire ErbB Receptor Ectodomains

Although the structure of subdomains within ErbB receptor extracellular regions was anticipated from their homology to the IGF1R subdomains, their arrangement in intact receptors was not. The structures of unliganded HER3 and unactivated EGFR extracellular regions revealed a ~15 Å beta-hairpin loop that extends from domain II to interact with a pocket at the C-terminus of domain IV (Fig. 4). This interaction, akin to a "snap" or "tether," constrains the EGFR and HER3 extracellular regions into a bracelet-like structure in which domains I and III are prevented from coming into close contact. The domain II/IV interaction is mediated by three sets of side-chain interactions including a tyrosine from domain II (Tyr 246 in HER3) that makes a hydrogen bond to a salt-bridged aspartate-lysine pair from domain IV (Lys 583 and Asp 562), a phenylalanine from domain II (Phe 251) that packs on a glycine from domain IV (Gly 563), and a glutamine from domain II (Gln 252) that makes a hydrogen bond to a histidine from domain IV (His 565) (Fig. 5). With the exception of Phe 251 and His 565, which are replaced by the similar residues tyrosine in EGFR and asparagine in HER4, respectively, each of these residues is conserved in all ErbB receptors except HER2.

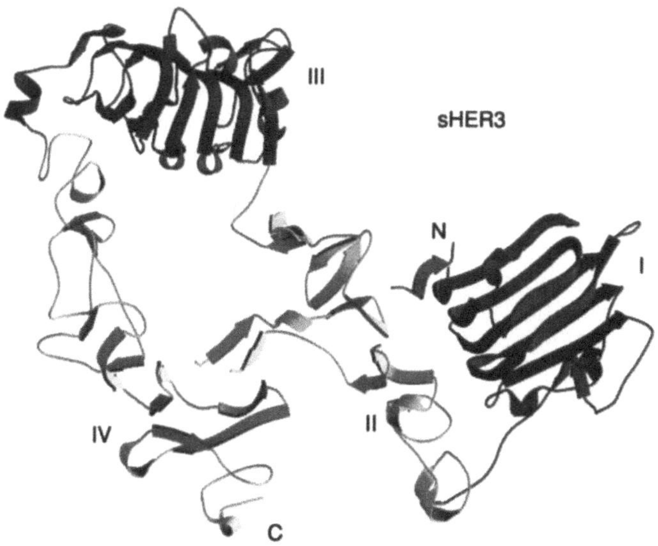

Fig. 4. Ribbon diagram of sHER3 (Cho and Leahy, 2002). Individual domains and the N- and C-termini are labeled. Domains I and III are colored dark gray; domains II and IV are colored light gray.

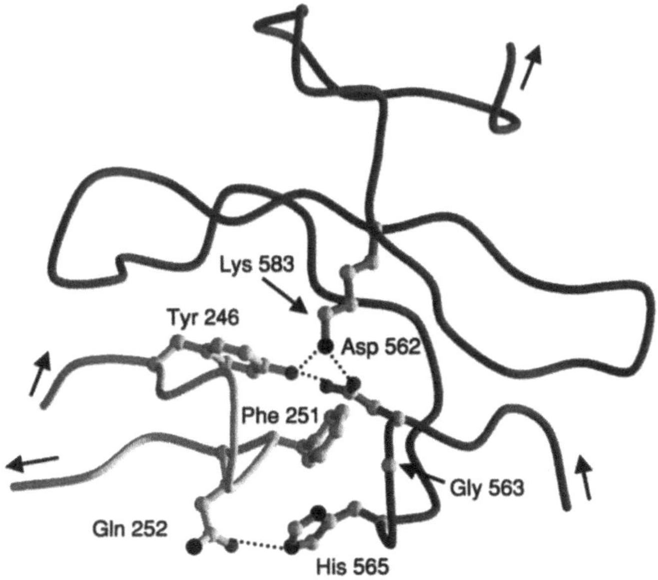

FIG. 5. Interactions in the HER3 domain II/IV "snap" contact region. Residues in the domain II loop (light gray) that contact residues in the domain IV pocket (dark gray) are indicated.

Structures of the first three subdomains of EGFR complexed with either EGF (Ogiso *et al.*, 2002) or TGFα (Garrett *et al.*, 2002) proved very similar to one another but revealed an entirely different arrangement of receptor subdomains than observed in unactivated forms of sEGFR and sHER3. Both EGF and TGFα bind to surfaces on domains I and III and mediate a close juxtaposition of these two domains (Fig. 6). The "snap-like" loop from domain II that mediates an interaction with domain IV in the unactivated state no longer contacts domain IV but instead mediates an inter-receptor dimer. The importance of the domain II loop for signaling has been demonstrated by site-directed mutagenesis (Garrett *et al.*, 2002; Ogiso *et al.*, 2002). In the case of the EGFR/EGF complex, domain IV was present in the crystal (although not visible in the electron density) indicating that participation of the domain II loop in mediating interreceptor dimers is favored over interactions with domain IV when ligand is present (Ogiso *et al.*, 2002).

A striking feature of ligand-induced dimers of sEGFR is that the dimer interface is mediated entirely by interreceptor contacts (Garrett *et al.*,

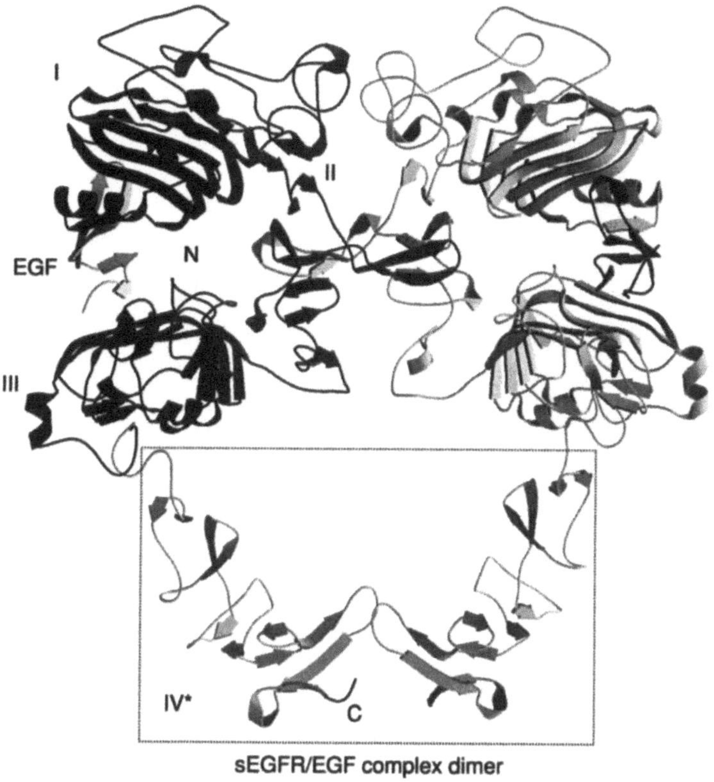

FIG. 6. Dimer of EGFR/EGF complex (Garrett *et al.*, 2002; Ogiso *et al.*, 2002). One EGFR subunit is colored dark gray and its bound EGF light gray; the other EGFR is colored light gray and its bound EGF dark gray. The relatively conserved interdomain orientation between domains III and IV of ErbB receptors has been used to model the position of domain IV, enclosed in a dashed box (Ferguson *et al.*, 2003).

2002; Ogiso *et al.*, 2002). In all previous structures of complexes of growth factor receptors with ligand, the ligand directly cross-links receptors (Schlessinger, 2000). An entirely receptor-mediated dimer allows formation of heteromeric receptor complexes without requiring ligand to bind simultaneously to two receptors. Co-receptors may thus evolve without requiring co-evolution of a ligand. Additionally, inter-receptor dimers mediated entirely by membrane-bound components may explain in part the weak association of ErbB extracellular regions that is observed in solution (Brown *et al.*, 1994; Lemmon *et al.*, 1997). The greatly increased local concentration that results from being confined to two dimensions may not require as strong an association to drive dimerization as is needed when one or more components is free to move in three dimensions (Grasberger *et al.*, 1986).

Comparison of structures of different ErbB receptors indicates that both the domain I/II and domain III/IV pairs retain a relatively fixed interdomain orientation (Fig. 7). Some spine-like movement of the submodules of domain II is observed, but differences between ErbB structures appear to arise primarily from rigid body movements of the domain I/II and III/IV pairs relative to one another. The relatively fixed orientation between domains III and IV allowed modeling of the position of domain IV in the ligand-induced dimers of sEGFR (Burgess *et al.*, 2003) (Fig. 6), which showed that the C-termini of subunits within the dimer are located adjacent to one another. Close proximity of the juxtamembrane regions of subunits of a signaling dimer is required if receptor autophosphorylation occurs in trans as generally believed, and this result provides added confidence that the dimer observed in the sEGFR crystals is in fact a physiological dimer.

Receptor dimerization does not appear sufficient for signaling, however. Introduction of interreceptor disulfide bonds at different sites within the juxtamembrane region of ErbB receptors results in both active and inactive dimers (Burke *et al.*, 1997). Requirement for a stereospecific dimer is also seen for cytokine receptors (Jiang and Hunter, 1999) and suggests that conformational information is transmitted across the plasma

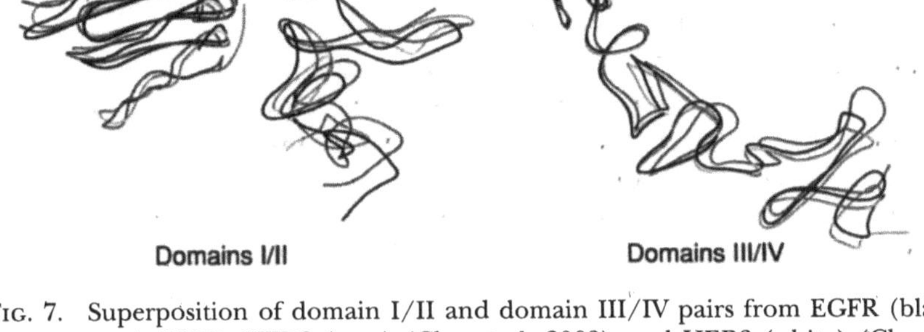

FIG. 7. Superposition of domain I/II and domain III/IV pairs from EGFR (black) (Ferguson *et al.*, 2003), HER2 (gray) (Cho *et al.*, 2003), and HER3 (white) (Cho and Leahy, 2002).

membrane during signaling. That is, simply inducing proximity of the cytoplasmic kinase regions is not sufficient to lead to activation.

A consequence of the recent structures of ErbB extracellular regions is the emergence of a simple model for ligand-induced dimerization of ErbB receptors (Burgess *et al.*, 2003) (Fig. 8). This model was as unexpected as it is satisfying in explaining several unusual features of ErbB receptors. In the unliganded state, the extracellular regions of EGFR, HER3, and—almost certainly—HER4 adopt a constrained structure in which an extended hairpin loop from domain II binds to a pocket at the C-terminus of domain IV. This conformation results in a large separation between domains I and III, which together comprise the ligand binding site. For ligand to bind, the domain II/IV connection must be broken and a large (~130°) rotation of the domain I/II pair relative to the domain III/IV pair must occur. This change brings together domains I and III to form a complete binding site and exposes the extended domain II loop, which is free to mediate inter-receptor dimerization and initiate signaling. The extended domain II loop

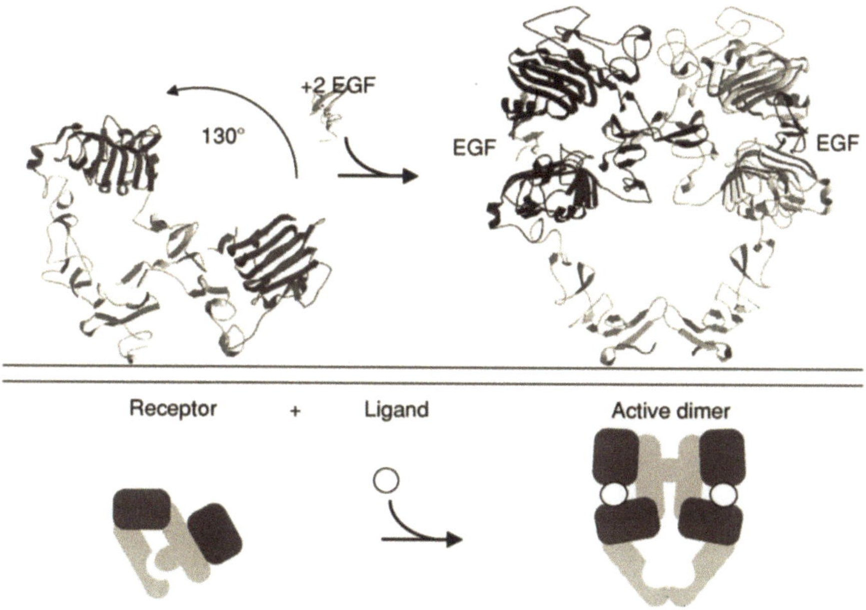

FIG. 8. Ligand-induced signaling mechanism. Structures and models of unliganded sHER3 (left) and ligand-bound sEGFR dimers (right) are shown. EGFR with bound ligand exhibits a ~130° rotation of the domain I/II pair relative to the domain III/IV pair when compared to unactivated sEGFR (Ferguson *et al.*, 2003) or sHER3 (Cho and Leahy, 2002). Adapted from a figure in Burgess *et al.* (2003).

has thus been called the "dimerization" loop. An essential feature of this model is a weak domain II/IV contact, which is consistent with the fact that its elimination by deletion or mutagenesis results in only a ~5–10 fold increase in the affinity of the receptor for ligand (Elleman *et al.*, 2001; Ferguson *et al.*, 2003).

The structure of a complex of the EGFR extracellular region and EGF determined at low pH provides a snapshot of a likely mechanism for release of bound ligand in the low pH environment of the endosome. EGFR is endocytosed after interacting with ligand, and at least some of this receptor is trafficked to endosomes where ligand dissociates. In crystals of a sEGFR/EGF complex grown at pH 5.0, EGF is bound only to the domain I binding surface, and the snap-like contact between domains II and IV is present (Ferguson *et al.*, 2003) (Fig. 9). Several histidines are conserved at the interface between EGF and domain III of EGFR, and protonation of

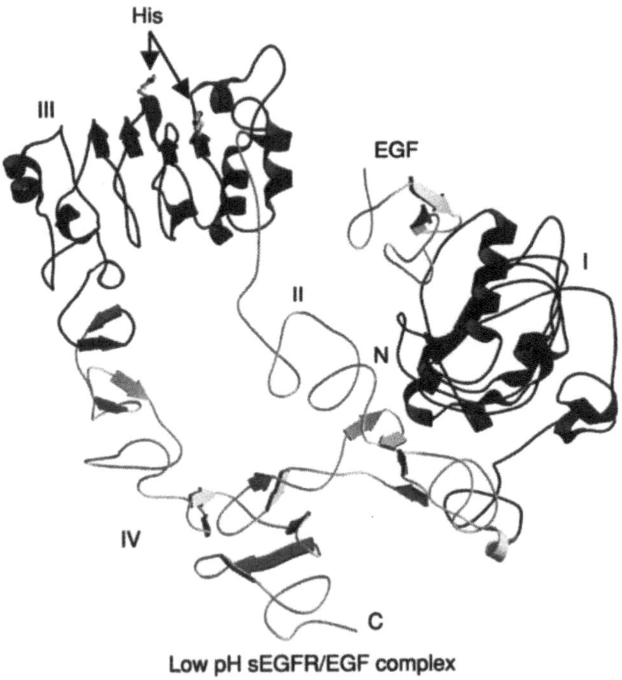

FIG. 9. Low pH form of an sEGFR/EGF complex (Ferguson *et al.*, 2003). Individual domains and the N- and C-termini are labeled; domains I and III are colored dark gray, and domains II and IV are colored light gray. EGF is colored light gray, and two histidines in domain III (His 346 and His 409) that contact EGF in the sEGFR/EGF complex formed at high pH (Garrett *et al.*, 2002; Ogiso *et al.*, 2002) are indicated.

these histidines at low pH appears to disrupt this interface (Ferguson *et al.*, 2003; Garrett *et al.*, 2002). Most of the interaction energy between EGFR and EGF is mediated by domain III (Kohda *et al.*, 1993). The interaction of EGF with domain I observed in these crystals is likely to be driven by the high concentrations of receptor and ligand used in crystallization trials. This structure thus appears to recapitulate events in the endosome—low pH results in protonation of histidines at the interface between EGF and domain III of EGFR leading to disruption of this interface and loss of high-affinity ligand binding.

VI. HER2

HER2 is unique among ErbB receptors in that no high-affinity HER2 ligand has been found, it functions as a co-receptor with each of the other ErbB receptors, and it is transforming when overexpressed (Di Fiore *et al.*, 1987; Klapper *et al.*, 1999). Much attention has been focused on HER2 because it is activated in many cancers (Holbro *et al.*, 2003; Slamon *et al.*, 1987; Tang and Lippman, 1998) and a HER2-targeted therapy, the monoclonal anti-HER2 antibody Herceptin, has demonstrated efficacy in a subset of breast cancers (Slamon *et al.*, 2001).

Structures of the entire extracellular region (Cho *et al.*, 2003) and the first three domains of HER2 (Garrett *et al.*, 2003) show it to adopt a very different conformation than unactivated forms of sEGFR or sHER3. The "snap-like" hairpin loop from domain II is present in HER2, but it is exposed to solvent and does not mediate a contact with domain IV (Fig. 10). Instead, an extensive ($\sim$1200 Å^2) and highly complementary contact is made between domains I and III, which appears to fix the orientation of the domain I/II pair relative to the domain III/IV pair. The interface between domains I and III is conserved in all three crystal forms of HER2 and appears to be a fixed feature of HER2 homologs. Two key hydrophobic residues buried at this interface, Leu 443 and Leu 472, are conserved in all HER2 homologs but not in other ErbB receptors. The HER2 domain I/II and III/IV pairs align well with the corresponding pairs from all other ErbB receptors (Fig. 7), suggesting that the interdomain orientations within these pairs are relatively rigid. By fixing the orientation of the domain I/II and III/IV pairs relative to one another, the HER2-specific interaction between domains I and III thus fixes the conformation of the entire HER2 extracellular region.

In light of the mechanism of ligand-induced signaling apparent from the sEGFR and sHER3 structures (Fig. 8), the structure of the HER2 extracellular domain explains several of its unique properties. The surface buried between domains I and III of HER2 overlaps significantly with

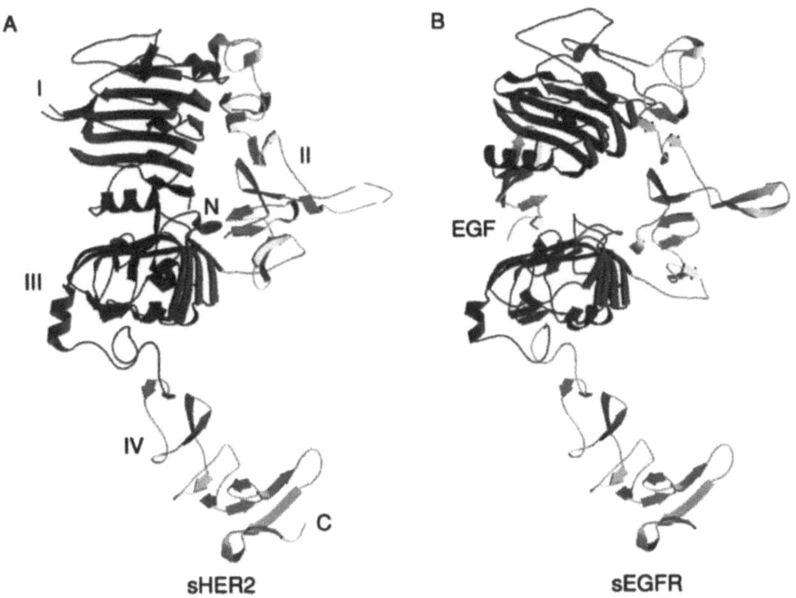

FIG. 10. Ribbon diagrams of (A) sHER2 (Cho *et al.*, 2003) and (B) a monomer of the sEGFR/EGF complex (Garrett *et al.*, 2002; Ogiso *et al.*, 2002). Individual domains and the N- and C-termini of sHER2 are labeled; domains I and III are colored dark gray, and domains II and IV are colored light gray. EGF is colored light gray. Domain IV of sEGFR is modeled based on the conserved interdomain orientation observed between domains III and IV of ErbB receptors (Ferguson *et al.*, 2003).

regions homologous to the ligand-binding site of EGFR (Garrett *et al.*, 2002; Ogiso *et al.*, 2002), effectively blocking canonical ligand binding. The presence of an intrinsic bridge between HER2 domains I and III appears not only to preclude normal ligand binding but also to substitute for it. The conformation of the HER2 extracellular region is very similar to the conformation of the EGFR extracellular region with either EGF or TGFα bound (Fig. 10). In each case domains I and III are brought close together and the domain II dimerization loop is exposed. If domain III of ligand-bound EGFR is superimposed with domain III of HER2, the domain II dimerization loops occupy nearly identical positions and the orientation of domain I of EGFR and HER2 differ only by a rotation of $\sim$30°. That is, HER2 adopts a constitutively "active-like" structure.

The active-like structure of HER2 provides a structural basis for its role as the preferred heterodimerization partner among ErbB receptors. HER2 does not adopt the autoinhibited conformation observed for EGFR and HER3 but is instead poised to interact with any ErbB receptor that becomes available—regardless of the stimulating ligand. This model assumes that during signaling the HER2 extracellular region interacts with other ErbB receptors in a manner similar to that observed for dimers of ligand-activated sEGFR. For the moment this assumption seems justified based on two arguments: it requires the simplest evolutionary path following receptor duplication and, as described later, an anti-HER2 antibody that binds to the HER2 dimerization loop (2C4) blocks formation of HER2-containing receptor heterodimers (Agus *et al.*, 2002; Franklin *et al.*, 2004). The development of HER2 as a "universal" ErbB co-receptor independent of stimulating ligand is facilitated by the entirely receptor-mediated signaling dimers, which do not require co-evolution of HER2-binding surfaces for each ligand.

The constitutively active-like conformation of HER2 raises the question of how inappropriate signaling through HER2 homodimers is avoided. In principle, subtle changes in interdomain relationships or differences at specific amino-acid positions could strongly disfavor homodimers without adversely affecting heteromeric receptor interactions, and it may not be possible to identify or quantitate all contributing factors. Nonetheless, three possible explanations have been put forward. Among dimerization loop residues involved in mediating EGFR inter-receptor dimers (Garrett *et al.*, 2002; Ogiso *et al.*, 2002), only a single site, Leu 291 in HER2, differs from its counterpart in EGFR (Arg 285). Mutation of Arg 285 in EGFR to serine reduces ligand-induced signaling, presumably by disfavoring formation of appropriate receptor dimers (Ogiso *et al.*, 2002). This single amino-acid change in the HER2 dimerization loop may be tolerated in HER2-containing heterodimers, which must accommodate only single change, but it may pose a higher barrier to formation of HER2 homodimers, which must accommodate this change in both dimer subunits (Cho *et al.*, 2003). It has also been noted that a surface encompassing the HER2 dimerization loop is predominantly negatively charged, which could serve to preferentially disfavor formation of HER2 homodimers (Garrett *et al.*, 2003). Finally, it has been observed that a loop on HER2 domain II at the canonical interreceptor interface is extended in such way as to give rise to a clash with itself in a modeled HER2 homodimer but not in a HER2-containing heterodimer (Franklin *et al.*, 2004). Each of these factors may be assessed through study of the properties of mutated or chimeric forms of HER2.

Interest in the mechanism by which HER2 homodimerization is disfavored is stimulated in part by a desire to understand how HER2 becomes activated when overexpressed (Brennan *et al.*, 2000; Di Fiore *et al.*, 1987). HER2 overexpression and activation contributes to the genesis and severity of human cancers (Blume-Jensen and Hunter, 2001; Holbro *et al.*, 2003; Slamon *et al.*, 1987; Tang and Lippman, 1998), and understanding HER2 activation in these circumstances may lead to new therapeutic strategies aimed at inhibiting HER2 activation. Although perhaps the most obvious hypothesis, it seems unlikely that overexpression of HER2 overcomes an inherently weak propensity of its extracellular region to mediate stereospecific homodimers. In crystals of HER2 extracellular regions, where the receptor concentration is 5–10 mM, no interreceptor interactions suggestive of a signaling complex are observed. Furthermore, an anti-HER2 monoclonal antibody ineffective at blocking the formation of HER2-containing receptor heterodimers, Herceptin, appears to be a comparably effective treatment for HER2-overexpressing tumors as 2C4, an anti-HER2 monoclonal antibody that does block heterodimer formation (Agus *et al.*, 2002).

Although formation of stereospecific dimers mediated by the extracellular region seems unlikely when HER2 is overexpressed, the absence of an autoinhibited conformation for HER2 appears to remove a barrier to transformation in HER-overexpressing cells. ErbB receptors lacking significant parts the extracellular region exhibit constitutive activation and demonstrate that extracellular regions normally inhibit activation in the absence of ligand (Humphrey *et al.*, 1991; Molina *et al.*, 2002). HER2 is the only ErbB receptor that both lacks an apparent autoinhibited conformation and is transforming when overexpressed—each of the other ErbB receptors requires the presence of ligand to become transforming (Tang and Lippman, 1998). Interestingly, mutations in EGFR that disrupt the domain II/IV interaction responsible for the autoinhibited conformation do not lead to constitutive activation (Ferguson *et al.*, 2003; Mattoon *et al.*, 2004). These results suggest that shielding of the dimerization loop in the autoinhibited conformation is not sufficient to account for the inhibitory effect of the EGFR extracellular region, and additional inhibitory mechanisms are present. The effects of these mutations on overexpressed receptors needs to be examined to fully justify this conclusion, however. An intriguing possibility is that preformed receptor dimers account for the inactivity of receptors in the absence of ligand (Gadella and Jovin, 1995; Moriki *et al.*, 2001; Sako *et al.*, 2000). It seems clear that ErbB receptors utilize multiple means to regulate receptor activity, and it remains an important challenge to understand how receptors are kept off as well as how they are turned on (Schlessinger, 2003).

VII. Therapeutic Anti-HER2 Antibodies

A major development in the treatment of breast cancer has been the success of Herceptin, an anti-HER2 monoclonal antibody, in treating the 20–25% of breast cancers in which HER2 is overexpressed (Slamon *et al.*, 2001). Although not a cure—average life expectancy is extended

4–5 months—Herceptin has demonstrated that targeting activated ErbB receptors in cancer can be beneficial and holds out the hope of improved therapies. A particular advantage of specific receptor-targeted therapies is that they frequently cause fewer side effects than standard cancer chemotherapies. Much interest has thus been generated in the related problems of understanding how Herceptin works and developing improved strategies for targeting activated ErbB receptors in cancer (Arteaga, 2003; Sliwkowski *et al.*, 1999).

Of the many mechanisms proposed for Herceptin action, unequivocal evidence exists for antibody-directed cellular cytoxicity (ADCC) playing a major role. Herceptin is not effective either as a Fab fragment or when used against tumors grown in mice lacking functional Fc receptors (Clynes *et al.*, 2000). ADCC cannot account for all of Herceptin's activity, however, as Herceptin has an antiproliferative effect on tumors cells grown in culture (Sliwkowski *et al.*, 1999). Another possible explanation for Herceptin's activity is suggested by the observation that the extracellular region of HER2 is normally cleaved off at a low level, leaving behind a truncated and activated receptor. It has been suggested that HER2 overexpression results in an increased level of this proteolysis and background activation (Baselga *et al.*, 2001; Molina *et al.*, 2001). Herceptin inhibits this cleavage, which may contribute to its antiproliferative effect. Herceptin does not appear to work by blocking HER2-mediated signaling, however (Agus *et al.*, 2002).

The crystal structure of a complex between HER2 and the Herceptin Fab shows Herceptin to bind to the juxtamembrane region of domain IV (Cho *et al.*, 2003). Binding at this site, which is consistent with earlier epitope mapping (Sliwkowski *et al.*, 1999), provides a steric block to proteolytic cleavage of the HER2 extracellular region explaining its ability to inhibit cleavage. The HER2/Herceptin structure is also entirely consistent with binding studies of Herceptin variants bearing binding site mutations introduced both systematically (Kelley and O'Connell, 1993) and through selection by phage display (Gerstner *et al.*, 2002). Together, these studies make the HER2/Herceptin interface among the most thoroughly characterized of protein-protein interactions. This level of characterization may be helpful for the design and selection of Herceptin variants with better binding but so far has added little insight into how Herceptin works.

The recent structure of a complex between the HER2 extracellular region and the Fab fragment of the 2C4 antibody (also known as pertuzumab or Omnitarg) has proven much more illuminating (Franklin *et al.*, 2004). Elucidation of the signaling mechanism of ErbB receptors combined with the HER2 structure immediately suggested that targeting the HER2 dimerization loop might interfere with HER2-mediated signaling

(Burgess *et al.*, 2003). The sHER2/2C4 structure shows that 2C4 binds to a region encompassing the dimerization loop and provides an immediate validation of this hypothesis as 2C4 blocks formation of HER2-containing receptor heterodimers, which Herceptin does not (Agus *et al.*, 2002; Franklin *et al.*, 2004). These results also provide the best evidence that the HER2 dimerization loop is involved in mediating inter-receptor dimers in a fashion analogous to the EGFR dimerization loop.

The different binding epitope and biochemical properties of Herceptin and 2C4 appear to translate into different clinical properties as well. Two classes of HER2 activation in cancer have been observed: ligand-independent activation, as seen for example when HER2 is overexpressed, and ligand-dependent forms that appear secondary to dysregulation of an ErbB ligand such as TGFα (Agus *et al.*, 2002; Sizeland and Burgess, 1992). HER2 activation in ligand-dependent cancers must occur through the simultaneous presence of the cognate ErbB receptor, but the HER2 activation appears to contribute significantly to proliferation of affected cells. Herceptin has no effect against cancers in which HER2 activation is ligand-dependent (and HER2 expression levels are normal), but 2C4 does appear to have an anti-proliferative effect in initial studies with these cancers. 2C4 is currently in phase II clinical trials for breast, lung, prostate, and ovarian cancers in which HER2 expression levels are normal (Agus *et al.*, 2002; Franklin *et al.*, 2004).

Comparison of structural, biochemical, and clinical results with Herceptin and 2C4 are adding up to a consistent picture with important implications for cancer therapy. 2C4 represents the ideal antibody with which to target HER2 function based on structural and biochemical considerations, and it appears to have wider application than Herceptin, which is currently approved for clinical use. Experience with anti-HER2 antibodies and knowledge of the ErbB activation mechanism is likely to influence selection of future antibodies targeting other ErbB receptors activated in specific cancers as well as guide development of entirely new therapeutics. For example, mutated ligands that bind receptor at only the domain I or domain III binding surface are likely to fail to induce the conformational change necessary for activation and function as competitive inhibitors of receptor activation. It is also conceivable that small molecules blocking or mimicking key regions of ErbB receptors may be developed and exhibit useful biological activities.

VIII. Remaining Questions

Recent structural studies have greatly advanced our understanding of signaling by members of the ErbB family of receptors, but many key questions remain. Principal among these is how events on the outside of the cell are communicated across the plasma membrane and lead to activation of the kinase activity. Structural and biochemical studies clearly point to receptor dimers as key elements of signaling complexes (Garrett *et al.*, 2002; Ogiso *et al.*, 2002; Yarden and Schlessinger, 1987). Not all dimers are capable of signaling, however, indicating that proximity of the C-terminal or transmembrane sequences is insufficient to trigger activation (Burke *et al.*, 1997). ErbB receptors share this property with cytokine receptors, which also appear to require a stereospecific dimer to trigger signaling (Jiang and Hunter, 1999). If proximity is insufficient to generate signals, some aspect of conformation must be communicated across the membrane to activate the kinase. Understanding the conformational features important for activation and how they are transmitted across the cell membrane presents an exciting challenge for the future.

Although dimerization must play an essential role in ErbB signaling, a role for higher-order oligomers cannot be ruled out. In fact, since the HER3 kinase domain is inactive and interreceptor phosphorylation appears to occur in trans, higher order oligomers are often invoked to explain the phosphorylation of HER2 observed when signaling is initiated through the HER2/HER3 pair. In addition to sorting out the conformation and oligomeric state of a minimal ErbB signaling complex, understanding the conformation and oligomeric state of inactive ErbB receptors also needs attention. Several mechanisms appear to conspire to keep ErbB receptors inactive in the absence of ligand, including the constrained conformation mediated by the interdomain tether and possible preformed dimers (Cho and Leahy, 2002; Ferguson *et al.*, 2003; Gadella and Jovin, 1995; Moriki *et al.*, 2001; Sako *et al.*, 2000; Schlessinger, 2003). At this point, advancing our understanding of the nature and importance of resting and active states of ErbB receptors is likely to depend on our ability to study intact receptors.

Deciphering the molecular details of ErbB receptor signaling is of great fundamental interest to biologists and biochemists and certain to prove of clinical interest as well. The structure-based model of the signaling cycle of the extracellular regions of ErbB receptors—and the unique role played by HER2 in this cycle—have rationalized different properties of current anti-cancer therapeutics and suggested new approaches to modulating dysfunctional ErbB receptors (Burgess *et al.*, 2003; Cho and Leahy, 2002; Ferguson *et al.*, 2003; Franklin *et al.*, 2004). Understanding the conformations and interactions of intact ErbB receptors is likely to reveal new interaction surfaces or states that when targeted by antibodies or small molecules may result in beneficial modulation of ErbB receptor function.

Acknowledgments

I thank Mark Lemmon, Mark Sliwkowski, and members of my lab for helpful discussions and HHMI and NIH for funding the work in my laboratory.

References

Agus, D. B., Akita, R. W., Fox, W. D., Lewis, G. D., Higgins, B., Pisacane, P. I., Lofgren, J. A., Tindell, C., Evans, D. P., and Maiese, K. *et al.* (2002). Targeting ligand-activated ErbB2 signaling inhibits breast and prostate tumor growth. *Cancer Cell* **2,** 127–137.

Arteaga, C. L. (2002). Overview of epidermal growth factor receptor biology and its role as a therapeutic target in human neoplasia. *Semin. Oncol.* **29,** 3–9.

Arteaga, C. L. (2003). ErbB-targeted therapeutic approaches in human cancer. *Exp. Cell Res.* **284,** 122–130.

Bajaj, M., Waterfield, M. D., Schlessinger, J., Taylor, W. R., and Blundell, T. (1987). On the tertiary structure of the extracellular domains of the epidermal growth factor and insulin receptors. *Biochim. Biophys. Acta* **916,** 220–226.

Bargmann, C. I., Hung, M. C., and Weinberg, R. A. (1986a). Multiple independent activations of the neu oncogene by a point mutation altering the transmembrane domain of p185. *Cell* **45,** 649–657.

Bargmann, C. I., Hung, M. C., and Weinberg, R. A. (1986b). The neu oncogene encodes an epidermal growth factor receptor-related protein. *Nature* **319,** 226–230.

Bargmann, C. I., and Weinberg, R. A. (1988a). Increased tyrosine kinase activity associated with the protein encoded by the activated neu oncogene. *Proc. Natl. Acad. Sci. USA* **85,** 5394–5398.

Bargmann, C. I., and Weinberg, R. A. (1988b). Oncogenic activation of the neu-encoded receptor protein by point mutation and deletion. *EMBO J.* **7,** 2043–2052.

Baselga, J., Albanell, J., Molina, M. A., and Arribas, J. (2001). Mechanism of action of trastuzumab and scientific update. *Semin. Oncol.* **28,** 4–11.

Blume-Jensen, P., and Hunter, T. (2001). Oncogenic kinase signalling. *Nature* **411,** 355–365.

Brennan, P. J., Kumagai, T., Berezov, A., Murali, R., Greene, M. I., and Kumogai, T. (2000). HER2/neu: Mechanisms of dimerization/oligomerization. *Oncogene* **19,** 6093–6101.

Brown, P. M., Debanne, M. T., Grothe, S., Bergsma, D., Caron, M., Kay, C., and O'Connor-McCourt, M. D. (1994). The extracellular domain of the epidermal growth factor receptor. Studies on the affinity and stoichiometry of binding, receptor dimerization and a binding-domain mutant. *Eur. J. Biochem.* **225,** 223–233.

Burgess, A. W., Cho, H. S., Eigenbrot, C., Ferguson, K. M., Garrett, T. P., Leahy, D. J., Lemmon, M. A., Sliwkowski, M. X., Ward, C. W., and Yokoyama, S. (2003). An open-and-shut case? Recent insights into the activation of EGF/ErbB receptors. *Mol. Cell* **12,** 541–552.

Burke, C. L., Lemmon, M. A., Coren, B. A., Engelman, D. M., and Stern, D. F. (1997). Dimerization of the p185neu transmembrane domain is necessary but not sufficient for transformation. *Oncogene* **14,** 687–696.

Carpenter, G. (1987). Receptors for epidermal growth factor and other polypeptide mitogens. *Annu. Rev. Biochem.* **56,** 881–914.

Carpenter, G. (2003). ErbB-4: Mechanism of action and biology. *Exp. Cell Res.* **284,** 66–77.

Carpenter, G., and Cohen, S. (1979). Epidermal growth factor. *Annu. Rev. Biochem.* **48,** 193–216.

Carpenter, G., King, L., Jr., and Cohen, S. (1978). Epidermal growth factor stimulates phosphorylation in membrane preparations *in vitro. Nature* **276,** 409–410.

Carpenter, G., King, L., Jr., and Cohen, S. (1979). Rapid enhancement of protein phosphorylation in A-431 cell membrane preparations by epidermal growth factor. *J. Biol. Chem.* **254,** 4884–4891.

Carpenter, G., Lembach, K. J., Morrison, M. M., and Cohen, S. (1975). Characterization of the binding of 125-I-labeled epidermal growth factor to human fibroblasts. *J. Biol. Chem.* **250,** 4297–4304.

Cho, H. S., and Leahy, D. J. (2002). Structure of the extracellular region of HER3 reveals an interdomain tether. *Science* **297,** 1330–1333.

Cho, H. S., Mason, K., Ramyar, K. X., Stanley, A. M., Gabelli, S. B., Denney, D. W., Jr., and Leahy, D. J. (2003). Structure of the extracellular region of HER2 alone and in complex with the Herceptin Fab. *Nature* **421,** 756–760.

Citri, A., Skaria, K. B., and Yarden, Y. (2003). The deaf and the dumb: The biology of ErbB-2 and ErbB-3. *Exp. Cell Res.* **284,** 54–65.

Clynes, R. A., Towers, T. L., Presta, L. G., and Ravetch, J. V. (2000). Inhibitory Fc receptors modulate *in vivo* cytoxicity against tumor targets. *Nat. Med.* **6,** 443–446.

Cohen, S. (1986). Nobel lecture. Epidermal growth factor. *Biosci. Rep.* **6,** 1017–1028.

Cohen, S. (1987). Epidermal growth factor. *In Vitro Cell Dev. Biol.* **23,** 239–246.

Cohen, S., Carpenter, G., and King, L., Jr. (1980). Epidermal growth factor-receptor-protein kinase interactions. Co-purification of receptor and epidermal growth factor-enhanced phosphorylation activity. *J. Biol. Chem.* **255,** 4834–4842.

Cohen, S., Ushiro, H., Stoscheck, C., and Chinkers, M. (1982). A native 170,000 epidermal growth factor receptor-kinase complex from shed plasma membrane vesicles. *J. Biol. Chem.* **257,** 1523–1531.

Das, M., Miyakawa, T., Fox, C. F., Pruss, R. M., Aharonov, A., and Herschman, H. R. (1977). Specific radiolabeling of a cell surface receptor for epidermal growth factor. *Proc. Natl. Acad. Sci. USA* **74,** 2790–2794.

de Vos, A. M., Ultsch, M., and Kossiakoff, A. A. (1992). Human growth hormone and extracellular domain of its receptor: Crystal structure of the complex. *Science* **255,** 306–312.

Di Fiore, P. P., Pierce, J. H., Kraus, M. H., Segatto, O., King, C. R., and Aaronson, S. A. (1987). erbB-2 is a potent oncogene when overexpressed in NIH/3T3 cells. *Science* **237,** 178–182.

Downward, J., Yarden, Y., Mayes, E., Scrace, G., Totty, N., Stockwell, P., Ullrich, A., Schlessinger, J., and Waterfield, M. D. (1984). Close similarity of epidermal growth factor receptor and v-erb-B oncogene protein sequences. *Nature* **307,** 521–527.

Elleman, T. C., Domagala, T., McKern, N. M., Nerrie, M., Lonnqvist, B., Adams, T. E., Lewis, J., Lovrecz, G. O., Hoyne, P. A., and Richards, K. M. *et al.* (2001). Identification of a determinant of epidermal growth factor receptor ligand-binding specificity using a truncated, high-affinity form of the ectodomain. *Biochemistry* **40,** 8930–8939.

Ferguson, K. M., Berger, M. B., Mendrola, J. M., Cho, H. S., Leahy, D. J., and Lemmon, M. A. (2003). EGF activates its receptor by removing interactions that autoinhibit ectodomain dimerization. *Mol. Cell* **11,** 507–517.

Franklin, M. C., Carey, K. D., Vajdos, F. F., Leahy, D. J., de Vos, A. M., and Sliwkowski, M. X. (2004). Insights into ErbB signaling from the structure of the ErbB2-pertuzumab complex. *Cancer Cell* **5,** 317–328.

Frykberg, L., Palmieri, S., Beug, H., Graf, T., Hayman, M. J., and Vennstrom, B. (1983). Transforming capacities of avian erythroblastosis virus mutants deleted in the erbA or erbB oncogenes. *Cell* **32,** 227–238.

Gadella, T. W., Jr., and Jovin, T. M. (1995). Oligomerization of epidermal growth factor receptors on A431 cells studied by time-resolved fluorescence imaging microscopy. A stereochemical model for tyrosine kinase receptor activation. *J. Cell Biol.* **129,** 1543–1558.

Garrett, T. P., McKern, N. M., Lou, M., Elleman, T. C., Adams, T. E., Lovrecz, G. O., Kofler, M., Jorissen, R. N., Nice, E. C., Burgess, A. W., and Ward, C. W. (2003). The crystal structure of a truncated ErbB2 ectodomain reveals an active conformation, poised to interact with other ErbB receptors. *Mol. Cell* **11,** 495–505.

Garrett, T. P., McKern, N. M., Lou, M., Elleman, T. C., Adams, T. E., Lovrecz, G. O., Zhu, H. J., Walker, F., Frenkel, M. J., and Hoyne, P. A. *et al.* (2002). Crystal structure of a truncated epidermal growth factor receptor extracellular domain bound to transforming growth factor alpha. *Cell* **110,** 763–773.

Garrett, T. P., McKern, N. M., Lou, M., Frenkel, M. J., Bentley, J. D., Lovrecz, G. O., Elleman, T. C., Cosgrove, L. J., and Ward, C. W. (1998). Crystal structure of the first three domains of the type-1 insulin-like growth factor receptor. *Nature* **394,** 395–399.

Gassmann, M., Casagranda, F., Orioli, D., Simon, H., Lai, C., Klein, R., and Lemke, G. (1995). Aberrant neural and cardiac development in mice lacking the ErbB4 neuregulin receptor. *Nature* **378,** 390–394.

Gerstner, R. B., Carter, P., and Lowman, H. B. (2002). Sequence plasticity in the antigen-binding site of a therapeutic anti-HER2 antibody. *J. Mol. Biol.* **321,** 851–862.

Grasberger, B., Minton, A. P., DeLisi, C., and Metzger, H. (1986). Interaction between proteins localized in membranes. *Proc. Natl. Acad. Sci. USA* **83,** 6258–6262.

Guy, P. M., Platko, J. V., Cantley, L. C., Cerione, R. A., and Carraway, K. L., 3rd. (1994). Insect cell-expressed p180erbB3 possesses an impaired tyrosine kinase activity. *Proc. Natl. Acad. Sci. USA* **91,** 8132–8136.

Haigler, H. T., McKanna, J. A., and Cohen, S. (1979). Direct visualization of the binding and internalization of a ferritin conjugate of epidermal growth factor in human carcinoma cells A-431. *J. Cell. Biol.* **81,** 382–395.

Heldin, C. H. (1995). Dimerization of cell surface receptors in signal transduction. *Cell* **80,** 213–223.

Hock, R. A., Nexo, E., and Hollenberg, M. D. (1979). Isolation of the human placenta receptor for epidermal growth factor-urogastrone. *Nature* **277,** 403–405.

Holbro, T., Civenni, G., and Hynes, N. E. (2003). The ErbB receptors and their role in cancer progression. *Exp. Cell Res.* **284,** 99–110.

Humphrey, P. A., Gangarosa, L. M., Wong, A. J., Archer, G. E., Lund-Johansen, M., Bjerkvig, R., Laerum, O. D., Friedman, H. S., and Bigner, D. D. (1991). Deletion-mutant epidermal growth factor receptor in human gliomas: Effects of type II mutation on receptor function. *Biochem. Biophys. Res. Commun.* **178,** 1413–1420.

Humphrey, P. A., Wong, A. J., Vogelstein, B., Zalutsky, M. R., Fuller, G. N., Archer, G. E., Friedman, H. S., Kwatra, M. M., Bigner, S. H., and Bigner, D. D. (1990). Anti-synthetic peptide antibody reacting at the fusion junction of deletion-mutant epidermal growth factor receptors in human glioblastoma. *Proc. Natl. Acad. Sci. USA* **87,** 4207–4211.

Jiang, G., and Hunter, T. (1999). Receptor signaling: When dimerization is not enough. *Curr. Biol.* **9,** R568–R571.

Kelley, R. F., and O'Connell, M. P. (1993). Thermodynamic analysis of an antibody functional epitope. *Biochemistry* **32,** 6828–6835.

Klapper, L. N., Glathe, S., Vaisman, N., Hynes, N. E., Andrews, G. C., Sela, M., and Yarden, Y. (1999). The ErbB-2/HER2 oncoprotein of human carcinomas may function solely as a shared coreceptor for multiple stroma-derived growth factors. *Proc. Natl. Acad. Sci. USA* **96,** 4995–5000.

Klapper, L. N., Kirschbaum, M. H., Sela, M., and Yarden, Y. (2000). Biochemical and clinical implications of the ErbB/HER signaling network of growth factor receptors. *Adv. Cancer Res.* **77,** 25–79.

Kohda, D., Odaka, M., Lax, I., Kawasaki, H., Suzuki, K., Ullrich, A., Schlessinger, J., and Inagaki, F. (1993). A 40-kDa epidermal growth factor/transforming growth factor alpha-binding domain produced by limited proteolysis of the extracellular domain of the epidermal growth factor receptor. *J. Biol. Chem.* **268,** 1976–1981.

Kraulis, P. J. (1991). A program to produce both detailed and schematic plots of protein structures. *Journal of Applied Crystallography* **24,** 946–950.

Lax, I., Bellot, F., Howk, R., Ullrich, A., Givol, D., and Schlessinger, J. (1989). Functional analysis of the ligand binding site of EGF-receptor utilizing chimeric chicken/human receptor molecules. *EMBO J.* **8,** 421–427.

Lax, I., Johnson, A., Howk, R., Sap, J., Bellot, F., Winkler, M., Ullrich, A., Vennstrom, B., Schlessinger, J., and Givol, D. (1988). Chicken epidermal growth factor (EGF) receptor: cDNA cloning, expression in mouse cells, and differential binding of EGF and transforming growth factor alpha. *Mol. Cell. Biol.* **8,** 1970–1978.

Lee, K. F., Simon, H., Chen, H., Bates, B., Hung, M. C., and Hauser, C. (1995). Requirement for neuregulin receptor erbB2 in neural and cardiac development. *Nature* **378,** 394–398.

Lemmon, M. A., Bu, Z., Ladbury, J. E., Zhou, M., Pinchasi, D., Lax, I., Engelman, D. M., and Schlessinger, J. (1997). Two EGF molecules contribute additively to stabilization of the EGFR dimer. *EMBO J.* **16,** 281–294.

Mattoon, D., Klein, P., Lemmon, M. A., Lax, I., and Schlessinger, J. (2004). The tethered configuration of the EGF receptor extracellular domain exerts only a limited control of receptor function. *Proc. Natl. Acad. Sci. USA* **101,** 923–928.

Miettinen, P. J., Berger, J. E., Meneses, J., Phung, Y., Pedersen, R. A., Werb, Z., and Derynck, R. (1995). Epithelial immaturity and multiorgan failure in mice lacking epidermal growth factor receptor. *Nature* **376,** 337–341.

Molina, M. A., Codony-Servat, J., Albanell, J., Rojo, F., Arribas, J., and Baselga, J. (2001). Trastuzumab (herceptin), a humanized anti-Her2 receptor monoclonal antibody, inhibits basal and activated Her2 ectodomain cleavage in breast cancer cells. *Cancer Res.* **61,** 4744–4749.

Molina, M. A., Saez, R., Ramsey, E. E., Garcia-Barchino, M. J., Rojo, F., Evans, A. J., Albanell, J., Keenan, E. J., Lluch, A., and Garcia-Conde, J. *et al.* (2002). NH(2)-terminal truncated HER-2 protein but not full-length receptor is associated with nodal metastasis in human breast cancer. *Clin. Cancer Res.* **8,** 347–353.

Moriki, T., Maruyama, H., and Maruyama, I. N. (2001). Activation of preformed EGF receptor dimers by ligand-induced rotation of the transmembrane domain. *J. Mol. Biol.* **311,** 1011–1026.

Morris, J. K., Lin, W., Hauser, C., Marchuk, Y., Getman, D., and Lee, K. F. (1999). Rescue of the cardiac defect in ErbB2 mutant mice reveals essential roles of ErbB2 in peripheral nervous system development. *Neuron* **23,** 273–283.

Muller, W. E., and Schacke, H. (1996). Characterization of the receptor protein-tyrosine kinase gene from the marine sponge Geodia cydonium. *Prog. Mol. Subcell Biol.* **17,** 183–208.

Ogiso, H., Ishitani, R., Nureki, O., Fukai, S., Yamanaka, M., Kim, J. H., Saito, K., Sakamoto, A., Inoue, M., Shirouzu, M., and Yokoyama, S. (2002). Crystal structure of the complex of human epidermal growth factor and receptor extracellular domains. *Cell* **110,** 775–787.

Olayioye, M. A., Neve, R. M., Lane, H. A., and Hynes, N. E. (2000). The ErbB signaling network: Receptor heterodimerization in development and cancer. *EMBO J.* **19,** 3159–3167.

Riethmacher, D., Sonnenberg-Riethmacher, E., Brinkmann, V., Yamaai, T., Lewin, G. R., and Birchmeier, C. (1997). Severe neuropathies in mice with targeted mutations in the ErbB3 receptor. *Nature* **389,** 725–730.

Sako, Y., Minoghchi, S., and Yanagida, T. (2000). Single-molecule imaging of EGFR signalling on the surface of living cells. *Nat. Cell Biol.* **2,** 168–172.

Schechter, Y., Hernaez, L., Schlessinger, J., and Cuatrecasas, P. (1979). Local aggregation of hormone-receptor complexes is required for activation by epidermal growth factor. *Nature* **278,** 835–838.

Schlessinger, J. (2000). Cell signaling by receptor tyrosine kinases. *Cell* **103,** 211–225.

Schlessinger, J. (2003). Signal transduction. Autoinhibition control. *Science* **300,** 750–752.

Sibilia, M., and Wagner, E. F. (1995). Strain-dependent epithelial defects in mice lacking the EGF receptor. *Science* **269,** 234–238.

Sizeland, A. M., and Burgess, A. W. (1992). Anti-sense transforming growth factor alpha oligonucleotides inhibit autocrine stimulated proliferation of a colon carcinoma cell line. *Mol. Biol. Cell* **3,** 1235–1243.

Slamon, D., Leyland-Jones, B., Shak, S., Fuchs, H., Paton, V., Bajamonde, A., Fleming, T., Eiermann, W., Wolter, J., and Pegram, M. *et al.* (2001). Use of chemotherapy plus a monoclonal antibody against HER2 for metastatic breast cancer that overexpresses HER2. *N. Engl. J. Med.* **344,** 783–792.

Slamon, D. J., Clark, G. M., Wong, S. G., Levin, W. J., Ullrich, A., and McGuire, W. L. (1987). Human breast cancer: Correlation of relapse and survival with amplification of the HER-2/neu oncogene. *Science* **235,** 177–182.

Sliwkowski, M. X., Lofgren, J. A., Lewis, G. D., Hotaling, T. E., Fendly, B. M., and Fox, J. A. (1999). Nonclinical studies addressing the mechanism of action of trastuzumab (Herceptin). *Semin. Oncol.* **26,** 60–70.

Stamos, J., Sliwkowski, M. X., and Eigenbrot, C. (2002). Structure of the epidermal growth factor receptor kinase domain alone and in complex with a 4-anilinoquinazoline inhibitor. *J. Biol. Chem.* **277,** 46265–46272.

Stein, R. A., and Staros, J. V. (2000). Evolutionary analysis of the ErbB receptor and ligand families. *J. Mol. Evol.* **50,** 397–412.

Tang, C. K., and Lippman, M. E. (1998). EGF family of receptors and their ligands in human cancer. *In* Hormones and Signaling (B. W. O'Malley, Ed.), pp. 113–165. Academic Press, San Diego.

Threadgill, D. W., Dlugosz, A. A., Hansen, L. A., Tennenbaum, T., Lichti, U., Yee, D., LaMantia, C., Mourton, T., Herrup, K., and Harris, R. C. *et al.* (1995). Targeted disruption of mouse EGF receptor: Effect of genetic background on mutant phenotype. *Science* **269,** 230–234.

Todaro, G. J., De Larco, J. E., and Cohen, S. (1976). Transformation by murine and feline sarcoma viruses specifically blocks binding of epidermal growth factor to cells. *Nature* **264,** 26–31.

Ullrich, A., Coussens, L., Hayflick, J. S., Dull, T. J., Gray, A., Tam, A. W., Lee, J., Yarden, Y., Libermann, T. A., and Schlessinger, J. *et al.* (1984). Human epidermal growth factor receptor cDNA sequence and aberrant expression of the amplified gene in A431 epidermoid carcinoma cells. *Nature* **309,** 418–425.

Ushiro, H., and Cohen, S. (1980). Identification of phosphotyrosine as a product of epidermal growth factor-activated protein kinase in A-431 cell membranes. *J. Biol. Chem.* **255,** 8363–8365.

Ward, C. W., Hoyne, P. A., and Flegg, R. H. (1995). Insulin and epidermal growth factor receptors contain the cysteine repeat motif found in the tumor necrosis factor receptor. *Proteins* **22,** 141–153.

Wrann, M., Linsley, P. S., and Fox, C. F. (1979). Identification of the EGF receptor on 3T3 cells by surface-specific iodination and gel electrophoresis. *FEBS Lett.* **104,** 415–419.

Yamamoto, T., Ikawa, S., Akiyama, T., Semba, K., Nomura, N., Miyajima, N., Saito, T., and Toyoshima, K. (1986). Similarity of protein encoded by the human c-erb-B-2 gene to epidermal growth factor receptor. *Nature* **319,** 230–234.

Yamamoto, T., Nishida, T., Miyajima, N., Kawai, S., Ooi, T., and Toyoshima, K. (1983). The erbB gene of avian erythroblastosis virus is a member of the src gene family. *Cell* **35,** 71–78.

Yarden, Y., and Schlessinger, J. (1987). Epidermal growth factor induces rapid, reversible aggregation of the purified epidermal growth factor receptor. *Biochemistry* **26,** 1443–1451.

Yarden, Y., and Sliwkowski, M. X. (2001). Untangling the ErbB signalling network. *Nat. Rev. Mol. Cell. Biol.* **2,** 127–137.

THE THREE-DIMENSIONAL STRUCTURE OF INTEGRINS AND THEIR LIGANDS, AND CONFORMATIONAL REGULATION OF CELL ADHESION

By TIMOTHY A. SPRINGER* AND JIA-HUAI WANG†

*CBR Institute for Biomedical Research, Department of Pathology, Harvard Medical School, Boston, Massachusetts 02115; †Dana-Farber Cancer Institute, Departments of Pediatrics, Biological Chemistry and Molecular Pharmacology, Harvard Medical School, Boston, Massachusetts 02115

Abstract

Integrins are a structurally elaborate family of adhesion molecules that transmit signals bidirectionally across the plasma membrane by undergoing large-scale structural rearrangements. By regulating cell-cell and cell-matrix contacts, integrins participate in a wide-range of biological interactions including development, tissue repair, angiogenesis, inflammation and hemostasis. From a therapeutic standpoint, integrins are probably the most important class of cell adhesion receptors. Structural investigations on integrin-ligand interactions reveal remarkable features in molecular detail. These details include the atomic basis for divalent cation-dependent ligand binding and how conformational signals are propagated long

0065-3233/04 $35.00

distances from one domain to another between the cytoplasm and the extracellular ligand binding site that regulate affinity for ligand, and conversely, cytosolic signaling pathways.

I. Introduction

Members of the integrin family of adhesion molecules are non-covalently associated α/β heterodimers that mediate cell-cell, cell-extracellular matrix, and cell-pathogen interactions by binding to distinct but often overlapping combinations of ligands. Eighteen different integrin α subunits and eight different β subunits are present in vertebrates, forming at least 24 $\alpha\beta$ heterodimers and perhaps making integrins the most structurally and functionally diverse family of cell adhesion molecules (Hynes, 1992; Springer, 1994) (Fig. 1). Half of integrin α subunits contain inserted (I) domains, which are the major ligand binding domains when present (Humphries, 2000; Shimaoka *et al.*, 2002). The structural and functional diversity and complexity of integrins allow this family of adhesion molecules to play pivotal roles in broad contexts of biology including inflammation, innate and antigen-specific immunity, hemostasis, wound healing, tissue morphogenesis, and regulation of cell growth and differentiation (Hynes, 1992; Springer, 1994). Conversely, dysregulation of integrins is involved in the pathogenesis of many disease states, from autoimmunity to thrombotic vascular diseases to cancer metastasis (Curley *et al.*, 1999). Therefore, extensive efforts to discover and develop integrin antagonists have been made for clinical applications.

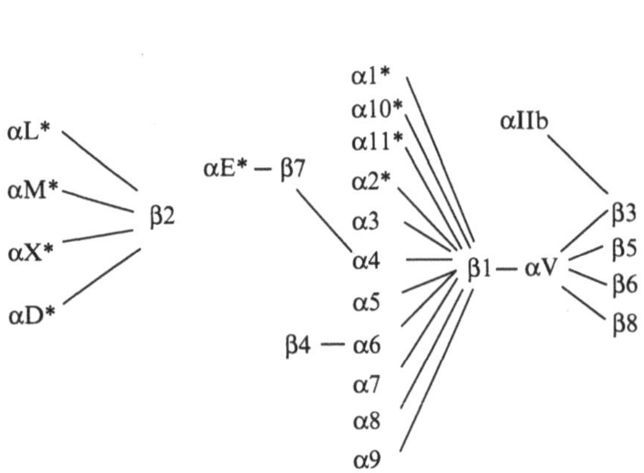

Fig. 1. Integrin α- and β-subunits form 24 heterodimers that recognize distinct but overlapping ligands. Half of the α subunits contain I domains (asterisks).

II. Conformational Regulation of Integrin Structure and Function

A. *Overall Picture of Integrin Heterodimers and Integrin Domains*

Integrins have a key role in organizing the cellular architecture of metazoan organisms. Furthermore, integrins are perhaps the only known family of adhesion molecules that can mediate migration of cells within tissues. Migration requires dynamic regulation both of integrin cytoplasmic domain association with cytoskeletal components, and extracellular domain binding to matrix components and cell surface glycoproteins. These functions are facilitated by the presence of two large, noncovalently associated subunits in integrins, each with a complex extracellular domain organization (Fig. 2). Each subunit spans the membrane, and the cytoplasmic domains with one exception (de Pereda *et al.*, 1999) are surprisingly short. Integrins are not found in prokaryotes, fungi, or plants but are present in the most primitive metazoans, including sponges and corals. By the time that *Drosophila* and *C. elegans* evolved, two of the main families of integrin α subunits were already present, which bind laminins and Arg-Gly-Asp (RGD)-containing matrix components (Hynes and Zhao, 2000). In the billion years of metazoan evolution, the number of extracellular domains and their organization in integrins has not changed, with the exception that in chordates, some integrin α subunits contain an inserted or I domain (Figs. 1 and 2).

The integrin α subunit ectodomain of >940 residues contains four domains (5 in I domain-containing integrins) and the β subunit of ~640 residues contains eight domains (Fig. 2). The crystal structure of the ectodomain of integrin $\alpha_V\beta_3$ revealed the structure, in a bent conformation, of eight of these domains, and a portion of a ninth (Fig. 2B) (Xiong *et al.*, 2001). The structures of integrin-EGF domains 2 and 3 of the β_2 subunit were determined by NMR, complementing a part of the missing portions of the $\alpha_V\beta_3$ structure (Beglova *et al.*, 2002). The N-terminal portions of the α and β subunits fold into the globular headpiece, which is connected through α and β tailpiece domains to the membrane (Du *et al.*, 1993; Takagi *et al.*, 2001; Weisel *et al.*, 1992; Xiong *et al.*, 2001). Dramatic rearrangements occur in the orientation of these domains during integrin activation (Fig. 3) (Takagi *et al.*, 2002). In this review we will focus primarily on the integrin headpiece, where ligands bind. However, it should be emphasized that in the bent conformation, extensive interfaces totaling over 4000 Å^2 of solvent accessible surface are buried between the headpiece and tailpiece and between the α tailpiece and β tailpiece. These interfaces stabilize the bent conformation, and are important in regulating

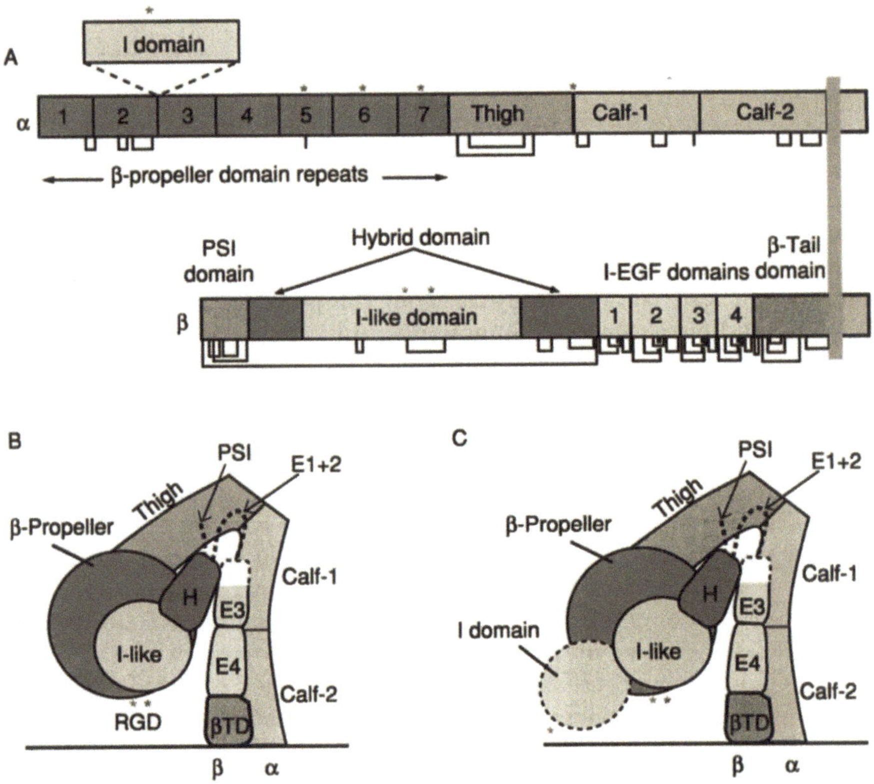

FIG. 2. Integrin architecture. (A) Organization of domains within the primary structure. Some α subunits contain an I domain inserted in the position denoted by the dotted lines. Cysteines and disulfide bonds are shown as lines below the stick figures. Red and blue asterisks denote Ca^{2+} and Mg^{2+} binding sites, respectively. (B) Arrangement of domains within the three-dimensional crystal structure of $\alpha_v\beta_3$ (Xiong *et al.*, 2001). Each domain is color coded as in A. (C) The structure in (B) with an I domain added. (See Color Insert.)

the equilibrium between the bent and extended integrin conformations (Fig. 3) (Luo *et al.*, 2003b; Takagi *et al.*, 2002).

B. vWF-Type A Domains: I Domain and I-Like Domain

All integrin β subunits and half of integrin α subunits contain von Willebrand factor-type A domains of about 200 amino acids, also referred to as the inserted (I) domain in the α-subunit and the I-like domain in the β subunit, respectively (Figs. 1, 2) (Humphries, 2000; Shimaoka *et al.*,

2002). Each domain adopts an α/β Rossmann fold with a metal ion dependent adhesion site (MIDAS) on the "top" of the domain, whereas its C and N-terminal connections are on the distal "bottom" face (Huang *et al.*, 2000; Lee *et al.*, 1995b; Shimaoka *et al.*, 2002; Xiong *et al.*, 2001) (Fig. 4A). Divalent cations are universally required for integrins to bind ligands and the metal at the MIDAS directly coordinates to a Glu or Asp residue in the ligand (Fig. 4B). This metal-dependent interaction through the MIDAS plays a central role in ligand recognition by the I and I-like domains.

C. I Domain

The I domain, which is inserted between blade 2 and 3 of the β-propeller domain of the α subunit (Fig. 2) (Springer, 1997), is a major ligand-binding domain and recognizes ligand directly when present (Diamond *et al.*, 1993; Michishita *et al.*, 1993). The ability of the I domain to bind ligand is controlled by conformational changes; the affinity of the I domain for its ligand is enhanced by downward axial displacement of its C-terminal α7-helix, which is conformationally linked to alterations of the MIDAS loops and Mg^{2+} coordination (Huth *et al.*, 2000; Shimaoka *et al.*, 2001, 2003b) (Fig. 4B, Fig. 5). In the case of $\alpha_L\beta_2$, compared to the default, low affinity conformation, downward displacements by one and two turns of α7-helix lead to intermediate- and high-affinity conformations with ~500 and 10,000-fold increased affinity, respectively (Shimaoka *et al.*, 2003b). A ratchet with a hydrophobic pocket stabilizes the β6-α7 loop in three alternative positions that correspond to downward movements of 0, 1, or 2 turns of 3_{10}-helix in the α7-helix (Fig. 5B, 5C). Disulfide bonds have been mutationally introduced that stabilize each of these three conformations and low, intermediate, and high affinity for ligand (Shimaoka *et al.*, 2003b) (Fig. 5A, 5C). Conversely, binding of ligand to the MIDAS of the I domain induces conformational change by stabilizing the high affinity conformation. This conformational change, which may be viewed as flowing from the ligand binding site to the α7-helix (Emsley *et al.*, 2000; Lee *et al.*, 1995b; Shimaoka *et al.*, 2003b) is identical to that flowing in reverse from the α7-helix to the ligand binding site (Shimaoka *et al.*, 2003b).

D. I-Like Domain

The β subunit I-like domain, which is inserted in the hybrid domain of the β-subunit (Fig. 2A), directly binds ligand in integrins that lack I domains in the α subunit (Fig. 3C, Fig. 6A). By contrast, when the I

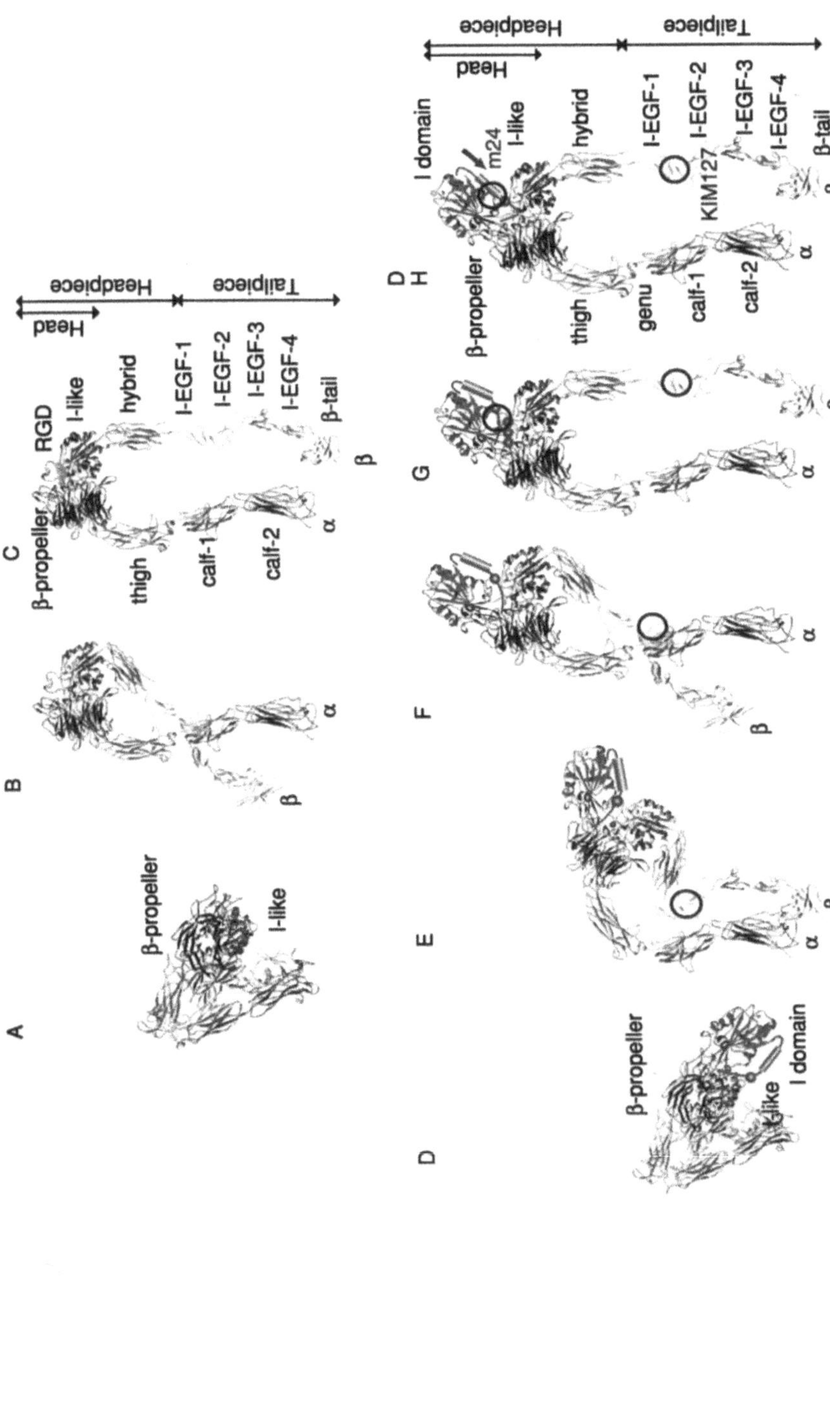
A
β-propeller
I-like
B
α
β
C
β-propeller
RGD
I-like
hybrid
thigh
calf-1
calf-2
α
β
I-EGF-1
I-EGF-2
I-EGF-3
I-EGF-4
β-tail
Head
Headpiece
Tailpiece
D
β-propeller
I-like
I domain
E
α
β
F
α
β
G
α
β
H
β-propeller
I domain
m24
I-like
hybrid
thigh
genu
calf-1
calf-2
α
β
KIM127
I-EGF-1
I-EGF-2
I-EGF-3
I-EGF-4
β-tail
Head
Headpiece
Tailpiece

domain is present the I-like domain functions indirectly by regulating the I domain (Fig. 3H, Fig. 6B). Compared to the I domain, the I-like domain contains two long loops, including one that is important for determining ligand specificity, and is referred to as the specificity-determining loop (SDL) (Takagi *et al.*, 1997). On either side of the MIDAS, the I-like domain contains two adjacent metal coordination sites, the ADMIDAS (*ad*jacent to *MIDAS*) and LIMBS (ligand-associated metal binding site) (Fig. 7A) (Xiong *et al.*, 2001, 2002).

The function of the I-like domain appears to be regulated by conformational changes similar to those observed in the I domain, in which a downward movement of the C-terminal α-helix allosterically alters the geometry of the MIDAS and increases the affinity for ligand (Shimaoka *et al.*, 2002; Takagi and Springer, 2002). Outward swing of the hybrid domain relative to the I-like domain (Figs. 3B, C and 6A) is thought to be coupled to the downward shift of the C-terminal α-helix of the I-like domain (Luo *et al.*, 2003b; Takagi *et al.*, 2002, 2003a).

E. *β-Propeller Domain*

The N-terminal region of the integrin α-subunit contains seven repeats of about 60 amino acids, which fold into a seven-bladed β-propeller domain (Springer, 1997; Xiong *et al.*, 2001) (Fig. 2A). A β-propeller domain with the same topology is also found in the trimeric G-protein β-subunit. The

---

FIG. 3. Conformational states for integrins. A–C. Model for $\alpha_v\beta_3$ integrin activation, with at least three conformations of the extracellular domain (Takagi *et al.*, 2002). (A) Bent, low affinity conformation. (B) Extended conformation with closed headpiece. (C) Extended conformation with open headpiece shown with bound RGD-mimetic peptide (green CPK). D–H. Model for $\alpha_L\beta_2$ integrin activation. (D) Bent conformation with low affinity. (E) and (F) $\alpha_L\beta_2$ with a closed headpiece and closed I domain in partially (E) or fully (F) extended states. (G) Extended conformation with open headpiece, and closed I domain, in the presence of α/β I-like allosteric antagonist, represented by three red spheres. (H) Extended conformation with open headpiece and open I domain. The models for all of the extracellular domains except for the I domain are based on conformational states of $\alpha_v\beta_3$ defined by negative stain electron microscopy (Takagi *et al.*, 2002), crystallography (Xiong *et al.*, 2002), NMR (Beglova *et al.*, 2002), and mapping of activation epitopes (Lu *et al.*, 2001a,c). The α_L I domain is a cartoon based on crystal structures (Shimaoka *et al.*, 2003b). The I domain is joined at the point of its insertion in the β-propeller domain but its orientation is arbitrary; it is shown at slightly larger scale for emphasis. The C-terminal I domain α-helix is represented by a red cylinder, and α_L Glu-310 in the linker as a blue sphere. The positions of activation epitopes m24 and KIM127 are circled only in the conformations in which they are thought to be exposed. (See Color Insert.)

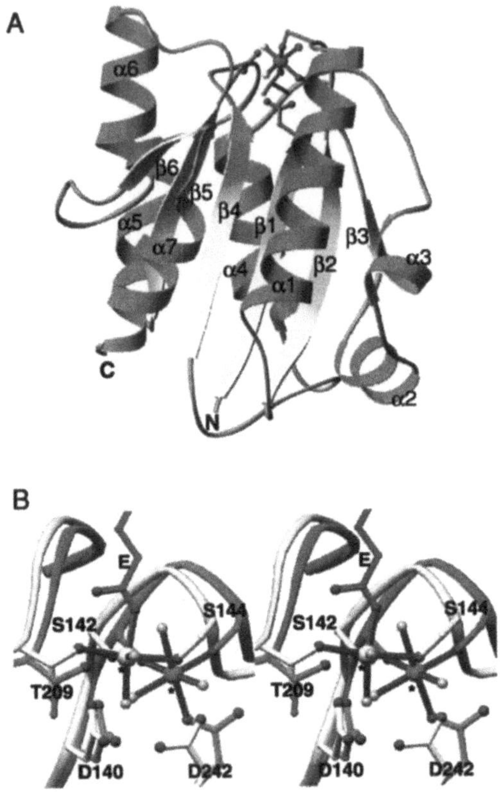

FIG. 4. α I domain structure and MIDAS conformational change. (A) Ribbon diagram of the α_M I domain in the open conformation (Lee *et al.*, 1995b). The β-strands (yellow), α-helices (cyan), and the N and C termini are labeled. The Mg ion is shown as a green sphere, and primary coordination bonds are blue. Side chains of residues that form primary or secondary coordinations to the metal ion are shown with grey bonds and carbon atoms and red oxygen atoms, and the oxygen of the ligand-mimetic Glu from another I domain is magenta. Coordinating water molecule oxygens are gold. Prepared with Ribbons (Carson, 1997). (B) Stereo view of alternative conformations of the α_M MIDAS. The backbone, coordinating side chain bonds, and metals (labeled with asterisks) are shown in yellow (open conformation) and cyan (closed conformation). The coordinating glutamate residue from the ligand-mimetic neighboring α_M I domain in α_M is in magenta. Primary coordination bonds to the metals are in blue. Oxygen atoms of the coordinating side chains and water molecules are red and gold, respectively. The 1IDO open α_M structure (Lee *et al.*, 1995a) was superimposed on the closed 1JLM α_M structure (Lee *et al.*, 1995b) using residues 132–141, 166–206, 211–241, 246–270, and 287–294. (See Color Insert.)

β-propeller domain directly participates in ligand recognition in those integrins that lack α I domains (Humphries, 2000).

F. *Structure of the Headpiece in Integrins that Lack I Domains*

The structure of $\alpha_v\beta_3$ reveals that the I-like domain makes extensive contact with the β-propeller domain, with the "top," ligand-binding faces of each domain oriented at about 90° to one another (Fig. 3A, B) (Xiong *et al.*, 2001). Loops in blades 2, 3, and 4 of the β-propeller domain are prominent in the ligand binding site. The structure of $\alpha_v\beta_3$ in complex with a cyclic peptide containing an Arg-Gly-Asp (RGD) sequence demonstrated binding to both the α and β subunits at the interface between the β-propeller and I-like domains (Fig. 3C). The Asp carboxylic acid side chain coordinates directly to the metal of the β subunit I-like domain MIDAS (Fig. 7A), while the Arg side chain binds to the α subunit β-propeller domain (Xiong *et al.*, 2002). Mapping by mutagenesis of residues important in binding to fibrinogen, the biological ligand of $\alpha_{IIb}\beta_3$, demonstrates a much larger interaction surface, centered on blades 2 to 4 of the α subunit β-propeller domain and the SDL loop of the β subunit I-like domain (Kamata *et al.*, 2001; Puzon-McLaughlin *et al.*, 2000).

Intriguing structural homology exists between the integrin β-propeller domain and the trimeric G protein β subunit, and between integrin I and I-like domains and G protein α subunits (Springer, 1997; Xiong *et al.*, 2001). Dissociation of these domains upon activation occurs in G proteins; however, in integrin activation there is little rigid body movement of the β-propeller domain relative to the I-like domain (Luo *et al.*, 2003a). Instead, ligand-binding affinity appears to be regulated primarily by conformational changes in loops of the I-like domain and possibly also in the β-propeller domain.

G. *Signal Transmission Across the Plasma Membrane*

Intracellular signaling pathways that are activated by other receptors (e.g., receptors coupled to G proteins or tyrosine kinases) impinge on integrin cytoplasmic domains and enhance the affinity of the extracellular headpiece for ligand. Recently the basis for bidirectional signal transmission across the membrane by integrins has been explained. The integrin α and β cytoplasmic tails associate with each other and constrain the integrin in its inactive form. Dissociation of the $\alpha\beta$ cytoplasmic tails by signals within the cell leads to the activation of the extracellular parts of the integrin (Kim *et al.*, 2003; Lu *et al.*, 2001d; Takagi *et al.*, 2001, 2002;

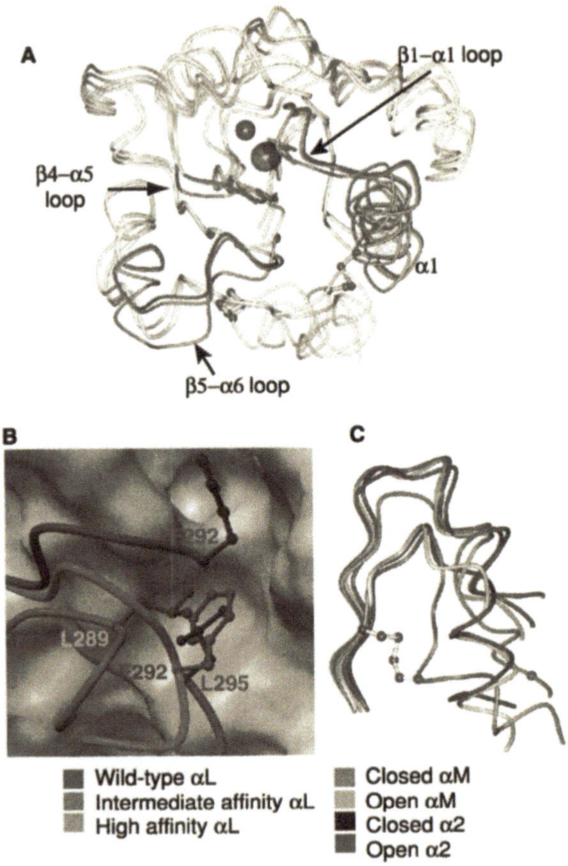

FIG. 5. Propagation of conformational change in the I domain. (A) The structures of the unliganded wild type, intermediate affinity, and high affinity αL I domains (Shimaoka *et al.*, 2003b). The three unliganded αL I domain backbones are shown superimposed and viewed centered on the MIDAS. Regions of the backbones that differ structurally are labeled and color keyed; other backbone regions are grey. The metal ions at the MIDAS and the atoms in the mutationally introduced, disulfide linked cysteine atoms are shown in the same colors as the backbone regions that differ; the cysteine sidechain bonds are yellow. The position of the missing metal ion in the high affinity structure is simulated with a smaller sphere. β6-α7 loop and α7 are shown in grey for clarity; differences in these regions are shown in (B) and (C). (B) The hydrophobic pocket that acts as a detent for the rachet-like movement of the β6-α7 loop. The backbone of the β6-α7 loop and the three residues that occupy the same hydrophobic pocket in the three different conformational states are color keyed. The pocket is shown as a GRASP van der Waals surface using the wild-type 1LFA structure with the residues from 287 to the C-terminus deleted. The upper hydrophobic pocket is also shown, which is occupied only in the closed conformation (by F292 which is shown in the wild type structure along with L295). For the closed conformation the β6-α7

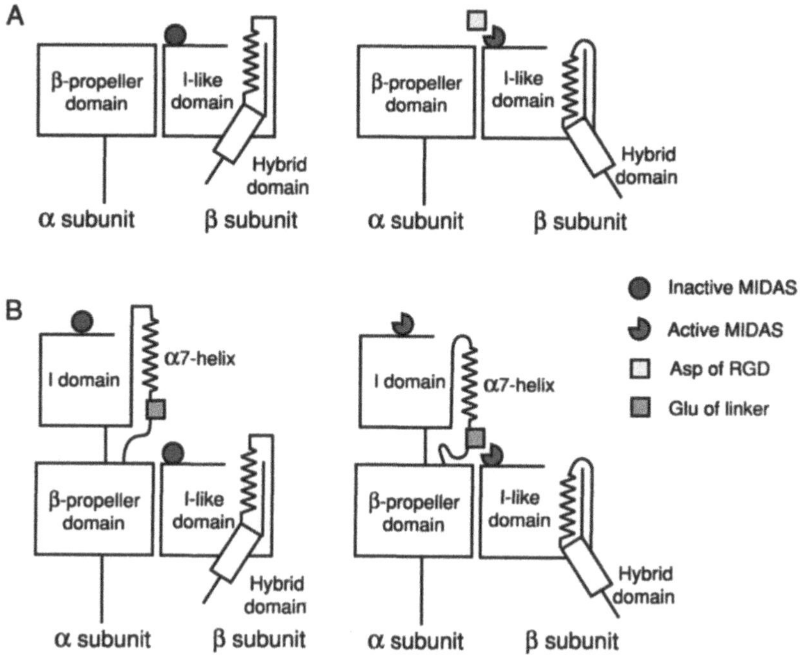

FIG. 6. Schematic drawing of conformational change in α I and β I-like domains and its coupling. (A) Integrins that lack α I domains. Swing-out of the hybrid domain is depicted as demonstrated for $\alpha_v\beta_3$ and $\alpha_5\beta_1$ (Takagi *et al.*, 2002, 2003a). Activation of the I-like domain is coupled to β I-like α7-helix downward displacement in both integrins that contain and lack α I domains (Luo *et al.*, 2004b; Yang *et al.*, 2004a), providing a mechanism for coupling activation to hybrid domain swing-out (Takagi *et al.*, 2002, 2003a). (B) In integrins that contain I domains, an invariant Glu in the linker following the α I α7-helix functions as an intrinsic ligand. Its binding to the β I-like MIDAS activates the α I MIDAS (Yang *et al.*, 2004b).

Vinogradova *et al.*, 2002). NMR studies reveal a weak association between the integrin α_{IIb} and β_3 subunit cytoplasmic domains, and that this association is disrupted by mutations that are known to activate integrins, and by binding of the β subunit cytoplasmic domain to the cytoskeletal

mainchain trace is broken between F292 and L295 for clarity. On the otherwise grey GRASP surface, hydrophobic residues are colored yellow. (C) The C-terminal fragments encompassing the β6-α7 loop for the three unligated α_L conformations and open and closed α_2 and α_M I domain structures (see color keys). The sidechain bonds of Cys-287 and Cys-294 in the designed disulfide bridge in the high affinity mutant are shown in yellow; the Cα atom of Cys-299 in the designed disulfide of the intermediate mutant is shown as a green sphere. (See Color Insert.)

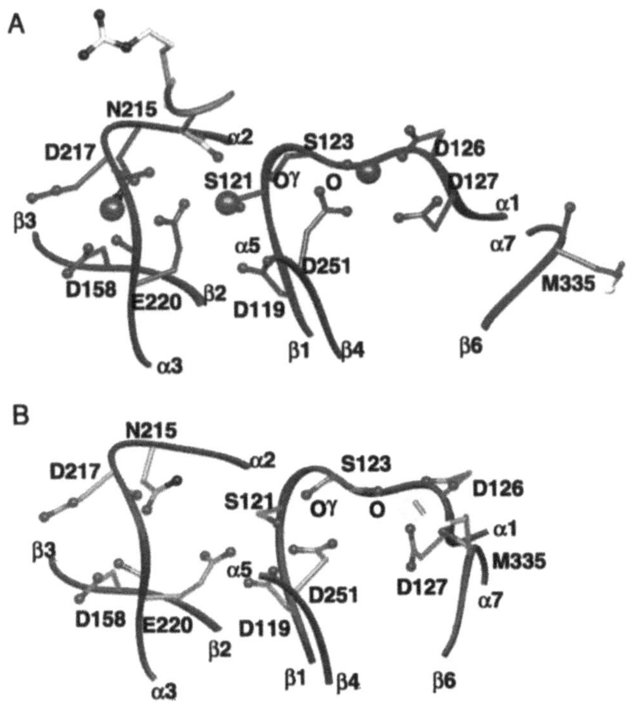

FIG. 7. The linear cluster of metal binding sites in the β I-like domain. (A) The liganded $\alpha_v\beta_3$ structure in Mn^{2+} with RGD (Xiong *et al.*, 2002) in gold. (B) The unliganded $\alpha_v\beta_3$ structure in Ca^{2+} (Xiong *et al.*, 2001). The structures were superimposed by using the I-like domain, so that equivalent positions in (A) and (B) are vertically aligned. The orientation is with the LIMBS, MIDAS, and ADMIDAS from left to right. Mn^{2+} and Ca^{2+} ions are large magenta and yellow spheres, respectively. All putative metal coordinating sidechains and backbone carbonyl groups are shown, with N, O, and S atoms in blue, red, and yellow, respectively. The carbonyl and sidechain oxygens of S123 are marked O and Oγ, respectively. (See Color Insert.)

protein talin, another known integrin activator (Garcia-Alvarez *et al.*, 2003; Vinogradova *et al.*, 2002; Weljie *et al.*, 2002). Fluorescence resonance energy transfer studies on $\alpha_L\beta_2$ demonstrate that the cytoplasmic domains separate upon activation of signaling pathways or talin activation within the cell, and upon binding of ligand to the extracellular domain (Kim *et al.*, 2003). Peptides containing integrin α and β subunit transmembrane domains form homodimers and homotrimers in detergent micelles (Li *et al.*, 2001, 2003); however, the physiologically relevant heterodimers fail to form under the same conditions, and homomeric interactions have not been demonstrated in lipid bilayers. Fluorescence resonance energy transfer (FRET) studies demonstrate that activation of $\alpha_L\beta_2$ does not lead to homomeric association (Kim *et al.*, 2004). Cysteine scanning and disulfide bond formation between the α_{IIb} and β_3 transmembrane domains demonstrates a specific α-helical interface between them in the resting state, and that the α_{IIb} and β_3 transmembrane domains separate after activation (Luo *et al.*, 2004a). Again in this system, heterodimer separation does not lead to homomeric interactions.

H. *Overall Conformational Change in the Ectodomain*

In the latent, low-affinity state, the integrin assumes a bent conformation (Fig. 3A) (Beglova *et al.*, 2002; Takagi *et al.*, 2002). The C-terminal residues in the α_v and β_3 subunits in the bent crystal structure are close together (Xiong *et al.*, 2001), and only a few residues thereafter the transmembrane domains begin. Therefore, separation of the transmembrane domains would destabilize the interface between the α subunit calf-1 and calf-2 leg domains and the β subunit I-EGF3, I-EGF4, and β-tail leg domains (Fig. 3). Destabilizing the interface between the α and β legs in the tailpiece would in turn destabilize the tailpiece-headpiece interface, and induce a switchblade-like opening to an extended conformation (Fig. 3). This extension re-orients the ligand binding face and exposes activation epitopes in the tailpiece. In the extended conformation, two different conformations of the headpiece, termed closed (Fig. 3B) and open (Fig. 3C), are seen (Takagi *et al.*, 2002). In the bent conformation only the closed conformation of the headpiece is present (Xiong *et al.*, 2001). Therefore, extension facilitates adoption of the open conformation of the headpiece, which corresponds to the ligand-bound and high affinity conformation (Takagi *et al.*, 2002, 2003a). Between the closed and open headpiece conformations, there is a marked change in orientation between the I-like domain and hybrid domain (Fig. 3B, C). Electron microscopic studies on both $\alpha_{IIb}\beta_3$ and $\alpha_5\beta_1$ directly demonstrate that binding induces the open headpiece with the swung-out hybrid domain (Takagi *et al.*, 2002, 2003a), and swing-out is also supported by solution X-ray scattering studies on ligand-bound $\alpha_5\beta_1$ (Mould *et al.*, 2003c) and exposure of epitopes on the hybrid domain (Mould *et al.*, 2003b). Notably, many antibody epitopes that are buried in the bent conformation become exposed in the extended conformation (Beglova *et al.*, 2002).

In solution and apparently on the cell surface as well, integrins are not fixed in a particular conformation, but equilibrate between them (Takagi *et al.*, 2002) (Fig. 3A–C). Whether the equilibrium favors the bent, low affinity conformation or the extended, high affinity conformation is affected by the presence of activating intracellular factors and the concentration of extracellular ligands. Activation by signals within the cell induces straightening and stabilizes the extended form. Binding of extracellular ligands also stabilizes the extended conformation and therefore enhances the separation of integrin tails, which transmits signals to the cytoplasm. Therefore, transition from the bent to the extended conformation is a bi-directional, allosteric mechanism for relaying conformational signals between the integrin headpiece and the cytoplasmic domains. All biological integrin ligands are multivalent, and therefore can also induce integrin clustering, which appears to be required, in addition to conformational change, for outside-in signaling.

I. Bistability of the β I-Like Domain

As described previously, EM studies show two distinct orientations between the I-like and hybrid domains. Ligand binding induces hybrid domain swing-out in solution as revealed by EM and apparently on cell surfaces as revealed by activation and "ligand-induced binding site" epitope exposure (Beglova *et al.*, 2002; Mould *et al.*, 2003b; Takagi *et al.*, 2002, 2003a). Consistent with these observations, it was suggested that downward movement of the β I-like α7-helix, analogous to that seen in α I domains, couples ligand binding affinity to hybrid domain swing-out (Takagi *et al.*, 2002). An N-glycosylation site introduced into the acute angle between the I-like and hybrid domains, designed to act as a wedge and stabilize the more obtuse angle in the open headpiece, activates ligand binding and integrin extension as predicted (Luo *et al.*, 2003b). An activating mutation in the β_1 I-like α7-helix supports the notion that this is an allosterically important region (Mould *et al.*, 2003b) and an activation epitope maps to the neighboring α1-helix (Mould *et al.*, 2002).

Two recent mutational studies provide direct evidence that downward movement of the α7-helix of the β I-like domain activates ligand binding. Mutagenic introduction of a disulfide bond between the β I-like β6-strand and α7-helix designed to stabilize the α7-helix in the conformation seen in the bent crystal structure abolished the ability to activate ligand binding by integrin $\alpha_{IIb}\beta_3$. By contrast, introduction of a disulfide bond into the β6-α7 loop that was designed to induce downward α7-helix displacement constitutively activated ligand binding (Luo *et al.*, 2004b). In another study, the effect of α7-helix displacement on the β6-α7 loop was mimicked by shortening the α7-helix by 4 residue deletions of ~1 turn of α7-helix. Two independent 4-residue deletions were tested in the β_2 and β_7 subunits. The $\alpha_L\beta_2$ mutants exhibit constitutive high affinity for ICAM-1, and full exposure of an activation epitope in the β_2 I-like domain. The $\alpha_4\beta_7$ mutants show the active phenotype of firm adhesion, rather than rolling adhesion, on the ligand mucosal addressin cell adhesion molecule-1 (MAdCAM-1) in a parallel plate flow chamber (Yang *et al.*, 2004a).

In the linear cluster of three metal ion binding sites in the β I-like domain (Fig. 7), the two outer sites regulate ligand binding by the middle MIDAS site (Chen *et al.*, 2003). These sites were mutated in the integrin $\alpha_4\beta_7$, and the effect studied on rolling adhesion on MAdCAM-1 substrates, which is hypothesized to be mediated by the extended conformation with the closed headpiece, with low or intermediate affinity, and firm adhesion on MAdCAM-1 substrates, which is hypothesized to be mediated by the extended conformation with the open headpiece, with high affinity (Chen *et al.*, 2003). Wild-type $\alpha_4\beta_7$ mediates rolling adhesion in Ca^{2+} and in contrast, firm adhesion in Mg^{2+} or Mn^{2+}. The middle MIDAS site is required for both rolling and firm adhesion. One polar site, the ADMIDAS, is required for rolling, because its mutation results in firm adhesion, no matter what divalent cation is present. The other polar site, the LIMBS, is required for firm adhesion, because its mutation results in rolling, no matter the divalent cation. The LIMBS mediates the positive regulatory effects of low Ca^{2+} concentrations, which synergize with Mg^{2+}, whereas the ADMIDAS mediates the negative regulatory effects of higher Ca^{2+} concentrations, which are competed by Mn^{2+}. The bipolar sites thus stabilize two alternative phases of adhesion. The higher affinity of Ca^{2+} than Mg^{2+} at the LIMBS is explained by the presence of three carbonyl O coordinations, for which Ca^{2+} has greater propensity than Mg^{2+} (Fig. 7A). The activating effect of Mn^{2+} and inhibitory effect of Ca^{2+} at the ADMIDAS may similarly be explained by the propensity of Ca^{2+} but not Mn^{2+} to form a carbonyl O coordination to the backbone of the $\beta6$-$\alpha7$ loop (Fig. 7B). This carbonyl coordination to Ca^{2+} seen in closed structures with Ca^{2+} (Fig. 7B) but with Mn^{2+} or Mn^{2+} plus ligand (Fig. 7A) (Xiong *et al.*, 2002) stabilizes the $\beta6$-$\alpha7$ loop against downward displacement with the $\alpha7$-helix in the postulated high affinity state of the β I-like domain. The ADMIDAS in some integrins appears required for downward movement of the $\alpha7$-helix, because in $\alpha_5\beta_1$ mutation of this site inhibits activation epitope exposure and ligand binding (Mould *et al.*, 2003a); this effect was not seen with $\alpha_4\beta_7$.

J. Activation of Integrins that Contain I Domains

In integrins that contain I domains, the I domain binds ligand, whereas the β-propeller and I-like domains have a regulatory role (Fig. 3D–H and Fig. 6B) (Lu *et al.*, 2001b,c). The bottom of the I domain is connected at its N- and C-termini to blades 2 and 3 of the β-propeller, respectively (Figs. 2A, 4A). A linker of ~15 residues C-terminal to the I domain must therefore locate near the β-propeller/I-like domain interface, i.e., to the ligand-binding face in integrins that lack I domains. This linker contains an invariant Glu residue, and its mutation abolishes ligand binding by both $\alpha_L\beta_2$ and $\alpha_M\beta_2$ (Alonso *et al.*, 2002; Huth *et al.*, 2000). It has been proposed that this universally conserved Glu residue is an "intrinsic ligand," and that binding of the activated β_2 I-like domain to this intrinsic ligand pulls the C-terminal α7-helix of the α I domain downward, and activates high affinity for ligand (Fig. 6B) (Alonso *et al.*, 2002; Shimaoka *et al.*, 2002; Takagi and Springer, 2002). Indeed, recent experiments demonstrate signal transmission between the α I linker and β I-like MIDAS by a receptor-like interaction (Yang *et al.*, 2004b). Second-site reversion with complementing Cys mutations was used to demonstrate the interaction. Mutation of the invariant Glu in the linker, α_L-Glu-310, to Cys or Ala, abolished ligand binding by $\alpha_L\beta_2$. Similarly, mutation of either of two non-metal-coordinating residues that flank the Mg^{2+} of the β_2 I-like MIDAS, β_2-Ala-210 or β_2-Tyr-115, to Cys also abolished ligand binding. By contrast, the double mutation of α_L-E310C with either β_2-A210C or β_2-Y115C forms a disulfide bond that constitutively activates ligand binding. Thus, the α7 helix and its linker function as a pull spring. The results suggest that in integrins that contain I domains, the Glu in the linker functions as an intrinsic ligand for the β I-like domain, and that when these integrins are activated, the β I-like MIDAS binds to the Glu, pulls the spring, and thereby activates the α I domain.

K. *Integrin Antagonists*

Three antagonists to integrin $\alpha_{IIb}\beta_3$, the platelet receptor for fibrinogen, have been approved by the FDA for acute treatment and prevention of thrombosis. These are an antibody Fab fragment, and two small molecule RGD mimetics, all of which competitively antagonize fibrinogen binding (Scarborough and Gretler, 2000). An antibody to the α I domain of $\alpha_L\beta_2$ which competitively blocks binding to ICAM-1 was recently approved for therapy of mild to severe plaque psoriasis, an autoimmune disease (Gordon *et al.*, 2003). Small molecule antagonists are also under development to $\alpha_v\beta_3$ and $\alpha_v\beta_5$ for blocking tumor metastasis, angiogenesis and bone resorption (Varner and Cheresh, 1996), and to β_2 integrins and α_4 integrins on leukocytes for treating autoimmune diseases and other inflammatory disorders (Giblin and Kelly, 2001; Yusuf-Makagiansar *et al.*, 2002).

Small molecule antagonists have been used to obtain important insights into the conformational regulation of integrin structure and function (Shimaoka and Springer, 2003). The ability of ligands and ligand-mimetic antagonists to induce conformational change in integrins has important clinical implications. $\alpha_{IIb}\beta_3$ antagonists expose LIBS epitopes on the surface of circulating platelets in treated patients. A small subset of patients have pre-existing LIBS antibodies, which cause thrombocytopenia after treatment (Billheimer *et al.*, 2002; Scarborough and Gretler, 2000). The current $\alpha_{IIb}\beta_3$ antagonists are designed for acute indications and have short *in vivo* half-lives. After drug dissociates from $\alpha_{IIb}\beta_3$ on live platelets, $\alpha_{IIb}\beta_3$ rapidly loses LIBS epitopes, and thus returns to the bent, low-affinity conformation (Kouns *et al.*, 1992). By contrast, if antagonist-treated platelets are fixed, and then antagonist is washed out, $\alpha_{IIb}\beta_3$ retains LIBS epitopes and has high affinity for fibrinogen (Du *et al.*, 1991). It is most important therapeutically that the antagonist-induced, high affinity conformation of $\alpha_{IIb}\beta_3$ is rapidly reversible *in vivo*; otherwise, antagonist dissociation and clearance would trigger thrombosis.

Two distinct classes of integrin antagonists have been discovered for the integrin $\alpha_L\beta_2$. In contrast to the competitive antagonists described previously, each acts allosterically (Shimaoka and Springer, 2003). α I allosteric antagonists bind between the α7-helix and the body of the I domain, and stabilize the domain in the closed, low affinity conformation (Kallen *et al.*, 1999) (Fig. 8). Mutant $\alpha_L\beta_2$ with the I domain locked in the high affinity, open conformation with a disulfide bond is completely resistant to inhibition (Lu *et al.*, 2001b). As assayed by epitope exposure, the conformation of the α_L I domain is coupled to the overall bent/extended integrin conformation through residue α_L-Glu-310 in the I domain linker; α I allosteric antagonists stabilize the bent conformation (Salas *et al.*, 2002, 2004; Shimaoka *et al.*, 2003a; Woska *et al.*, 2001).

α/β I-like allosteric antagonists appear to bind to a site in I domain-containing integrins (Fig. 3G) very similar to the site to which RGD-mimetics bind in integrins that lack I domains (Fig. 3C) (Shimaoka *et al.*, 2003a). These antagonists require the β_2 I-like MIDAS but not the α I domain for binding. Integrins that are stabilized in high affinity states either by disulfides within the α I domain, or by disulfides between the α I domain linker and residues adjacent to the β I-like MIDAS, are resistant to inhibition (Lu *et al.*, 2001b; Shimaoka *et al.*, 2003a; Yang *et al.*, 2004b). α/β I-like allosteric antagonists, like RGD-mimetic competitive antagonists, induce integrin extension as assayed by epitope exposure. However, they uncouple extension from I domain activation, revealing their binding site at the β I-like MIDAS to be a critical site for relaying conformational signals within I domain-containing integrins (Fig. 3G).

β_2 integrins primarily function in firm adhesion; they also can contribute to rolling adhesion, although markedly less well than α_4 integrins or selectins. Activated wild-type $\alpha_L\beta_2$, mutant high affinity $\alpha_L\beta_2$, and mutant

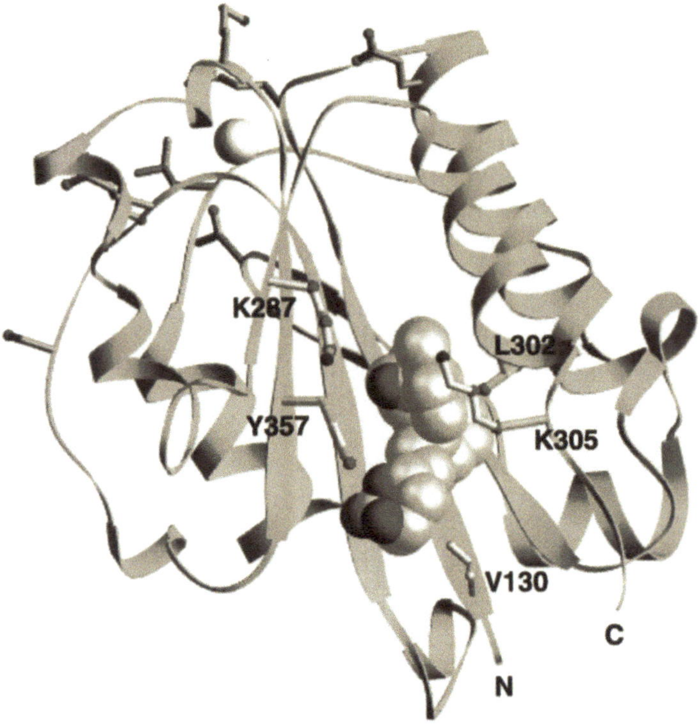

FIG. 8. Ribbon diagram of the α_L I domain in complex with an α I allosteric antagonist (Kallen *et al.*, 1999). The α I allosteric antagonist (shown by CPK with silver carbon atoms and red oxygen atoms) binds in the hydrophobic pocket underneath the C-terminal α-helix and stabilizes the I domain in the closed conformation. The side chains within the antagonist-binding pocket are shown with gold bonds and carbon atoms, red oxygen atoms, and blue nitrogen atoms. The residues critical for binding to ICAM-1 and ICAM-2 are shown with purple side chains and yellow sulfur, red oxygen, and blue nitrogen atoms. Note that these residues are located around the MIDAS, distal from the antagonist binding site. (See Color Insert.)

high affinity isolated, surface-expressed α_L I domains mediate firm adhesion. By contrast, resting, wild type $\alpha_L\beta_2$ and isolated surface-expressed wild type I domains mediate rolling adhesion (Salas *et al.*, 2002). α/β I-like allosteric antagonists inhibit firm adhesion, but because they induce the extended integrin conformation, they actually enhance rolling adhesion through $\alpha_L\beta_2$ (Salas *et al.*, 2004). This remarkable effect is seen both *in vitro* and *in vivo*.

III. Integrin/Ligand Interactions

The integrin counter-receptors on the cell surface include some members of immunoglobulin superfamily (IgSF) and the cadherin family. Ligands in ECM include collagen, laminin, fibronectin, fibrinogen, vitronectin, thrombospondin, and osteopontin. Structural investigations on integrin/ligand interactions have revealed intriguing features of these interactions with molecular detail. Figure 9 shows the integrin-binding domain structures of some representative ligands.

A. *IgSF Protein as Counter-Receptor*

Many of IgSF members are important cell adhesion molecules. What we will be discussing here is a subset of the IgSF proteins that share a greater sequence homology within the subset than to the other IgSF members and serve as counter-receptors for integrins. These are the intercellular adhesion molecules (ICAMs, including ICAM-1, ICAM-2, ICAM-3, ICAM-4 and ICAM-5, etc.), vascular adhesion molecule-1 (VCAM-1), and mucosal addressin cell adhesion molecule-1 (MAdCAM-1). As transmembrane glycoproteins, they are known to be composed of 2–9 IgSF domains in tandem

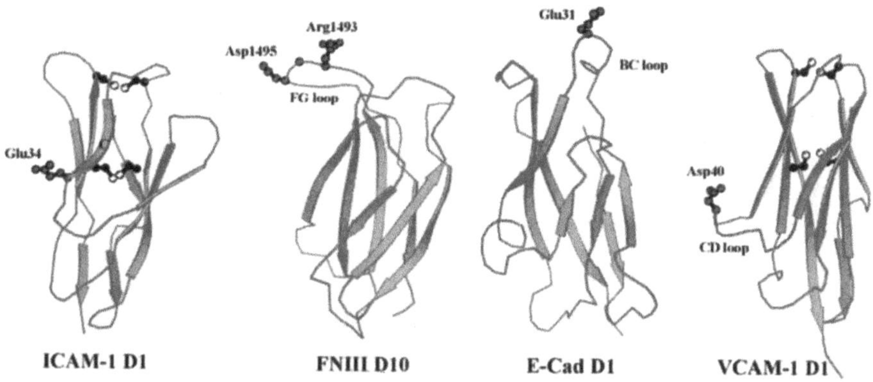

Fig. 9. Ribbon diagram of integrin-binding domain structures of four representative integrin ligands: ICAM-1 domain 1 (ICAM-1 D1), 10th fibronectin type III repeat (FNIII D10), E-cadherin domain 1 (E-Cad D1), and VCAM-1 domain 1 (VCAM-1 D1). In the figure, those key integrin-binding acidic residues are shown in red ball-and-stick model. They are located at the end of the C strand of IgSF domain (ICAM-1 D1), at the tip of the FG loop of FNIII domain (FNIII D10), at the tip of the BC loop of cadherin domain (E-Cad D1), and at the tip of the CD loop of IgSF domain (VCAM-1 D1), respectively. The figure demonstrates the diversity of integrin-binding modes to structurally very similar domains. The figure was prepared with MOLSCRIPT. (See Color Insert.)

on cell surface. The crystal structures of biologically the most important N-terminal two-domain fragment of ICAM-1 (Fig. 9), ICAM-2, VCAM-1 (Fig. 9) and MAdCAM-1 are known and reviewed (Wang and Springer, 1998). Here we offer a brief summary on the structural features of these molecules in the context of being integrin counter-receptors and then discuss most recent progress in structural studies on ICAM-1/$\alpha_L \beta_2$ interactions.

1. Both domain 1 and 2 (D1 and D2) belong to the IgSF fold, which consists of two anti-parallel β sheets linked together by an inter-sheet disulfide bond. The Ig fold has been grouped into three major sets: V set, C set and I set. The V and C sets originally were derived from antibody structure's variable and constant domains, respectively. The V set's two β sheets contain ABED and A′GFCC′C″ β strands, respectively, whereas the edge strands A′, C″ and C′ or D are missing in C set. I set is referred to as an intermediate set between V and C sets. Although, like a C set, topologically an I set is a truncation of V set, structurally it is close to V set. The key feature a V set and an I set share is that the A strand runs halfway down the domain in one sheet and kinks over at a *cis*-proline position to become A' strand, joining the opposite sheet by forming mainchain hydrogen bonds to the G strand in a parallel fashion (Harpaz and Chothia, 1994). Both D1 and D2 of these CAM molecules discussed here fall into I set of Ig fold category.

2. There are two unique characteristics of D1 structures of these CAM molecules. The first is an "extra" disulfide bond at the very tip of the molecule, between the BC loop and FG loop, not seen in any other IgSF structures. This disulfide bond brings these two otherwise widely open loops closed up and makes them impossible to serve as ligand-binding sites like their counterparts CDR1 and CDR3 antigen-combining loops in antibody do. The second is that an acidic residue that is known as the key integrin-binding residue is located either on the protruding CD loop (VCAM-1 and MAdCAM-1) or at the end of the C strand (ICAMs) near the bottom of the D1. Therefore, in contrast to antibodies and T cell receptors, the ligand-binding region of these CAMs is centered on the side of molecules' domain 1, but not at the top.

3. Structure-based mutation experiments have mapped the integrin-binding site on ICAM-2 domain 1 structure. The mapping applies to other ICAM family members as well. The site runs from the CD edge of the domain diagonally across the GFC β sheet (Casasnovas *et al.*, 1999). In the case of VCAM-1 and MAdCAM-1 the binding seems also to involve domain 2.

4. Although VCAM-1 and MAdCAM-1 differ from ICAMs in ligand-binding elements used and the type of integrins they interact (receptors for ICAMs are integrins with I domain, whereas those for VCAM-1 and MAdCAM-1 with no I domain), they all have an elaborate hydrogen bond networks around the key integrin-binding residues maintaining local conformation to facilitate binding.

5. Structure data suggested that each CAM molecule seems to evolve a special way to present its integrin-binding site in a proper orientation for recognition by an integrin on an opposing cell. ICAM-1, for example, appears to benefit from being a dimer on the cell surface. The dimeric ICAM-1 structure orients the key integrin-binding residue, Glu34, of domain 1 upwards for binding (Casasnovas *et al.*, 1998). On the other hand, ICAM-2 takes advantage of having three glycans on domain 2 forming a tripod-like architecture and retaining a relatively rigid D1-D2 bend to facilitate the key binding residue, Glu37, exposed (Casasnovas *et al.*, 1997).

Recently the first complex structure between binding domains of a CAM molecule and an integrin has revealed important features of how an IgSF molecule interacts with an integrin. This is the ICAM-1 D1–D3 fragment in complex with the $\alpha_L\beta_2$ I domain (Shimaoka *et al.*, 2003b) (Fig. 10, left panel). In the structure, the I domain has been engineered to introduce a disulfide bond that locks the I domain in an intermediately higher affinity compared to the wild type one. Several essential points are immediately apparent from the structure. First, domain 1 of ICAM-1 is exclusively engaged in binding, and there are no contacts between the ICAM-1 D2 and I domain. The ICAM-1 D1 docks onto I domain "top" face's shallow groove, with β strands on the D1 domain's CD edge parallel to the groove and roughly perpendicular to the central β sheet of the I domain. If one views the complex structure from ICAM-1 toward the I domain, the ICAM-1 D1's shadow would just cover the I domain. The key integrin-binding residue Glu34 at the end of C strand directly coordinates to the Mg^{2+} ion on the MADAS site at the groove's center. The binding does not induce any significant conformational changes in ICAM-1 molecule. Surrounding this Glu34, there is a ring of hydrophobic contact area between ICAM-1 and $\alpha_L\beta_2$. On the ICAM-1 side, the contributing residues include Pro36, Met64 and Tyr66, while on $\alpha_L\beta_2$ side these are Leu204, Leu205 and Met140. This non-polar environment around Glu34 should enforce the interaction between the Glu34 of ICAM-1 and Mg^{2+} ion on MIDAS, and provide major binding energy. Outside this hydrophobic ring are hydrophilic interactions. Apparently these interactions optimize the docking orientation. Of particular interests are many hydrogen bonds

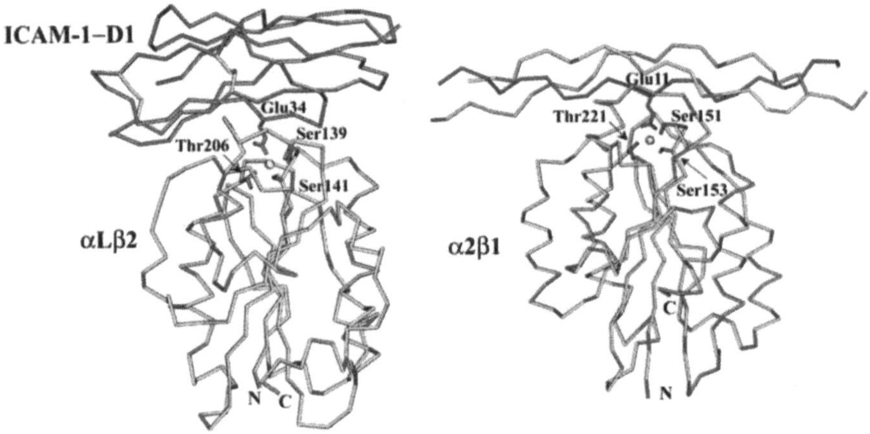

FIG. 10. Backbone diagram of two integrin/ligand complex structures. On the left panel is an intermediate affinity I domain from $\alpha_L\beta_2$ (in blue) in complex with ICAM-1 (in red). For clarity, only ICAM-1 D1 is shown. The sidechains of the key integrin-binding residue, Glu34, of ICAM-1 and MIDAS residues that directly coordinate the metal (in yellow) are illustrated. On the right panel is an I domain from $\alpha_2\beta_1$ (in blue) in complex with synthetic collagen-originated 21-mer triple-peptides (the major interacting middle strand is in red, and the other two strands are in yellow and green, respectively). Also, sidechains of the key integrin-binding residue, Glu11, of collagen and MIDAS residues of I domain are illustrated. The figure was prepared with SETOR. (See Color Insert.)

involving backbonds or sidechains that are conserved in ICAM family members. For example, the sidechain of the conserved Gln75 of ICAM-1 appears to make a pair of hydrogen bonds to sidechain of Asn207 of $\alpha_L\beta_2$ I domain. Another extremely interesting interaction is a salt bridge between Lys39 of ICAM-1 and Glu241 of I domain. The interaction can not take place without reorientation of Glu241 upon conversion of I domain from close to open conformation as described earlier. The Glu241 and the preceding Gly240 sit at the tip of a flexible $\beta4$-$\alpha5$ loop. Upon ligation, the I domain's Gly240 changes its dihedral angles from the β conformation to the area in Ramachadran map that only glycine is allowed. This glycine is conserved. The observation strongly suggests that the Gly240 facilitated swing of $\beta4$-$\alpha5$ loop and hence the sidechain of Glu241 appears to be crucial in switching the I domain into a high affinity state. The swing of the Glu241 and the resulting formation of the salt bridge of Glu241–Lys39 may be a general mechanism for regulating ligand binding by I domains.

B. E-Cadherin as Counter-Receptor

Cadherins are primary cell adhesion molecules responsible for Ca^{2+}-dependent cell–cell adhesion in vertebrate tissues. These transmembrane glycoproteins form a large cadherin superfamily. Depending on where they are mainly expressed, classical cadherins comprise E-, P-, N-, M- and VE-cadherin from epithelia, placenta, neurons, muscle and endothelial cells, respectively. Their extracellular portion consists of 5–6 cadherin folds, each being structurally related to an Ig fold (reviewed in Takeichi, 1990; Yagi and Takeichi, 2000). Crystal structures of domains of N- and E-cadherins are known (reviewed in Shapiro and Colman, 1998). More recently, the structure of an entire extracellular fragment of C-cadherin was reported (Boggon *et al.*, 2002). These structures showed how Ca^{2+} contributes to domain junction stability, and suggested a mechanism of homophilic adhesion.

E-cadherin is unique in that it not only, like other cadherin family members, mediates homophilic adhesion to establish and maintain cell-cell contacts, it also serves as a counter-receptor for integrins $\alpha_E\beta_7$ (Cepek *et al.*, 1994) and $\alpha_2\beta_1$ (Whittard *et al.*, 2002) in heterophilic adhesion. In fact, the interaction between E-cadherin on mucosal epithelial cells and $\alpha_E\beta_7$ on intraepithelial lymphocytes has been the best characterized tissue-specific interaction for lymphocyte retention. Although structure of binding domains between E-cadherin and $\alpha_E\beta_7$ is not available, mutagenesis data are informative of how these two molecules may interact (Higgins *et al.*, 2000; Taraszka *et al.*, 2000). Distinct from IgSF molecule described previously, E-cadherin's key integrin-binding residue, Glu31, is located on the N-terminal domain's BC loop at the very tip of the molecule (Fig. 9). Whereas this acidic residue coordinates metal ion on MADAS of $\alpha_E\beta_7$'s I domain, the sidechain of Phe298 in $\beta4$-$\alpha5$ loop from the I domain may extend into a hydrophobic pocket formed at E-cadherin's top part to strengthen binding.

C. Fibronectin as Counter-Receptor

Fibronectin (FN) is a large extracellular glycoprotein. It exists both as a soluble plasma protein and an aggregated fibril. The soluble plasma FN consists of two similar but not identical chains with molecular weight of 220–250 kDa each, joined together at their C-termini via two disulfide bonds. Each FN chain folds into 5–6 concatenate domains, each of which is specialized for binding to a particular molecule, such as collagen, heparin and integrin family member. Through these interactions, FN plays crucial roles in cell adhesion, cell morphology, cell migration, thrombosis, hemostasis, etc. An individual domain can in turn be divided into smaller homologous modules of 40–90 residues, designated as type I, II, and III repeats. The 90-residue long FN-III repeats are the most common in FN, and are building units for many other proteins as well. In fact, the FN-III repeat is among the most common of all protein domains in vertebrates (for review, see Hynes, 1990; Schwarzbauer, 1991).

Structures of all three type FN modules are known. NMR data showed that FN-I and FN-II are both a small module of only 45 and 55 residues, respectively, with a few short β strands connected together by two disulfide bonds. The FN-I structure may suggest its binding surface to fibrin (Williams *et al.*, 1994). The known structures of FN-II modules were from metalloproteinase-2 (MMP-2) (Briknarova *et al.*, 2001; Gehrmann *et al.*, 2002). The structures have revealed enzyme's possible binding site for substrates (gelatin, laminin and other ECM proteins). FN-III has received much more attention from the structural biology community for the exploration of its integrin and other ligand binding properties. The first serious effort was from Erickson and his colleagues in their X-ray analyses of the third FN-III repeat alone (Leahy *et al.*, 1992) and later a four-repeat FN-III, the FN7-10 (Leahy *et al.*, 1996). Although with distinction, these structures have clearly illustrated the module's similar topology to the C set IgSF domain, having two facing β sheets of ABE and C′CFG but without the disulfide bond in between. A triplet sequence motif Arg-Gly-Asp (RGD) from the tenth FN-III was first discovered to be the key integrin-binding site for fibronectin (Pierschbacher and Ruoslahti, 1984).

Subsequently this RGD tripeptide was found in numerous adhesive proteins (fibrinogen, collagen, vitronectin, osteoponin, thrombospondin, etc.) present in ECM and blood, and identified to be a common integrin-recognition motif for these proteins (Ruoslahti and Pierschbacher, 1987). Crystal structure of FN7-10 depicted that the RGD motif is located on the tip of an unusually long and protruding FG loop of the 10th domain such that despite a rod-like contiguous abutting arrangement of the four FN-III modules, the RGD motif is 10 Å away from the molecule body and still well accessible for integrin-binding (Fig. 9). Apparently, having the glycine in place favors the RGD motif to be in a type II′ β turn, which may represent a general integrin-binding conformation (Leahy *et al.*, 1996). Another feature of the structure FN7-10 is that it suggests a synergy region on domain 9 aligning well with the RGD site on domain 10 on the same face of molecule. The two sites are separated by 30–40 Å, which should allow a single integrin molecule to interact simultaneously with both the RGD and the synergy region. More recently structure of FN12-14 suggests a new class of integrin-binding sites (Sharma *et al.*, 1999). This is a sequence motif of Pro-Arg-Ala-Arg-Ile (PRARI) on FN14. The authors' speculation was based on: (1) This sequence PRARI in FN14 is at the same region as PHSRN sequence motif in FN9 that interact with integrins $\alpha_5\beta_1$ and $\alpha_{IIb}\beta_3$ in synergy with the RGD site in FN10; (2) Mutation of either arginine in PRARI motif to alanine impaired the integrin $\alpha_4\beta_1$ binding; (3) The PRARI motif is essentially conserved across all FN species. Since another ligand, heparin, binds to FN13 on a positively charged surface, which is on opposite face of integrin-binding site, the authors further hypothesized that loading or unloading heparin may facilitate integrin binding or *vice versa.*

Very recently, molecular electron microscopic image of a complex between the integrin $\alpha_5\beta_1$ headpiece and its physiological ligand, the FN7-10 fragment, for the first time has provided a view of how a protein ligand binds to an integrin (Takagi *et al.*, 2003b or #16024). The results convincingly demonstrate that the unliganded integrin is in a close conformation, and upon ligation the integrin transforms to an open conformation. The conformation changes are dramatic: the integrin's hybrid domain swings by 80° relative to the I-like domain in the β chain to let the large integrin molecule "stand up" from its bent state. The FN binding site is between the β-propeller and I-like domains of integrin headpiece. One interesting observation from this study is that it does not support the two-site binding mode. Their EM images and their kinetic analysis of $\alpha_5\beta_1$ binding to both wild type and mutant FN7-10 appear to suggest that the contribution of FN9's synergy site on to integrin-binding may only be an indirect one. In other words, the synergy site may just help orient the RGD loop on FN10 for binding and/or provide long-range electrostatic steering.

D. *Collagen as Counter-Receptor*

Collagens are a family of fibrous proteins present in all multicellular organisms. As a major component of skin and bone, they are the most abundant proteins in mammals, constituting about 25% of total protein mass. Their polypeptide chains have a regular $(\text{Gly-X-Y})_n$ sequence repeats, in which proline and 4-hydroxyproline frequently occupy the X and Y positions. This allows for three chains winding into a unique collagen helical structure with every third residue, the Gly, meeting in the center of three-helix bundle. Slightly different collagen chains assemble to create various type of collagens that either further polymerize to become collagen fibrillar or facilitate network-formation with other ECM proteins (see review Eyre, 1980).

Collagens as the major ECM proteins serve as counter-receptors for $\beta 1$ and $\beta 3$ integrins (Hynes, 1992). A specific collagen sequence motif Gly-Phe-HydroxyPro-Gly-Glu-Arg (GFOGER) has been identified as the integrin $\alpha_2\beta_1$ recognition site (Knight *et al.*, 2000). This finding stimulated the structure determination of a complex that contains the I domain from $\alpha_2\beta_1$ and a synthetic 21-mer peptide with the sequence of [Ac-(GPO)$_2$ GFOGER(GPO)$_3$-NH$_2$] that was shown to be in a triple helical conformation (Fig. 10, right panel) (Emsley *et al.*, 2000). In the structure, the authors named three collagen strands as "leading," "middle," and "trailing." The middle strand makes the majority of contacts with the I domain, whereas the trailing strand contributes fewer and the leading strand has no interactions. The key feature of the structure is the coordination of a glutamate sidechain of the middle strand to the metal ion on I domain's MADAS site. Like elaborated in preceding sections, three I domain metal-coordinating residues, Ser153, Ser155 and Thr221, lack a formal negative charge, which enables the metal ion to have strong bond to the collagen glutamate. One extremely interesting observation is that the collagen triple helical bundle also lies on a shallow groove of I domain with the collagen helical axis roughly perpendicular to the I domain's central β sheet, in a similar fashion as described for ICAM-1/$\alpha_L\beta_2$ complex structure. Aside from the key binding residue Glu, the neighboring residue Arg of the middle strand also participates in the specific interaction to Asp219 of I domain. The Arg of the trailing strand extends into a negatively charged well, close to a Glu256 of I domain. This Glu256 of $\alpha_2\beta_1$ is in an equivalent position as Glu241 of $\alpha_L\beta_2$ on the $\beta 4$-$\alpha 5$ loop, which undergoes the conformational flip upon ligation facilitated by a conserved Gly just one residue upstream from the Glu. It is worth mentioning again that in ICAM-1/$\alpha_L\beta_2$ binding domain complex structure, there is a salt bridge between this Glu from I domain to a Lys39 from ICAM-1. The Arg of the trailing strand, therefore, may well play a similar part as Lys39 does in ICAM-1/$\alpha_L\beta_2$ complex formation. It is not very clear what role another collagen residue, Phe, might play. Phe from the middle strand and trailing strand may just make hydrophobic contact to the I domain surface.

E. *Fibrinogen as Counter-Receptor*

Fibrinogen is a 340 kDa soluble glycoprotein found in the blood plasma of all vertebra animals. It functions *in vivo* as the precursor to an insoluble fibrin clot. Fibrinogen is a dimeric protein, with each protomer consisting of three chains, Aα, Bβ, and γ. The dimer of hetero-trimer forms a long molecule of sigmoidal shape (Yang *et al.*, 2001). The β and γ chains each have a globular β structured domain at C-terminus. They combine to make up a D nodule. The two protomers meet at their N-termini with six chains interwinding into a small E nodule. D and E nodules are connected by a long coiled coil, which has four helices, two from hairpin-shaped α chain and one each from β and γ chains. That gives the molecule a size about 470Å long (Yang *et al.*, 2001).

The multi-functional fibrinogen also serves as ligand for β_2 and β_3 integrins. For instance, fibrinogen is the most abundant ligand for integrin $\alpha_{IIb}\beta_3$ on platelets, and the binding results in the formation of platelet-fibrin thrombi *in vivo* (Hawiger, 1994). The fibrinogen-binding sites on integrin $\alpha_{IIb}\beta_3$ map at the edge of the top and on the side of the β-propeller domain. The experiments were done by swapping predicted loops between α_{IIb} and other integrins such as α_4 or α_5, alanine scanning mutagenesis and molecular modeling (Kamata *et al.*, 2001). On the other hand, the integrin-binding site of fibrinogen has been mapped to the C-terminus of γ chain. The site encompasses residues γ400–γ411. The same segment also bears donor and acceptor sites for factor XIIIa-catalyzed cross-linking of fibrin (Chen and Doolittle, 1971). A crystal structure of a chimeric protein having this segment attached to the C-terminus of chicken egg white lysozyme gave a view of how this tail's conformation might look (Donahue *et al.*, 1994). A noticeable feature of the segment is to have two wide turns. In particular, one loop is stabilized by four hydrogen bonds through residue γGln399's sidechain to mainchain atoms contributed from γHis400–γHis401-γ Leu402–γGly403. It is likely that at least this part of structure may well represent the native structure in fibrinogen γ chain as a ligand for $\alpha_{IIb}\beta_3$, but not an artifact from the chimeric molecule.

Mutagenesis data have implied that the β_2 integrins $\alpha_M\beta_2$ and $\alpha_X\beta_2$ may share a common recognition site on fibrinogen, also located at γ chain on the D nodule. There are two sequence motifs on γ chain that are suggestive of involving β_2-binding: γ190–γ202 (P1 site) and γ377–γ395 (P2 site) (reviewed in Ugarova and Yakubenko, 2001). We noticed that within the D nodule, the γ chain C-terminal domain is at the two extremes of the elongated fibrinogen molecule, which may explain its functional importance for ligand-binding of both β_2 and β_3 integrins. There is no doubt that a complex structure between domains of integrin and fibrinogen is urgently needed to reveal this extremely important interaction.

F. Laminin as Counter-Receptor

Laminins are a family of large extracellular matrix glycoproteins. The best known family member is laminin-1, or classic laminin. Each laminin is a heterotrimer assembled from α, β and γ chains, secreted and incorporated with other matrix proteins (nidogen, type IV collagen and perlecan) into basement membranes. Basement membranes are thin layers of specialized extracellular matrices surrounding cells or separating layer of cells of different lineages. The basement membranes are fundamental to tissue organization and physiology in all metazoans. As a key component of basement membrane, laminins also bind to cell surface receptors, particularly integrins and dystroglycan, and the interactions control cellular activities (for review, see Colognato and Yurchenco, 2000).

A laminin heterotrimer appears in an asymmetric cross shape. Like many other matrix proteins, laminins' each chain has a modular architecture with tandem arrays of globular N-terminal domain (LN), rod-like EGF repeats (LE), globular domain IV (L4) and coiled coil region. In addition, the C-terminus of α chain has five G domain (LG) repeat (Engel *et al.*, 1981; Hohenester *et al.*, 1999). The receptor- and matrix-bindings of laminins are exerted by different regions of the molecules. The high-affinity interactions with nidogen within matrix is mediated through the LE modules 3–5 from γ1 chain domain III (Gerl *et al.*, 1991), particularly the module 4 (Mayer *et al.*, 1993). The crystal structures of this laminin fragment (Stetefeld *et al.*, 1996) and its complex to nidogen G3 domain, a β-propeller domain (Takagi *et al.*, 2003b), have been reported. Integrins that serve for laminin receptor fall into β_1 and β_4 subgroups. While $\alpha_1\beta_1$ and $\alpha_2\beta_1$ integrins' binding sites were found within the N-terminal LN domain of laminin's α chain short arm (Colognato *et al.*, 1997), other integrins, such as $\alpha_6\beta_1$, $\alpha_6\beta_4$ and $\alpha_7\beta_1$ bind to laminin's C-terminal LG domain (Aumailley *et al.*, 1990; Lee *et al.*, 1992; Song *et al.*, 1992). Crystal structure of LG5 from laminin α2 was determined (Hohenester *et al.*, 1999). This domain of about 200 residues folds into an anti-parallel β sandwich, reminiscent of a lectin fold. However, it was believed that the multi-functional ligand-binding sites of LG domain are distinct from lectins' carbohydrate-binding sites. Whereas the carbohydrate-binding sites of a lectin cluster on the concave surface of the β sandwich, the LG domain may offer the rim of the β sandwich as its ligand-binding region, similar to antibody's antigen combining loops (Rudenko *et al.*, 2001). It is interesting to observe from the crystal structure that a calcium cation is bound to one edge of the sandwich by conserved acidic residues, and a sulfate ion has been identified in the structure to take part in calcium coordination, a mimic of ligand-binding. The cation-binding edge is on the opposite side of where the N- and C-termini of the LG domain meet (Hohenester *et al.*, 1999), which appears ideal for binding to ligands, probably also including integrins. So far there are no further experimental data available about how laminin and integrin interact.

References

Alonso, J. L., Essafi, M., Xiong, J. P., Stehle, T., and Arnaout, M. A. (2002). Does the integrin αA domain act as a ligand for its βA domain? *Curr. Biol.* **12,** R340–R342.

Aumailley, M., Timpl, R., and Sonnenberg, A. (1990). Antibody to integrin $\alpha 6$ subunit specifically inhibits cell-binding to laminin fragment 8. *Exper. Cell Res.* **188,** 55–60.

Beglova, N., Blacklow, S. C., Takagi, J., and Springer, T. A. (2002). Cysteine-rich module structure reveals a fulcrum for integrin rearrangement upon activation. *Nat. Struct. Biol.* **9,** 282–287.

Billheimer, J. T., Dicker, I. B., Wynn, R., Bradley, J. D., Cromley, D. A., Godonis, H. E., Grimminger, L. C., He, B., Kieras, C. J., and Pedicord, D. L. *et al.* (2002). Evidence that thrombocytopenia observed in humans treated with orally bioavailable glycoprotein IIb/IIIa antagonists is immune mediated. *Blood* **99,** 1–7.

Boggon, T. J., Murray, J., Chappuis-Flament, S., Wong, E., Gumbiner, B. M., and Shapiro, L. (2002). C-cadherin ectodomain structure and implications for cell adhesion mechanisms. *Science* **296,** 1308–1313.

Briknarova, K., Gehrmann, M., Banyai, L., Tordai, H., Patthy, L., and Llinas, M. (2001). Gelatin-binding region of human matrix metalloproteinase-2: Solution structure, dynamics, and function of the COL-23 two-domain construct. *J. Biol. Chem.* **276,** 27613–27621.

Carson, M. (1997). Ribbons. *Methods Enzymol.* **277,** 493–505.

Casasnovas, J., Pieroni, C., and Springer, T. A. (1999). Lymphocyte function-associated antigen-1 binding residues in intercellular adhesion molecule-2 (ICAM-2) and the integrin binding surface in the ICAM subfamily. *Proc. Natl. Acad. Sci. USA* **96,** 3017–3022.

Casasnovas, J. M., Springer, T. A., Liu, J.-h., Harrison, S. C., and Wang, J.-h. (1997). The crystal structure of ICAM-2 reveals a distinctive integrin recognition surface. *Nature* **387,** 312–315.

Casasnovas, J. M., Stehle, T., Liu, J.-h., Wang, J.-h., and Springer, T. A. (1998). A dimeric crystal structure for the N-terminal two domains of ICAM-1. *Proc. Natl. Acad. Sci. USA* **95,** 4134–4139.

Cepek, K. L., Shaw, S. K., Parker, C. M., Russell, G. J., Morrow, J. S., Rimm, D. L., and Brenner, M. B. (1994). Adhesion between epithelial cells and T lymphocytes mediated by E-cadherin and the $\alpha E\beta 7$ integrin. *Nature* **372,** 190–193.

Chen, J. F., Salas, A., and Springer, T. A. (2003). Bistable regulation of integrin adhesiveness by a bipolar metal ion cluster. *Nat. Struct. Biol.* **10,** 995–1001.

Chen, R., and Doolittle, R. F. (1971). Cross-linking sites in human and bovine fibrin. *Biochemistry* **10,** 4487–4491.

Colognato, H., MacCarrick, M., O'Rear, J. J., and Yurchenco, P. D. (1997). The laminin $\alpha 2$-chain short arm mediates cell adhesion through both the $\alpha 1\beta 1$ and $\alpha 2\beta 1$ integrins. *J. Biol. Chem.* **272,** 29330–29336.

Colognato, H., and Yurchenco, P. D. (2000). Form and function: The laminin family of heterotrimers. *Dev. Dyn.* **218,** 213–234.

Curley, G. P., Blum, H., and Humphries, M. J. (1999). Integrin antagonists. *Cell Mol. Life Sci.* **56,** 427–441.

de Pereda, J. M., Wiche, G., and Liddington, R. C. (1999). Crystal structure of a tandem pair of fibronectin type III domains from the cytoplasmic tail of integrin $\alpha 6\beta 4$. *EMBO J.* **18,** 4087–4095.

Diamond, M. S., Garcia-Aguilar, J., Bickford, J. K., Corbi, A. L., and Springer, T. A. (1993). The I domain is a major recognition site on the leukocyte integrin Mac-1 (CD11b/CD18) for four distinct adhesion ligands. *J. Cell. Biol.* **120,** 1031–1043.

Donahue, J. P., Patel, H., Anderson, W. F., and Hawiger, J. (1994). Three-dimensional structure of the platelet integrin recognition segment of the fibrinogen γ chain obtained by carrier protein-driven crystallization. *Proc. Natl. Acad. Sci. USA* **91,** 12178–12182.

Du, X., Gu, M., Weisel, J. W., Nagaswami, C., Bennett, J. S., Bowditch, R., and Ginsberg, M. H. (1993). Long range propagation of conformational changes in integrin $\alpha_{IIb}\beta_3$. *J. Biol. Chem.* **268,** 23087–23092.

Du, X., Plow, E. F., Frelinger, A. L., III, O'Toole, T. E., Loftus, J. C., and Ginsberg, M. H. (1991). Ligands "activate" integrin αIIbβ3 (platelet GPIIb-IIIa). *Cell* **65,** 409–416.

Emsley, J., Knight, C. G., Farndale, R. W., Barnes, M. J., and Liddington, R. C. (2000). Structural basis of collagen recognition by integrin $\alpha 2\beta 1$. *Cell* **101,** 47–56.

Engel, J., Odermatt, E., Engel, A., Madri, J. A., Furthmayr, H., Rohde, H., and Timpl, R. (1981). Shapes, domain organizations and flexibility of laminin and fibronectin, two multifunctional proteins of the extracellular matrix. *J. Mol. Biol.* **150,** 97–120.

Eyre, D. R. (1980). Collagen: Molecular diversity in the body's protein scaffold. *Science* **207,** 1315–1322.

Garcia-Alvarez, B., de Pereda, J. M., Calderwood, D. A., Ulmer, T. S., Critchley, D., Campbell, I. D., Ginsberg, M. H., and Liddington, R. C. (2003). Structural determinants of integrin recognition by talin. *Mol. Cell* **11,** 49–58.

Gehrmann, M., Briknarova, K., Banyai, L., Patthy, L., and Llinas, M. (2002). The col-1 module of human matrix metalloproteinase-2 (MMP-2): Structural/functional relatedness between gelatin-binding fibronectin type II modules and lysine-binding kringle domains. *Biol. Chem.* **383,** 137–148.

Gerl, M., Mann, K., Aumailley, M., and Timpl, R. (1991). Localization of a major nidogen-binding site to domain III of laminin B2 chain. *Eur. J. Biochem.* **202,** 167–174.

Giblin, P. A., and Kelly, T. A. (2001). Antagonists of $\beta 2$ integrin-mediated cell adhesion. *Annu. Rep. Med. Chem.* **36,** 181–190.

Gordon, K. B., Papp, K. A., Hamilton, T. K., Walicke, P. A., Dummer, W., Li, N., Bresnahan, B. W., and Menter, A. (2003). Efalizumab for patients with moderate to severe plaque psoriasis: A randomized controlled trial. *Jama* **290,** 3073–3080.

Harpaz, Y., and Chothia, C. (1994). Many of the immoglobulin superfamily domains in cell adhesion molecules and surface receptors belong to a new structural set which is close to that containing variable domains. *J. Mol. Biol.* **238,** 528–539.

Hawiger, J. (1994). *In* "Hemostasis and Thrombosis: Basic Principles and Clinical Practice", (R. W. Colman, J. Hirsch, V. J. Marder, and E. W. Salzman, Eds.), pp. 762. Lippincot, PA.

Higgins, J. M., Cernadas, M., Tan, K., Irie, A., Wang, J., Takada, Y., and Brenner, M. B. (2000). The role of α and β chains in ligand recognition by $\beta 7$ integrins. *J. Biol. Chem.* **275,** 25652–25664.

Hohenester, E., Tisi, D., Talts, J. F., and Timpl, R. (1999). The crystal structure of a laminin G-like module reveals the molecular basis of α-dystroglycan binding to laminins, perlecan, and agrin. *Mol. Cell* **4,** 783–792.

Huang, C., Zang, Q., Takagi, J., and Springer, T. A. (2000). Structural and functional studies with antibodies to the integrin β2 subunit: A model for the I-like domain. *J. Biol. Chem.* **275,** 21514–21524.

Humphries, M. J. (2000). Integrin structure. *Biochem. Soc. Trans.* **28,** 311–339.

Huth, J. R., Olejniczak, E. T., Mendoza, R., Liang, H., Harris, E. A., Lupher, M. L., Jr., Wilson, A. E., Fesik, S. W., and Staunton, D. E. (2000). NMR and mutagenesis evidence for an I domain allosteric site that regulates lymphocyte function-associated antigen 1 ligand binding. *Proc. Natl. Acad. Sci. USA* **97,** 5231–5236.

Hynes, R. (1990). Fibronectins. Springer-Verlag, New York.

Hynes, R. O. (1992). Integrins: Versatility, modulation, and signaling in cell adhesion. *Cell* **69,** 11–25.

Hynes, R. O., and Zhao, Q. (2000). The evolution of cell adhesion. *J. Cell. Biol.* **150,** F89–F96.

Kallen, J., Welzenbach, K., Ramage, P., Geyl, D., Kriwacki, R., Legge, G., Cottens, S., Weitz-Schmidt, G., and Hommel, U. (1999). Structural basis for LFA-1 inhibition upon lovastatin binding to the CD11a I-domain. *J. Mol. Biol.* **292,** 1–9.

Kamata, T., Tieu, K. K., Springer, T. A., and Takada, Y. (2001). Amino acid residues in the αIIb subunit that are critical for ligand binding to integrin αIIbβ3 are clustered in the β-propeller model. *J. Biol. Chem.* **276,** 44275–44283.

Kim, M., Carman, C. V., and Springer, T. A. (2003). Bidirectional transmembrane signaling by cytoplasmic domain separation in integrins. *Science* **301,** 1720–1725.

Kim, M., Carman, C. V., Yang, W., Salas, A., and Springer, T. A. (2004). LFA-1 clustering occurs as a consequence of, rather than a prelude to, cell adhesion. In preparation.

Knight, C. G., Morton, L. F., Peachey, A. R., Tuckwell, D. S., Farndale, R. W., and Barnes, M. J. (2000). The collagen-binding A-domains of integrins $\alpha1\beta1$ and $\alpha2\beta1$ recognize the same specific amino acid sequence, GFOGER, in native (triple-helical) collagens. *J. Biol. Chem.* **275,** 35–40.

Kouns, W. C., Kirchhofer, D., Hadvary, P., Edenhofer, A., Weller, T., Pfenninger, G., Baumgartner, H. R., Jennings, L. K., and Steiner, B. (1992). Reversible conformational changes induced in glycoprotein IIb-IIIa by a potent and selective peptidomimetic inhibitor. *Blood* **80,** 2539–2547.

Leahy, D. J., Aukhil, I., and Erickson, H. P. (1996). 2.0 angstrom crystal structure of a four-domain segment of human fibronectin encompassing the RGD Loop and synergy region. *Cell* **84,** 155–164.

Leahy, D. J., Hendrickson, W. A., Aukhil, I., and Erickson, H. P. (1992). Structure of a fibronectin type III domain from tenascin phased by MAD analysis of the selenomethionyl protein. *Science* **258,** 987–991.

Lee, E. C., Lotz, M. M., Steele, G. D. J., and Mercurio, A. M. (1992). The integrin $\alpha_6\beta_4$ is a laminin receptor. *J. Cell Biol.* **117,** 671–678.

Lee, J.-O., Bankston, L. A., Arnaout, M. A., and Liddington, R. C. (1995a). Two conformations of the integrin A-domain (I-domain): A pathway for activation? *Structure* **3,** 1333–1340.

Lee, J.-O., Rieu, P., Arnaout, M. A., and Liddington, R. (1995b). Crystal structure of the A domain from the α subunit of integrin CR3 (CD11b/CD18). *Cell* **80,** 631–638.

Li, R., Babu, C. R., Lear, J. D., Wand, A. J., Bennett, J. S., and Degrado, W. F. (2001). Oligomerization of the integrin αIIbβ3: Roles of the transmembrane and cytoplasmic domains. *Proc. Natl. Acad. Sci. USA* **98,** 12462–12467.

Li, R., Mitra, N., Gratkowski, H., Vilaire, G., Litvinov, S. V., Nagasami, C., Weisel, J. W., Lear, J. D., DeGrado, W. F., and Bennett, J. S. (2003). Activation of integrin αIIbβ3 by modulation of transmembrane helix associations. *Science* **300,** 795–798.

Lu, C., Ferzly, M., Takagi, J., and Springer, T. A. (2001a). Epitope mapping of antibodies to the C-terminal region of the integrin β2 subunit reveals regions that become exposed upon receptor activation. *J. Immunol.* **166,** 5629–5637.

Lu, C., Shimaoka, M., Ferzly, M., Oxvig, C., Takagi, J., and Springer, T. A. (2001b). An isolated, surface-expressed I domain of the integrin αLβ2 is sufficient for strong adhesive function when locked in the open conformation with a disulfide. *Proc. Natl. Acad. Sci. USA* **98,** 2387–2392.

Lu, C., Shimaoka, M., Zang, Q., Takagi, J., and Springer, T. A. (2001c). Locking in alternate conformations of the integrin αLβ2 I domain with disulfide bonds reveals functional relationships among integrin domains. *Proc. Natl. Acad. Sci. USA* **98,** 2393–2398.

Lu, C., Takagi, J., and Springer, T. A. (2001d). Association of the membrane-proximal regions of the α and β subunit cytoplasmic domains constrains an integrin in the inactive state. *J. Biol. Chem.* **276,** 14642–14648.

Luo, B.-H., Springer, T. A., and Takagi, J. (2003a). High affinity ligand binding by integrins does not involve head separation. *J. Biol. Chem.* **278,** 17185–17189.

Luo, B.-H., Springer, T. A., and Takagi, J. (2003b). Stabilizing the open conformation of the integrin headpiece with a glycan wedge increases affinity for ligand. *PNAS* **100,** 2403–2408.

Luo, B.-H., Springer, T. A., and Takagi, J. (2004a). A specific interface between integrin transmembrane helices and affinity for ligand. *Pub. Lib. Sci.* **2,** E113.

Luo, B.-H., Takagi, J., and Springer, T. A. (2004b). Locking the β3 integrin I-like domain into high and low affinity conformations with disulfides. *J. Biol. Chem.* **279,** 10215–10221.

Mayer, U., Nischt, R., Poschl, E., Mann, K., Fukuda, K., Gerl, M., Yamada, Y., and Timpl, R. (1993). A single EGF-like motif of laminin is responsible for high affinity nidogen binding. *EMBO J.* **12,** 1879–1885.

Michishita, M., Videm, V., and Arnaout, M. A. (1993). A novel divalent cation-binding site in the A domain of the β2 integrin CR3 (CD11b/CD18) is essential for ligand binding. *Cell* **72,** 857–867.

Mould, A. P., Askari, J. A., Barton, S., Kline, A. D., McEwan, P. A., Craig, S. E., and Humphries, M. J. (2002). Integrin activation involves a conformational change in the α1 helix of the β subunit A-domain. *J. Biol. Chem.* **277,** 19800–19805.

Mould, A. P., Barton, S. J., Askari, J. A., Craig, S. E., and Humphries, M. J. (2003a). Role of ADMIDAS cation-binding site in ligand recognition by integrin α5β1. *J. Biol. Chem.* **278,** 51622–51629.

Mould, A. P., Barton, S. J., Askari, J. A., McEwan, P. A., Buckley, P. A., Craig, S. E., and Humphries, M. J. (2003b). Conformational changes in the integrin βA domain provide a mechanism for signal transduction via hybrid domain movement. *J. Biol. Chem.* **278,** 17028–17035.

Mould, A. P., Symonds, E. J., Buckley, P. A., Grossmann, J. G., McEwan, P. A., Barton, S. J., Askari, J. A., Craig, S. E., Bella, J., and Humphries, M. J. (2003c). Structure of an integrin-ligand complex deduced from solution x-ray scattering and site-directed mutagenesis. *J. Biol. Chem.* **278,** 39993–39999.

Pierschbacher, M. D., and Ruoslahti, E. (1984). Cell attachment activity of fibronectin can be duplicated by small synthetic fragments of the molecule. *Nature* **309,** 30–33.

Puzon-McLaughlin, W., Kamata, T., and Takada, Y. (2000). Multiple discontinuous ligand-mimetic antibody binding sites define a ligand binding pocket in integrin αIIbβ3. *J. Biol. Chem.* **275,** 7795–7802.

Rudenko, G., Hohenester, E., and Muller, Y. A. (2001). LG/LNS domains: Multiple functions—one business end? *Trends Biochem. Sci.* **26,** 363–368.

Ruoslahti, E., and Pierschbacher, M. D. (1987). New perspectives in cell adhesion: RGD and integrins. *Science* **238,** 491–497.

Salas, A., Shimaoka, M., Chen, S., Carman, C. V., and Springer, T. A. (2002). Transition from rolling to firm adhesion is regulated by the conformation of the I domain of the integrin LFA-1. *J. Bio. Chem.* **277,** 50255–50262.

Salas, A., Shimaoka, M., Kogan, A. N., Harwood, C., von Andrian, U. H., and Springer, T. A. (2004). Rolling adhesion through an extended conformation of integrin αLβ2 and relation to α I and β I-like domain interaction. *Immunity* In press.

Scarborough, R. M., and Gretler, D. D. (2000). Platelet glycoprotein IIb-IIIa antagonists as prototypical integrin blockers: Novel parenteral and potential oral antithrombotic agents. *J. Med. Chem.* **43,** 3453–3473.

Schwarzbauer, J. E. (1991). Fibronectin: From gene to protein. *Curr. Opin. Cell Biol.* **3,** 786–791.

Shapiro, L., and Colman, D. R. (1998). Structural biology of cadherins in the nervous system. *Curr. Opin. Neurobiol.* **8,** 593–599.

Sharma, A., Askari, J. A., Humphries, M. J., Jones, E. Y., and Stuart, D. I. (1999). Crystal structure of a heparin- and integrin-binding segment of human fibronectin. *EMBO J.* **18,** 1468–1479.

Shimaoka, M., Lu, C., Palframan, R., von Andrian, U. H., Takagi, J., and Springer, T. A. (2001). Reversibly locking a protein fold in an active conformation with a disulfide bond: Integrin αL I domains with high affinity and antagonist activity *in vivo. Proc. Natl. Acad. Sci. USA* **98,** 6009–6014.

Shimaoka, M., Salas, A., Yang, W., Weitz-Schmidt, G., and Springer, T. A. (2003a). Small molecule integrin antagonists that bind to the β2 subunit I-like domain and activate signals in one direction and block them in another. *Immunity* **19,** 391–402.

Shimaoka, M., and Springer, T. A. (2003). Therapeutic antagonists and the conformational regulation of integrin structure and function. *Nat. Rev. Drug. Disc.* **2,** 703–716.

Shimaoka, M., Takagi, J., and Springer, T. A. (2002). Conformational regulation of integrin structure and function. *Annu. Rev. Biophys. Biomol. Struct.* **31,** 485–516.

Shimaoka, M., Xiao, T., Liu, J.-H., Yang, Y., Dong, Y., Jun, C.-D., McCormack, A., Zhang, R., Joachimiak, A., and Takagi, J. *et al.* (2003b). Structures of the αL I domain and its complex with ICAM-1 reveal a shape-shifting pathway for integrin regulation. *Cell* **112,** 99–111.

Song, W. K., Wang, W., Foster, R. F., Bielser, D. A., and Kaufman, S. J. (1992). H36-α7 is a novel integrin α chain that is developmentally regulated during skeletal myogenesis. *J. Cell. Biol.* **117,** 643–657.

Springer, T. A. (1994). Traffic signals for lymphocyte recirculation and leukocyte emigration: The multi-step paradigm. *Cell* **76,** 301–314.

Springer, T. A. (1997). Folding of the N-terminal, ligand-binding region of integrin α-subunits into a β-propeller domain. *Proc. Natl. Acad. Sci. USA* **94,** 65–72.

Stetefeld, J., Mayer, U., Timpl, R., and Huber, R. (1996). Crystal structure of three consecutive laminin-type epidermal growth factor-like (LE) modules of laminin γ1 chain harboring the nidogen binding site. *J. Mol. Biol.* **257,** 644–657.

Takagi, J., Erickson, H. P., and Springer, T. A. (2001). C-terminal opening mimics "inside-out" activation of integrin $\alpha5\beta1$. *Nature Struct. Biol.* **8,** 412–416.

Takagi, J., Kamata, T., Meredith, J., Puzon-McLaughlin, W., and Takada, Y. (1997). Changing ligand specificities of $\alpha v\beta 1$ and $\alpha v\beta 3$ integrins by swapping a short diverse sequence of the β subunit. *J. Biol. Chem.* **272,** 19794–19800.

Takagi, J., Petre, B. M., Walz, T., and Springer, T. A. (2002). Global conformational rearrangements in integrin extracellular domains in outside-in and inside-out signaling. *Cell* **110,** 599–611.

Takagi, J., and Springer, T. A. (2002). Integrin activation and structural rearrangement. *Immunological Rev.* **186,** 141–163.

Takagi, J., Strokovich, K., Springer, T. A., and Walz, T. (2003a). Structure of integrin $\alpha5\beta1$ in complex with fibronectin. *EMBO J.* **22,** 4607–4615.

Takagi, J., Yang, Y., Liu, J.-h., Wang, J.-h., and Springer, T. A. (2003b). Complex between nidogen and laminin fragments reveals a paradigmatic β-propeller interface. *Nature* **424,** 969–974.

Takeichi, M. (1990). Cadherins: A molecular family important in selective cell-cell adhesion. *Annu. Rev. Biochem.* **59,** 237–252.

Taraszka, K. S., Higgins, J. M., Tan, K., Mandelbrot, D. A., Wang, J. H., and Brenner, M. B. (2000). Molecular basis for leukocyte integrin $\alpha E\beta 7$ adhesion to epithelial (E)-cadherin. *J. Exp. Med.* **191,** 1555–1567.

Ugarova, T. P., and Yakubenko, V. P. (2001). Recognition of fibrinogen by leukocyte integrins. *Ann. NY Acad. Sci.* **936,** 365–385.

Varner, J. A., and Cheresh, D. A. (1996). Tumor angiogenesis and the role of vascular cell integrin $\alpha v\beta 3$. *Important Adv. Oncol.* 69–87.

Vinogradova, O., Velyvis, A., Velyviene, A., Hu, B., Haas, T. A., Plow, E. F., and Qin, J. (2002). A structural mechanism of integrin $\alpha_{IIb}\beta_3$ "inside-out" activation as regulated by its cytoplasmic face. *Cell* **110,** 587–597.

Wang, J.-h., and Springer, T. A. (1998). Structural specializations of immunoglobulin superfamily members for adhesion to integrins and viruses. *Immunol. Rev.* **163,** 197–215.

Weisel, J. W., Nagaswami, C., Vilaire, G., and Bennett, J. S. (1992). Examination of the platelet membrane glycoprotein IIb-IIIa complex and its interaction with fibrinogen and other ligands by electron microscopy. *J. Biol. Chem.* **267,** 16637–16643.

Weljie, A. M., Hwang, P. M., and Vogel, H. J. (2002). Solution structures of the cytoplasmic tail complex from platelet α IIb- and β 3-subunits. *Proc. Natl. Acad. Sci. USA* **99,** 5878–5883.

Whittard, J. D., Craig, S. E., Mould, A. P., Koch, A., Pertz, O., Engel, J., and Humphries, M. J. (2002). E-cadherin is a ligand for integrin $\alpha2\beta1$. *Matrix Biol.* **21,** 525–532.

Williams, M. J., Phan, I., Harvey, T. S., Rostagno, A., Gold, L. I., and Campbell, I. D. (1994). Solution structure of a pair of fibronectin type 1 modules with fibrin binding activity. *J. Mol. Biol.* **235,** 1302–1311.

Woska, J. R., Jr., Shih, D., Taqueti, V. R., Hogg, N., Kelly, T. A., and Kishimoto, T. K. (2001). A small-molecule antagonist of LFA-1 blocks a conformational change important for LFA-1 function. *J. Leukoc. Biol.* **70,** 329–334.

Xiong, J.-P., Stehle, T., Diefenbach, B., Zhang, R., Dunker, R., Scott, D. L., Joachimiak, A., Goodman, S. L., and Arnaout, M. A. (2001). Crystal structure of the extracellular segment of integrin $\alpha V\beta 3$. *Science* **294,** 339–345.

Xiong, J. P., Stehle, T., Zhang, R., Joachimiak, A., Frech, M., Goodman, S. L., and Arnaout, M. A. (2002). Crystal structure of the extracellular segment of integrin $\alpha V\beta 3$ in complex with an Arg-Gly-Asp ligand. *Science* **296,** 151–155.

Yagi, T., and Takeichi, M. (2000). Cadherin superfamily genes: Functions, genomic organization, and neurologic diversity. *Genes Dev.* **14,** 1169–1180.

Yang, W., Shimaoka, M., Chen, J. F., and Springer, T. A. (2004a). Activation of integrin β subunit I-like domains by one-turn C-terminal α-helix deletions. *Proc. Natl. Acad. Sci. USA* **101,** 2333–2338.

Yang, W., Shimaoka, M., Salas, A., Takagi, J., and Springer, T. A. (2004b). Inter-subunit signal transmission in integrins by a receptor-like interaction with a pull spring. *Proc. Natl. Acad. Sci. USA* **101,** 2906–2911.

Yang, Z., Kollman, J. M., Pandi, L., and Doolittle, R. F. (2001). Crystal structure of native chicken fibrinogen at 2.7 A resolution. *Biochemistry* **40,** 12515–12523.

Yusuf-Makagiansar, H., Anderson, M. E., Yakovleva, T. V., Murray, J. S., and Siahaan, T. J. (2002). Inhibition of LFA-1/ICAM-1 and VLA-4/VCAM-1 as a therapeutic approach to inflammation and autoimmune diseases. *Med. Res. Rev.* **22,** 146–167.

STRUCTURES OF AXON GUIDANCE MOLECULES AND THEIR NEURONAL RECEPTORS

By WILLIAM A. BARTON, JUHA-PEKKA HIMANEN, ALEXANDER ANTIPENKO, AND DIMITAR B. NIKOLOV

Structural Biology Program, Memorial Sloan-Kettering Cancer Center, New York, New York 10021

I. Introduction

The establishment of proper neuronal connectivity during development proceeds through a progressive series of choices made by the growing axons (Dickson, 2002; Goodman, 1996; Yu and Bargman, 2001). These choices are made as the extending axons navigate through the developing embryo to their synaptic targets. The direction of neurite growth is determined by the growth cone at the tip of the axon. This is a highly motile structure that is responsible for sensing and integrating multiple signals present in its environment. Furthermore, it translates these signals into cytoskeletal changes that determine the rate and direction of extension (Goodman, 1996). Guidance cues, also called *axon guidance molecules*, which provide directional information, are especially important for the navigation of growth cones. Intensive *in vitro* and *in vivo* studies in both

ADVANCES IN PROTEIN CHEMISTRY, Vol. 68

0065-3233/04 $35.00

invertebrate and vertebrate model systems resulted in the identification of several families of guidance cues, including the membrane-bound ephrins, the secreted netrin and slit proteins, and the semaphorins that contain both membrane-bound and secreted members. Their neuronal cell-surface receptors that recognize the guidance signals and transduce them to the inside of the cell were also identified. The axon guidance ligand/receptor families seem to share several common features (Fiore and Püschel, 2003). One such feature is the bi-functional nature of the axon guidance molecules that can act as attractive or repulsive agents depending on the composition of receptors and downstream adapter molecules present in the cells. In addition, different guidance molecules control axonal outgrowth in a cooperative fashion. For example, the commissural axons of the vertebrate spinal cord are guided by the coordinated action of members of the netrin family, acting as chemoattractant, and members of slit and semaphorin families acting as chemorepellents. Finally, the guidance signals are evolutionarily conserved. Members of all axon guidance families are found from nematodes to humans where they play similar roles during neuronal development.

One important scientific question in regard to the molecular mechanism of axon guidance is how does ligand binding by the growth cone receptors results in the initiation of intracellular signaling cascades. Structural biology, and in particular X-ray crystallography, in combination with other biophysical and biochemical methods, is best suited to address this question since it can provide direct visualization of the conformational rearrangements in the receptors that lead to the transduction of the signal through the cell membrane. In the past several years many of the genetically well characterized axon guidance molecules and their receptors have indeed become the focus of intensive structural studies. The Eph/ephrin ligand/receptor family is by far the best structurally characterized system and is, therefore, the main focus of this review. In addition we review recent structural and biochemical studies of semaphorins and of the Nogo receptor, a neuronal receptor that mediates the repulsive signals of several myelin-associated inhibitory proteins.

II. Eph Receptors and Ephrins

The Eph receptors represent the largest class of receptor tyrosine kinases (RTK) with 15 members (Boyd and Lackmann, 2001; Dodelet and Pasquale, 2000; Flanagan and Vanderhaeghen, 1998; Himanen and Nikolov, 2003b; Kullander and Klein, 2002). Initially described as important regulators of axon pathfinding (Drescher *et al.*, 1995; Henkemeyer *et al.*, 1996), they are now known to have roles in controling many other cell-cell interactions, including those of vascular endothelial cells and specialized epithelia (Adams *et al.*, 1999; Gerety *et al.*, 1999; Wang *et al.*, 1998). A number of membrane-attached ligands that are collectively termed ephrins were subsequently identified on the basis of their ability to bind Eph family receptors and activate their tyrosine kinase catalytic domain (Flanagan and Vanderhaeghen, 1998).

The Eph receptors and the ephrins are divided into two subclasses (A and B) based on sequence conservation and their binding affinities (Flanagan and Vanderhaeghen, 1998; Gale *et al.*, 1996). In general, the nine different EphA RTKs (EphA1-A9) promiscuously interact with five A-ephrins (ephrin-A1-A6), that are attached to the cell via a glycosylphosphatidylinositol (GPI) linkage. The EphB subclass receptors (EphB1-B6) interact with three different B-ephrins (ephrin-B1-B3), which are attached to the cell by a hydrophobic transmembrane region and a short cytoplasmic domain.

Eph receptors and ephrins mediate bi-directional signaling (Holland *et al.*, 1996; Kullander and Klein, 2002). Concomitant with activation of the Eph tyrosine kinase domain and transduction of the typical receptor forward signal into the receptor-expressing cell, the interactions between Ephs and ephrins also lead to transduction of a reverse signal into the ephrin-expressing cell. For the B-ephrins, the reverse signal involves phosphorylation of conserved tyrosine residues located in the cytoplasmic domain, while A-ephrins are suggested to signal through unidentified co-receptors.

The membrane attachment of both Ephs and ephrins limits their functional interactions to sites of cell-cell contact. Ligand binding leads to the multimerization of both molecules to distinct clusters within their respective plasma membranes, resulting in the formation of signaling centers restricted to zones of cell-cell and axon-cell contact (Davis *et al.*, 1994; Henkemeyer *et al.*, 1994). Most RTKs activate signaling pathways whose ultimate targets are the transcription patterns in the nucleus that regulate cell proliferation and/or differentiation. The Ephs and ephrins, on the other hand, like the other axon guidance ligand/receptor systems, regulate cell migration, repulsion, adhesion, or attachment to the extracellular matrix. The signaling cascades, which they initiate, therefore, ultimately converge on targets such as integrins and small Rho-family GTP-ases.

A. *Expression Profiles and Biological Roles of Ephs and Ephrins*

Analysis of Eph and ephrin protein expression has been carried out in several species including zebrafish, Xenopus, mouse, rat, and chicken embryos (reviewed in Boyd and Lackmann, 2001). They are broadly expressed during development but most abundant in the nervous system and, to a lesser extent, in vascular endothelium and specialized epithelia. Tight spatial and temporal regulation of expression is crucial because the ligand-receptor interactions can be quite promiscuous. The presence of multiple family members in the same place and time is nevertheless observed and may account for the lack of strong phenotypes in most single-molecule knockouts (Kullander and Klein, 2002).

Eph-ephrin interactions induce repulsive bi-directional signals and, therefore, ligands and receptors are often expressed on adjacent but non-overlapping cell groups, thus defining and maintaining developmental tissue boundaries, an example being the formation of hindbrain rhombomeres. In a similar way, ephrin-B2 and Eph-B3 are expressed in developing arteries and veins respectively thus preventing the intermixing of these two cell types (Adams *et al.*, 1999; Gerety *et al.*, 1999; Wang *et al.*, 1998).

While Eph-mediated signaling is critical for the correct development of many tissues and organs, their function is best understood in the nervous system. The formation of the retinocollicular topographic map represents the best-characterized example of gradient expression of Ephs and ephrins (Flanagan and Vanderhaeghen, 1998). Retinal neurons are guided via establishment of precise spatial expression patterns of the ephrin-A2, ephrin-A5, EphA2, EphA5, and EphA6 proteins in the retina and the superior colliculus (or tectum in chicken) leading to proper development of the visual system. In addition to the role of class-A molecules in the formation of the retinocollicular map, class-B Ephs and ephrins are also expressed in the retina and guide the movement of newly formed axons to the optic disc to form the optic nerve.

At the midline, ephrin-B3, prevents EphA4 expressing corticospinal neurons from recrossing. Furthermore, the complementary expression of class-B Ephs and ephrins on longitudinally projecting commissural neurons suggests a role in positioning longitudinal fiber tracts. During neuromuscular development, Ephs and ephrins guide both limb and carnial motor neurons to their appropriate target muscles. They also direct neuromuscular synaptogenesis, where similarly to the retinocollicular system, motor axons innervate muscles with some topographic order maintained by ephrin-A2 and ephrin-A5 signals (Kullander and Klein, 2002). Recent studies demonstrate that class-B Ephs and ephrins regulate the formation and function of CNS synapses. More specifically, the extracellular domain of EphB2 directly interacts with NMDA receptors. This association is triggered by ephrin-B ligand stimulation, and ultimately leads to increases in the density of NMDA receptor synaptic clusters and the number of postsynaptic release sites (Dalva *et al.*, 2000; Ethell *et al.*,

2001; Takasu *et al.*, 2002). Thus, lack of EphB2 impairs long-term potentiation (LTP) and long-term depression (Grunwald *et al.*, 2001; Henderson *et al.*, 2001). Interestingly, during early neural tube closure, expression of alternative splice isoforms of the ephrin-A5 receptor, EphA7, can switch the signaling from repulsion to attraction raising the possibility that suppression of the repulsive signaling of Eph receptors could convert them into cell adhesion molecules (Boyd and Lackmann, 2001).

B. Eph Receptor Structure

The Eph receptors (Fig. 1) contain a highly conserved N-terminal extracellular domain that is both necessary and sufficient for ligand recognition and binding (Labrador *et al.*, 1997). It is followed by a cysteine-rich region and two fibronectin-type III repeats that are suggested to be involved in receptor-receptor dimerization interactions (Lackmann *et al.*, 1998) and/or interactions with other proteins (e.g., NMDA receptors: Dalva *et al.*, 2000). The Eph intracellular side contains a juxtamembrane region (JM), a conserved kinase domain, a sterile alpha motif (SAM) domain, and a PDZ-binding motif (Kalo and Pasquale, 1999).

Structural studies reveal that the ligand-binding domain of the Eph receptors folds into a compact globular structure (Fig. 2A) with a β–sandwich jelly roll folding topology (Himanen *et al.*, 1998). The β-strands are connected by loops of varying length, including a long, well-ordered loop

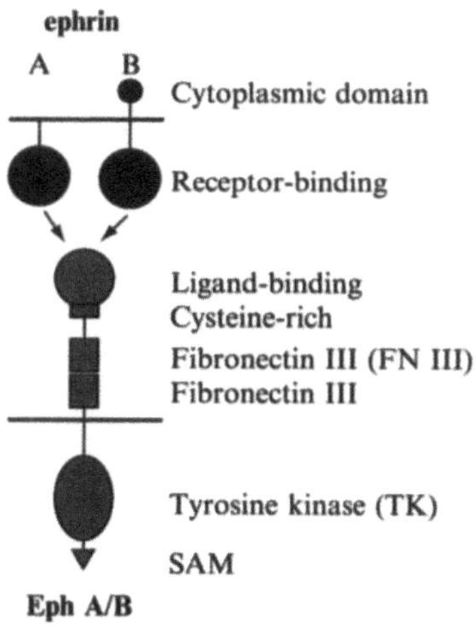

FIG. 1. Schematic representation of the domain organization of Eph receptors and ephrins. SAM, sterile alpha motif.

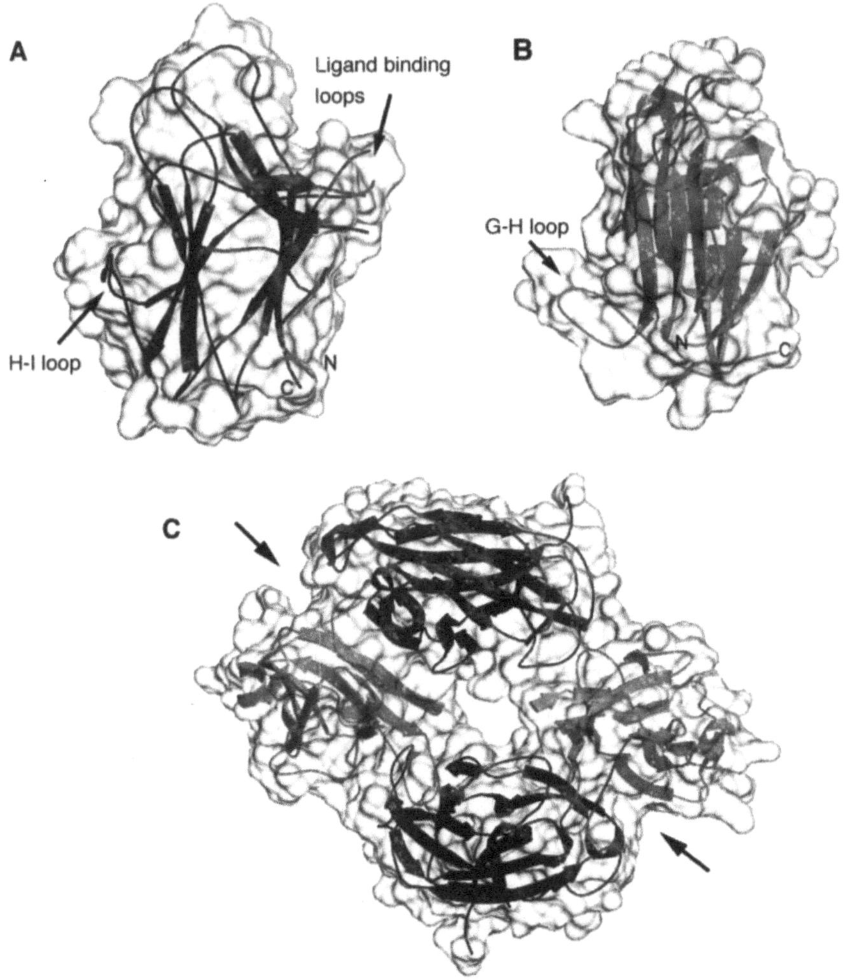

FIG. 2. Structures of the extracellular domains of Ephs and ephrins. The molecular surfaces (semi-transparent) are also indicated. (A) Structure of the ligand-binding domain of EphB2. The N- and C-termini of the molecule are labeled, as are the class-specificity loop (H-I) and the ligand-binding loops that are largely disordered in the absence of bound ephrin. (B) Structure of the extracellular receptor-binding domain of ephrin-B2. Indicated is the location of the receptor-binding G-H loop. (C) Structure of the EphB2/ephrin-B2 tetramer. Eph receptors are blue and ephrins are green. The high-affinity dimerization interfaces are indicated by arrows. (See Color Insert.)

(H-I), which packs against the concave β-sheet, and two partially disordered loops, which protrude from the middle of the convex β-sheet. These loops, (indicated on Fig. 2A) are involved in ligand recognition and binding (see further later).

The Eph ligand-binding domain shares no significant sequence homology to other known proteins. Nevertheless, they are weak structural homologues to some carbohydrate binding proteins (Himanen *et al.*, 1998; Rini, 1995). This observation suggested that the homology in molecular architecture might include the location of the ligand-binding site. Indeed, it is now known that the EphB2 tetramerization surface region (or "low-affinity" ligand-interaction region) is localized to the concave β-sheet around the H-I loop, in a position similar to the location of the carbohydrate binding site in lectins. Furthermore, the only sequence feature that is conserved within, but differs between the two Eph receptor subclasses, is the length of the H-I loop. This observation, as well as structure-based mutagenesis experiments demonstrating that a chimeric EphB2 receptor with an EphA3 H-I loop recognizes both A- and B-class receptors (Himanen *et al.*, 1998), has prompted the loop to be named "class specificity loop." Another intriguing possibility is that the carbohydrate moieties at the putative ephrin glycosylation sites may be directly involved in the ligand-receptor interactions. Indeed, it is interesting that in the crystal structure of the EphB2/ephrin-B2 complex (Himanen *et al.*, 2001) a conserved ephrin glycosylation site (Asn-39 in ephrin-B2) is located near the tetramerization ligand-receptor interface and the H-I loop. Of course, since bacterially-produced, non-glycosylated ephrin-B2 binds cognate receptors with high-affinity and specificity (Himanen *et al.*, 1998), carbohydrate-mediated interactions are unlikely to occur at the "high-affinity" ligand-receptor dimerization interface (see *Eph/Ephrin Recognition*), but could play a possible role at "lower-affinity" interfaces mediating ligand-receptor tetramerization and/or clustering.

C. *Ephrin Structure*

The ephrins possess a unique N-terminal receptor-binding domain (RBD) (Fig. 1), which is separated from the membrane via a linker of approximately forty amino acids. A-ephrins are attached to the cell via a GPI linkage. B-type ephrins have a transmembrane region and short but conserved 80-amino-acid cytoplasmic domain, which harbors a C-terminal PDZ-binding motif.

X-ray crystallographic studies revealed that the ephrin RBD has a globular β-barrel structure (Fig. 2B) with a Greek key folding topology

(Himanen *et al.*, 2001; Toth *et al.*, 2001). In the crystals of uncomplexed ephrin-B2, the molecule forms homodimers by burying the hydrophobic surface regions around the G-H loop. Since this same loop is involved in receptor binding, the ephrin molecules might exhibit significant rearrangement when their homodimers are displaced upon interaction with the Eph receptors.

As with the Eph receptors, the primary sequence of ephrins has no similarities to that of any other proteins. Unexpectedly though, the ephrins are structurally homologous to the cupredoxin/phytocyanin family of copper-binding proteins. In addition, there is a plant nodulin subfamily that shares sequence similarity with the phytocyanins indicating that they are also structurally similar to the ephrins. Interestingly, the nodulins are extracellular signaling proteins, which can be GPI-anchored to the cell surface. The significance of the ephrin-nodulin structural homology is yet unclear but may point to a more widely phylogenetically conserved role of this receptor-ligand architecture (Dresher, 2002).

The solution structure of the cytoplasmic domain of human ephrin-B2 was also determined (Song *et al.*, 2002) and revealed that the 48 N-terminal ephrin residues are unstructured and are prone to aggregation. The highly conserved 33 C-terminal residues, on the other hand, form a well-packed hairpin structure followed by the flexible PDZ-binding tail.

D. *Stoichiometry of the Eph/Ephrin Complex*

The first step in the initiation of Eph-mediated signaling is the recognition and binding of Eph receptors and ligands located on the opposed cell surfaces. Biophysical studies indicate that in solution, isolated Eph and ephrin ectodomains interact with each other via a two-step process. Initially, they form high-affinity heterodimers that are the predominant form of the complex throughout a large concentration range (Himanen *et al.*, 2002; Lackmann *et al.*, 1997). Dimer pairs can then tetramerize with a much lower K_D to form 2:2 heterotetramers. *In vivo*, at the cell membrane, the Eph/ephrin tetramers become further arranged in higher-order aggregates.

Regarding the stoichiometry of ligand-receptor complexes, the functional unidirectional signaling complexes are usually trimers composed of two receptors and one ligand, since this is sufficient to bring two intracellular signaling domains (e.g., kinase) together. On the other hand, a functional bi-directional signaling complex is expected to be at least tetrameric with two molecules initiating signaling in each direction. Crystallographic analysis of the EphB2/ephrin-B2 complex (Himanen *et al.*,

2001) indeed reveals a tetrameric complex forming a ring-like structure where each receptor interacts with two ligands and each ligand with two receptors (Fig. 2C). One of the ligand-receptor interfaces is very extensive and is presumably responsible for the initial 1:1 high-affinity binding. The second interface is much smaller and is suggested to be responsible for the assembly of the dimers into functional tetrameric 2:2 complexes. In the tetramer, the molecules are positioned with the C-termini of both ligands located on one side and the C-termini of the receptors on the other. This molecular arrangement allows the membrane-associated ephrins and Eph receptors to interact between the surfaces of adjacent cells.

Detailed structural analysis of the EphB2/ephrin-B2 interactions (see further later) revealed that the lower-affinity Eph/ephrin tetramerization interface displays a clear structural basis for subclass discrimination. This interface is centered around an Eph surface loop (called the "class-specificity" loop), whose length is invariant within each of the subclasses but differs by 4 residues between the two subclasses (Himanen *et al.*, 1998). By contrast, the high-affinity dimerization interface, although revealing structural features that could mediate subclass selectivity, does not provide a clear view into Eph class discrimination, and suggests the possibility of cross-subclass Eph/ephrin interactions (Himanen *et al.*, 2001). Indeed, it has been well established that the EphA4 receptor can interact with both A- and B-subclass ligands (Flanagan and Vanderhaeghen, 1998). In addition it was recently documented that two other molecules, ephrin-A5 and EphB2 also bind each other with significant affinity (Himanen *et al.*, 2004). Furthermore, exposure of EphB2-expressing cells to ephrin-A5 induced receptor autophosphorylation, and initiation of downstream signaling resulting in neurite retraction in a model system (Himanen *et al.*, 2004).

The crystal structure of the complex of the interaction domains of ephrin-A5 and EphB2 was then determined and revealed that this cross-class ligand receptor complex is a heterodimer, architecturally distinct from the tetramer observed in the EphB2/ephrin-B2 structure. Indeed, the minimal interaction domains of ephrin-A5 and EphB2 form only heterodimers not only in the crystals, but also in solution, even at millimolar concentrations. Nevertheless, ephrin-A5 activates and signals via the EphB2 receptor. This unexpected observation suggests that receptor activation either does not require precise positioning of nearby Eph receptors (and that just bringing the receptors together by the pre-clustered ligand is sufficient for activation of their kinase domains), or alternatively, that the orientation of the receptors with respect to each other is important, but can be maintained by interacting domains outside of the minimal ligand/receptor binding modules. The latter hypothesis is supported by studies implicating both extracellular (Cys-rich) (Lackmann *et al.*, 1998), and intracellular (SAM) (Stapleton *et al.*, 1999; Thanos *et al.*, 1999a) domains of Eph receptors in ligand-independent receptor-receptor interactions.

Interestingly, the high-affinity, intra-subclass interactions between A-class Ephs and ephrins, although not yet structurally characterized, also seem to involve ligand/receptor heterotetramers, similar to those observed in the EphB2/ephrin-B2 structure. Specifically, a random mutagenesis approach was recently employed to study the molecular determinants of the EphA3/ephrin-A5 recognition (Smith *et al.*, 2004). Selection and functional characterization of EphA3 point mutants with impaired ephrin-A5 binding from a yeast expression library defined three EphA3 surface areas that are essential for the EphA3/ephrin-A5 interaction. Two of these interfaces coincide with the dimerization and tetramerization interfaces identified in the EphB2/ephrin-B2 complex structure. The third falls within the Eph cys-rich region, outside of the structurally characterized interaction domains. Furthermore, functional analysis of selected EphA3 mutants revealed that all three Eph/ephrin contact areas are essential for the *in vivo* assembly of signaling-competent, receptor-ligand complexes. These results suggest that structural studies involving the complete Eph and ephrin ectodomains are required in order to fully understand the architecture of their functional signaling assembly.

E. *Eph/Ephrin Recognition*

Comparison of the structures of the bound and free EphB2, ephrin-B2, and ephrin-A5 molecules indicates that their recognition proceeds via an induced fit mechanism (or ligand-induced receptor folding). More specifically, the ligand-receptor dimerization interface centers around the long ephrin G-H loop, which inserts in a channel on the Eph surface (Fig. 3). The Eph loops forming the sides of this channel are unstructured in the unbound receptor, but fold upon ligand binding to generate an extensive interaction surface. The thermodynamic driving force for complex formation derives from the preference of the hydrophobic G-H loop to be away from the polar solvent. The secondary structure rearrangements in the interacting proteins are strictly localized to the interaction interface, and therefore downstream signaling is most likely triggered in the cytoplasmic sides not through conformational changes in the membrane-associated ligands and receptors but through their translocational rearrangements and repositioning relative to each other. However, it is yet to be determined how the other extracellular domains and regions of Ephs and ephrins, which are not visualized in the current crystal structures, participate in signal-initiating interactions.

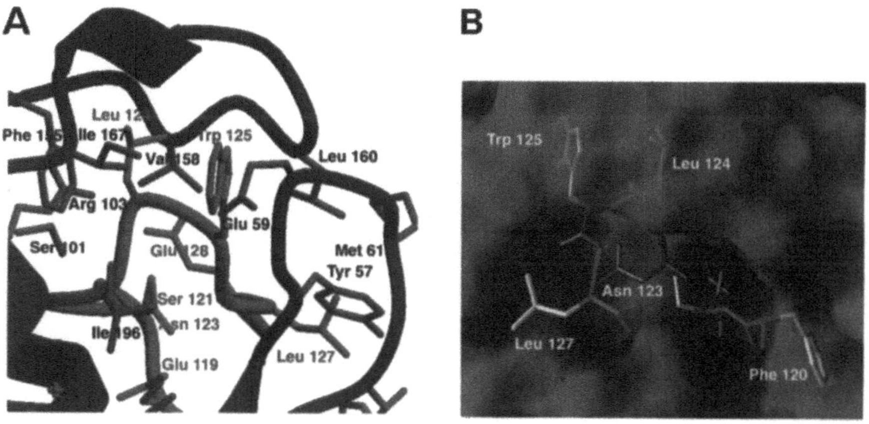

FIG. 3. Structure of the Eph/ephrin interface. (A) The channel on the surface of EphB2 viewed from the distal end of the ligand, showing the receptor (blue) and the incoming ephrin-B2 ligand (red). (B) Molecular surface representation of EphB2, viewed from position of the incoming ligand, showing the entrance of the hydrophobic channel with deeply inserted ephrin-B2 loop. The molecular surface of the receptor is color-coded according to curvature with concave regions in grey and convex – in green. Reproduced with permission from (Himanen *et al.*, 2001). (See Color Insert.)

In the case of the lower affinity cross-subclass ephrin-A5/EphB2 complex, the crystallographically observed dimer is architecturally similar to the high-affinity EphB2/ephrin-B2 dimer (Fig. 4), and its formation would also proceed via a similar induced-fit mechanism. Interestingly, the EphB2/ephrin-B2 high-affinity dimerization interface includes two distinct regions, only one of which is observed in the inter-class EphB2/ephrin-A5 complex. Specifically, in the EphB2/ephrin-B2 complex, in addition to the Eph-channel/ephrin-G–H_L-loop interactions, a second, structurally separate, contact area encompasses the ephrin docking site along the upper convex surface of the receptor (Fig. 4A). In this region the concave ephrin-B2 β-sandwich interacts with three EphB2 β-strands. Several ligand-receptor interactions likely to mediate subclass-specificity are located within this contact area and are described in detail in Himanen *et al.* (2001). For example key interacting side-chain pairs are composed of bulky polar residues, conserved in B-subclass ligands, positioned against small polar residues, conserved in the B-subclass receptors. In A-subclass members, the corresponding side chains are either hydrophobic or polar but with switched positions of the bulky and the small side chains. As a result, ligand-receptor combinations containing mixed subclasses would have either two bulky residues facing each other or polar residues facing

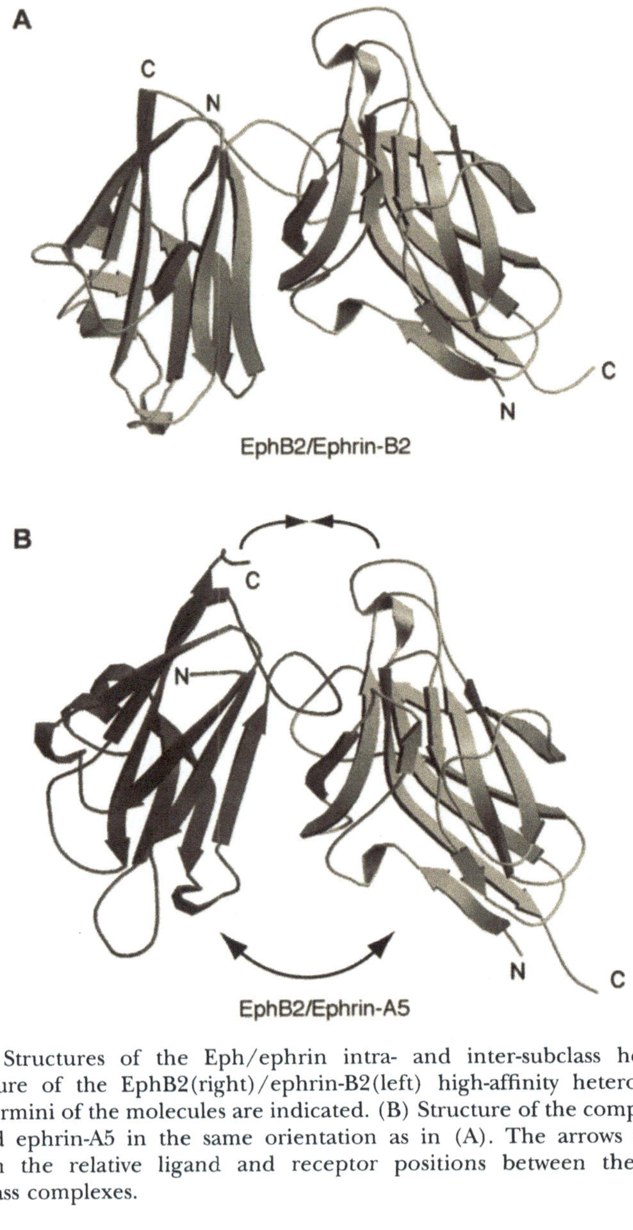

FIG. 4. Structures of the Eph/ephrin intra- and inter-subclass heterodimers. (A) Structure of the EphB2(right)/ephrin-B2(left) high-affinity heterodimer. The N- and C-termini of the molecules are indicated. (B) Structure of the complex between EphB2 and ephrin-A5 in the same orientation as in (A). The arrows indicate the changes in the relative ligand and receptor positions between the intra- and inter-subclass complexes.

hydrophobic ones, and would therefore be energetically unfavorable. Not surprisingly, the corresponding surface regions do not interact with each other in the EphB2/ephrin-A5 complex (bottom arrow on Fig. 4B), resulting in the much smaller overall inter-class ligand-receptor interface. Indeed, the EphB2/ephrin-A5 interface buries only 670 Å^2 from the total of ~10,000 Å^2 receptor surface area and 700 Å^2 from the total of ~8500 Å^2 ligand surface (1370 Å^2 total buried area). In contrast, formation of the EphB2/ephrin-B2 dimer buries 1150 Å^2 from the surface of the receptor and 1200 Å^2 from the surface of the ligand (2350 Å^2 total buried area).

Analysis of the EphB2/ephrin-A5 structure also suggests that while high-affinity cross-subclass Eph/ephrin interactions are likely to be involved in important steps during development, they are not a general phenomenon. Rather, they are probably constrained to a relatively small but strategic subset of Ephs and ephrins. Indeed only ephrin-A5 out of all A-class ephrins can bind to and activate EphB2. The structural explanation for this observation lies in the presence of several compensatory stabilizing interactions between ephrin-A5 and EphB2 that could not form between EphB2 and most other A-subclass ephrins. Specifically, the separation of ligand and receptor indicated with the lower arrow on Fig. 4B leads to slight pivoting around the receptor-bound G–H_L loop, and the approximation of ligand and receptor residues at the top of the G–H_L loop of ephrin-A5 and the G–H_R, and E–F_R loops of EphB2 (upper arrow on Fig. 4B). A critical ephrin residue in this region that participates in interactions observed only in the cross-class EphB2/ephrin-A5 structure is Leu120. Of the A-subclass ephrins, only ephrin-A5 and ephrin-A2 contain Leu, rather than Arg, at position 120 and, therefore, have the highest binding affinities for EphB2.

F. Eph Receptor Activation

In general, receptor tyrosine kinases are under tight regulation because they control critical cellular events including proliferation, differentiation, and death. Activation of all RTKs follows some general rules (Hubbard and Till, 2000; Pawson and Nash, 2000; Schlessinger, 2000; Simon, 2000): Ligand binding serves to bring together two catalytically-repressed kinase domains and hold them in an orientation favoring phosphorylation *in trans.* One of the monomers then phosphorylates regulatory sequences on the other monomer, leading to activation of its catalytic domain. The active kinase can then phosphorylate other molecules, initiating the downstream signaling cascade.

The usual mechanism for kinase activation involves the phosphorylation of the so-called activation loop within the kinase domain, which in its non-phosphorylated form blocks the kinase active site. The structural basis for this has been well established (Hubbard, 2002; Huse and Kuriyan, 2002; Johnson, 1996) In addition, it was recently suggested that in some RTKs, including the Eph, Kit, Flt, PDGF-β and TrkB receptors, the juxtamembrane region is also involved in regulation of the kinase activity (Hubbard, 2001; Kullander and Klein, 2002). The crystal structure of the intracellular region of EphB2 containing both the kinase and juxtamembrane domains revealed the molecular basis for this type of receptor activation (Wybenga-Groot *et al.*, 2001). The unphosphorylated JM region forms a well-ordered, mostly helical structure (Fig. 5A, blue) which interacts intimately with the N-terminal lobe of the kinase (orange) and weakly with the C-terminal lobe (green). The interactions cause the distortion of a key α-helix, leading to the displacement of a catalytic glutamate residue away from the active site and to kinase inactivation. Phosphorylation of the JM tyrosines (red) would lead to kinase activation since electrostatic forces will push the phosphorylated JM region away from the kinase, relieving the structural constraints that distort the active site. In addition, when phosphorylated, the solvent exposed JM region presents interaction sites for signaling proteins containing phosphotyrosine-binding motifs. This model is consistent with experiments, where mutations in the two

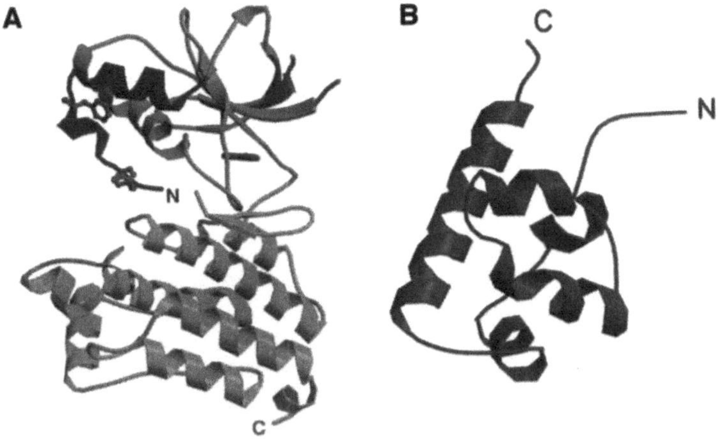

FIG. 5. Structures of the intracellular domains of Eph receptors. (A) Structure of the EphB2 kinase domain autoinhibited by the nonphosphorylated juxtamembrane region. The C-terminal kinase lobe is green and the N-terminal lobe is brown. The two regulatory tyrosine residues in the blue juxtamembrane region are shown in red. The ATP location in the active site is indicated in gray. (B) Structure of the monomeric sterile α motif (SAM) domain of EphA4. Reproduced with permission from (Himanen *et al.*, 2003b). (See Color Insert.)

juxtamembrane tyrosines to glutamic acids did not cause loss in catalytic activity but abolished SH2 binding (Holland *et al.*, 1997; Zisch *et al.*, 2000).

It is interesting that the JM region of the type I TGFβ receptor serine/threonine kinase plays a similar functional role, elucidated by the crystal structures of the molecule in its phosphorylated (Huse *et al.*, 2001) and non-phosphorylated (Huse *et al.*, 1999) states. In this case, the non-phosphorylated JM region acts together with another protein, FKBP12, to suppress the kinase activity again via a distortion of the active site. JM serine/threonine phosphorylation by the type II TGFβ receptor creates a binding site for the Smad2 protein, which displaces FKBP12, initiating TGFβ signaling. Although the structural details of kinase regulation of the Eph and the TGFβ receptors are different, they illustrate an emerging general property of receptor JM regions to perform double duty during signaling. In their unphosphorylated state they repress the kinase catalytic activity, and upon phosphorylation serve as docking sites for downstream signaling proteins (Hubbard, 2001).

G. *Initiation of Bi-Directional Signaling*

The recent structural studies of Eph receptors, ephrins, and their complex suggest a likely mechanism for initiation of bi-directional signaling (Himanen *et al.*, 2001, 2003a,b; Smith, 2004; Toth *et al.*, 2001) (Fig. 6): Prior to cell-cell contact, Ephs and ephrins are loosely pre-clustered at the cell surface, most likely in cholesterol-rich lipid rafts (Bruckner *et al.*, 1999). Following cell-cell contact, the Ephs and ephrins bind each other with 1:1 stoichiometry and nanomolar affinity via an extensive heterodimerization interface. The heterodimerization creates complementary interaction surfaces, which facilitate the formation of tetrameric complexes (Fig. 2C) thereby allowing the receptor kinase domains to phosphorylate each other and to initiate forward signaling. Importantly, the disruption of the potential ephrin-ephrin homodimers also results in a repositioning of the ephrin transmembrane and cytoplasmic domains, converting them from an inactive to an active configuration. This promotes tyrosine phosphorylation of the ephrin cytoplasmic tail by a yet unidentified kinase and initiation of reverse signaling. The tetramers are further arranged into the higher-order clusters observed *in vivo* at the sites of cell-cell contact. It should be noted that while Eph-ephrin tetramerization is likely to be sufficient for receptor phosphorylation, the formation of ligand/receptor clusters is absolutely essential for physiological signaling. Moreover the size of the aggregates may define the nature of the generated signals (Boyd and Lackmann, 2001; Stein *et al.*, 1998). Other Eph and ephrin regions are probably also involved in the final positioning of the bidirectional

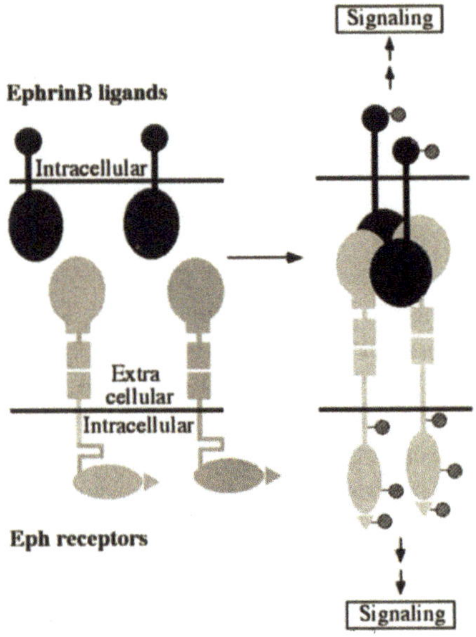

FIG. 6. Activation of the bi-directional Eph/ephrin signaling pathway. Upon binding, the molecules form a 2:2 circular heterotetramer causing the phosphorylation of their cytoplasmic domains. The domains symbols are the same as in Fig. 1. The phosphate groups are depicted as small circles.

signaling complexes, including the extracellular cysteine rich linker (Lackmann *et al.*, 1998) and intracellular SAM domain of the receptors (Stapleton *et al.*, 1999; Thanos *et al.*, 1999a,b) as well as the C-terminal PDZ domain binding sites of both receptors and ligands (Lin *et al.*, 1999; Torres *et al.*, 1998).

The Eph SAM domain was suggested to be involved in the clustering process because in many other proteins it mediates protein-protein interactions including homo- and hetero-oligomerization. All Eph receptors contain a conserved SAM domain but surprisingly its removal does not appear to disrupt Eph signaling (Boyd and Lackmann, 2001). Crystallographic (Stapleton *et al.*, 1999; Thanos *et al.*, 1999a,b) and NMR (Smalla *et al.*, 1999) studies reveal that the ~70 amino acid domain has a compact helical structure (Fig. 5B), but their implications for understanding the function of the Eph SAM domain during signaling are not straightforward because the different structures reveal different modes of homotypic interactions including dimerization (Stapleton *et al.*, 1999), multimerization (Thanos *et al.*, 1999a) or no interactions at all (Thanos *et al.*, 1999b). Moreover, biophysical studies reveal that isolated SAM domains are mostly monomeric in solution, and form dimers only at very high concentrations (Behlke *et al.*, 2001).

It is well established that Eph receptors accomplish their principle function of directing axonal growth by monitoring and responding to ephrin gradients. The formation of Eph/ephrin clusters may indeed be necessary for high sensitivity and wide dynamic range of the response to graded ligand densities as observed, for example, with the bacterial chemotaxis receptors (Kim *et al.*, 2002). It is interesting that a dynamic model of chemotaxis receptor clustering suggests that in the clusters the receptors adopt an arrangement where they interact with different neighboring receptors via their cytoplasmic and periplasmic domains (Kim *et al.*, 2002). One can thus speculate that in the Eph/ephrin clusters the cytoplasmic domains (kinase or SAM) of an activated receptor could also functionally interact with the intracellular domains of receptors from neighboring tetramers.

The Eph-ephrin interactions result in stable, long-lived complexes, and until recently it was not clear how these molecules could mediate repulsive cell-cell interactions. An elegant recent study by Flanagan and colleagues (Hattori *et al.*, 2000) reveals that ephrin-A2 co-localizes with a membrane-bound metalloprotease, ADAM10 (a.k.a. Kuzbanian). The authors suggest a general mechanism where, after a ligand- and a receptor-expressing cell interact, ADAM10 is activated and cleaves the Eph-bound ephrins from the cell surface allowing for the ligand-receptor complex to be internalized and degraded and the two cells to move apart.

Ephrin-A2 is cleaved by the ADAM protease domain in the membrane-proximal linker region, but the formation of a stable ADAM10-ephrin complex involves the ephrin receptor-binding domain and the ADAM10 disintegrin and/or cys-rich domains (Hattori *et al.*, 2000). Interestingly, a 15 amino-acid ephrin-derived peptide causes a concentration-dependent, synergistic activation of ADAM10-mediated ephrin-A2 cleavage in the presence of the EphA3 receptor. Since the peptide covers part of a putative motif found also in other ADAM10 substrates, it was suggested that it might directly bind to and activate the proteinase (Hattori *et al.*, 2000; Toth *et al.*, 2001). The structures of uncomplexed and Eph-bound ephrin reveal that this peptide corresponds to a region around the G-H loop, which is directly involved both in ephrin homodimerization and in receptor binding, and is unlikely to be exposed for direct interactions with ADAM10. An alternative model based on the assumption that ephrin homodimers inhibit ADAM10 cleavage while Eph/ephrin complexes activate it, can be proposed (Himanen and Nikolov, 2003b), which could explain the observation that the peptide-induced ephrin cleavage is suppressed at high peptide concentrations. At relatively low concentrations, the peptide might target the ephrin homodimerization interface inducing monomerization of the cleavage-resistant ephrin homodimers and their subsequent shedding. At high concentrations, the peptide would target the Eph/ephrin interface causing dissociation of their complexes and inhibition of ephrin shedding.

The ability of Ephs and ephrins to form highly-ordered multimeric assemblies might be responsible for an emerging novel, architectural (non-signaling) role they might play. More specifically, Eph receptors and ephrins could facilitate the assembly of stable membrane-associated platforms necessary for the organization of various cellular structures. The best-studied example is their role in the formation, organization, and function of some excitatory CNS synapses (Murai and Pasquale, 2002). Interestingly, based on the Eph/ephrin structures, and assuming that the extracellular regions of the molecules adopt relatively extended conformations, the distance between the membranes of two interacting cells can be estimated at approximately 20 nm, which corresponds to the size of the synaptic cleft.

III. Semaphorins, Neuropilins, and Plexins

A. *The Semaphorins and Their Receptors*

The semaphorins are, overall, the largest family of guidance molecules identified so far. Semaphorin 3A (Sema3A, also called collapsin), was the first molecularly characterized chemorepellent, originally described as an activity in membrane preparations from embryonic or adult chick brain that induced the collapse of sensory growth cones (Kapfhammer and Raper, 1987; Luo *et al.*, 1995; Raper and Kapfhammer, 1990). The molecular cloning of Sema3A revealed that it shares significant sequence homology with Fasciclin IV (later renamed Sema-1a), a gene identified in a screen for neuronal glycoproteins in grasshoppers (Kolodkin *et al.*, 1992, 1993). All semaphorins are characterized by a conserved N-terminal domain, named the semaphorin (or sema) domain (Messersmith *et al.*, 1995; Püschel *et al.*, 1995), which contains approximately 500 amino acids, including 14 highly conserved cysteine residues, and two potential N-glycosylation sites (Fig. 7).

More than 30 semaphorins have been identified to date that can be divided into 8 classes based on sequence similarity and domain organization. (Fiore and Püschel, 2003; Mark *et al.*, 1997; Semaphorin Nomenclature Committee, 1999). Classes 1 and 2 are found only in invertebrates, while classes 3 to 7 are found in vertebrates. Class "V" members

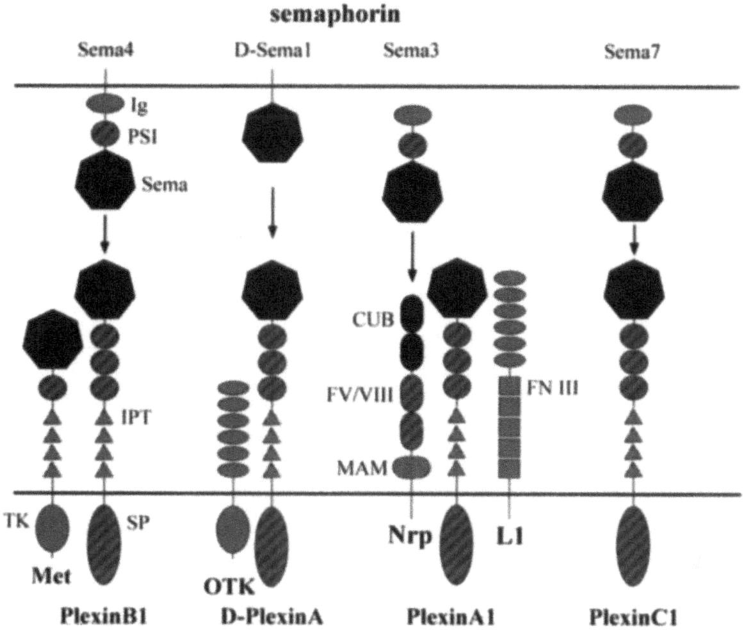

FIG. 7. Schematic representation of the domain organization, as well as ligand-, receptor-, and co-receptor-binding preferences of the semaphorins and their neuronal receptors. The individual domains are labeled. PSI, *P*lexin/*S*emaphorin/*I*ntegrin domain; IPT, *I*mmunoglobulin-like domain found in *P*lexins (and Met) and in some *T*ranscription factors; CUB, domain homologous to complement-binding factors C1r and C1s; FV/VIII, domain homologous to coagulation factor V and VIII (also known as F5/8 type C or discoidin domain); MAM, *M*eprin/*A*5/μ domain; SP, *S*ex-*P*lexin domain. Semaphorins, plexins and scatter-factor receptors (MET) share a common semaphorin domain (black heptagon).

are encoded by viral genomes. Class 5 is the only one that includes both vertebrate and invertebrate members (Adams *et al.*, 1996; Bahri *et al.*, 2001; Khare *et al.*, 2000). The semaphorin family includes both secreted (class 2 and 3) and membrane-bound proteins that are anchored in the plasma membrane by a transmembrane domain (classes 1, 4, 5, and 6), or a GPI-anchor (class 7). At their C-terminal regions most semaphorins contain immunoglobulin-like domain (classes 2–4 and 7). The C-terminal regions of class 5 semaphorins contain 7 type I thrombospondin (TSP) repeats that are also found in extracellular matrix proteins, such as Thrombospondin 1 and -2 (Adams *et al.*, 1996).

Class 3 semaphorins are the best functionally characterized group. The presence of a sema domain and dimerization are the minimal requirements for their activity. Indeed, the semaphorin-3A sema domain alone is sufficient for induction of growth cone collapse when produced in dimer form (Klostermann *et al.*, 1998; Koppel and Raper, 1998). Interestingly, the specific effects of class 3 semaphorins on different types of axons are determined by a relatively short region within the sema domain (Koppel *et al.*, 1997).

Class 3 semaphorins are synthesized as inactive precursors that require furin proteolytic processing for activation. Several highly conserved clusters of basic amino acid residues are located in the sema domain and the C-terminus of the protein that contain consensus recognition sites for furin-like proteases. The proteolytic processing results in the generation of several isoforms that differ in their repulsive activity depending on the combination of cleaved sites (Adams *et al.*, 1997). Since the relative amount of the isoforms changes during embryogenesis, the proteolytic processing of class 3 semaphorins may be important for the modulation of their chemorepellent activity during development.

Sema3A acts via a receptor complex that contains Neuropilin 1 (Nrp-1) as the ligand-binding subunit and an A-Plexin as the signal-transducing subunit (Fiore and Püschel, 2003) (Fig. 7). Nrp-1 and the closely related Nrp-2 appear to bind all class 3 semaphorins, but differ in their affinity for individual members (Fiore and Püschel, 2003; He and Tessier-Lavigne, 1997; Kolodkin *et al.*, 1997). Nrp-1 contains a large extracellular domain, a single transmembrane region and a short cytoplasmic tail. The neuropilin extracellular domain includes two CUB motifs (domain A: A1, A2), followed by two domains with similarity to coagulation Factor V/VIII (domain B: B1, B2), and one MAM domain (domain C). Both the CUB and V/VIII domains are essential for binding the semaphorin domain of Sema3A while only the N-terminal V/VIII domain is required for the interaction with VEGF165 or the basic carboxy terminus of Sema3A (Giger *et al.*, 1998; Gu *et al.*, 2002; Lee *et al.*, 2003; Nakamura *et al.*, 1998; Renzi *et al.*, 1999). While the C-terminus does not contribute to the biological specificity of semaphorins it is a major determinant of their affinity for neuropilins (Giger *et al.*, 1998; Nakamura *et al.*, 1998).

Vertebrate A-Plexins are the signaling subunits for secreted class 3 semaphorins, while other plexins interact with semaphorins from different classes (Fiore and Püschel, 2003). In contrast to Sema3A, which is not able to bind directly to A-Plexins, Sema4D and Sema7A directly interact with Plexin-B1 and Plexin-C1, respectively (Takahashi *et al.*, 1999; Tamagnone *et al.*, 1999). Interestingly, all Plexins contain an extracellular semaphorin domain that is strictly required for their function (Fig. 7). A semaphorin domain is also present in the scatter-factor receptor MET (Tamagnone *et al.*, 1999).

B. *Semaphorin Structure*

Two crystal structures of semaphorin family members were recently reported including the structure of an observed *in vivo* proteolytic form of semaphorin-3A (Sema3A-65K, including the complete sema domain and part of the PSI domain) (Antipenko *et al.*, 2003) and of the nearly full-length semaphorin-3D including the sema, PSI, and Ig domains (Love *et al.*, 2003).

Unexpectedly, the crystallographic studies revealed that the semaphorin fold is a variation of the β propeller topology with seven blades radially arranged around a central axis (Fig. 8A). Each blade contains a four-stranded (strands A to D) antiparallel β sheet. In contrast to most known β-propeller structures with approximately 40–45 residues per blade (Fulop and Jones, 1999) the semaphorin blades average 70 amino acids. The large size of the semaphorin domain results from the presence of additional secondary structure elements inserted in most of the blades. Blade 5 has the largest insertion (~70 amino acids), also termed "the extrusion," that is composed of 3 helixes and 2 β strands. Some of the loops connecting the secondary structure elements are also unusually long, including the four loops that form the proposed neuropilin-binding site. The two faces of the β propeller are by convention designated top and bottom, and in

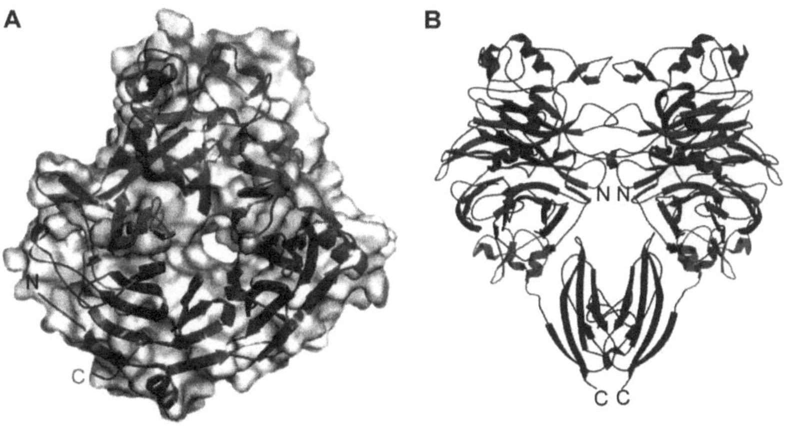

FIG. 8. Structures of Sema3A-65K and Sema4D. (A) The structure of Sema3A-65K viewed from the "top" face of the molecule. The molecular surface (semi-transparent) is also indicated. The individual Sema3A-65K pseudo-repeats corresponding to the individual β propeller blades are colored (from -N to -C terminus) in red (1), orange (2), yellow (3), green (4), cyan (5), blue (6) and magenta (7). (B) The structure of Sema4D homodimer. The sema domain is in red, the PSI – in green, and the Ig – in blue. (See Color Insert.)

semaphorins, as in other members of the β propeller family, they differ markedly in nature: the bottom surface is relatively smooth and polar, whereas the top surface presents a series of loops bearing several prominent hydrophobic residues.

Like most of the other known propeller structures, semaphorin uses a "Velcro" system to close the circle between the first and the last blades (Fulop and Jones, 1999; Paoli, 2001). The Sema's Velcro (Fig. 8A) is stronger than usually observed. Indeed, the N-terminus of the molecule, in addition to contributing the fourth antiparallel β strand of the 7th blade (purple), wraps further around and contributes a fifth, this time parallel strand to the 6th blade (blue).

The C-terminal β strand of the semaphorin propeller leads directly into the PSI domain, which nestles against the side of blade 6 (Fig. 8B). As predicted from sequence analysis, this small domain (approximately 50 residues—green on Fig. 8B) forms a compact cysteine knot structure, with a core stabilized by three disulfide bridges. Interestingly, the pattern of disulfide linkages differs from that predicted by sequence comparisons for the PSI domains of integrins (reviewed in Bork *et al.*, 1999), in which the linkages were assigned on the basis of biochemical data (Calvete *et al.*, 1991). Sequence analysis (Behar *et al.*, 1999) had suggested a structural homology with hanatoxin (a peptide toxin inhibitor of voltage-gated K+ channels) and indeed they contain a marked, although distant, resemblance in fold topology but no equivalence in the arrangement of disulfide bridges. The C-terminal Ig-like domain of Sema4D abuts the PSI domain and blade 6. It conforms most closely to the I set of Ig-like folds, with one three-stranded and one four-stranded β sheets.

Semaphorins, plexins, and scatter-factor receptors were not recognized until recently (Gherardi *et al.*, 2003) as β propeller proteins because they lack any detectable repeating motif in their sequences. Even in light of the determined structures, a superposition of the seven semaphorin blades fails to reveal any consensus sequence repeat (Antipenko *et al.*, 2003). In contrast, most other β propellers are stabilized by different sets of interactions that result in sequence repeats (or motifs) shared by all or most individual blades. For example, in the WD repeats of Gβ the residues Trp-Ser/Thr-His-Asp form an electrostatic tetrad reoccurring throughout the circular array. Other such motifs include the YWTD, TolB, RCC1, and Tachylectin-2 repeats, the aspartate box, and the (WD), *kelch* and Tryptophan-docking motifs (Fulop and Jones, 1999). Instead, the semaphorin fold is partially stabilized by the conserved disulfide bonds, which contribute to maintaining the protein architecture, while allowing divergence in the sequence. In this respect the semaphorins are similar to viral neuraminidase, the six-bladed propeller of which has 9 disulfide bonds and

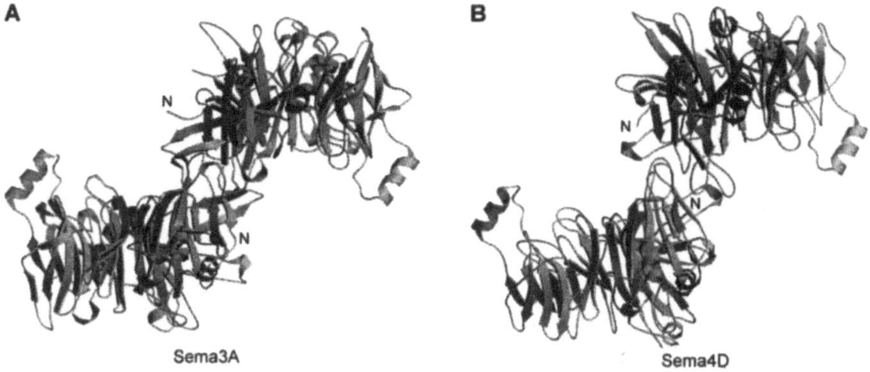

FIG. 9. Non-covalent semaphorin dimerization mediated by the sema domains. (A) The Sema3A-65K dimer in the asymmetric unit of the crystals. (B) The interacting sema domains in the Sema4D dimer viewed in the same orientation as in (A). The expansive dimerization interface (total buried area of approximately 3000 $Å^2$) is generated by the approximation of four protruding loops from each monomer. These loops, located at the "top" face of the β propeller are also implicated in interactions with the semaphorin receptors.

very little inter-repeat sequence similarity (Paoli, 2001; Varghese *et al.*, 1983).

Both structurally characterized semaphorin constructs are dimeric in solution and in the crystals. In the Sema3A and Sema4D structures the monomers bind each other in a very similar fashion (Fig. 9) by using four protruding surface loops, which intertwine with each other to form an intimate and extensive interface burying a total of approximately 3000 $Å^2$. The interface contains both hydrophobic and polar residues and the molecular interactions between them include hydrogen bonds, salt bridges and Van der Waals contacts. In Sema4A, which functions as a membrane-attached axon guidance molecule, the dimerization mode places the Ig-like domains in an approximately parallel orientation consistent with attachment to the same cell surface (Fig. 8B). Analytical ultracentrifugation experiments documented that Sema3A-65K is a dimer in solution (the estimated Kd of dimerization is 3 μM) and undergoes a dimer-to-monomer transition upon binding to monomeric Nrp-1AB (Kd of 0.23 μM), thus forming a 1:1 complex. The neuropilin complex formed by Sema3A-90K (or "full length" Sema3A), which is the predominant Sema3A form *in vivo* (Adams *et al.*, 1997), is expected to have a 2:2 stoichiometry, since Sema3A-90K is a preformed covalent dimer via a disulfide bridge located outside of the receptor-binding domain (see Fig. 10A).

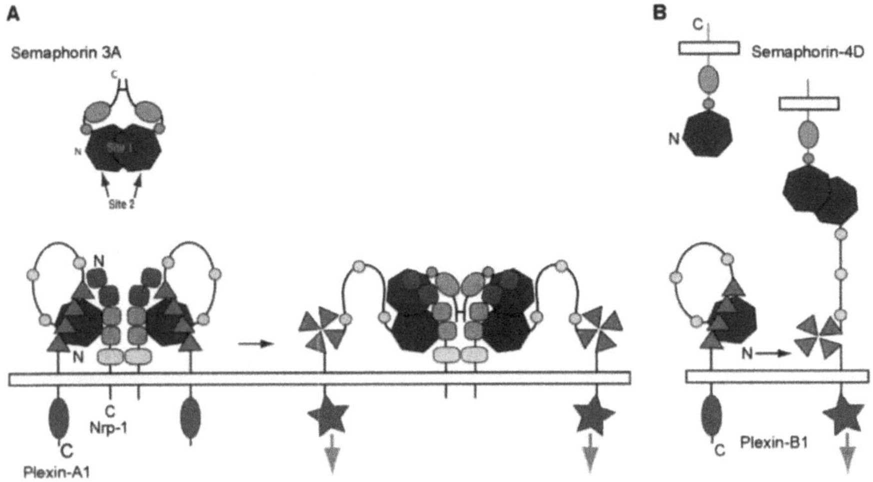

FIG. 10. Model for initiation of semaphorin signaling. (A) Initiation of Sema3A-mediated signaling via the Nrp-1/Plexin-A1 complex (see discussion in text). Sema3A is in magenta (Sema domain – heptagon; PSI domain – circle; Ig domain – oval); In plexin, the Sema domain is in dark blue, the IPT domains in green, and the PSI regions in light blue. The intracellular Sex-Plexin domain is in brown and the star corresponds to the activated form (the red arrow indicates signaling directing growth cone collapse). In Nrp-1, the A1 and A2 domains are in red, B1 and B2 in orange, and the MAM domain is in yellow. Sema3A binding results in a 2:2:2 ligand/receptor/co-receptor complex formation and the release of the plexin membrane-proximal extracellular region which in the absence of constrains adopts an active conformation (right panel). (B) Initiation of Sema4D signaling via the Plexin-B1 receptor. In this case no co-receptors are required. As with Sema3A, ligand binding disrupts the inhibitory semaphorin conformation resulting in activation of downstream signaling. Reproduced with permission from (Antipenko *et al.*, 2003). (See Color Insert.)

The fact that neuropilin binding competes with semaphorin dimer formation indicates that the neuropilin-binding region of Sema3A overlaps with or is very close to the dimerization interface. Structure-based deletion mutagenesis experiments (Antipekno *et al.*, 2003) confirmed that the neuropilin-binding surface of the semaphorins is indeed localized to the upper part of the top. Furthermore, a Sema3A-neutralizing antibody has also been shown to target one of the loops at the dimerization interface (Shirvan *et al.*, 2002). In addition, peptides corresponding to this Sema3A region mimick the biological activity of full-length Sema3A in cell-based assays (Shirvan *et al.*, 2002). Interestingly, the closest structural homologues of Sema3A, namely the transducin β subunit and integrin, utilize corresponding surfaces on their β propellers to bind their respective interaction partners (Sondek *et al.*, 1996; Xiong *et al.*, 2001). In all cases, the interaction interfaces are located on the top face of the propeller, slightly off-center, with most interface residues belonging to blades 2–5. Analyses of the Sema3A structure also suggested other potential surfaces (on the side of the β propeller) that might be involved in interactions with the plexin co-receptors but these interactions have not yet been confirmed.

An interesting result of the semaphorin crystallographic analysis is the suggestion, based on the presence and the specific localization of the ligand-binding β propeller domains in plexins and the LDL receptor (LDLR), that they might share a similar autoinhibition mechanism (Antipenko *et al.*, 2003; Rudenko *et al.*, 2002; Takahashi and Strittmatter, 2001). In both cases the β propeller domain functions as an alternate substrate and binds in *cis* other receptor domains (the LA4/R4 and LA5/R5 cysteine-rich modules in LDLR and the IPT domains of plexins) thus inhibiting the molecules. The inhibition of the LDLR is relieved at high pH allowing the β propeller domain to be displaced by LDL. In plexins, on the other hand, interactions with the semaphorin domain of their ligands presumably dissociate the inhibitory complex resulting in activation of downstream signaling. Interestingly, the LA4/LA5 binding site on the LDL β propeller domain is centered on blades 4 and 5 at the "top" face of the molecule, occupying the same general location as the proposed neuropilin-interaction site on the Sema3A semaphorin domain.

C. *Initiation of Semaphorin-Mediated Signaling*

The crystallographic (Antipenko *et al.*, 2003; Love *et al.*, 2003), and other studies (Giger *et al.*, 1998; Gu *et al.*, 2002; Nakamura *et al.*, 1998; Renzi *et al.*, 1999; Takahashi and Strittmatter, 2001; Tamagnone *et al.*, 1999) suggest that the following molecular events might occur during the initiation of semaphorin signaling (Fig. 10). In the absence of ligand, plexins assume an autoinhibited state where their N-terminal semaphorin domain intramolecularly interacts with the C-terminal part (IPT sequences) of the ectodomain, constraining the juxtamembrane and intercellular domains in an inactive conformation. Nrp-1 interacts with both the semaphorin domain and the IPT region of Plexin-A1 (Takahashi and Strittmatter, 2001) stabilizing the autoinhibited conformation of the molecule. Binding of Sema3A to Nrp-1 leads to a conformational rearrangement in Plexin-A1 that is transmitted to the cytosolic domain. It was also suggested that Sema3A might contain two distinct interaction surfaces. One is more extensive, overlaps with the Sema3A-65K dimerization interface (Fig. 10A, site 1), and is responsible for the initial high-affinity binding to the AB domain of Nrp-1. The second site (Fig. 10A, site 2) is smaller and located at the side of the β-propeller. The binding of Sema3A to Nrp-1, presents site 2 for binding to the Plexin-A1 semaphorin domain. The resulting Sema3A/plexinA1 interaction displaces the bound IPT region and releases the plexin juxtamembrane and intercellular domains from their inactive conformation, initiating downstream signaling. In the activated receptor, Nrp-1 remains associated with Plexin-A1 through a direct interaction with its semaphorin domain. The interaction with the C-terminal half of the Plexin-A1 ectodomain might also be preserved (Takahashi and Strittmatter, 2001). The two co-receptors interact also indirectly, through the bound Sema3A ligand. Some plexins, such as Plexin-B1 are directly activated by semaphorins (Tamagnone *et al.*, 1999). In this case, the interaction between the semaphorin domains of the ligand and receptor is of high affinity and the binding does not require assistance from a neuropilin (Fig. 10B).

Interestingly, the large conformational change in plexins that likely accompanies the release of autoinhibition is somewhat reminiscent of the rearrangement of the integrin ectodomains after their activation. Furthermore, the interaction of Rac with the cytoplasmic domain of Plexin-B1 leads to an increase in its affinity for Sema4D and its transport to the plasma membrane (Vikis *et al.*, 2002) suggesting a similar mechanism of inside-out signaling that is used by integrins (Hynes, 2002; Takagi *et al.*, 2002). It was also suggested that plexins and integrins may both utilize a novel activation mechanism (Fig. 10) where their intracellular signaling domains are held in close proximity to each other in the inactive state and undergo spatial separation upon receptor activation (Kim *et al.*, 2003; Takahashi and Strittmatter, 2001).

IV. Myelin-Associated Inhibitors of Axonal Regeneration

The past 10 years have yielded considerable progress toward the identification and characterization of several myelin-associated inhibitors of axonal growth. These include Nogo, MAG, and OMgp (Fig. 11), which have been shown to limit axonal regeneration after injury of the spinal cord and brain. These cell-surface proteins signal through multi-subunit neuronal receptors that contain a common ligand-binding subunit termed the Nogo-66 Receptor (NgR).

A. *Nogo and NgR*

Recent advances into the identification of axon-regrowth inhibitors included the characterization of a monoclonal antibody (IN-1) that was

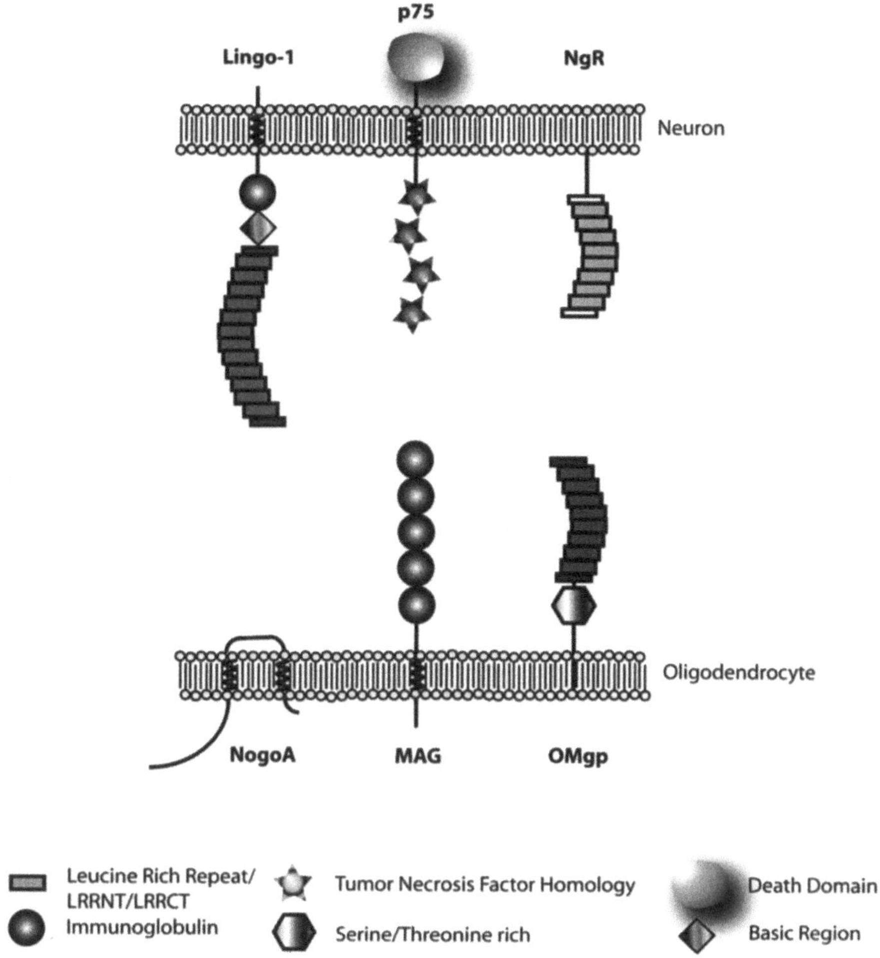

FIG. 11. Schematic representation of the myelin-associated glycoproteins and their receptors and co-receptors involved in inhibition of neurite outgrowth. The Nogo receptor (NgR) and associated co-receptors are expressed on neurons while its ligands, NogoA, OMgp, and MAG, are expressed on the oligodendrocyte cell surface.

capable of partially restoring the ability of injured axons to extend processes following key injuries (Caroni and Schwab, 1988). Several groups used this antibody to identify its target, a protein that was named Nogo-A, or Nogo (Chen *et al.*, 2002; GrandPre *et al.*, 2002; Prinjha *et al.*, 2000). Its C-terminal region of approximately 200 amino acids shares ~70% homology with various members of the reticulin (Rtn) family, and the protein has therefore been classified as Reticulon-4A (GrandPre *et al.*, 2000). Reticulons are expressed primarily, if not exclusively, in neuroendocrine cells and are localized to the endoplasmic reticulum as a result of a dilysine ER retention signal within their C-terminus (van de Velde *et al.*, 1994). As a result of alternative mRNA splicing, Nogo exists as three main isoforms A, B, and C, all of which retain the C-terminal reticulon homology region but differ at their amino terminus. NogoA transcripts can be found in the brain, but not in the peripheral nervous system. Nogo-B and Nogo-C are expressed both in the brain and in various other adult tissues (Chen *et al.*, 2000; GrandPre *et al.*, 2000; Hunt *et al.*, 2003). The largest isoform, Nogo-A is ~1200 amino acids and contains two potential transmembrane regions. Its large amino terminal domain and its C-terminus are predicted to have a cytoplasmic localization, while an internal 66 amino-acid domain (named Nogo-66) is predicted to reside in the extracellular space. While immunohistochemistry experiments seem to support such transmembrane topology, there have been some recent speculations that both the N-terminal region and Nogo-66 might be located on the luminal/extracellular side of the membrane. The majority of Nogo is found in the endoplasmic reticulum but some is also exported to the cell surface (GrandPre *et al.*, 2000).

Studies designed at identifying the inhibitory regions within Nogo have revealed that both the amino terminal segment as well as the Nogo-66 domain can, by themselves, inhibit axonal growth in *in vitro* experiments. Nogo-66 shares some sequence similarity with the other members of the reticulon family but none of these displays growth cone collapse activity. Interestingly, peptides derived from this region possess neurite retraction capability (GrandPre *et al.*, 2002). It was documented, for example, that a peptide corresponding to amino acids 31–55 of Nogo-66, although less potent than the entire Nogo-66 domain, can induce axonal inhibition. In addition, the Strittmatter group has identified a different peptide, NEP1-40 (containing the first 40 amino acids of Nogo-66), which is a potent antagonist of the growth cone collapse activity of Nogo-66. This peptide has a high binding affinity to the neuronal Nogo receptor but even at high concentrations (μM range) lacks any neurite inhibition activity. NEP1-40 prevents Nogo-66 from inducing neuronal collapse *in vitro* and promotes neuron sprouting in rats following intrathecal application at the site of a mid-thoracic dorsal hemisection, thus leading to improved functional recovery (Li and Strittmatter, 2003).

Considerable effort has been directed at the identification of cellular receptors responsible for the Nogo-mediated signal transduction. While the receptor for N-terminal Nogo region is yet unknown, the Strittmatter's group has identified the receptor for the Nogo-66 ligand (Fournier *et al.*,

2001). The Nogo-66 receptor, NgR, is a leucine-rich, GPI anchored, extracellular molecule. Consistent with its function, NgR is expressed mainly on neurons while its Nogo ligand is found on oligodendrocytes. Biophysical experiments with recombinant proteins indicate high-affinity binding between NgR and Nogo-66 with dissociation constants in the 1–3 nM range (GrandPre *et al.*, 2002). *In vitro* pull-down assays confirm the high affinity interaction, which can be competed by the NEP1-40 peptide. Importantly, NEP1-40 not only binds NgR with high affinity *in vitro*, but also inhibits the Nogo-66-induced axonal retraction in cell-based assays, confirming that NgR mediates the *in vivo* neurite outgrowth inhibition signal of Nogo-66 (GrandPre *et al.*, 2002).

The N-terminal region of NgR harbors eight canonical leucine rich repeats (LRR) that contain the LRR-signature sequence LxxLxLN/CxL. The NgR LRRs are flanked by a leucine rich repeat N-terminal subdomain (LRRNT) and a leucine rich repeat C-terminal subdomain (LRRCT), which are small protein motifs frequently found next to LRR domains. Binding studies reveal that the leucine rich domains are necessary and sufficient for ligand recognition (Fournier *et al.*, 2002).

Two groups recently crystallized and determined the structure of the ligand-binding amino terminal NgR region (Barton *et al.*, 2003; He *et al.*, 2003) revealing an elongated molecule with banana shaped curvature (Fig. 12). The overall structure is most similar to the LRR containing proteins internalin B (InlB) (Marino *et al.*, 1999) and platelet glycoprotein 1Bα (Huizinga *et al.*, 2002). The predominant secondary structure elements are short β strands present within each LRR, which form a large parallel β sheet on the concave surface of the molecule. The molecular convex side is composed of loops with exposed charged amino acids and several small helices. The flanking LRRNT and LRRCT subdomains each contribute one small β strand to the large parallel β sheet. In addition, the LRRNT subdomain forms a three-stranded hydrophilic cap for the hydrophobic core of the LRRs. Four cysteines are involved in two disulfide bonds that serve to properly position the LRRNT subdomain within the NgR molecule. The LRRCT subdomain is slightly larger, containing two helices and five short β strands, and also functions to shield the NgR hydrophobic core. Two disulfide bonds within this domain further stabilize the structure. Kobe and Kajava (2001) refer to this particular disulfide bond arrangement as type CF1 and it is the one most commonly observed in LRRCT domains. Unlike the platelet glycoprotein 1Bα, NgR lacks large protruding fingers in its LRRNT and LRRCT subdomains, and thus does not rely on such structural features for ligand recognition.

The primary biological function of LRRs is to provide a structural framework for protein-protein interactions. Crystal structures of several

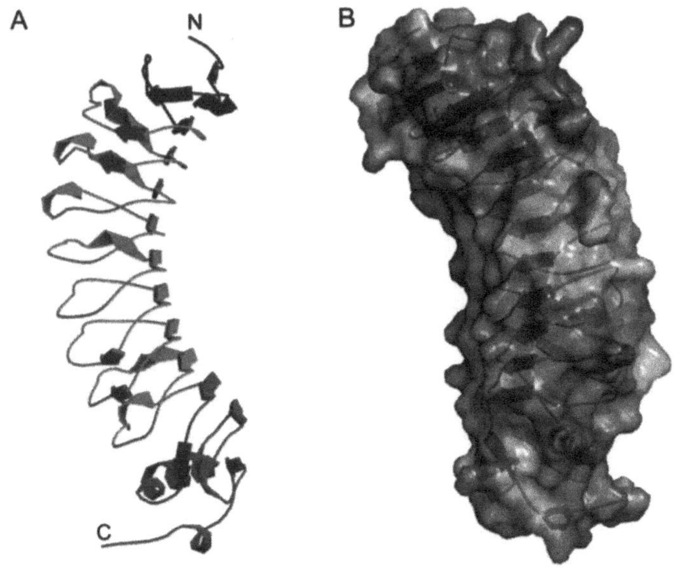

FIG. 12. (A) Side view of the NgR structure. The LRRNT domain is shown in blue, the LRRs in green, and the LRRCT domain in red. (B) Surface representation of the NgR structure showing the electrostatic potential in red and blue. Note the two large basic (blue) patches near the N- and C-termini and the diffuse acidic (red) region on the concave surface. (See Color Insert.)

LRR-containing proteins in complex with ligands, suggest that the LRR concave β sheet is the surface most often utilized for ligand binding (Kobe and Deisenhofer, 1995; Papageorgiou *et al.*, 1997). However, in some cases, the majority of the interactions involve the LRRNT and LRRCT domains (Huizinga *et al.*, 2002).

NgR presents an extensive molecular surface of over 12,000 Å^2 that could serve as a scaffold for multiple molecular recognition events. Overall, NgR has a slightly positive charge with a theoretical pI of 8.6. The LRRCT subdomain is even more electrostatically positive (theoretical pI value of 9.4 if isolated alone) and contains several surface-exposed basic residues. Interestingly, two particularly large positively charged patches are present on the NgR molecular surface (Fig. 12B). One of these comprises residues from the LRRCT subdomain and from the C-terminal LRRs. Interestingly this region is involved in making a crystal contact with a slightly acidic patch located on the N-terminal region of the concave LRR surface (see later). The other patch comprises residues on the concave surface of the mid to N-terminal LRRs. Comparison of the NgR molecular surface with models of the related NgR2 and NgR3 shows that these electrostatic features are not conserved (Barton *et al.*, 2003). Interestingly, while NgR2 and NgR3 are close sequence homologues of NgR (~50% identity), they display unique binding characteristics, and neither interacts with any of the NgR ligands (Barton *et al.*, 2003). Since Nogo-66 is negatively charged, with two consecutive glutamic acids that are required for efficient binding to NgR (GrandPre, 2002), the presence of a large number of NgR basic residues that are not conserved in its non-Nogo66-interacting homologues, suggests that they may be involved in ligand or co-receptor recognition.

B. MAG

Almost a decade ago, the Myelin-Associated Glycoprotein (MAG) was shown to inhibit neurite outgrowth *in vitro* (McKerracher *et al.*, 1994). MAG is a member of the immunoglobulin (Ig) superfamily, and of its sialic-acid binding (siglec) subfamily. The extracellular MAG region contains five tandem immunoglobulin repeats (Fig. 11). The sialic acid binding region includes the first three Ig domains, and most likely centers around Arg 118 of the first Ig domain. The last two Ig domains, on the other hand, contain the protein neurite inhibitory activity. Using a truncated MAG construct, Filbin and colleagues demonstrated that Ig repeats four and five are likely involved in interactions with NgR. In agreement with these studies, the sialic-acid binding function of MAG was shown to be independent from its neurite outgrowth inhibitory activity (Domeniconi *et al.*, 2002; Tang *et al.*, 1997).

Like Nogo, MAG is proposed to signal through the NgR/p75 receptor/co-receptor complex (Domeniconi *et al.*, 2002; Liu *et al.*, 2002; Wang *et al.*, 2002b). Subsequent experiments with knockout mice, though, have somewhat clouded the *in vivo* role of MAG. For example, mice deficient in MAG show no detectable differences from wild type mice in neuronal regenerative effects following optical or corticospinal damage. In addition, myelin preparations from either wild-type or MAG deficient mice are similar in their neurite outgrowth extension inhibiton activity or growth cone collapse potential (Bartsch *et al.*, 1995). Further studies are obviously required to define the precise biological functions of MAG during development and in the adult organisms.

C. OMgp

The latest identified myelin-specific inhibitor of axonal regeneration is the oligodendrocyte-myelin glycoprotein (OMgp). Like NgR, OMgp is a GPI anchored LRR protein containing eight LRRs and a LRRNT subdomain

(Figure 11). The C-terminal region of OMgp includes a serine and threonine rich domain. Interestingly both OMgp domains appear to be capable of independent binding to NgR on overexpressing cell surfaces but the LRR domains mediate the higher affinity interactions (Wang *et al.*, 2002a). OMgp expression is mostly limited to the oligodendrocyte cell surface and can be easily detected in axon-associated myelin. Exposure of NgR-expressing retinal ganglion cells to OMgp results in a growth cone collapse response, providing further evidence that OMgp is a physiologic ligand for NgR. As Nogo and MAG, OMgp is proposed to signal through the NgR/p75 complex (Wang *et al.*, 2002a,b).

D. The p75 Co-Receptor

The neurotrophin receptor p75 was first identified as a nerve growth factor (NGF)-binding protein and was subsequently shown to interact with each of the other neurotrophic factors, BDNF, neurotrophin-3, and neurotrophin-4/-5. It also modulates the activity of several members of the tropomyosin-related receptor tyrosine kinase family (Trk) (reviewed in Chao, 2003). p75, a member of the tumor necrosis factor superfamily, is a type I transmembrane protein with four cysteine-rich domains in its extracellular region and a Death domain in its cytoplasmic protein (Fig. 11).

Since NgR is a GPI anchored protein and lacks any cytoplasmic domains, it was suggested that it signals through the action of a co-receptor. Recently p75 was shown to associate with NgR via interactions within the LRR region, forming a functional complex capable of transmitting growth-inhibitory information inside the cell (Wang *et al.*, 2002b). The association of NgR and p75 was demonstrated in experiments utilizing overexpressed proteins in CHO cells and by immunoprecipitation of endogenous proteins from rat postnatal cerebellar granule neurons. Furthermore, dorsal root ganglion (DRG) neurons derived from p75 knockout mice were shown to be impervious to the growth cone collapsing effects of MAG, OMgp, and Nogo. Truncated, dominant negative, forms of p75, lacking the intracellular domain, compete effectively for NgR binding on the cell surface and thus block NgR-mediated growth cone collapse providing further evidence for a p75-NgR functional interaction *in vivo* (Wang *et al.*, 2002b).

In contrast to its mechanism of action in neurotrophic factor signaling, p75 apparently does not interact independently with any of the NgR ligands prior to its association with NgR (Wang *et al.*, 2002b). It is also interesting that addition of MAG stimulates the binding of p75 to NgR both in immunoprecipitation assays and in cell labeling experiments. From a structural perspective it is surprising that NgR and p75 form a receptor complex responsible for mediating the inhibitory signals of such structurally diverse molecules as MAG, OMgp, and Nogo. Thus further crystallographic investigations into how these molecules interact with each other are urgently needed.

While considerable amount of data points to p75 as the functional neuronal NgR co-receptor, the overall weak affinity between p75 and NgR combined with a few recent reports raises questions about the exact role of p75 in NgR signaling. Song *et al.* (2004), for example, studied the *in vivo* role of p75 in nerve regeneration using p75 knockout mice and found that the absence of p75 had little effect on neuronal regeneration following injury. Additionally, a dominant negative p75-Fc fusion protein that binds NgR but does not signal, failed to block neurite outgrowth inhibitors. Interestingly, researchers at Biogen have recently discovered a second novel NgR co-receptor, Lingo-1, that appears to have affinity for both NgR and p75 (Mi *et al.*, 2004). Like NgR, Lingo-1 is a LRR protein containing 12 repeats flanked by small LRRNT and LRRCT subdomains. The LRRCT domain is followed in the Lingo-1 sequence by a basic region, an Ig domain, a transmembrane helix and a short intracellular region. Lingo-1 expression is limited, as expected for a NgR co-receptor, to the nervous system. In a neuronal tissue culture model system, a dominant negative Lingo-1 variant blocked the inhibition caused by a soluble OMgp. Further studies are needed to address the precise roles of the NgR co-receptors p75 and Lingo-1 *in vivo.*

E. NgR Activation and Downstream Signaling Events

Deletion analysis of NgR suggests that each of the leucine rich repeats as well as the LRRNT and LRRCT subdomains are required for binding of all three ligands (Fournier *et al.*, 2002). Given the diverse structural nature of the NgR-binding proteins, it is likely, that multiple and disparate surfaces are involved in ligand recognition, and that the different NgR ligands might utilize distinct modes of receptor activation.

Although the ligand-binding region of NgR is monomeric in solution, the molecule is likely to form multimers (aggregates) *in vivo* prior to ligand-binding and initiation of signaling. Indeed, recombinant NgR associates with the endogenous NgR expressed at the cell surface of COS-7 cells (Fournier *et al.*, 2002). Interestingly, in the crystals, the NgR molecules pack in a fashion that is proposed to correspond to the lower-affinity ligand-independent receptor-receptor association observed at the cell surface. Upon ligand approximation, the electrostatic receptor-receptor interaction

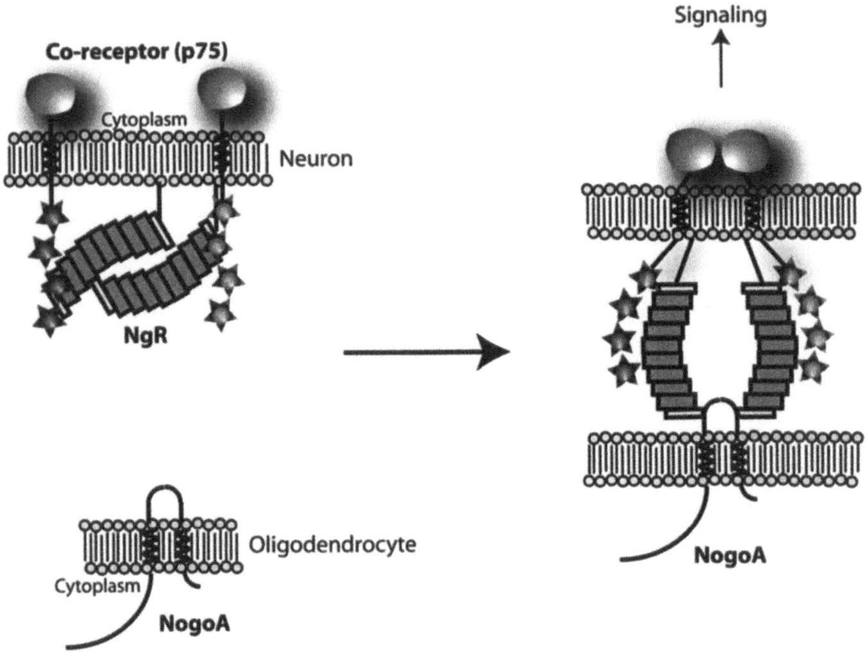

FIG. 13. A model for initiation of NgR signaling. While NgR exists primarily as a monomer in solution, studies have shown both self association and interactions with the co-receptor p75 on the cell surface. Currently, the stoichiometry of ligand and co-receptor binding is unknown.

would be disrupted in favor of the higher-affinity van der Waals ligand-receptor contacts, leading to the reorientation of receptor/co-receptor complexes and to the transduction of the repulsive signal to the inside of the cell (Fig. 13).

V. Conclusion and Perspectives

The past several years have seen dramatic advances in our understanding of the molecular events leading to the generation of correct neuronal connectivity during development. Several axon guidance ligand/receptor families were identified and their biological functions and mechanism of action have been investigated in detail by using genetic and biochemical methods. These molecules are of particular interest to the structural biologist since they present excellent model systems to study transmembrane signal transduction. Indeed the limited currently available structural and biophysical data demonstrate that the different axon guidance ligand/

receptor families utilize unique modes of ligand-induced receptor activation and downstream signaling initiation. In addition, the integration of multiple axon-guidance signals at the level of cell-surface receptor crosstalk provides an additional level of complexity that could best be addressed using the methods of structural biology.

It is now also clear that many axon guidance molecules and their receptors are involved not only in early developmental events, but also in the function of the adult organism, and the ability to modulate their signaling could have important medical applications. In the adult brain, for example, EphB signaling regulates neural stem cell migration and possibly proliferation, and therefore, Eph agonists could have therapeutic potential (Conover *et al.*, 2000). The semaphorins and the myelin-associated glycoproteins play crucial roles in inhibiting neuronal regeneration in the CNS, and are therefore becoming the focus of efforts directed towards the development of novel strategies to treat brain and spinal cord injuries and to promote axonal repair in diseases such as multiple sclerosis. Increasing evidence also implicates several of the axon guidance ligands and receptors in angiogenesis and cancer (Dodelet and Pasquale, 2000). The recent crystallographic studies of Ephs and ephrins demonstrate how structural biology can not only yield invaluable insight into the mechanistic signaling properties of these cell-surface receptors and ligands, but could also reveal potential drug-binding targets (Fig. 3). These can then be targeted by using computational structure-based screens for small-molecule antagonists or, as in the case of Eph receptors, by rational peptide design. Finally, the availability of high-resolution structural information should allow for structure-based engineering of axon guidance molecules and receptors with altered affinities and specificities (Himanen *et al.*, 1998), which can serve as important tools for understanding the precise biological function of these important signaling proteins.

References

Adams, R. H., Betz, H., and Püschel, A. W. (1996). A novel class of murine semaphorins with homology to thrombospondin is differentially expressed during early embryogenesis. *Mech. Dev.* **57,** 33–45.

Adams, R. H., Lohrum, M., Klostermann, A., Betz, H., and Püschel, A. W. (1997). The chemorepulsive activity of secreted semaphorins is regulated by furin-dependent proteolytic processing. *EMBO J.* **16,** 6077–6086.

Adams, R. H. *et al.* (1999). Roles of ephrinB ligands and EphB receptors in cardiovascular development: demarcation of arterial/venous domains, vascular morphogenesis, and sprouting angiogenesis. *Genes Dev.* **13,** 295–306.

Antipenko, A., Himanen, J. P., van Leyen, K., Nardi-Dei, V., Lesniak, J., Barton, W. A., Rajashankar, K. R., Lu, M., Hoemme, C., Puschel, A. W., and Nikolov, D. B. (2003). Structure of the semaphorin-3A receptor binding module. *Neuron* **39,** 589–598.

Bahri, S. M., Chia, W., and Yang, X. (2001). Characterization and mutant analysis of the Drosophila sema 5c gene. *Dev. Dyn.* **221,** 322–330.

Barton, W. A., Liu, B. P., Tzvetkova, D., Jeffrey, P. D., Fournier, A. E., Sah, D., Cate, R., Strittmatter, S. M., and Nikolov, D. B. (2003). Structure and axon outgrowth inhibitor binding of the Nogo-66 receptor and related proteins. *EMBO J.* **22,** 3291–3302.

Bartsch, U., Bandtlow, C. E., Schnell, L., Bartsch, S., Spillmann, A. A., Rubin, B. P., Hillenbrand, R., Montag, D., Schwab, M. E., and Schachner, M. (1995). Lack of evidence that myelin-associated glycoprotein is a major inhibitor of axonal regeneration in the CNS. *Neuron* **15,** 1375–1381.

Behar, O., Mizuno, K., Badminton, M., and Woolf, C. J. (1999). Semaphorin 3A growth cone collapse requires a sequence homologous to tarantula hanatoxin. *Proc. Natl. Acad. Sci. USA* **96,** 13501–13505.

Behlke, J. *et al.* (2001). Self-association studies on the EphB2 receptor SAM domain using analytical ultracentrifugation. *Eur. Biophys. J.* **30,** 411–415.

Bork, P., Doerks, T., Springer, T. A., and Snel, B. (1999). Domains in plexins: Links to integrins and transcription factors. *Trends Biochem. Sci.* **24,** 261–263.

Boyd, A. W., and Lackmann, M. (2001). Signals from Eph and ephrin proteins: A developmental tool kit. *Science's STKE* **112,** RE20.

Bruckner, K. *et al.* (1999). EphrinB ligands recruit GRIP family PDZ adaptor proteins into raft membrane microdomains. *Neuron* **22,** 511–524.

Calvete, J. J., Henschen, A., and Gonzalez-Rodriguez, J. (1991). Assignment of disulphide bonds in human platelet GPIIIa. A disulphide pattern for the beta-subunits of the integrin family. *Biochem. J.* **274,** 63–71.

Caroni, P., and Schwab, M. E. (1988). Antibody against myelin-associated inhibitor of neurite growth neutralizes nonpermissive substrate properties of CNS white matter. *Neuron* **1,** 85–96.

Chao, M. V. (2003). Neurotrophins and their receptors: A convergence point for many signalling pathways. *Nature Rev. Neurosci.* **4,** 299–309.

Chen, M. S., Huber, A. B., van der Haar, M. E., Frank, M., Schnell, L., Spillmann, A. A., Christ, Franziska, and Schwab, M. E. (2000). Nogo-A is a myelin-associated neurite outgrowth inhibitor and an antigen for monoclonal antibody IN-1. *Nature* **403,** 434–439.

Conover, J. C. *et al.* (2000). Disruption of Eph/ephrin signaling affects migration and proliferation in the adult subventricular zone. *Nat. Neurosci.* **3,** 1091–1097.

Dalva, M. B. *et al.* (2000). EphB receptors interact with NMDA receptors and regulate excitatory synapse formation. *Cell* **103,** 945–956.

Davis, S. *et al.* (1994). Ligands for EPH-related receptor tyrosine kinases that require membrane attachment or clustering for activity. *Science* **266,** 816–819.

Dodelet, V. C., and Pasquale, E. B. (2000). Eph receptors and ephrin ligands: Embryogenesis to tumorigenesis. *Oncogene* **19,** 5614–5619.

Domeniconi, M., Cao, Z., Spencer, T., Sivasankaran, R., Wang, K., Nikulina, E., Kimura, N., Cai, H., Deng, K., Gao, Y., He, Z., and Filbin, M. (2002). Myelin-associated glycoprotein interacts with the Nogo66 receptor to inhibit neurite outgrowth. *Neuron* **35,** 283–290.

Drescher, U. *et al.* (1995). In vitro guidance of retinal ganglion cell axons by RAGS, a 25 kDa tectal protein related to ligands for Eph receptor tyrosine kinases. *Cell* **82,** 359–370.

Drescher, U. (2002). Eph family functions from an evolutionary perspective. *Curr. Opin. Gen. Dev.* **12,** 397–402.

Dickson, B. J. (2002). Molecular mechanisms of axon guidance. *Science* **298,** 1959–1964.
Ethell, I. M. *et al.* (2001). EphB/syndecan-2 signaling in dendritic spine morphogenesis. *Neuron* **31,** 1001–1013.
Fiore, R., and Püschel, A. W. (2003). The function of semaphorins during nervous system development. *Front Biosci.* **8,** S484–S499.
Flanagan, J. G., and Vanderhaeghen, P. (1998). The ephrins and Eph receptors in neural development. *Annu. Rev. Neurosci.* **21,** 309–345.
Fournier, A. E., GranPre, T., and Strittmatter, S. M. (2001). Identification of a receptor mediating Nogo-66 inhibition of axonal regeneration. *Nature* **409,** 341–346.
Fournier, A. E., Gould, G. C., Liu, B. P., and Strittmatter, S. M. (2002). Truncated soluble Nogo receptor binds Nogo-66 and blocks inhibition of axon growth by myelin. *J. Neurosci.* **22,** 8876–8883.
Fulop, V., and Jones, D. T. (1999). Beta propellers: Structural rigidity and functional diversity. *Curr. Opin. Struct. Biol.* **9,** 715–721.
Gale, N. W. *et al.* (1996). Eph receptors and ligands comprise two major specificity subclasses and are reciprocally compartmentalized during embryogenesis. *Neuron* **17,** 9–19.
Gerety, S. S. *et al.* (1999). Symmetrical mutant phenotypes of the receptor EphB4 and its specific transmembrane ligand ephrin-B2 in cardiovascular development. *Mol. Cell* **4,** 403–414.
Gherardi, E., Youles, M. E., Miguel, R. N., Blundell, T. L., Iamele, L., Gough, J., Bandyopadhyay, A., Hartmann, G., and Butler, P. J. (2003). Functional map and domain structure of MET, the product of the c-met protooncogene and receptor for hepatocyte growth factor/scatter factor. *Proc. Natl. Acad. Sci. USA* **100,** 12039–12044.
Giger, R. J., Urquhart, E. R., Gillespie, S. K., Levengood, D. V., Ginty, D. D., and Kolodkin, A. L. (1998). Neuropilin-2 is a receptor for semaphorin IV: Insight into the structural basis of receptor function and specificity. *Neuron* **21,** 1079–1092.
Goodman, C. S. (1996). Mechanisms and molecules that control growth cone guidance. *Ann. Rev. Neurosci.* **19,** 341–377.
GrandPre, T., Nakamura, F., Vartanian, T., and Strittmatter, S. M. (2000). Identification of the Nogo inhibitor of axon regeneration as a Reticulon protein. *Nature* **403,** 439–444.
GrandPre, T., Li, S., and Strittmatter, S. M. (2002). Nogo-66 receptor antagonist peptide promotes axonal regeneration. *Nature* **417,** 547–551.
Grunwald, I. C. *et al.* (2001). Kinase-independent requirement of EphB2 receptors in hippocampal synaptic plasticity. *Neuron* **32,** 1027–1040.
Gu, C., Limberg, B. J., Whitaker, G. B., Perman, B., Leahy, D. J., Rosenbaum, J. S., Ginty, D. D., and Kolodkin, A. L. (2002). Characterization of neuropilin-1 structural features that confer binding to semaphorin 3A and vascular endothelial growth factor 165. *J. Biol. Chem.* **277,** 18069–18076.
Hattori, M. *et al.* (2000). Regulated cleavage of a contact-mediated axon repellent. *Science* **289,** 1360–1365.
He, Z., and Tessier-Lavigne, M. (1997). Neuropilin is a receptor for the axonal chemorepellent Semaphorin III. *Cell* **90,** 739–751.
He, X. L., Bazan, F., McDermott, G., Park, J. B., Wang, K., Tesier-Lavigne, M., He, Z., and Garcia, K. C. (2003). Structure of the Nogo receptor ectodomain: A recognition module implicated in myelin inhibition. *Neuron* **38,** 177–185.
Henderson, J. T. *et al.* (2001). The receptor tyrosine kinase EphB2 regulates NMDA-dependent synaptic function. *Neuron* **32,** 1041–1056.

Henkemeyer, M. *et al.* (1994). Immunolocalization of the Nuk receptor tyrosine kinase suggests roles in segmental patterning of the brain and axonogenesis. *Oncogene* **9,** 1001–1014.

Henkemeyer, M. *et al.* (1996). Nuk controls pathfinding of commissural axons in the mammalian central nervous system. *Cell* **86,** 35–46.

Himanen, J.-P. *et al.* (1998). Crystal structure of the ligand-binding domain of the receptor tyrosine kinase EphB2. *Nature* **396,** 486–491.

Himanen, J.-P. *et al.* (2001). Crystal structure of an Eph receptor-ephrin complex. *Nature* **414,** 933–938.

Himanen, J.-P., and Nikolov, D. B. (2002). Purification, crystallization and preliminary characterization of an Eph-B2/ephrin-B2 complex. *Acta Crystallogr. D* **58,** 533–535.

Himanen, J.-P., and Nikolov, D. B. (2003a). Molecules in focus: Eph receptors and ephrins. *Int. J. Biochem. Cell Biol.* **35,** 130–134.

Himanen, J. P., and Nikolov, D. B. (2003b). Eph signaling: A structural view. *Trends Neurosci.* **26,** 46–51.

Himanen, J. P. *et al.* (2004). Repelling Class Discrimination: Ephrin-A5 Binds to and Activates EphB2 Receptor Signaling. *Nat. Neurosc.* **7,** 501–509.

Holland, S. J. *et al.* (1996). Bidirectional signalling through the EPH-family receptor Nuk and its transmembrane ligands. *Nature* **383,** 722–725.

Holland, S. J. *et al.* (1997). Juxtamembrane tyrosine residues couple the Eph family receptor EphB2/Nuk to specific SH2 domain proteins in neuronal cells. *EMBO J.* **16,** 3877–3888.

Hubbard, S. R., and Till, J. H. (2000). Protein tyrosine kinase structure and function. *Annu. Rev. Biochem.* **69,** 373–398.

Hubbard, S. R. (2001). Theme and variations: Juxtamembrane regulation of receptor protein kinases. *Mol. Cell* **8,** 481–487.

Hubbard, S. R. (2002). Autoinhibitory mechanisms in receptor tyrosine kinases. *Front. Biosci.* **7,** d330–d340.

Huizinga, E. G., Tsuji, S., Romijn, R. A. P., Schiphorst, M. E., de Groot, P. G., Sixma, J. J., and Gros, P. (2002). Structures of glycoprotein Ibalpha and its complex with von Willebrand factor A1 domain. *Science* **297,** 1176–1179.

Hunt, D., Coffin, R. S., Prinjha, R. K., Campbell, G., and Anderson, P. N. (2003). Nogo-A expression in the intact and injured nervous system. *Mol. Cellular Neurosci.* **24,** 1083–1102.

Huse, M. *et al.* (1999). Crystal structure of the cytoplasmic domain of the type I TGFβ receptor in complex with FKBP12. *Cell* **96,** 425–436.

Huse, M. *et al.* (2001). The TGF? receptor activation process: An inhibitor- to substrate-binding switch *Mol. Cell* **8,** 671–682.

Huse, M., and Kuriyan, J. (2002). The conformational plasticity of protein kinases. *Cell* **109,** 275–282.

Hynes, R. O. (2002). Integrins: Bidirectional, allosteric signaling machines. *Cell* **110,** 673–687.

Johnson, L. N. (1996). Active and inactive protein kinases: Structural basis for regulation. *Cell* **85,** 149–158.

Kalo, M. S., and Pasquale, E. B. (1999). Signal transfer by Eph receptors. *Cell Tissue Res.* **298,** 1–9.

Kapfhammer, J. P., and Raper, J. A. (1987). Interactions between growth cones and neurites from different neural tissues in culture. *J. Neurosci.* **7,** 1595–1600.

Khare, N., Fascetti, N., DaRocha, S., Chiquet-Ehrismann, R., and Baumgartner, S. (2000). Expression patterns of two new members of the Semaphorin family in Drosophila suggest early functions during embryogenesis. *Mech. Dev.* **91,** 393–397.

Kim, S.-H. *et al.* (2002). Dynamic and clustering model of bacterial chemotaxis receptors: Structural basis for signaling and high sensitivity. *Proc. Natl. Acad. Sci.* **99,** 11611–11615.

Kim, M., Carman, C. V., and Springer, T. A. (2003). Bidirectional transmembrane signaling by cytoplasmic domain separation in integrins. *Science* **301,** 1720–1725.

Klostermann, A., Lohrum, M., Adams, R. H., and Püschel, A. W. (1998). The chemorepulsive activity of the axonal guidance signal semaphorin D requires dimerization. *J. Biol. Chem.* **273,** 7326–7331.

Kobe, B., and Deisenhofer, J. (1995). A structural basis of the interactions between leucine-rich repeats and protein ligands. *Nature* **374,** 183–186.

Kobe, B., and Kajava, A. V. (2001). The leucine-rich repeat as a protein recognition motif. *Curr. Opin. Struct. Biol.* **11,** 725–732.

Kolodkin, A. L., Matthes, D. J., O'Connor, T. P., Patel, N. H., Admon, A., Bemtley, D., Goodman, C. S., and Fasciclin, IV (1992). Sequence, expression, and function during growth cone guidance in the grasshopper embryo. *Cell* **9,** 831–845.

Kolodkin, A. L., Matthes, D. J., and Goodman, C. S. (1993). The semaphorin genes encode a family of transmembrane and secreted growth cone guidance molecules. *Cell* **75,** 1389–1399.

Kolodkin, A. L., Levengood, D. V., Rowe, E. G., Tai, Y. T., Giger, R. J., and Ginty, D. D. (1997). Neuropilin is a semaphorin III receptor. *Cell* **90,** 753–762.

Koppel, A. M., and Raper, J. A. (1998). Collapsin-1 covalently dimerizes, and dimerization is necessary for collapsing activity. *J. Biol. Chem.* **273,** 15708–15713.

Koppel, A. M., Feiner, L., Kobayashi, H., and Raper, J. A. (1997). A 70 amino acid region within the semaphorin domain activates specific cellular response of semaphorin family members. *Neuron* **19,** 531–537.

Kullander, K., and Klein, R. (2002). Mechanism and function of Eph and ephrin signaling. *Nature Rev. Mol. Cell. Biol.* **3,** 475–486.

Labrador, J. P. *et al.* (1997). The N-terminal globular domain of Eph receptors is sufficient for ligand binding and receptor signaling. *EMBO J.* **16,** 3889–3897.

Lackmann, M. *et al.* (1997). Ligand for EPH-related kinase (LERK) 7 is the preferred high affinity ligand for the HEK receptor. *J. Biol. Chem.* **272,** 16521–16530.

Lackmann, M. *et al.* (1998). Distinct subdomains of the EphA3 receptor mediate ligand binding and receptor dimerization. *J. Biol. Chem.* **273,** 20228–20237.

Lee, C. C., Kreusch, A., McMullan, D., Ng, K., and Spraggon, G. (2003). Crystal Structure of the Human Neuropilin-1 b1 Domain. *Structure* **11,** 99–108.

Lin, D. *et al.* (1999). The carboxyl terminus of B class ephrins constitutes a PDZ domain binding motif. *J. Biol. Chem.* **274,** 3726–3733.

Love, C. A., Harlos, K., Mavaddat, N., Davis, S. J., Stuart, D. I., Jones, E. Y., and Esnouf, R. M. (2003). The ligand-binding face of the semaphorins revealed by the high-resolution crystal structure of SEMA4D. *Nature Struct. Biol.* **10,** 843–848.

Luo, Y., Shepherd, I., Li, J., Renzi, M. J., Chang, S., and Raper, J. A. (1995). A family of molecules related to collapsin in the embryonic chick nervous system. *Neuron* **14,** 1131–1140.

Mark, M. D., Lohrum, M., and Püschel, A. W. (1997). Patterning neuronal connections by chemorepulsion: The semaphorins. *Cell Tissue Res.* **290,** 299–306.

Messersmith, E. K., Leonardo, E. D., Shatz, C. J., Tessier-Lavigne, M., Goodman, C. S., and Kolodkin, A. L. (1995). Semaphorin III can function as a selective chemorepellent to pattern sensory projections in the spinal cord. *Neuron* **14,** 949–959.

Mi, S., Lee, X., Shao, Z., Thill, G., Li, B., Relton, J., Levesque, M., Allaire, N., Perrin, S., Sands, B., Crowell, T., Cate, R. L., McCoy, J. M., and Pepinsky, R. B. (2004). Lingo-1 is a component of the Nogo-66 receptor/p75 signaling complex. *Nature Neurosci.* **7,** 221–228.

Murai, K. K., and Pasquale, E. B. (2002). Can Eph receptors stimulate the mind? *Neuron* **33,** 159–162.

Nakamura, F., Tanaka, M., Takahashi, T., Kalb, R. G., and Strittmatter, S. M. (1998). Neuropilin-1 extracellular domains mediate semaphorin D/III-induced growth cone collapse. *Neuron* **21,** 1093–1100.

Paoli, M. (2001). Protein folds propelled by diversity. *Biophys. Molec. Biol.* **76,** 103–130.

Papageorgiou, A. C., Shapiro, R., and Acharya, K. R. (1997). Molecular recognition of human angiogenin by placental ribonuclease inhibitor—an X-ray crystallographic study at 2.0 A resolution. *EMBO J.* **16,** 5162–5177.

Pawson, T., and Nash, P. (2000). Protein-protein interactions define specificity in signal transduction. *Genes Dev.* **14,** 1027–1047.

Prinjha, R., Moore, S. E., Vinson, M., Blake, S., Morrow, R., Christie, G., Michalovich, D., Simmons, D. L., and Walsh, F. S. (2000). Inhibitor of neurite outgrowth in humans. *Nature* **403,** 383–384.

Püschel, A. W., Adams, R. H., and Betz, H. (1995). Murine semaphorin D/collapsin is a member of a diverse gene family and creates domains inhibitory for axonal extension. *Neuron* **14,** 941–948.

Raper, J. A., and Kapfhammer, J. P. (1990). The enrichment of a neuronal growth cone collapsing activity from embryonic chick brain. *Neuron* **2,** 12–29.

Renzi, M. J., Feiner, L., Koppel, A. M., and Raper, J. A. (1999). A dominant negative receptor for specific secreted semaphorins is generated by deleting an extracellular domain from neuropilin-1. *J. Neurosci.* **19,** 7870–7880.

Rini, J. M. (1995). Lectin structure. *Annu. Rev. Biophys. Biomol. Struct.* **24,** 551–577.

Rudenko, G., Henry, L., Henderson, K., Ichtchenko, K., Brown, M. S., Goldstein, J. L., and Deisenhofer, J. (2002). Structure of the LDL receptor extracellular domain at endosomal pH. *Science* **298,** 2353–2358.

Schlessinger, J. (2000). Cell signaling by receptor tyrosine kinases. *Cell* **103,** 211–225.

Semaphorin Nomenclature Committee. (1999). Unified nomenclature for the semaphorins/collapsins. *Cell* **97,** 551–552.

Shirvan, A., Kimron, M., Holdengreber, V., Ziv, I., Ben-Shaul, Y., Melamed, S., Melamed, E., Barzilai, A., and Solomon, A. S. (2002). Anti-semaphorin 3A antibodies rescue retinal ganglion cells from cell death following optic nerve axotomy. *J. Biol. Chem.* **277,** 49799–49807.

Simon, M. (2000). Receptor tyrosine kinases: Specific outcomes from general signals. *Cell* **103,** 13–15.

Smalla, M. *et al.* (1999). Solution structure of the receptor tyrosine kinase EphB2 SAM domain and identification of two distinct homotypic interaction sites. *Protein Sci.* **8,** 1954–1961.

Smith, F. M. *et al.* (2004). Dissecting the EphA3/ephrin-A5 interactions using a novel functional mutagenesis screen. *J. Biol. Chem.* **279,** 9522–9531.

Sondek, J., Bohm, A., Lambright, D. G., Hamm, H. E., and Sigler, P. B. (1996). Crystal structure of a G-protein beta gamma dimer at 2.1A resolution. *Nature* **397,** 369–374.

Song, J. *et al.* (2002). Solution structure and backbone dynamics of the functional cytoplasmic subdomain of human ephrin b2, a cell-surface ligand with bi-directional signaling properties. *Biochemistry* **41,** 10942–10949.

Song, X.-Y., Zhong, J., Wang, X., and Zhou, X.-F. (2004). Suppression of p75NTR does not promote regeneration of injured spinal cord in mice. *J. Neurosci.* **24,** 542–546.

Stapleton, D. *et al.* (1999). The crystal structure of an Eph receptor SAM domain reveals a mechanism for modular dimerization. *Nature Struct. Biol.* **6,** 44–49.

Stein, E. *et al.* (1998). Eph receptors discriminate specific ligand oligomers to determine alternative signaling complexes, attachment, and assembly responses. *Genes Dev.* **12,** 667–678.

Takagi, J., Petre, B. M., Walz, T., and Springer, T. A. (2002). Global conformational rearrangements in integrin extracellular domains in outside-in and inside-out signaling. *Cell* **110,** 599–611.

Takahashi, T., Fournier, A., Nakamura, F., Wang, L. H., Murakami, Y., Kalb, R. G., Fujisawa, H., and Strittmatter, S. M. (1999). Plexin-neuropilin-1 complexes form functional semaphorin-3A receptors. *Cell* **99,** 59–69.

Takahashi, T., and Strittmatter, S. M. (2001). PlexinA1 autoinhibition by the plexin sema domain. *Neuron* **29,** 429–439.

Takasu, M. A., Dalva, M. B., Zigmond, R. E., and Greenberg, M. E. (2002). Modulation of NMDA receptor-dependent calcium influx and gene expression through EphB2 receptors. *Science* **295,** 491–495.

Tamagnone, L., Artigiani, S., Chen, H., He, Z., Ming, G. I., Song, H., Chedotal, A., Winberg, M. L., Goodman, C. S., Poo, M., Tessier-Lavigne, M., and Comoglio, P. M. (1999). Plexins are a large family of receptors for transmembrane, secreted, and GPI-anchored semaphorins in vertebrates. *Cell* **99,** 71–80.

Tang, S., Shen, Y. J., DeBellard, M. E., Mukhopadhyay, G., Salzer, J. L., Crocker, P. R., and Filbin, M. T. (1997). Myelin-associated glycoprotein interacts with neurons via a sialic acid binding site at ARG118 and a distinct neurite inhibition site. *J. Cell Biol.* **138,** 1355–1366.

Thanos, C. D. *et al.* (1999a). Oligomeric structure of the human EphB2 receptor SAM domain. *Science* **283,** 833–836.

Thanos, C. D. *et al.* (1999b). Monomeric structure of the human EphB2 sterile motif domain. *J. Biol. Chem.* **274,** 37301–37306.

Torres, R. *et al.* (1998). PDZ proteins bind, cluster, and synaptically colocalize with Eph receptors and their ephrin ligands. *Neuron* **21,** 1453–1463.

Toth, J. *et al.* (2001). Crystal structure of an ephrin ectodomain. *Dev. Cell* **1,** 83–92.

Van de Velde, H. J. K., Roebroek, A. J. M., Senden, N. H. M., Ramaekers, F. C. S., and Van de Ven, W. J. M. (1994). NSP-encoded reticulons, neuroendocrine proteins of a novel gene family associated with membranes of the endoplasmic reticulum. *J. Cell Sci.* **107,** 2403–2416.

Varghese, J. N., Laver, W. G., and Colman, P. M. (1983). Structure of the influenza virus glycoprotein antigen neuraminidase at 2.9 A resolution. *Nature* **303,** 35–40.

Vikis, H. G., Li, W., and Guan, K. L. (2002). The plexin-B1/Rac interaction inhibits PAK activation and enhances Sema4D ligand binding. *Genes Dev.* **16,** 836–845.

Wang, H. U. *et al.* (1998). Molecular distinction and angiogenic interaction between embryonic arteries and veins revealed by ephrin-B2 and its receptor Eph-B4. *Cell* **93,** 741–753.

Wang, K. C., Koprivica, V., Kim, J. A., Sivasankaran, R., Guo, Y., Neve, R. L., and He, Z. (2002a). Oligodendrocyte-myelin glycoprotein is a Nogo receptor ligand that inhibits neurite outgrowth. *Nature* **417,** 941–944.

Wang, K. C., Kim, J. A., Sivasankaran, R., Segal, R., and He, Z. (2002b). P75 interacts with the Nogo receptor as a co-receptor for Nogo, MAG and OMgp. *Nature* **420,** 74–78.

Wybenga-Groot, L. E. *et al.* (2001). Structural basis for autoinhibition of the EphB2 receptor tyrosine kinase by the unphosphorylated juxtamembrane region. *Cell* **106,** 745–757.

Xiong, J. P., Stehle, T., Diefenbach, B., Zhang, R., Dunker, R., Scott, D. L., Joachimiak, A., Goodman, S. L., and Arnaout, M. A. (2001). Crystal structure of the extracellular segment of integrin alpha Vbeta3. *Science* **294,** 339–345.

Yu, T. W., and Bargmann, C. I. (2001). Dynamic regulation of axon guidance. *Nat. Neurosci.* **4,** 1169–1176.

Zisch, A. H. *et al.* (2000). Replacing two conserved tyrosines of the EphB2 receptor with glutamic acid prevents binding of SH2 domains without abrogating kinase activity and biological responses. *Oncogene* **19,** 177–187.

SHARED CYTOKINE SIGNALING RECEPTORS: STRUCTURAL INSIGHTS FROM THE GP130 SYSTEM

By MARTIN J. BOULANGER AND K. CHRISTOPHER GARCIA

Departments of Microbiology and Immunology, and Structural Biology,
Stanford University School of Medicine,
Stanford, California 94305-5124

Abstract

The vast majority of cytokine signaling is mediated by "shared" receptors that form central signaling components of higher-order complexes incorporating ligand-specific receptors. These include the common γ chain (γ_c), common β chain (β_c), and gp130, as well as others. These receptors have the dual tasks of cross-reactive cytokine recognition, and formation of precisely oriented multimeric signaling assemblies. Currently, detailed structural information on a shared receptor complex exists only for gp130, which is a highly pleiotropic shared cytokine signaling receptor essential for mammalian cell growth and homeostasis. To date, more than 10 different four-helix bundle ligands have been identified that incorporate gp130, or one of its close relatives such as LIF receptor, into functional oligomeric signaling complexes. In this review we summarize our current knowledge of shared receptor recognition and activation, with a focus on gp130. We discuss recent structural and functional information to analyze overall architectural assemblies of gp130

ADVANCES IN PROTEIN CHEMISTRY, Vol. 68

0065-3233/04 $35.00

cytokine complexes and probe the basis for the extreme cross-reactivity of gp130 for its multiple cytokine ligands.

I. Cross-Reactive Signaling Receptors in Biology

The formation of high-affinity, ligand-mediated extracellular signaling complexes has emerged as a central paradigm in biology. This initial step in cellular communication is followed by the activation of intracellular signaling cascades, which are directly related to the composition and architectural assembly of the extracellular complexes. Although highly specific ligand/receptor interactions can provide the tight regulation necessary to maintain control of physiological responses, many receptor systems exhibit an extraordinary degree of cross-reactivity, which can result in a redundancy beneficial for survival of the organism. In fact, one might argue that with the emergence of the human genome sequence, we are finding that most receptor systems have the capacity to recognize and respond to more than one ligand. There are many examples of degenerate, shared receptors with central roles in signaling. In neurobiology, the p75 neurotrophin receptor (Dechant and Barde, 2002), the RET receptor for GDNF (Anders *et al.*, 2001), the Nogo receptor (GrandPre *et al.*, 2002), and others each have a variety of diverse ligands that can evoke distinct signaling pathways. In the immune system, shared receptors exist in both adaptive (T cell receptors, co-stimulatory receptors B7/CD28) and innate immunity (NKG2D natural killer receptor, scavenger, and pattern recognition receptors such as RAGE and Toll) (Strong, 2002). The cross-reactivity of T-cell receptors is an essential feature of their dual ability to both recognize and avoid self antigens as well as destroy foreign pathogens (Mason, 1998). Chemokine G-protein coupled receptors (GPCR) can exhibit an extraordinary degeneracy in chemokine recognition (Gong *et al.*, 1996). However, perhaps in no other system are shared receptors as integral to mammalian biology as seen for cytokines (Nicola, 1994; Taga and Kishimoto, 1995). Given its emerging importance, a clearer understanding of the biochemical and structural factors associated with receptor-ligand degeneracy is needed for a more comprehensive understanding of cellular communication. Gp130 is one of the most ubiquitous and important shared signaling receptors for cytokines, mediating the actions of more than 10 known ligands so far, each of which initiates both redundant and unique signaling outcomes (Bravo and Heath, 2000; Grotzinger *et al.*, 1999; Rose-John, 2002; Simpson *et al.*, 1997; Taga and Kishimoto, 1997). The past year has seen significant progress in obtaining structural information of gp130 in complex with different cytokine ligands (Boulanger *et al.*, 2003a,b). Also, new members of the gp130-class of cytokines are emerging from bioinformatics cloning approaches (Trinchieri *et al.*, 2003). In this review we take advantage of this recent structural information to study the use of a common assembly template among the higher order gp130 signaling complexes. We also probe in detail the structural basis for the extraordinary degree of cross-reactivity of gp130. The principles derived from structural analysis of gp130-ligand interactions may have some translational relevance to understanding the cross-reactive properties of other shared receptors.

At the outset of this review, we apologize that we are unable to cite all contributors to our understanding of this field in our review because of space constraints. We refer the reader to many excellent reviews on various aspects of cytokine structure, receptor interaction, and signaling (Bravo *et al.*, 1998; Davies and Wlodawer, 1995; de Vos *et al.*, 1992; Heinrich *et al.*, 2003; Kossiakoff and De Vos, 1998; Rose-John, 2002; Wells and de Vos, 1996; Wilson and Jolliffe, 1999).

II. Cytokines and Receptors

Hematopoietic cytokines comprise one of the largest groups of extracellular regulatory molecules and are key mediators of a diverse spectrum of physiological functions that include induction of immune responses, cell proliferation, and differentiation (Cosman, 1993; Cosman *et al.*, 1990; Kishimoto *et al.*, 1994; Nicola and Hilton, 1998; Taga and Kishimoto, 1992). The defining structural feature of the hematopoietic class of cytokines is a four-helix bundle motif organized into four anti-parallel helices that adopt an up-up-down-down motif (Bazan, 1990a; Sprang and Bazan, 1993). Structural predictions, later confirmed by several crystal and NMR structures, indicated that the four-helix bundle cytokines could be further sub-classified based on lengths of the helices (Davies and Wlodawer, 1995). The most common group is the long chain cytokines (10 to 20 residues), or type I cytokines, which include human growth hormone (GH), erythropoietin (EPO), granulocyte colony-stimulating factor (G-CSF) and the gp130 cytokines (Simpson *et al.*, 1997). The second class is the short chain cytokines with helices of 8 to 10 residues in length that include interleukins 2, 3, and 4. The third and final group of cytokines are formed by tandem four-helix bundle motifs to generate an eight-helix bundle architecture (Sprang and Bazan, 1993) and includes interleukin 5 and interferon gamma (Davies and Wlodawer, 1995; Walter, 2002). While this latter group is not strictly classified as "hematopoietic," they share highly related structural features and can be grouped as a whole with the hematopoietic cytokines to compose a large family of four-helix bundle cytokines with many redundant functions. In this review, we restrict our focus to the hematopoietic members.

Cytokines mediate their functional effects through a wide variety of signaling receptors that are generally classified into one of six major families based on common structural features that include secondary structure and conserved amino acid motifs (Bazan, 1990a,b; Cosman, 1993). The largest of these families is the hematopoietic receptor superfamily that is characterized by a cytokine binding homology region, or CHR, formed from tandem fibronectin repeats (Wells and de Vos, 1996). The amino terminal CHR domain contains four conserved cysteine resides and the carboxy terminal domain encodes a conserved WSXWS tryptophan motif (Bazan, 1990b; de Vos *et al.*, 1992) (Fig. 1). Mutagenesis studies have shown a structural role for these amino acids in maintaining the ternary structure and ligand recognition properties of the receptor (Yawata *et al.*, 1993). These structural features have been used successfully to identify and clone new hematopoietic receptors (Elson *et al.*, 1998; Mosley *et al.*, 1996; Sprecher *et al.*, 1998). Together, the four-helix cytokine and the corresponding CHR in the receptors comprise the basic building blocks of the cytokine world.

III. The Growth Hormone Paradigm: Generalities and Limitations

The structural basis for cytokine ligand-receptor association was initially defined from extensive structural, biochemical, and functional studies of human growth hormone (de Vos *et al.*, 1992; Kossiakoff and De Vos, 1998; Wells, 1996). The crystal structure of GH in complex with the human growth hormone receptor (GHR) (de Vos *et al.*, 1992) represented the first structure of a four-helix bundle cytokine in complex with its receptor (Fig. 1). Briefly, the assembly of the GH/GHR complex is mediated by two topologically distinct epitopes, termed *sites I* and *II*, on GH that engage the CHR domains of two identical receptor subunits. Sites I and II are located on opposite sides of GH, with site I formed by the B and D helical faces while site II is located on the A and C helical faces. The homodimerization of two receptors by a single cytokine, using different epitopes on the cytokine, but largely similar residues on the receptor, was completely unexpected. This complex established a structural linkage between the four-helix bundle cytokine, and the receptor CHR that can be found in all other cytokine receptor complexes. Following this assembly, specific tyrosine residues on the cytoplasmic domain of the receptor are

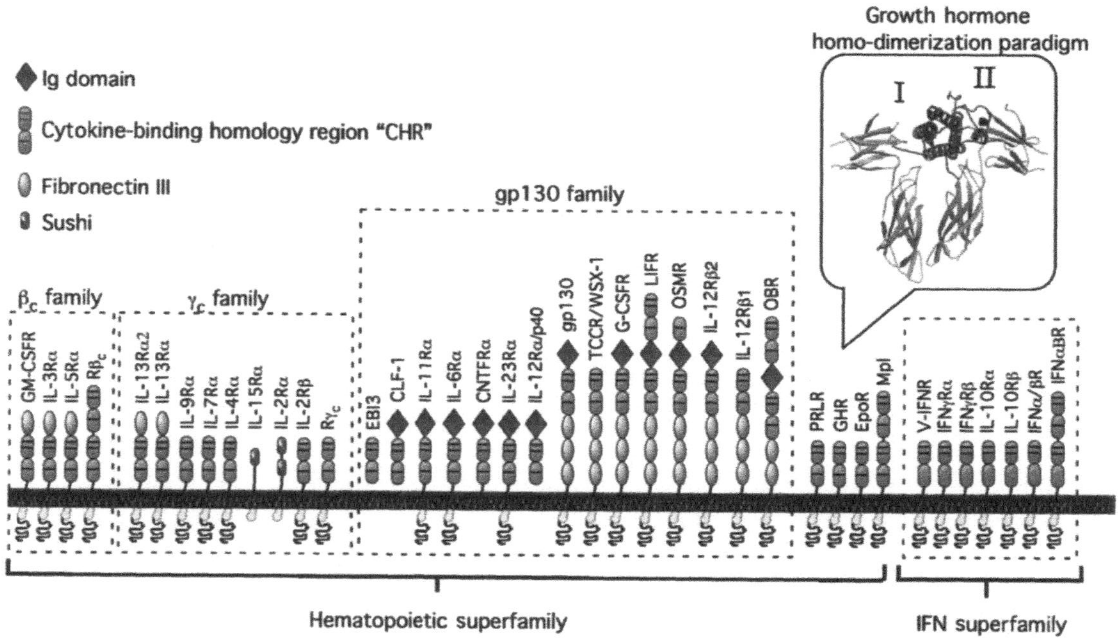

FIG. 1. Grouping of hematopoietic cytokine receptors by shared receptor usage. The majority of the hematopoietic cytokine receptors incorporate one of three shared signaling receptors, either the common beta (β_c) chain, the common gamma (γ_c) chain or gp130. The crystal structure of the growth hormone (red) complex (inset) was the first structure of a four helix bundle cytokine in complex with its receptor (green) (de Vos *et al.*, 1992) and established the paradigm of cytokine/receptor complex formation. (See Color Insert.)

phosphorylated and become docking sites for transcriptional regulatory molecules such as the STAT family of transcription factors (Heinrich *et al.*, 1998; Ihle, 1995). The nature of the transcriptional regulation is dictated directly by the identity of the cytokine and receptor and the orientational geometry of the final signaling complex (Constantinescu *et al.*, 2001). This has been most clearly observed with the EPO system (Livnah *et al.*, 1996), where mimics of EPO, and EPO itself, are able to homodimerize the erythropoietin receptor (EPOR) but have altered signaling capacities that seem to be correlated with the dimeric "angular orientation" of the receptors in the different complexes (Livnah *et al.*, 1998; Syed *et al.*, 1998). The implication is that the variable dimerization angles are relayed through the membrane to induce varied intracellular signaling outcomes (Seubert *et al.*, 2003; Wells, 1996; Wells and de Vos, 1996).

Although GH represents the simplest cytokine system, containing only a single class of receptor (similar to EPO and thrombopoietin [TPO], etc.), it established ligand-induced receptor homodimerization as the cytokine receptor signaling mechanism. At a gross structural level, all of the receptors shown in Fig. 1 use either a site I, site II, or combined I/II interaction mode with their cytokines as part of the higher-order signaling complexes (Davies and Wlodawer, 1995; de Vos *et al.*, 1992; Grotzinger, 2002). While the cytokine receptor complexes to date (IL-4, IL-10, IL-12, EPO, γ-IFN, etc.) recapitulate aspects of the modular architecture originally seen in the GH complex, similarities at the atomic level are less apparent. In other words, the GH complex revealed a gross docking topology, but each cytokine receptor pair uses divergent receptor-cytokine contacts, which prevent us from simply modeling the complexes based on the GH example (Walter, 2002). In fact, the other cytokine-receptor complexes have revealed significant variations in the docking geometries and interaction chemistries, which highlight the structural richness remaining in examining other complexes. The relationship of cytokine-receptor complexes to the GH example is similar in nature to the situation of T cell receptor/MHC interactions. The first complex structures established a roughly "diagonal" docking topology of the TCR on the MHC, but subsequent complex structures have revealed many unpredictable variations on this theme (Rudolph and Wilson, 2002). Detailed information on cytokine-receptor interactions still requires direct structural analysis. It is also important to remember that only a minority of cytokines signal by the GH homodimerization mechanism. Most other cytokine systems require a more complicated signaling assembly that incorporates different receptors, and binding epitopes to form a hetero-oligomeric signaling complex.

IV. Shared Signaling Receptors for Hematopoietic Cytokines

The vast majority of cytokines utilize combinations of monogamous and promiscuous receptors (Bagley *et al.*, 1995; Leonard *et al.*, 1994; Nicola, 1994; Taga and Kishimoto, 1995). Cytokines and their receptors can be classified based on their ligand-receptor pairing and usage of a limited set of shared receptors (Grotzinger, 2002) (Fig. 1). When grouped by shared receptor usage, we see three (or four, when the IL-10/IFN systems are added) major classes of hematopoietic cytokines: those that utilize the common gamma chain (γ_c), those that utilize common beta chain (β_c), and those within the largest group, the IL-6/IL-12 family that use gp130 and gp130-like shared receptors. These shared receptors are sometimes colloquially referred to as the "tall" (gp130 class) and the "short" (γ_c and β_c) receptors, and each have major structural differences from one another. In most cases, the shared receptors are used in concert with ligand specific α and/or β receptors. The shared receptor has the task of recognizing groups of cytokines with vastly different structural epitopes and converting this recognition event into an intracellular signal. The shared receptor then is required to be promiscuous for different ligand surface chemistries and structures yet specific enough to not cross-react with inappropriate cytokines. Thus these receptors have both recognition and assembly tasks in coordinating with the ligand-specific components of the signaling complexes. It is convenient to discuss these topics separately, but of course both recognition and assembly are intimately interrelated.

A. *The Common Gamma Chain:* γ_c

The glycoprotein γ_c is a type I transmembrane glycoprotein that serves as a shared signaling subunit for the receptors of interleukin (IL)-2, IL-4, IL-7, IL-9, IL-15, and IL-21 (He and Malek, 1998; [Kishimoto, 1995] Leonard *et al.*, 1994; Taga and Kishimoto, 1995) (Fig. 1, partially modeled as part of IL-2 ternary signaling complex in Fig. 2A (Bamborough *et al.*, 1994). The biological importance of γ_c is most dramatically illustrated by the fact that mutations of either γ_c or JAK3 are the primary causes of human X-linked severe combined immunodeficiency (X-SCID), characterized by a failure in T and natural killer cell development (Theze *et al.*, 1996). The IL-2 receptor (IL-2R) and IL-15R are heterotrimers comprised of unique α-chains and shared IL-2Rβ and γ_c subunits, whereas IL-4R, IL-7R, IL-9R, and IL-21R are heterodimers comprised of unique α-subunits and γ_c (Lai *et al.*, 1996). For each of these cytokine receptors, γ_c directly contributes to ligand binding through its extracellular domain and to

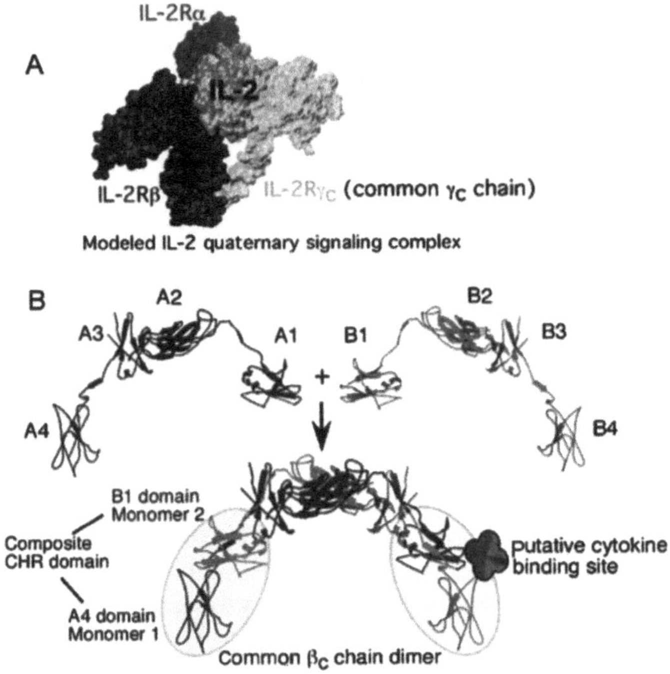

FIG. 2. Common gamma (γ_c) and beta (β_c) receptors. (A) Modeled IL-2 ternary complex based on the growth hormone complex showing the partially modeled common gamma (γ_c) chain, (yellow) (Bamborough *et al.*, 1994). In this model the γ_c chain docks against the long axis of IL-2 and also forms contacts with the D2 domain of IL-2Rβ receptor in a similar fashion to the growth hormone structure (de Vos *et al.*, 1992). (B) Structure of the extracellular homodimer of the common β_c chain. (Carr *et al.*, 2001). Recent mutagenesis studies have identified the cytokine binding homology region (CHR) to be formed by the A4 domain of one monomer (blue) and the B1 domain of the second monomer (green) to form a composite epitope (shaded cyan region) (Murphy *et al.*, 2003). The cytokine ligand is drawn in red and docked onto the CHR domain. (See Color Insert.)

signal transduction through the association of JAK3 with its cytoplasmic tail (Sugamura *et al.*, 1995).

The structural basis by which γ_c functions in binding six distinct cytokines is still not defined, although mutagenesis has localized sites of interaction (Olosz and Malek, 2000, 2002). The only member of this family for which a complex structure exists is IL-4 complex, with its alpha receptor (Hage *et al.*, 1999; Mueller *et al.*, 2002). Molecular modeling of the IL-2, IL-4, and IL-7 complexes with γ_c have been constructed based on assumed analogies with the GH paradigm of a site I and site II epitope on the cytokine (Bamborough *et al.*, 1994; Gustchina *et al.*, 1995; Kroemer and Richards, 1996; Zurawski *et al.*, 1993). In this case, site I is presumed to interact with the cytokine-specific α-receptor component, whereas site II interacts with γ_c. For IL-7, which uses a single alpha chain, the trimolecular complex with γ_c was modeled as being highly similar to the GH homodimer topology, with IL-7Rα bound to site I and γ_c bound to site II (Kroemer and Richards, 1996). The same conclusion was reached for IL-4, which also uses a single Rα. However, for IL-2 and IL-15, the situation is more complicated, since they use a unique Rα as well as an Rβ to form a quaternary signaling complex (Giri *et al.*, 1995; Leonard *et al.*, 1994; Liparoto *et al.*, 2002). Therefore for IL-2 and IL-15, the Rβ has been modeled as the site I binding receptor, keeping site II interaction with γ_c (Bamborough *et al.*, 1994). The unique Rα, which does not fall into the classical CHR fold of other hematopoietic receptors, has been placed as a "cap" on top of the four-helix bundle (Bamborough *et al.*, 1994). Why IL-2 and IL-15 utilize this structurally unique third receptor component is not known, although precomplexation with Rα is necessary for the subsequent transitions into the quaternary signaling complex. Numerous biochemical and cellular studies, particularly in the IL-2 system, have demonstrated cooperativity in the formation of various intermediate and higher-order forms of the multimeric signaling complexes (Liparoto *et al.*, 2002).

The interactions of γ_c with its ligands have been probed by mutagenesis based on the crude models of the complexes. Panels of γ_c mutants in combination with antagonistic antibodies have narrowed down the γ_c binding site (Olosz and Malek, 2000, 2002; Zhang *et al.*, 2003). γ_c utilizes a common mechanism for its interactions with multiple cytokines, but the binding sites are largely overlapping yet not identical (Olosz and Malek, 2000, 2002). The functional region of γ_c, the classical cytokine-binding homology region (CHR), is composed of the EF1, BC2, and FG2 loops projecting out from the intersection between top and bottom β-sandwiches, which is highly similar to other type I cytokine receptors. Analogous ligand binding sites have been found in the corresponding loops of GHR, EPOR, gp130, granulocyte CSFR (G-CSFR), and granulocyte-macrophage (GM-CSFR). From the mutagenesis studies, Asn^{128} and Tyr^{103} likely act as contact residues with cytokine. Importantly, Tyr^{103} of γ_c is homologous to critical ligand binding sites located in the hydrophobic clusters of GHR (Trp^{104}) and EPOR (Phe^{93}). Asn128 is located in a "linker" region between the D1 and D2 domains of γ_c that is not found in other CHR-containing receptors. This linker may allow inter-domain flexibility as an adaptive mechanism to diverse cytokine and receptor components. It has been noted from the structure-function studies that the magnitude of the mutational effects differs for different members of the γ_c cytokine family (Olosz and Malek, 2000, 2002; Zhang *et al.*, 2003). For example, mutations of Tyr^{103} appeared to have a more severe effect on the binding of IL-15 than on IL-2. Similarly, the single mutation Asn-128 had a more noticeable effect on IL-7 versus IL-2 and IL-15 binding. The IL-21 epitope comprises γ_c residues Asn^{44}, Tyr^{103}, Asn^{128}, Leu^{161}, Glu^{162}, and Leu^{208}, which are only partially overlapping with the previously established γ_c epitope for IL-4 binding. Perhaps Tyr^{103} functions as a structurally conserved anchor residue for cytokine binding, whereas surrounding amino acids such as Asn^{128} may be related to fine specificity in binding for the different cytokines. Such a structural makeup would be consistent with the "hotspot" idea from Wells and colleagues (Clackson and Wells, 1995). The primacy of a central aromatic in enabling cross-reactivity has emerged as a critical feature in gp130 recognition of structurally diverse cytokines (Boulanger *et al.*, 2003a), thus γ_c and gp130 likely share this property.

In summary, the γ_c class of cytokines clearly utilizes the site I/II architectural assembly of the GH paradigm as the signaling complex, with some important exceptions such as the role of the sushi-domain in the IL-2 and IL-15Rα. The precise definition of the chemistry of cross-reactive cytokine recognition in this family remains an important future structural challenge.

B. *The Common Beta Chain:* β_c

The β_c is one of the most enigmatic shared cytokine receptors and is used by GM-CSF, IL-3, and IL-5 (Fig. 1), which are related cytokines involved in the regulation of hematopoiesis and inflammation (Bagley *et al.*, 1995; Nicola *et al.*, 1997; Tavernier *et al.*, 1991). The receptors for the three cytokines consist of cytokine-specific α receptors essential to the activation of the shared β_c receptor subunit that is believed to be the main signaling entity. Although β_c does not bind cytokines by itself, its coexpression with the α-chains enhances the affinity of cytokine binding. Such a characteristic "affinity conversion" effect is utilized by cytokine shared receptors as a means of imposing tissue specificity. The stoichiometry of the activated receptor complexes and the mechanism of β_c activation are not definitively known. Crystal structures exist for GM-CSF (Diederichs *et al.*, 1991) and IL-5 (Milburn *et al.*, 1993) but until recently, no X-ray structures were available for the complete extracellular domains of any of the α subunits, the β_c subunit, or their ligand complexes. In 2001 there was a great achievement in this system with the report of crystal structure of the full-length extracellular domain of β_c (Carr *et al.*, 2001) (Fig. 2B). Together with a previously determined D4-mAb complex structure

(Rossjohn *et al.*, 2000) and mutagenesis information (Woodcock *et al.*, 1994), new insight has been gained into the recognition and signaling mechanisms in this system.

From sequence analysis, the extracellular part of β_c comprised four FNIII-type domains, forming two contiguous CHR modules, with features especially conserved among cytokine receptors (e.g., the conserved cysteines and WSXWX box). Mutational analysis has localized the cytokine-binding region to residues in the fourth extracellular domain (D4), and in particular the residues Tyr^{365}, His^{367}, Ile^{368} are crucial for receptor activation (Woodcock *et al.*, 1994). The complex of the D4 with an antagonistic mAb localized the positions of these residues to be in surface exposed strands and loops on the FNIII domain (Rossjohn *et al.*, 2000). Unexpectedly, the structure of the full-length β_c ectodomain revealed it to be a domain-swapped, head-to-tail intertwined dimer (Carr *et al.*, 2001) (Fig. 2B). Domains 1 and 3 have a very novel feature: their G strands of the beta sheets are unraveled and extended away from its own domain to form the G strand of a fibronectin domain in the other protein chain. The unusual fold in the β_c receptor structure generates a stable intertwined dimer that is consistent with cellular studies showing the β_c is a dimer on the cell surface (Carr *et al.*, 2001; Woodcock *et al.*, 1997). In the dimeric structure, the membrane proximal domains are positioned far from one another as a result of the antiparallel dimer orientation forcing the long axis of the dimer to lie almost parallel to the membrane surface (Fig. 2B). The completely unexpected structure of β_c is without precedent in other cytokine receptors.

Despite the fascinating structure of the unliganded β_c, the mode of interaction with cytokine or α receptor ligands is still not known. The intricate structure of the dimer presents an array of possibilities (Carr *et al.*, 2001). The two domains closest to the membrane on either side of the dimer (domains A1-B4 and B1-A4) adopt the characteristic L-shaped topology of the classical CHR module seen in the growth hormone receptor and related two-domain receptors. However, the interfaces between domains A2 and B3 are also similar to those of cytokine receptors that are activated by homodimerization. Thus the familiar "elbow" of CHR receptors is present in two places in the dimer. However, in both cases the putative interaction side would be formed by joining of two different receptors in the dimer, which would be unprecedented. A recent mutagenesis study appears to solidify the unusual A1-B4 module as the ligand interaction site (Murphy *et al.*, 2003). This would represent the first example of a class I cytokine receptor interface to be composed of two noncontiguous fibronectin III domains. Direct visualization of the architecture of the higher-order signaling assembly and the structural basis for cross-reactive recognition of three cytokines remains to be determined. Clearly, though, the β_c suggests the possibility of some very interesting structural surprises.

V. gp130 Family of Cytokines and Receptors

The only structural information for complex formation by a shared cytokine receptor is for gp130, which is activated by four-helix cytokines known as the ''gp130-cytokines'' or ''IL-6-type cytokines'' (Simpson *et al.*, 1997) (Fig. 3). These cytokines generally share a very low sequence homology (10–20%) and so pose an interesting question of how gp130 can recognize each with high affinity. Gp130 is the founding member of a group of ''tall'' receptors that includes LIF-R, OSM-R, OB-R, G-CSF-R, and others (Fig. 1). These receptors have sequence homology to one another that places them as genetic relatives. They also share roughly similar domain architecture, particularly in the use of the top-mounted Ig-domain, the FNIII spacer domains, and the location of the cytokine binding regions (Fig. 3). However, these tall receptors may not all be shared. For instance, G-CSF is the only known ligand for G-CSFR, and leptin is the only known ligand for OB-R.

Despite the ubiquitous expression of gp130, cellular responsiveness mediated by gp130 cytokines is largely determined the half-lives of the locally secreted cytokines and by regulated expression of ligand specific chains. The biological importance of gp130 is clear from deletion studies in which the targeted disruption of gp130 gene in mice is lethal during embryogenesis. Collectively, gp130 signaling complexes mediate a wide variety and often overlapping biological functions, yet each member of the family yields a unique phenotype on inactivation. Deletion of LIF, for example, results in female infertility caused by failure of embryo implantation (Stewart *et al.*, 1992), while deletion of IL-6 results in impaired acute phase and anti-viral response (Kopf *et al.*, 1994; Poli *et al.*, 1994). We refer the reader to a number of excellent comprehensive reviews on the gp130 system (Bravo and Heath, 2000; Chow *et al.*, 2002; Grotzinger *et al.*, 1997, 1999; Heinrich *et al.*, 1998, 2003; Hirano *et al.*, 1997; Kishimoto *et al.*, 1995; Simpson *et al.*, 1997; Taga and Kishimoto 1997).

The family of gp130 cytokines can be subdivided into two main groups based on the identity of the signaling receptors incorporated into the final complex (Fig. 3). The first group of gp130 cytokines includes Interleukin-6 (IL-6) and Interleukin-11 (IL-11), both of which require association with a specific α-receptor as the first step in the assembly pathway. Although dispensable for complex formation and signaling, the intracellular domain of IL-6Rα appears to function in sorting of the α receptor to the

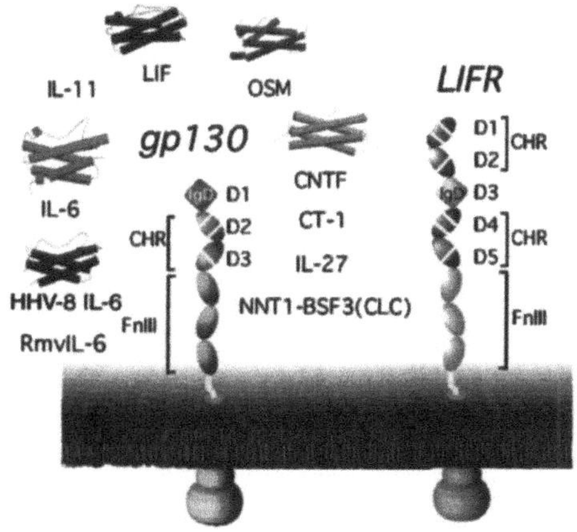

Cytokine	Site I	Site II	Site III
HHV-8 IL-6		gp130	gp130
Rm IL-6		gp130	gp130
IL-6	IL-6Rα	gp130	gp130
IL-11	IL-11Rα	gp130	gp130
LIF		gp130	LIFR
OSM		gp130	LIFR(*OSMR*)
CNTF	CNTFRα	gp130	LIFR
CT-1	CNTFRα	gp130	LIFR
NNT1-BSF3(CLC)	CNTFRα	gp130	LIFR
IL-27		TCCR	gp130

FIG. 3. The gp130 family of cytokines and receptors. Schematic representation of gp130 and leukemia inhibitory factor receptor (LIFR) oriented in a cell membrane. Of the four-helix bundle gp130 cytokines, structural information currently exists for human interleukin 6 (IL-6) (green) (Somers *et al.*, 1997), human herpes virus interleukin 6 (HHV-8 IL-6) (purple) (Chow *et al.*, 2001a), ciliary neurotrophic factor (CNTF) (orange) (McDonald *et al.*, 1995), leukemia inhibitory factor (LIF) (blue) (Robinson *et al.*, 1994), and oncostatin-M (OSM) (red) (Deller *et al.*, 2000). Lower panel is a detailed list of gp130 cytokines and the associated receptors incorporated into the final signaling complex. (See Color Insert.)

basolateral membrane of polarized cells (Martens *et al.*, 2000). A subgroup of this "homodimerization" class of gp130 cytokines is comprised of two viral homologues of IL-6, one from human herpes virus (HHV-8 IL-6) and the second from the Rhesus macaque rhadinovirus (Rm IL-6). These viral homologs short-circuit the traditional assembly pathway established for

IL-6 and IL-11 by dispensing with the requirement for an α receptor. A recent study suggests that "hydrophobic enhancement" of the receptor binding epitopes of HHV-8 IL-6 may be responsible for obviating the requirement for a specific α receptor (Boulanger *et al.*, 2004). The second group of gp130 cytokines mediate the heterodimerization of gp130 with a second signaling receptor known as LIF receptor (LIFR) (Benigni *et al.*, 1996; Hibi *et al.*, 1990; Taga and Kishimoto, 1997) (Fig. 3). This group represents the largest of the gp130 cytokines and includes leukemia inhibitory factor (LIF), oncostatin (OSM), cardiotrophin (CT-1), ciliary neurotrophic factor (CNTF), and NNT-1/BSF3, also known as cardiotrophin-like cytokine (CLC). The common theme between both groups is that the gp130 cytokines mediate the complexation of two signaling receptors. Interestingly, artificial homodimerization of the cytoplasmic domains of LIFR resulted in only weak signal activation highlighting the functional requirement of gp130 (Starr *et al.*, 1997).

The family of gp130 cytokines and receptors continues to grow with several recent additions (Trinchieri *et al.*, 2003). A third receptor, OSM receptor (OSM-R) (Mosley *et al.*, 1996) has also been identified as a potential signaling receptor for OSM. OSM-R is structurally similar to LIF-R but encodes only a partial N-terminal domain of the upper (N-terminal) CHR motif (Fig. 1). Functionally, Cosman and coworkers have shown that OSM is not capable of binding to either LIF-R or OSM-R except in the presence of gp130 (Mosley *et al.*, 1996). Using isothermal titration calorimetry of the soluble ectodomains, minus the three membrane proximal fibronectin repeats, of LIF-R, OSM-R, and gp130, we find that OSM can bind directly to LIF-R with nanomolar affinity in the absence of gp130 but shows no detectable binding to OSM-R even following precomplexation of OSM with gp130 (our unpublished data). Clearly, further studies are required to address these inconsistencies. Very recently a new gp130 cytokine, IL-27 (Trinchieri *et al.*, 2003), has been discovered that is secreted as a preformed complex consisting of the four-helix bundle cytokine p28 and an α receptor EB13. IL-27 appears to mediate the assembly of a fourth type of gp130 signaling complex, which consists of one gp130 receptor and a second signaling receptor known as TCCR or gp132 that is similar architecturally to gp130 but lacks the N-terminal Ig domain (Pflanz *et al.*, 2002; Sprecher *et al.*, 1998). For the sake of this review we will focus on the two primary groups of the higher order signaling complexes that incorporate gp130 and LIFR.

The three-dimensional structures are known for many gp130 cytokines including murine LIF (Robinson *et al.*, 1994), human LIF (Hinds *et al.*, 1998), CNTF (McDonald *et al.*, 1995), human Interleukin 6 (IL-6) (Somers *et al.*, 1997), HHV-8 IL-6 (Chow *et al.*, 2001a) and oncostatin M (OSM)

(Deller *et al.*, 2000) (Fig. 3). While each gp130 cytokine adopts the classical four-helix bundle motif, the lengths of the helices and interhelical loops differ with the loops even showing variation between homologs from different species (Bravo and Heath, 2000). The most interesting structural difference between the gp130 cytokines is observed in the topology of the A and D helices, which are straight in IL-6, IL-11 and HHV-8 IL-6 but exhibit a distinct kink in LIF, OSM and CNTF. The fact that these two structural features coincide with the receptor composition in the final signaling complexes—straight helical cytokines incorporate only gp130 as a signaling receptor while the kinked cytokines use both gp130 and an additional signaling receptor such as LIFR—has led to the suggestion that this structural feature plays an important role in molecular recognition (Bravo and Heath, 2000).

Architecturally, the ectodomains of gp130 and LIFR are composed of six and eight contiguous β-sandwich domains containing one or two signature CHR domains (Bazan, 1990a), respectively, and a single Ig-domain (IgD) (Hammacher *et al.*, 1998) (Fig. 3). Both gp130 and LIFR encode three membrane-proximal fibronectin type-III domains that do not play a specific role in cytokine recognition (Bravo *et al.*, 1998) but have recently been implicated in signal activation by correctly orienting and positioning the intracellular domains (Timmermann *et al.*, 2002). The non-signaling α receptors of IL-6, IL-11, CT-1, CNTF, and CLC are similar in structure to gp130 with an N-terminal Ig domain followed by a CHR domain yet differ in that they do not incorporate the three membrane proximal fibronectin repeats. Instead, the functional extracellular domains of these α receptors are linked to the membrane through amino acid stalks (IL-6, IL-11, CT-1) or GPI anchors (CNTF). The D1 or IgD domain of gp130 forms an essential component of the higher-order signaling complex with human IL-6, HHV-8, IL-6, and IL-11 (Boulanger *et al.*, 2003b; Chow *et al.*, 2001a) yet appears to be dispensable for assembly of the hetero complexes of gp130 with LIFR (Boulanger *et al.*, 2003a). The Ig domain is, however, thought to be involved in stabilizing the receptor during intracellular trafficking through the secretory pathway (Vollmer *et al.*, 1999). The "top mounted" Ig domain is a signature feature of the gp130 family that allows us to predict the architecture of all gp130-related signaling assemblies based on which receptor components in the complex carry this domain (discussed later).

In a similar fashion to the classical GH system, gp130 cytokines mediate the assembly of higher-order signaling complexes through distinct epitopes. The sites I and II, originally identified in GH, formed by distinct helical faces of the four-helix bundle are conserved in the gp130 cytokines. In addition, gp130 cytokines also include a third epitope, termed *site III*, that is formed by the tip of the four-helix bundle, formed by the C-D loops and the end of the D helix, and the N-terminus of the cytokine. Site I is used by IL-6, CNTF, IL-11, CT-1, and CLC to engage ligand-specific non-signaling α receptors. Site II is used by all gp130 cytokines to bind the CHR domain of gp130. Site III was originally identified as a functional epitope by mutagenesis and is responsible for driving the transition of the higher-order gp130 signaling complex through the interaction with the Ig domain of gp130 or LIFR (Hammacher *et al.*, 1998; Inoue *et al.*, 1995; Kallen *et al.*, 1999; Kurth *et al.*, 1999). Using a combination of these sites, and in some cases all three sites, the functional role of gp130 cytokines is to oligomerize the extracellular domains of the signaling receptors into a high-affinity complex, thereby promoting the association of the intracellular domains.

A. *Current Structural Knowledge*

Despite significant structural information detailing the features of the unliganded gp130 cytokines, far less is known about the structures of the receptors alone or in complex with a ligand. The CHR domain of gp130 expressed in *E. coli* was solved in 1998 by Jones and co-workers (Bravo *et al.*, 1998). The 2.0 angstrom resolution structure of the gp130 CHR revealed a similar topology to other class I receptors of the hematopoietic cytokine receptor superfamily. The basic structural scaffold consists of two fibronectin, seven-stranded β sandwich repeats that are related by a ~90-degree elbow angle connected by a proline rich linker (Fig. 4). Mutagenesis studies have localized the cytokine binding interface on gp130 to the elbow region of the CHR domain at the interface between the N and C-terminal fibronectin domains. Structural information describing gp130 in complex with a cytokine, however, remained elusive for several more years in large part because of the technical difficulties of assembling the larger, hetero-oligomeric receptor complexes. We have now solved three different crystal structures of gp130/cytokine complexes (Boulanger *et al.*, 2003a,b; Chow *et al.*, 2001a).

The first complex structure of gp130 was obtained with HHV-8 IL-6 that mediates the homodimerization of two gp130 molecules (Chow *et al.*, 2001a) (Fig. 4). Oligomerization of gp130 by the HHV-8 IL-6 protein is independent of an α receptor and as a result, the site I of HHV-8 IL-6 is unoccupied. Biochemical studies showed that HHV-8 IL-6 did not form a stable complex with the gp130 CHR only, but addition of the gp130 Ig-domain resulted in the formation of a highly stable "tetramer" containing two copies each of cytokine and gp130. In tandem with the reconstitution of the viral tetramer, we also expressed human IL-6 and its α receptor

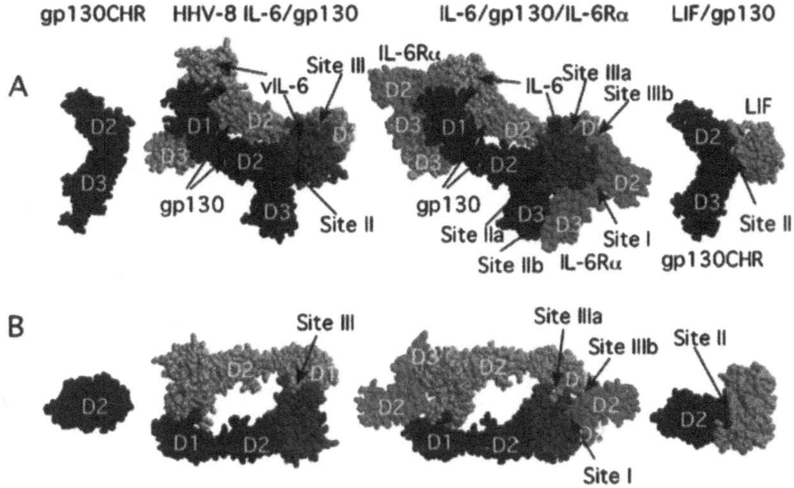

FIG. 4. Current structural knowledge of gp130 cytokine/receptor complexes. Four crystal structures are shown in the same relative orientations based on the gp130 CHR, as seen from a tilted side view (A) and a view onto the top of the complexes (B). CPK models are shown for the crystal structures of gp130 (D2-D3–CHR domain) (Bravo *et al.*, 1998), HHV-8 IL-6/gp130(D1-D3) (Chow *et al.*, 2001a), IL-6/IL-6Rα /gp130(D1-D3) (Boulanger *et al.*, 2003b) and LIF/gp130(D2-D3–CHR domain) (Boulanger *et al.*, 2003a). In transitioning from the HHV-8 IL-6/gp130 complex to the IL-6/IL-6Rα/gp130 complex, the additional IL-6Rα docks onto the site I. CPK models were generated with MOLSCRIPT (Kraulis, 1991) and RASTER3D (Merritt and Murphy, 1994). (See Color Insert.)

along with the gp130 D1D2D3. This stable solution complex was shown to be a "hexamer" containing two copies of each component (Chow *et al.*, 2001b). Thus it was clear from biochemical studies that the major difference between the human and viral complexes was the presence of the α receptor in the human version. We crystallized and determined the structure of the viral complex (Chow *et al.*, 2001a). The complex assumes a tetrameric "hammock-like" configuration with two parallel gp130 molecules (D1-D3) bridged by two HHV-8 IL-6 cytokines (Fig. 4). The complex is tethered together through the interaction of one face of HHV-8 IL-6 (site II) with a gp130 CHR (D2D3) and a second epitope (site III) at the tip of the HHV-8 IL-6 four-helix bundle interacting with the IGD (D1) domain of a different gp130 from the other half of the tetramer. The cytokine, then, uses spatially distinct epitopes to enforce dimerization on gp130 in an antiparallel orientation. The veracity of the viral complex to the mammalian cytokines was clear to us, especially seeing that a conserved aromatic residue (Trp^{144}) in HHV-8 IL-6 forms the major site III interface contact. In human IL-6, a Trp^{157} superimposes in the identical location as found on the viral cytokine. This structure provided the first accurate view of the site III epitope that appears to be unique to gp130 cytokines, which rationalized functional and mutagenesis data highlighting the importance of the D1 domain of gp130 in formation of the higher order signaling complex.

The HHV-8 IL-6/gp130 crystal structure also help to resolve structural inconsistencies of a putative site III epitope observed in complex of granulocyte colony-stimulating factor (G-CSF) with its receptor G-CSFR. G-CSFR is the most closely related receptor to gp130 with better than 45% homology and the same extracellular domain structure. Functionally, G-CSFR differs from gp130 in that it is not a shared receptor but instead is specific for G-CSF. The crystal structure of G-CSF in complex with G-CSFR was solved by Morikawa and coworkers (Aritomi *et al.*, 1999) and showed the conserved site II interaction between G-CSF and G-CSFR consistent with previous mutagenesis data (Layton *et al.*, 1999). No non-signaling α receptor is required, and thus the site I is unoccupied. The G-CSF receptor does, however, possess an Ig domain that is known to be required for signaling, suggesting the possibility of a functional site III on G-CSF (Layton *et al.*, 2001). In the crystal structure of the G-CSF/G-CSFR CHR complex, there is a "side-by-side" dimer of 1:1 complexes to form an overall 2:2, which is the predicted stoichiometry of the active extracellular signaling complex for G-CSF with the G-CSFR. Although this dimer of CHR complexes lacks the D1, a case was made for this assembly representing the active signaling complex even though it was topologically unlikely that a D1 domain could fold back and make further interactions with the complex (Aritomi *et al.*, 1999). The putative site III epitope suggested by this model was at odds with the HHV-8 IL-6/gp130 structure; however, epitope-mapping studies of G-CSF binding to G-CSFR defined an interaction between a "site III" epitope of GCSF with the D1 domain of G-CSFR, resulting in a model identical in topology to the HHV-8 IL-6/gp130 framework (Chow *et al.*, 2001a; Layton *et al.*, 2001).

More recently we have determined the crystal structure of the IL-6/gp130/IL-6Rα hexamer, which provides the first comprehensive view of a hematopoietic cytokine that incorporates all three binding epitopes—site I, II, and III—into the final assembled complex (Fig. 4). The core of the hexamer (gp130 D1D2D3 and IL-6) is reminiscent of the HHV-8 IL-6 tetramer complex, where two IL-6 cytokines bridge two gp130 molecules forming a tetramer. However, the addition of the IL-6Rα, not required for HHV-8 IL-6 signaling (Mullberg *et al.*, 2000; Osborne *et al.*, 1999), results in a much more complicated network of interfaces than seen in the viral complex. Rather than simply "decorating" the exposed site I of the human cytokine, IL-6Rα forms a series of complex composite binding interfaces with gp130. Overall, the hexamer is held together by 10 two-fold-related protein–protein interfaces, five of which are unique to each half of the hexamer (i.e., a dimer of five interfaces: sites I, sites IIa and b, sites IIIa and b), that results in a total buried surface area of nearly 5500 Å^2 (Fig. 4).

In the IL-6 hexamer, the site I binding epitope of IL-6 is localized to the A and D helices and interacts with IL-6Rα to form the initial IL-6/IL-6Rα binary complex. The D3 domain of IL-6Rα provides the majority of contact surface with IL-6, contributing more that 70% of the total buried surface area. The remaining four unique (or eight 2-fold related) protein-protein interfaces exist as composite epitopes (Fig. 4). The composite site II interaction of the IL-6/IL-6Rα complex with gp130 is separated into two spatially distinct interfaces: site IIa is between the IL-6 A and C helical faces and the "elbow" region at the boundary between the D2 and D3 domains of gp130 CHR, and site IIb is between the IL-6Rα D3 domain and the gp130 D3 domain. In site IIb there is overall shape complementarity with the D3 domain of IL-6Rα presenting a convex bulge that docks into an extended cavity on the D3 domain of gp130. The composite site III epitope that is unique to gp130 cytokines is formed by Site IIIa, which is a broad and discontinuous interface between IL-6 and gp130, where the tip of the IL-6 four-helical bundle (A/B loop and N-terminal region of D helix) abuts into the bottom β-sheet of the D1 domain of gp130. Site IIIb is formed by the interface between N-terminal portion of the D1 domain of gp130 and the D2 domain of IL-6Rα. The unexpected "composite epitopes" revealed the basis for the required pre-complexation between IL-6 and IL-6Rα. The additional surfaces provided by the sites IIb and IIIb interface enhances the overall binding affinity. Since neither IL-6 nor IL-6Rα have measurable affinity for gp130 alone, the composite interfaces appear to act as a molecular brace to hold IL-6 against the gp130 CHR and D1 binding site of two distinct gp130 receptors.

B. *A New Cytokine Receptor Recognition Module: The Logic of Site III*

The growth hormone structure established the architecture of the site I and II interactions in a homodimeric receptor complex. The viral and human IL-6 complexes with gp130 extend this paradigm to heterodimeric complexes but also elaborate the modular epitope concept by adding the site III. It is clear from the gp130 complexes that in the case of the HHV-8 IL-6, a site II is used for the gp130 CHR contact. In the human structure, a site I is used for the α receptor and a site II is used for the gp130 CHR interaction. Similar to GH, in the human IL-6 structure, precomplexation to site I is required for subsequent engagement of site II. The structural basis for this is extensive receptor-receptor contact, as is also seen in the GH homodimer, and is the basis for the sequential binding model where site I is first engaged followed by the second GH receptor to site II. One may ask, then, why site III? In GH, each receptor polypeptide contains the full complement of adaptor motifs to engage JAK/STAT components. In the gp130 system, the α receptors do not contain intracellular signaling domains, which are only present on gp130. Thus the trimolecular complexes of α receptor/cytokine/gp130 need a mechanism to themselves dimerize to bring together the signaling components of gp130. The protruding tip of the cytokine, as it sits sandwiched on both flanks (or one for the HHV-8 IL-6), presents a convenient interaction point to complex with the protruding Ig-domain of gp130 to then "double" the complexes (Figs. 4 and 5), thereby initiating signaling. Therefore the logic of a site III results from the fact that gp130, as a shared receptor, forms heterodimers with non-signaling α receptors. An additional binding epitope is required to bring two competent intracellular signaling motifs into proximity.

C. LIF Complex with the CHR of gp130

To date there is no structural information describing LIFR by itself or in complex with a gp130 cytokine. However, we have recently determined the crystal structure of the binary complex of gp130 with LIF (Boulanger *et al.*, 2003a), where LIF is the representative member of the second group of gp130 cytokines that incorporate LIFR into the final signaling complex (Fig. 4). The assumption in the literature was that LIF required precomplexation with LIFR prior to binding gp130. However, we expressed soluble forms of the gp130 CHR, the gp130 DID2D3 domains, and the top five domains of LIFR and tested them for interaction with LIF. We found that LIF is capable of forming complexes with both gp130 and LIF independent of one another: there was no requirement for precomplexation (discussed later). In fact, LIF formed a moderate affinity complex with both the gp130 CHR and the three domain gp130 CHR+IgD, indicating that the top mounted Ig domain of gp130 was not playing a role in the LIF signaling complex. This result provided a strong clue that the Ig domain of gp130 was not acting as an activation module in gp130/LIFR heterocomplexes. We crystallized and determined the structure of LIF complexed with the gp130 CHR. The interaction between LIF and the CHR domain of gp130 differs somewhat from that observed for HIV-8 IL-6 and human IL-6. Overall, LIF interacts primarily with the D2 domain of gp130 (Fig. 4) rather than within the "elbow" formed by the D2D3 bend,

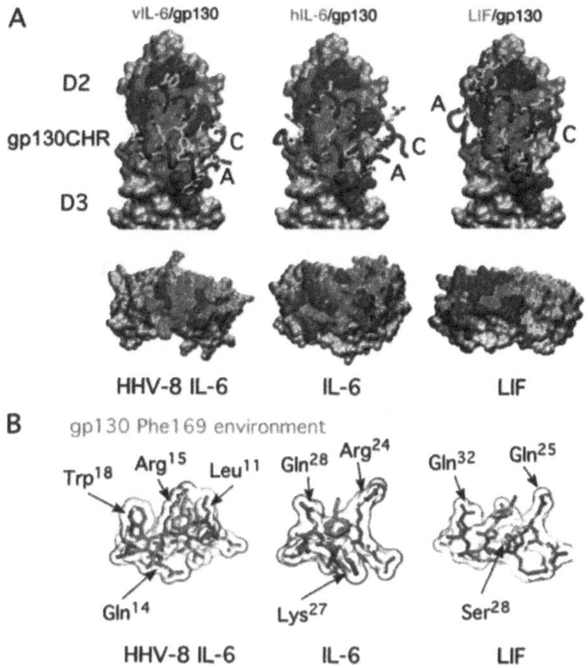

FIG. 5. Cross-reactivity of the cytokine-binding homology region (CHR) of gp130. (A) Upper panel: Contact surface of gp130 when bound to HHV-8 IL-6, IL-6 and LIF. Residues contributing to the buried surface area of the interface, as calculated using the Protein-Protein Interaction (PPI) server http://www.biochem.ucl.ac.uk/bsm/PP/server. Hydrophobic residues are colored as green surface and hydrophilic residues as red surface. Note the core hydrophobic region surrounded by discontinuous patches of polar residues. The A and C helices (blue) and the associated contact residues for each of the three cytokines are shown docked onto the surface of gp130. Lower panel: Surface representations showing the hydrophobic properties of the site II gp130 CHR binding epitope of HHV-8 IL-6, IL-6, and LIF. Note the significant hydrophilic contact surface of LIF relative to the primarily hydrophobic contact surface of HHV-8 IL6. (B) The packing of the conserved gp130 residue Phe^{169} (green) into surface pockets on HHV-8 IL-6, IL-6 and LIF. The binding affinity of the complexes appear to be consistent with the polarity of the Phe^{169} binding pocket (i.e., the more hydrophobic the pocket the higher the associated binding affinity) (HHV8 IL-6 > IL-6 > LIF). VMD (Humphrey *et al.*, 1996) was used to prepare MSMS surface representations (Sanner *et al.*, 1996). (See Color Insert.)

which was the binding strategy observed both in the HHV-8 IL-6/gp130 structure (Chow *et al.*, 2001a) and the human IL-6 hexameric complex structure (Boulanger *et al.*, 2003b). Site II of LIF is positioned towards the extreme N-terminal end of the four-helix bundle where the site II of HHV-8 IL-6 and IL-6 are more centrally located. The translational shift of the gp130 binding epitope along the long helical axis of LIF is likely due to an N-terminal "flap" preceding helix A that is unique among gp130 cytokines to LIF. Structurally this flap is tethered at both ends by disulfide bonds (Cys^{12}–Cys^{134} and Cys^{18}–Cys^{131}) and forms a protruding lip at the base of the four helix bundle. The flap effectively buttresses gp130 on one side, rigidifying the complex and also creates an obvious shape complementarity in the interface through formation of a concave pocket that packs snugly against the convex CD loop of gp130CHR.

VI. A Structural Basis of gp130 Cross-Reactivity: Site II

The basis for the extreme cross-reactivity that permits gp130 to bind with more than 10 known ligands is a fundamental question in gp130 signaling. The site II epitope is used by all gp130 cytokines to engage the CHR domain of gp130. We analyzed the protein-protein interfaces of gp130 in complex with human IL-6 (Boulanger *et al.*, 2003b), HHV-8 IL-6 (Chow *et al.*, 2001a), and LIF (Boulanger *et al.*, 2003a) (Fig. 5A). Each of the three cytokines use entirely different site II residues to contact gp130. A structural comparison of the contact surfaces on the CHR of gp130 in complex with HHV-8 IL-6, LIF, and IL-6 revealed a core shared region that dominated ligand binding (Fig. 5A). Surrounding this region are discontinuous patches that show ligand specificity and which we have termed "specificity islands." By characterizing the chemical nature of the residues that contribute to binding, we found that the core shared region is largely hydrophobic while the residues that comprise the specificity islands are largely polar. Despite the clear parsing of polar and apolar residues into distinct regions, gp130 contributes an approximately equal mix of polar and apolar residues to the binding interface likely necessary for the chemically diverse ligand surfaces with which it interacts. The site II residues of HHV-8 IL-6 are strikingly hydrophobic with two bulky tryptophan residues (Trp^{18} and Trp^{21}) contributing 30% of the total buried surface area in complex with gp130, and forms a broad hydrophobic patch on helix A (Fig. 5A). The site II epitope of hIL-6 is significantly more polar with a single, centrally disposed tyrosine (Tyr^{31}), which from mutational studies is known to play an important role in receptor binding (Savino *et al.*, 1994). The contact surface of LIF is the most polar consistent, with four well-defined solvent atoms that participate in an inter-molecular hydrogen bond network observed in the crystal structure of LIF in complex with gp130 (Boulanger *et al.*, 2003a). A noteworthy observation was that the binding surface on gp130 is extremely rigid between the unliganded and liganded forms, showing almost no rotameric flexibility in the side-chains (Boulanger *et al.*, 2003a). Together, these observations suggest that the extreme cross-reactivity of gp130 relies on the chemical and thermodynamic properties of the different cytokine epitopes rather than the more commonly observed structural plasticity (Atwell *et al.*, 1997; Sundberg and Mariuzza, 2000).

Extensive structure function studies have identified Phe^{169}, which is centrally disposed on the gp130 CHR interface as critical for ligand engagement for all human gp130-cytokines (Fig. 5) (Bravo *et al.*, 1998; Horsten *et al.*, 1997; Kurth *et al.*, 1999; Li and Nicholas, 2002). The use of a bulky hydrophobic solvent exposed residue at site II is also observed in the GH/GHR complex (de Vos *et al.*, 1992) and the EPO/EPOR complex (Syed *et al.*, 1998) suggesting a more fundamental role in mediating protein-protein interactions. In these cases, a bulky tyrosine docks into a hydrophobic pocket on the cytokine formed by the $C\alpha$ backbone of helix C.

Phe^{169} contributes the largest fraction of buried surface area in the LIF-gp130 interface (113 Å^2) the HHV-8 IL-6-gp130 interface (127 Å^2) (Chow *et al.*, 2001a), as well as the human IL-6 hexamer site II (128 Å^2) (Boulanger *et al.*, 2003b). In each case, Phe^{169} docks into a pocket formed on the site II epitope of the cytokine, although the topology and hydrophobic nature of the pocket varies greatly between cytokine (Fig. 5B). The surface of HIV-8 IL-6 represents one extreme with a deep pocket formed by Trp^{18} on one side and Leu^{11} on the other. The depth of the pocket is dictated largely by the protruding side-chain of Trp^{18}, which packs orthogonally against the side-chain of Phe^{169}. Two large polar residues, Glu^{14} and Arg^{15}, are positioned in the middle of the pocket but are flattened out so that their polar head groups are directed toward solvent and the hydrophobic methylene part of the side-chains form the primary contact with Phe^{169}. The surface pockets on human IL-6 and LIF that accommodate Phe^{169} are more similar in overall topology and are not as deep as observed for HHV-8 IL-6. On human IL-6, Phe^{169} of gp130 does not really sit in a pocket on the cytokine but instead packs directly against the splayed out methylene side-chain of Lys^{27}. Forming the backside of the binding region on human IL-6 are Arg^{24} and Gln^{28}, with the polar head group of Gln^{28} packing against the $C\beta$ atom of Phe^{169}. The gp130 binding surface on LIF is the most polar of all three cytokines. The polar head groups of Gln^{32} and Gln^{25} form the walls of the pocket and direct the aromatic ring of Phe^{169} to pack against the $C\beta$ atom of Ser^{28} in a similar fashion to the packing of Phe^{169} against Lys^{27} on human IL-6.

The binding affinity of the cytokine ligand for gp130 then appears to be commensurate with the polarity of the Phe^{169} docking site on the cytokine. From this observation, we suggest that the shared region on gp130, anchored by Phe^{169}, forms the primary molecular determinant for complex formation, while the surrounding residues utilize an extreme chemical flexibility, involving both main-chain and side-chain, to form interactions appropriate to the surface chemistry of each of the ten cytokines which bind gp130.

VII. Thermodynamic Basis for a Universal Binding Solution

Given the relative rigidity observed in the gp130 CHR binding surface, it is apparent that structural adaptation is not used as a means of cross-reactivity but suggests that the basis of gp130 cross-reactivity lies in the unique chemistry of the CHR epitope. To test this we used isothermal titration calorimetry to measure the energetics of the interaction between the CHR domain of gp130 and individually LIF, OSM, CNTF, and human IL-6 (Table I). In all cases, the interactions are primarily entropy-driven desolvation processes (LIF: 5 cal/[molK]; human IL-6: 45 cal/[molK]; OSM: 30 cal/[molK]; and CNTF: 62 cal/[molK]), albeit to varying extents commensurate with the surface polarity of the cytokine in the site II. For instance, the highly polar LIF showed the least desolvation and largest enthalpy. The ITC measurements, then, vividly reveal the underlying basis of cross-reactivity, which was suggested by crystal structures. We interpret these data to mean that gp130 is utilizing desolvation as a structurally insensitive means of cross-reacting with structurally unique surfaces. We suggest that as a cross-reactive receptor, gp130 has an extraordinary ability to order water at its CHR binding site. The multiple solvent exposed aromatic residues are likely covered with immobilized water clathrates in

Table I

A Thermodynamic Solution for the Cross-Reactivity of gp130

	LIF	OSM	CNTF/CNTFRα	IL-6/IL-6Rα
ΔH kcal/mol	-7.7 ± 0.07	-1.9 ± 0.02	8.3 ± 0.54	3.3 ± 0.3
ΔS cal/(molK)	5.3	30	62	45
ΔCp cal/mol/K	-245	-212	-340	-490
Δ kcal/mol	-9.2	-10.3	-9.0	-9.6
K_DnM	80 ± 6.2	11.8 ± 2.8	103 ± 20	35 ± 5.0
T(K)	283	283	280	281
n	1.1 ± 0.006	1.0 ± 0.005	0.9 ± 0.037	1.0 ± 0.016

The measured and calculated thermodynamic parameters obtained from isothermal titration calorimetry of the binary complexes of LIF–gp130CHR, oncostatin M (OSM)–gp130CHR, CNTF/CNTFRα–gp130CHR and IL-6/IL-6Rα–gp130CHR. Note that each titration shows favorable entropy (ΔS), albeit to varying degrees.

the unbound state, setting up a situation in which expulsion of the water into bulk solvent would be extremely entropically favorable. The subsequent formation of protein–protein contacts, either polar or apolar, provide an additional driving force for complex formation with cytokines whose surface structure and chemistry may not be ideal. The necessity for gp130 to complex with multiple ligands means that gp130 must achieve a balance between elimination of unfavorable interactions that would prevent certain cytokines from binding but preservation of favorable interactions with each ligand. However, the favorable interactions may have limitations, since the formation of energetically ideal interactions with one cytokine may result in disfavoring interactions with another cytokine. Specificity is likely not achieved without a loss of cross-reactivity. When the affinities of the gp130 CHR for cytokines are examined, they are in the high nanomolar range. Such "moderate" affinities probably represent a ceiling in which too high affinity for a cytokine would result in a loss of cross-reactivity for others. In fact, the true energetic basis of gp130 cross-reactivity may be very subtle. Recently a shotgun alanine scan of a high-affinity GH variant revealed the basis for the affinity enhancement was the elimination of negative bystander residues in the phage selection and not necessarily the introduction of more favorable interactions (Sidhu *et al.*, 2000). Certainly natural evolution has probably utilized such a mechanism on the gp130 CHR to clear the binding site of residues not favorable for cross-reactivity and create a fairly inoffensive, generic binding site. Overall, there appears to be divergent structural solutions for gp130 recognition of cytokine but a convergence in the overall thermodynamic properties underlying the recognition.

One might ask if the energetic and structural mechanism by which gp130 cross-reacts could also be used by other shared receptors such as the γ_c and β_c. The structural analogies are very clear. The γ_c extracellular domain is a two-domain CHR module similar to the site II CHR module of gp130. Thus γ_c clearly recognizes a range of diverse cytokines using the same "elbow" bend in the CHR carrying the amino acid contacts on the loops connecting the β-sheets of the individual FNIII-style domains. Mutagenesis studies have implicated Try^{103} of γ_c as being a major contact point for essentially most γ_c class cytokines. This Tyr^{103} appears to be analogous to the Phe^{169} of gp130. Hence, γ_c may utilize Tyr^{103} as an anchor point that is surrounded by either chemically amphipathic or structurally flexible peripheral contacts as a means of cross-reactivity. One major difference between γ_c and gp130 is the presence of a linker region between the two CHR domains in γ_c. Some residues in this region have been shown to affect cytokine recognition. The gp130 CHR does not have such a linker, instead showing a close abutment between the two CHR domains and essentially no interdomain flexibility. Based on the presence of this linker, we suggest that the γ_c receptor probably has a significant degree of interdomain flexibility and also is likely quite structurally plastic in its binding site, as has been seen in the GH receptor, for instance, as a means of accommodating mutations in GH (Atwell *et al.*, 1997). Therefore γ_c appears more likely to utilize structural flexibility as a means of cytokine cross-reactivity than gp130. Direct structural information will be required to answer these questions.

The basis of the cross-reactivity by β_c is a more difficult question for many of the reasons discussed earlier when reviewing the structure of the entire ectodomain. At the present time we are not exactly sure which regions of the β_c dimer are the cytokine interaction module. However, the recent mutagenesis study by Murphy *et al.* (2003) seems to indicate that the CHR is formed by the convergence of the two receptors to form a CHR-like module with protruding loops similar to that used in the site II interaction by gp130. A further refinement of this finding, as well as a more precise definition of the cytokine structural epitope, is needed before making any structural or energetic predictions regarding cross-reactivity.

VIII. Cross-Reactivity of the gp130 IGD with Cytokine Site III

While the gp130 CHR is clearly the "gateway" entry point for all gp130-cytokines, and is the focal point of this shared receptors cross-reactivity, the top-mounted Ig domain also interacts with the IL-6, IL-11 and HHV-8 IL-6 site III surfaces. At this point we only have two structures with intact site III/gp130 interfaces to examine (Boulanger *et al.*, 2003b; Chow *et al.*, 2001a), but some obvious features emerge (Fig. 6). The site III binding patch on the gp130 IGD contains both degenerate and specific structural features relevant to its function as a shared signal-transducer (Fig. 6). In both the human and viral complexes, the site III interaction is comprised of an extensive interface between the tips of the cytokine four-helix bundle (A/B loop and start of D-helix) and the edge (GF strands) of the upper 3-stranded β-sheet of the gp130 IGD (D1) (Fig. 6) (Boulanger *et al.*, 2003b; Chow *et al.*, 2001a). The IGD is a rigid structural framework that does not appear to utilize conformational plasticity to cross-react with multiple different cytokine site III surfaces. Rather, the site III interaction surface of the cytokine is composed primarily of flexible inter-helical loops and appears to adapt its structure to the surface of the IGD in an induced-fit type of interaction (Fig. 6, top). One very curious feature of the site III interface in both human and viral complexes is the four N-terminal peptide residues of gp130 insert into a groove between the C helix and

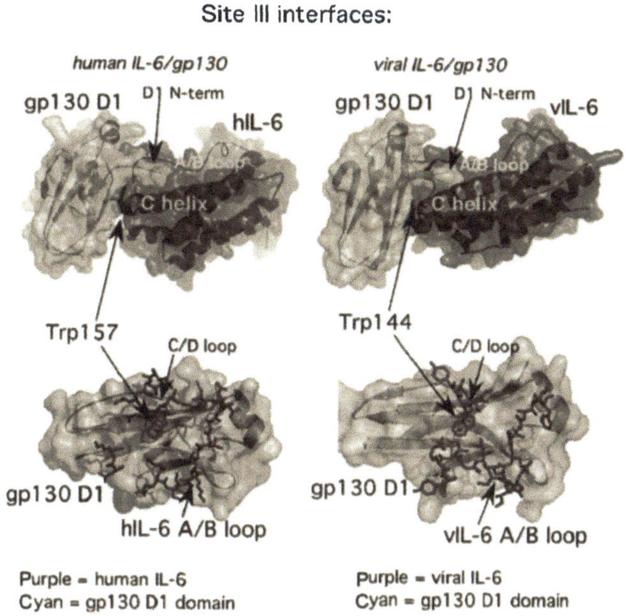

FIG. 6. Structural anatomy of the site III interface. The Ig domain of gp130 (blue) from the human and viral receptor complexes is shown in isolation with the respective cytokines (purple) at the site III interfaces. At the top is a "top view" of a transparent molecular surface with the ribbons of the structure within. In this view the convex tip of the cytokine is abutting into the bottom sheet of the gp130 Ig-domain. The penetrating N-terminal peptide from gp130 is also seen to insert into a groove on the cytokine surface. At the bottom of the figure, the view is directly into the cytokine binding sheet of the gp130 Ig-domain with the interacting loops of the cytokines shown as sticks, so that we are looking through the cytokine into the receptor. The centrally-disposed Trp residue at the heart of both site III interfaces is indicated in both views. Figure generated with PyMol (W. Delano). (See Color Insert.)

A/B loop of the cytokine, forming a sheet-like structure with the A/B loop (Fig. 6, top). In both the viral and human IL-6 site II complexes, the A/B loop is ordered, whereas this loop I disordered in most unliganded cytokines. In both the viral IL-6 and human IL-6 site II interfaces, a Trp residue (Trp^{144} and Trp^{157} respectively) forms the heart of the interfaces, burying more surface than any other residues (Fig. 6, Fig. 7A). The placement of this aromatic residue is a conserved feature of all gp130-cytokines and so likely represents the site III hotspot. This central Trp is surrounded by a chemically diverse array of amino acids to make up a very large interaction surface with the gp130 D1 domain. In fact, many of the contacts are main chain Van der Waals and hydrogen bonds with the

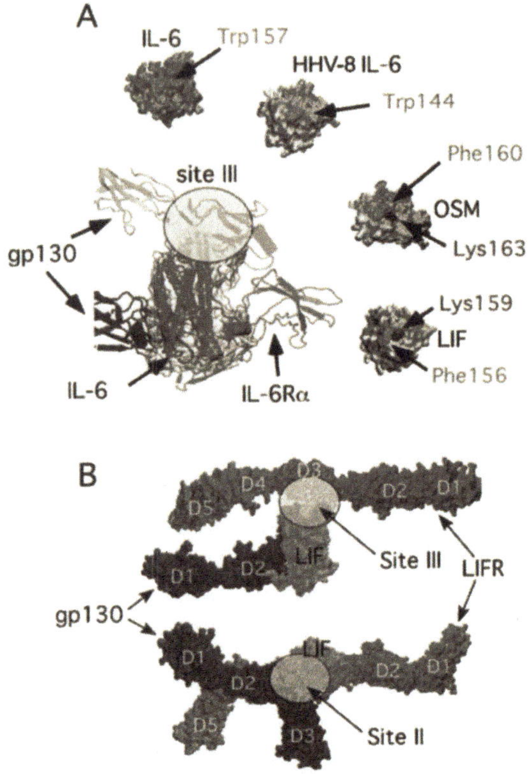

FIG. 7. The role of site III in the higher order assembly of gp130 signaling complexes. (A) Secondary structure depiction of one half of the hexameric IL-6/IL-6Rα/gp130 complex with the site III highlighted in yellow formed by the tip of the four helix bundle abutting into the D1 (IgD) domain of gp130 (Boulanger *et al.*, 2003b). Surface representations show the site III localization of the conserved tryptophan residue on IL-6 and HHV-8 IL-6 and the conserved phenylalanine/lysine pair on LIF and OSM. (B) Model of the extracellular signaling complex of LIF/gp130/LIFR where the site II of LIF (green) engages the CHR domain of gp130 (blue) (Boulanger *et al.*, 2003a) and the site III of LIF contacts the Ig domain of LIFR (purple). (See Color Insert.)

cytokine A/B loop, which suggests a relative lack of discrimination as to the amino acid content of the protein surface surrounding the Trp (Fig. 6). On the D1 side, the contact surface is a depression in a β-sheet of the Ig domain, which receives the bulky Trp residues (Fig. 6, bottom). The G β-strand, where the IGD receives the conserved aromatic (Trp^{144} in HHV-8 IL6, Trp^{157} in human IL-6) present on all gp130-cytokines only exposes main chain atoms and a glycine (Gly^{95}) to the cytokine, reducing the possibility of side-chain-specific interactions that would diminish cross-reactivity. The site III binding face of the Ig domain is distinctly lacking in prominent charge or structural features, with no large protruding side chains, perhaps, then, tuned to "absorb" a range of complementary surfaces on the cytokine. In fact, one might describe the site III interface as the reverse of the site II interface in that the protruding aromatic is provided by the cytokine in the site III, but provided by the receptor in site II, with both jutting into rather vanilla depressions on the opposing surface. Similar to site II, the D1 binding site uses amphipathicity to broaden the range of ligand surface chemistries with which it can interact, with an abundance of residues such as Tyr and Asn participating in both polar and non-polar interactions. Overall, we would suggest that, similar to the site II Phe^{169} situation, the site III interface is focused on the burial of the aromatic from the cytokine, with surrounding complementarity provided by a degree of chemical amphipathicity. Thus, as long as the site III contains the conserved aromatic, the surrounding interactions appear relatively devoid of highly specific interactions.

LIF receptor also uses its Ig domain (D3) to receive site III epitopes of CNTF, LIF, CT-1, OSM, and others. So the Ig domain of LIF receptor is, in fact, more cross-reactive than gp130. The site III structures of all gp130-cytokines maintain a conserved aromatic residue at the tip of the D helix (Fig. 7) that is certainly the structural analog to the Trp residues we see in the center of the gp130 site III interface. Therefore we predict that the LIFR site III interface cross-reactivity is achieved by using similar structural features at the gp130 IgD.

IX. Predicting the Higher-Order Assemblies of the Asymmetric Complexes (gp130/LIFR)

The architecture of the signaling assembly provided by the viral and human IL-6 complexes with gp130 allows us to predict the structure of all other gp130-cytokine complexes. The key information necessary for this prediction is to know which receptor carries the "active" CHR and which carries the top-mounted Ig domain used in the site II interaction (Grotzinger *et al.*, 1999; Kallen *et al.*, 1999). A prescient example is the LIF/LIFR/gp130 complex. One of the key questions regarding the cytokines, which use both gp130 and LIFR, is how they manage to break the symmetry of the core tetramer template we see in the IL-6 complexes. The gp130/LIFR cytokines use two different receptors, so they cannot assemble a symmetric complex. In the absence of structural information on the complete LIF/gp130/LIFR complex, biophysical (Boulanger *et al.*, 2003a), mutagenesis, and functional studies (Plun-Favreau *et al.*, 2003) have been used to probe the higher order assembly (Aasland *et al.*, 2002, 2003).

Multi-angle light scattering (MALS) of the LIF/gp130/LIFR complex showed the molecular mass to be 118 kDa, consistent with a heterotrimeric assembly including one copy each of LIF, gp130, and LIFR (Boulanger *et al.*, 2003a). Furthermore, pre-complexation of LIF with either gp130 or LIFR does not influence binding to the other receptor, indicating that in the final complex the five N-terminal domains of LIFR and the three N-terminal domains of gp130 do not interact.

With the overall heterotrimeric stoichiometry of the LIF signaling complex established, the next step was to characterize the binding epitope on LIFR. From the crystal structures of viral and human IL-6, which mediate the homodimerization of two gp130 receptors, a centrally disposed tryptophan at the tip of the four-helix bundle (site III) abuts into the D1 or Ig domain of gp130 (Fig. 6). A sequence comparison of LIF, OSM and CT-1—all of which incorporate one copy of gp130 and one copy of LIFR into the final signaling complex—highlight a conserved phenylalanine and lysine that form part of the FXXK motif (Fig. 7A), as also noted previously (Deller *et al.*, 2000). Crystal structures of LIF and OSM show that this conserved motif is located at the tip of the four-helix bundle forming the site III binding epitope (Deller *et al.*, 2000). Although no structure exists for LIFR, recent modeling and docking studies of the Ig domain of LIFR, which is located at the D3 position, with LIF revealed that the conserved Phe/Lys pair on LIF (Fig. 7A) are complemented by a Phe/Asp pair on LIFR (Plun-Favreau *et al.*, 2003). Mutagenesis of these residues to alanines, while not adversely affecting the expression or topology of LIFR, severely impaired LIF-dependent activation in cell assays (Plun-Favreau *et al.*, 2003), consistent with the Ig domain of LIFR being the docking site for the site III of LIF. Furthermore, Baca and coworkers, using a phage display selection process, showed that increasing the hydrophobicity of residues proximal to the Phe/Lys pair on the site III of LIF dramatically increased the binding affinity between LIF and LIFR (Fairlie *et al.*, 2003). Based on these observations, we have generated a theoretical model of the LIF/gp130/LIFR complex in which the site II of LIF engages the gp130 CHR domain as observed in the crystal structure (Boulanger *et al.*, 2003a) and the site III of LIF engages the D3 or Ig domain of LIFR (Fig. 7B). The topology of the LIF/gp130/LIFR is likely conserved in the gp130 heterodimeric complexes that include LIFR and represents a blueprint distinct from the "core tetramer" motif used in the homodimeric gp130 signaling complexes. The key feature of this heterotrimeric complex is that the interaction of the LIF site III with the LIFR Ig domain is very high affinity and therefore does not require a site II interaction to enhance the overall affinity, as seen in the gp130 situation. Thus LIFR is stable simply "hanging" off the end of LIF without any other support. The gp130 site III interaction is too weak to remain bound to site III unless an opposing site II was constraining it to be in that orientation. The conclusion, then, is that the symmetry of the homodimer template is broken through a high-affinity site III, negating its own site II interaction. This model predicts that neither CHR of LIF is, in fact, a bona fide cytokine binding module and that the only functional cytokine binding site on LIFR is the Ig domain located at the D3 position. The functional data regarding the activity of these CHRs is not clear (Aasland *et al.*, 2003).

This logic can be extended to some of the new gp130-class cytokines recently identified. For instance, TCCR (or WSX-1, or gp130b) is a signaling receptor used in tandem with gp130 by IL-27 (Pflanz *et al.*, 2002; Sprecher *et al.*, 1998). TCCR does not have a top Ig domain but does have a CHR module at D1D2. This tells us that IL-27 uses the CHR of TCCR but then must have a very high affinity for the Ig domain of gp130. The structure of the IL-27 complex with TCCR and gp130 would resemble the LIF/LIFR/gp130 complex where TCCR would replace gp130, and gp130 would replace LIFR.

X. Translating Ligand Recognition into Signaling

The solution of this problem remains the holy grail for any receptor system. How is ligand recognition coupled to the activation of intracellular signaling? For gp130, we have seen that the obvious initiating event occurs distal to the cell surface. The complicated architecture of the top three domains of gp130 complexed with cytokine positions of the intracellular domains in a precise geometry required for signaling. One might say the hexamer and tetramer ligand-receptor complexes form a "cap." However, we do not have any structural information for the three FNIII domains, which we refer to as "legs," leading from the ligand binding CHR and Ig domains to the membrane (Fig. 8). A very important question then is what the relationship is between the cap and the legs. Examination of our structures reveals that the D3 domains of gp130, as they lead downward, actually project away from each other. If the FNIII domains continued along this direction in a straight path, by the time they reached the cell surface each gp130 would enter the membrane hundreds of angstroms away from each other (Fig. 8). If the role of gp130 dimerization is to bring the intracellular segments into close relative proximity, it is hard to imagine how the intracellular domains could communicate over such a long distance. Our ITC data on the six-domain gp130 complex indicates that there is likely an interaction between the legs of gp130 in the full-length complex (Boulanger *et al.*, 2003b). The indicator is that the six-domain gp130 extracellular domain has an overall less favorable free

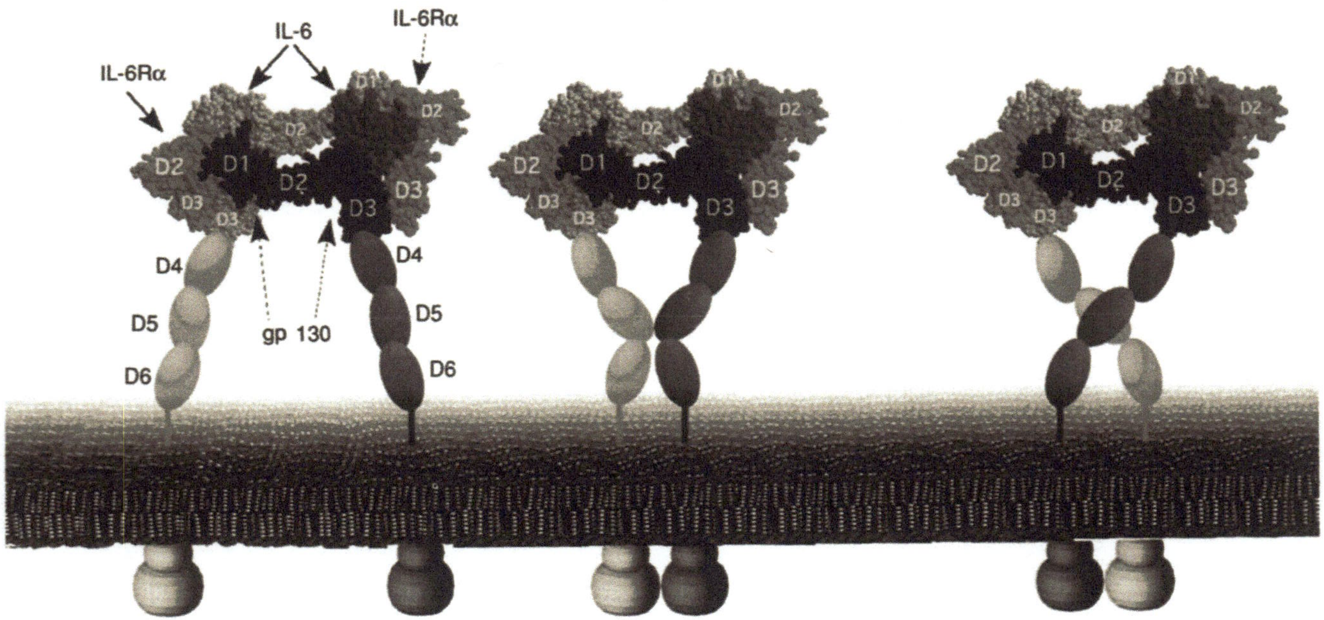

FIG. 8. Possible orientations of the "legs" of gp130. The CPK model of the IL-6/IL-6Rα/gp130 hexamer structure (Boulanger *et al.*, 2003b) is drawn along with schematic model of the three membrane proximal fibronectin repeats that connect the cap to the membrane. The two models to the right incorporate data suggesting an interaction between the gp130 legs (Boulanger *et al.*, 2003b), possibly the D5 domain (Kurth *et al.*, 2000). To date, no structural information exists describing the position and orientation of the three membrane proximal domains. (See Color Insert.)

energy for hexamer formation with IL-6 than the three-domain soluble gp130-ligand complex. We interpret this to mean that the six-domain gp130 is in some steric strain, preventing a collapse into the lower free energy state of the three-domain version. Since the difference in the two complexes is the presence of the extra three FNIII domains, we suggest that these "legs" are in contact with one another and that this contact prevents the top-mounted complex from relaxing into the lowest energy state. In fact this further suggests that a role of complex formation is to drive the legs together. These thermodynamic data are supported by cellular studies showing that the D5 domain of gp130 form a cross-link in the final signaling complex (Kurth *et al.*, 2000). Collectively, these data suggest two surprising features. One is that the gp130 extracellular domain must be "bent" at an angle as it exits the D3 domain at the base of the cap (Fig. 8). The only way that the legs could be in contact is if the receptor breaks off at an almost 90-degree angle at the base of the D3 domains and projects back towards the other half of the dimer. The second implication is that the six-domain gp130 ectodomain is a rigid structure that utilizes this rigidity as a means of compressing the legs together and maintain a close proximity of the receptors as they enter the cell membrane. The disposition of the gp130 FNIII spacer domains is an important future question of this family.

Acknowledgment

The authors acknowledge Mathias Rickert and the members of the Garcia lab for helpful discussions. The authors also acknowledge the NIH, the Keck Foundation, and the Cancer Research Institute for support. M.J.B. is supported by a Natural Sciences and Engineering Research Council of Canada post-doctoral fellowship.

References

Aasland, D., Oppmann, B., Grotzinger, J., Rose-John, S., and Kallen, K. J. (2002). The upper cytokine-binding module and the Ig-like domain of the leukaemia inhibitory factor (LIF) receptor are sufficient for a functional LIF receptor complex. *J. Mol. Biol.* **315,** 637–646.

Aasland, D., Schuster, B., Grotzinger, J., Rose-John, S., and Kallen, K. J. (2003). Analysis of the leukemia inhibitory factor receptor functional domains by chimeric receptors and cytokines. *Biochemistry* **42,** 5244–5252.

Anders, J., Kjar, S., and Ibanez, C. F. (2001). Molecular modeling of the extracellular domain of the RET receptor tyrosine kinase reveals multiple cadherin-like domains and a calcium-binding site. *J. Biol. Chem.* **276,** 35808–35817.

Aritomi, M., Kunishima, N., Okamoto, T., Kuroki, R., Ota, Y., and Morikawa, K. (1999). Atomic structure of the GCSF-receptor complex showing a new cytokine-receptor recognition scheme. *Nature* **401,** 713–717.

Atwell, S., Ultsch, M., De Vos, A. M., and Wells, J. A. (1997). Structural plasticity in a remodeled protein–protein interface. *Science* **278,** 1125–1128.

Bagley, C. J., Woodcock, J. M., Hercus, T. R., Shannon, M. F., and Lopez, A. F. (1995). Interaction of GM-CSF and IL-3 with the common beta-chain of their receptors. *J. Leukoc. Biol.* **57,** 739–746.

Bamborough, P., Hedgecock, C. J., and Richards, W. G. (1994). The interleukin-2 and interleukin-4 receptors studied by molecular modelling. *Structure* **2,** 839–851.

Bazan, J. F. (1990a). Haemopoietic receptors and helical cytokines. *Immunol. Today* **11,** 350–354.

Bazan, J. F. (1990b). Shared architecture of hormone binding domains in type I and II interferon receptors. *Cell* **61,** 753–754.

Benigni, F., Fantuzzi, G., Sacco, S., Sironi, M., Pozzi, P., Dinarello, C. A., Sipe, J. D., Poli, V., Cappelletti, M., and Paonessa, G. *et al.* (1996). Six different cytokines that share GP130 as a receptor subunit, induce serum amyloid A and potentiate the induction of interleukin-6 and the activation of the hypothalamus-pituitary-adrenal axis by interleukin-1. *Blood* **87,** 1851–1854.

Boulanger, M. J., Bankovich, A., Kortemme, T., Baker, D., and Garcia, K. C. (2003a). Convergent mechanisms for recognition of divergent cytokines by the shared signaling receptor gp130. *Mol. Cell* **12,** 577–589.

Boulanger, M. J., Chow, D., Brevnova, E., Martick, M., Sandford, G., Nicholas, J., and Garcia, K. C. (2004). Molecular Mechanisms for Viral Mimicry of a Human Cytokine: Activation of gp130 by HHV-8 Interleukin-6. *J. Mol. Biol.* **335,** 641–654.

Boulanger, M. J., Chow, D. C., Brevnova, E. E., and Garcia, K. C. (2003b). Hexameric structure and assembly of the interleukin-6/IL-6 alpha-receptor/gp130 complex. *Science* **300,** 2101–2104.

Bravo, J., and Heath, J. K. (2000). Receptor recognition by gp130 cytokines. *EMBO J.* **19,** 2399–2411.

Bravo, J., Staunton, D., Heath, J. K., and Jones, E. Y. (1998). Crystal structure of a cytokine-binding region of gp130. *EMBO J.* **17,** 1665–1674.

Carr, P. D., Gustin, S. E., Church, A. P., Murphy, J. M., Ford, S. C., Mann, D. A., Woltring, D. M., Walker, I., Ollis, D. L., and Young, I. G. (2001). Structure of the complete extracellular domain of the common beta subunit of the human GM-CSF, IL-3, and IL-5 receptors reveals a novel dimer configuration. *Cell* **104,** 291–300.

Chow, D., Brevnova, L., He, X., Martick, M. M., Bankovich, A., and Garcia, K. C. (2002). A structural template for gp130-cytokine signaling assemblies. *Biochim. Biophys. Acta* **1592,** 225–235.

Chow, D., He, X., Snow, A. L., Rose-John, S., and Garcia, K. C. (2001a). Structure of an extracellular gp130 cytokine receptor signaling complex. *Science* **291,** 2150–2155.

Chow, D., Ho, J., Nguyen Pham, T. L., Rose-John, S., and Garcia, K. C. (2001b). *In vitro* reconstitution of recognition and activation complexes between interleukin-6 and gp130. *Biochemistry* **40,** 7593–7603.

Clackson, T., and Wells, J. A. (1995). A hot spot of binding energy in a hormone-receptor interface. *Science* **267,** 383–386.

Constantinescu, S. N., Huang, L. J., Nam, H., and Lodish, H. F. (2001). The erythropoietin receptor cytosolic juxtamembrane domain contains an essential, precisely oriented, hydrophobic motif. *Mol. Cell* **7,** 377–385.

Cosman, D. (1993). The hematopoietin receptor superfamily. *Cytokine* **5,** 95–106.

Cosman, D., Lyman, S. D., Idzerda, R. L., Beckmann, M. P., Park, L. S., Goodwin, R. G., and March, C. J. (1990). A new cytokine receptor superfamily. *Trends. Biochem. Sci.* **15,** 265–270.

Davies, D. R., and Wlodawer, A. (1995). Cytokines and their receptor complexes. *FASEB J.* **9,** 50–56.

de Vos, A. M., Ultsch, M., and Kossiakoff, A. A. (1992). Human growth hormone and extracellular domain of its receptor: Crystal structure of the complex. *Science* **255,** 306–312.

Dechant, G., and Barde, Y. A. (2002). The neurotrophin receptor p75(NTR): Novel functions and implications for diseases of the nervous system. *Nat. Neurosci.* **5,** 1131–1136.

Deller, M. C., Hudson, K. R., Ikemizu, S., Bravo, J., Jones, E. Y., and Heath, J. K. (2000). Crystal structure and functional dissection of the cytostatic cytokine oncostain M. *Structure Fold Des.* **8,** 863–874.

Diederichs, K., Boone, T., and Karplus, P. A. (1991). Novel fold and putative receptor binding site of granulocyte-macrophage colony-stimulating factor. *Science* **254,** 1779–1782.

Elson, G. C., Graber, P., Losberger, C., Herren, S., Gretener, D., Menoud, L. N., Wells, T. N., Kosco-Vilbois, M. H., and Gauchat, J. F. (1998). Cytokine-like factor-1, a novel soluble protein, shares homology with members of the cytokine type I receptor family. *J. Immunol.* **161,** 1371–1379.

Fairlie, W. D., Uboldi, A. D., McCoubrie, J. E., Wang, C. C., Lee, E. F., Yao, S., De Souza, D. P., Mifsud, S., Metcalf, D., and Nicola, N. A. *et al.* (2003). Affinity maturation of leukemia inhibitory factor and conversion to potent antagonists of signalingI. *J. Biol. Chem.* **279,** 2125–2134.

Giri, J. G., Anderson, D. M., Kumaki, S., Park, L. S., Grabstein, K. H., and Cosman, D. (1995). IL-15, a novel T cell growth factor that shares activities and receptor components with IL-2. *J. Leukoc. Biol.* **57,** 763–766.

Gong, J. H., Uguccioni, M., Dewald, B., Baggiolini, M., and Clark-Lewis, I. (1996). RANTES and MCP-3 antagonists bind multiple chemokine receptors. *J. Biol. Chem.* **271,** 10521–10527.

GrandPre, T., Li, S., and Strittmatter, S. M. (2002). Nogo-66 receptor antagonist peptide promotes axonal regeneration. *Nature* **417,** 547–551.

Grotzinger, J. (2002). Molecular mechanisms of cytokine receptor activation. *Biochim. Biophys. Acta* **1592,** 215–223.

Grotzinger, J., Kernebeck, T., Kallen, K. J., and Rose-John, S. (1999). IL-6 type cytokine receptor complexes: Hexamer, tetramer or both? *Biol. Chem.* **380,** 803–813.

Grotzinger, J., Kurapkat, G., Wollmer, A., Kalai, M., and Rose-John, S. (1997). The family of the IL-6-type cytokines: Specificity and promiscuity of the receptor complexes. *Proteins* **27,** 96–109.

Gustchina, A., Zdanov, A., Schalk-Hihi, C., and Wlodawer, A. (1995). A model of the complex between interleukin-4 and its receptors. *Proteins* **21,** 140–148.

Hage, T., Sebald, W., and Reinemer, P. (1999). Crystal structure of the interleukin-4/receptor alpha chain complex reveals a mosaic binding interface. *Cell* **97,** 271–281.

Hammacher, A., Richardson, R. T., Layton, J. E., Smith, D. K., Angus, L. J., Hilton, D. J., Nicola, N. A., Wijdenes, J., and Simpson, R. J. (1998). The immunoglobulin-like module of gp130 is required for signaling by interleukin-6, but not by leukemia inhibitory factor. *J. Biol. Chem.* **273,** 22701–22707.

He, Y. W., and Malek, T. R. (1998). The structure and function of gamma c-dependent cytokines and receptors: Regulation of T lymphocyte development and homeostasis. *Crit. Rev. Immunol.* **18,** 503–524.

Heinrich, P. C., Behrmann, I., Haan, S., Hermanns, H. M., Muller-Newen, G., and Schaper, F. (2003). Principles of interleukin (IL)-6-type cytokine signalling and its regulation. *Biochem. J.* **374,** 1–20.

Heinrich, P. C., Behrmann, I., Muller-Newen, G., Schaper, F., and Graeve, L. (1998). Interleukin-6-type cytokine signalling through the gp130/Jak/STAT pathway. *Biochem. J.* **334**(Pt 2), 297–314.

Hibi, M., Murakami, M., Saito, M., Hirano, T., Taga, T., and Kishimoto, T. (1990). Molecular cloning and expression of an IL-6 signal transducer, gp130. *Cell* **63,** 1149–1157.

Hinds, M. G., Maurer, T., Zhang, J. G., Nicola, N. A., and Norton, R. S. (1998). Solution structure of leukemia inhibitory factor. *J. Biol. Chem.* **273,** 13738–13745.

Hirano, T., Nakajima, K., and Hibi, M. (1997). Signaling mechanisms through gp130: A model of the cytokine system. *Cytokine Growth Factor Rev.* **8,** 241–252.

Horsten, U., Muller-Newen, G., Gerhartz, C., Wollmer, A., Wijdenes, J., Heinrich, P. C., and Grotzinger, J. (1997). Molecular modeling-guided mutagenesis of the extracellular part of gp130 leads to the identification of contact sites in the interleukin-6 (IL-6). IL-6 receptor.gp130 complex. *J. Biol. Chem.* **272,** 23748–23757.

Humphrey, W., Dalke, A., and Schulten, K. (1996). *J. Mol. Graphics* **14,** 33.

Ihle, J. N. (1995). Cytokine receptor signalling. *Nature* **377,** 591–594.

Inoue, M., Nakayama, C., Kikuchi, K., Kimura, T., Ishige, Y., Ito, A., Kanaoka, M., and Noguchi, H. (1995). D1 cap region involved in the receptor recognition and neural cell survival activity of human ciliary neurotrophic factor. *Proc. Natl. Acad. Sci. USA* **92,** 8579–8583.

Kallen, K. J., Grotzinger, J., Lelievre, E., Vollmer, P., Aasland, D., Renne, C., Mullberg, J., Myer zum Buschenfelde, K. H., Gascan, H., and Rose-John, S. (1999). Receptor recognition sites of cytokines are organized as exchangeable modules. Transfer of the leukemia inhibitory factor receptor-binding site from ciliary neurotrophic factor to interleukin-6. *J. Biol. Chem.* **274,** 11859–11867.

Kishimoto, T., Akira, S., Narazaki, M., and Taga, T. (1995). Interleukin-6 family of cytokines and gp130. *Blood* **86,** 1243–1254.

Kishimoto, T., Taga, T., and Akira, S. (1994). Cytokine signal transduction. *Cell* **76,** 253–262.

Kopf, M., Baumann, H., Freer, G., Freudenberg, M., Lamers, M., Kishimoto, T., Zinkernagel, R., Bluethmann, H., and Kohler, G. (1994). Impaired immune and acute phase responses in interleukin-6 deficient mice. *Nature* **368,** 339–342.

Kossiakoff, A. A., and De Vos, A. M. (1998). Structural basis for cytokine hormone-receptor recognition and receptor activation. *Adv. Protein. Chem.* **52,** 67–108.

Kraulis, P. (1991). MOLSCRIPT: A Program to Produce Both Detailed and Schematic Plots of Protein Structures. *J. Appl. Cryst.* **24,** 946–950.

Kroemer, R. T., and Richards, W. G. (1996). Homology modeling study of the human interleukin-7 receptor complex. *Protein Eng.* **9,** 1135–1142.

Kurth, I., Horsten, U., Pflanz, S., Dahmen, H., Kuster, A., Grotzinger, J., Heinrich, P. C., and Muller-Newen, G. (1999). Activation of the signal transducer glycoprotein 130 by both IL-6 and IL-11 requires two distinct binding epitopes. *J. Immunol.* **162,** 1480–1487.

Kurth, I., Horsten, U., Pflanz, S., Timmermann, A., Kuster, A., Dahmen, H., Tacken, I., Heinrich, P. C., and Muller-Newen, G. (2000). Importance of the membrane-proximal extracellular domains for activation of the signal transducer glycoprotein 130. *J. Immunol.* **164,** 273–282.

Lai, S. Y., Xu, W., Gaffen, S. L., Liu, K. D., Longmore, G. D., Greene, W. C., and Goldsmith, M. A. (1996). The molecular role of the common gamma c subunit in signal transduction reveals functional asymmetry within multimeric cytokine receptor complexes. *Proc. Natl. Acad. Sci. USA* **93,** 231–235.

Layton, J. E., Hall, N. E., Connell, F., Venhorst, J., and Treutlein, H. R. (2001). Identification of ligand-binding site III on the immunoglobulin-like domain of the granulocyte colony-stimulating factor receptor. *J. Biol. Chem.* **276,** 36779–36787.

Layton, J. E., Shimamoto, G., Osslund, T., Hammacher, A., Smith, D. K., Treutlein, H. R., and Boone, T. (1999). Interaction of granulocyte colony-stimulating factor (G-CSF) with its receptor. Evidence that Glu19 of G-CSF interacts with Arg288 of the receptor. *J. Biol. Chem.* **274,** 17445–17451.

Leonard, W. J., Noguchi, M., and Russell, S. M. (1994). Sharing of a common gamma chain, gamma c, by the IL-2, IL-4, and IL-7 receptors: Implications for X-linked severe combined immunodeficiency (XSCID). *Adv. Exp. Med. Biol.* **365,** 225–232.

Li, H., and Nicholas, J. (2002). Identification of amino acid residues of gp130 signal transducer and gp80 alpha receptor subunit that are involved in ligand binding and signaling by human herpesvirus 8-encoded interleukin-6. *J. Virol.* **76,** 5627–5636.

Liparoto, S. F., Myszka, D. G., Wu, Z., Goldstein, B., Laue, T. M., and Ciardelli, T. L. (2002). Analysis of the role of the interleukin-2 receptor gamma chain in ligand binding. *Biochemistry* **41,** 2543–2551.

Livnah, O., Johnson, D. L., Stura, E. A., Farrell, F. X., Barbone, F. P., You, Y., Liu, K. D., Goldsmith, M. A., He, W., and Krause, C. D. *et al.* (1998). An antagonist peptide-EPO receptor complex suggests that receptor dimerization is not sufficient for activation. *Nat. Struct. Biol.* **5,** 993–1004.

Livnah, O., Stura, E. A., Johnson, D. L., Middleton, S. A., Mulcahy, L. S., Wrighton, N. C., Dower, W. J., Jolliffe, L. K., and Wilson, I. A. (1996). Functional mimicry of a protein hormone by a peptide agonist: The EPO receptor complex at 2.8 A. *Science* **273,** 464–471.

Martens, A., Bode, J. G., Heinrich, P. C., and Graeve, L. (2000). The cytoplasmic domain of the interleukin-6 receptor gp80 mediates its basolateral sorting in polarized madin-darby canine kidney cells. *J. Cell Sci.* **113,** 3593–3602.

Mason, D. (1998). A very high level of crossreactivity is an essential feature of the T-cell receptor. *Immunol. Today* **19,** 395–404.

McDonald, N. Q., Panayotatos, N., and Hendrickson, W. A. (1995). Crystal structure of dimeric human ciliary neurotrophic factor determined by MAD phasing. *EMBO J.* **14,** 2689–2699.

Merritt, E. A., and Murphy, E. P. M. (1994). Raster3D Version 2.0 A Program for Photorealistic Molecular Graphics. *Acta Cryst. D* **50,** 869–873.

Milburn, M. V., Hassell, A. M., Lambert, M. H., Jordan, S. R., Proudfoot, A. E., Graber, P., and Wells, T. N. (1993). A novel dimer configuration revealed by the crystal structure at 2.4 A resolution of human interleukin-5. *Nature* **363,** 172–176.

Mosley, B., Imus, C., Friend, D., Boiani, N., Thoma, B., Park, L. S., and Cosman, D. (1996). Cloning and characterization of an alternate signaling subunit conferring OSM specific receptor activation. *J. Biol. Chem.* **271,** 32635–32643.

Mueller, T. D., Zhang, J. L., Sebald, W., and Duschl, A. (2002). Structure, binding, and antagonists in the IL-4/IL-13 receptor system. *Biochim. Biophys. Acta* **1592,** 237–250.

Mullberg, J., Geib, T., Jostock, T., Hoischen, S. H., Vollmer, P., Voltz, N., Heinz, D., Galle, P. R., Klouche, M., and Rose-John, S. (2000). IL-6 receptor independent stimulation of human gp130 by viral IL-6. *J. Immunol.* **164,** 4672–4677.

Murphy, J. M., Ford, S. C., Wiedemann, U. M., Carr, P. D., Ollis, D. L., and Young, I. G. (2003). A novel functional epitope formed by domains 1 and 4 of the human common beta-subunit is involved in receptor activation by granulocyte macrophage colony-stimulating factor and interleukin 5. *J. Biol. Chem.* **278,** 10572–10577.

Nicola, N. A. (1994). Cytokine pleiotropy and redundancy: A view from the receptor. *Stem Cells* **12**(Suppl. 1), 3–12; discussion 12–14.

Nicola, N. A., and Hilton, D. J. (1998). General classes and functions of four-helix bundle cytokines. *Adv. Protein. Chem.* **52,** 1–65.

Nicola, N. A., Smith, A., Robb, L., Metcalf, D., and Begley, C. G. (1997). The structural basis of the biological actions of the GM-CSF receptor. *Ciba Found Symp.* **204,** 19–27; discussion 27–32.

Olosz, F., and Malek, T. R. (2000). Three loops of the common gamma chain ectodomain required for the binding of interleukin-2 and interleukin-7. *J. Biol. Chem.* **275,** 30100–30105.

Olosz, F., and Malek, T. R. (2002). Structural basis for binding multiple ligands by the common cytokine receptor gamma-chain. *J. Biol. Chem.* **277,** 12047–12052.

Osborne, J., Moore, P. S., and Chang, Y. (1999). KSHV-encoded viral IL-6 activates multiple human IL-6 signaling pathways. *Hum. Immunol.* **60,** 921–927.

Pflanz, S., Timans, J. C., Cheung, J., Rosales, R., Kanzler, H., Gilbert, J., Hibbert, L., Churakova, T., Travis, M., and Vaisberg, E. *et al.* (2002). IL-27, a heterodimeric cytokine composed of EBI3 and p28 protein, induces proliferation of naive CD4(+) T cells. *Immunity* **16,** 779–790.

Plun-Favreau, H., Perret, D., Diveu, C., Froger, J., Chevalier, S., Lelievre, E., Gascan, H., and Chabbert, M. (2003). Leukemia inhibitory factor (LIF), cardiotrophin-1, and oncostatin M share structural binding determinants in the immunoglobulin-like domain of LIF receptor. *J. Biol. Chem.* **278,** 27169–27179.

Poli, V., Balena, R., Fattori, E., Markatos, A., Yamamoto, M., Tanaka, H., Ciliberto, G., Rodan, G. A., and Costantini, F. (1994). Interleukin-6 mice are protected from bone loss caused by estrogen depletion. *EMBO J.* **13,** 1189–1196.

Robinson, R. C., Grey, L. M., Staunton, D., Vankelecom, H., Vernallis, A. B., Moreau, J. F., Stuart, D. I., Heath, J. K., and Jones, E. Y. (1994). The crystal structure and biological function of leukemia inhibitory factor: Implications for receptor binding. *Cell* **77,** 1101–1116.

Rose-John, S. (2002). Cytokines come of age. *Biochim. Biophys. Acta* **1592,** 213–214.

Rossjohn, J., McKinstry, W. J., Woodcock, J. M., McClure, B. J., Hercus, T. R., Parker, M. W., Lopez, A. F., and Bagley, C. J. (2000). Structure of the activation domain of the GM-CSF/IL-3/IL-5 receptor common beta-chain bound to an antagonist. *Blood* **95,** 2491–2498.

Rudolph, M. G., and Wilson, I. A. (2002). The specificity of TCR/pMHC interaction. *Curr. Opin. Immunol.* **14,** 52–65.

Sanner, M. F., Spehner, J.-C., and Olson, A. J. (1996). *Biopolymers* **38,** 305–320.

Savino, R., Ciapponi, L., Lahm, A., Demartis, A., Cabibbo, A., Toniatti, C., Delmastro, P., Altamura, S., and Ciliberto, G. (1994). Rational design of a receptor superantagonist of human interleukin-6. *EMBO J.* **13,** 5863–5870.

Seubert, N., Royer, Y., Staerk, J., Kubatzky, K. F., Moucadel, V., Krishnakumar, S., Smith, S. O., and Constantinescu, S. N. (2003). Active and inactive orientations of the transmembrane and cytosolic domains of the erythropoietin receptor dimer. *Mol. Cell* **12,** 1239–1250.

Sidhu, S. S., Weiss, G. A., and Wells, J. A. (2000). High copy display of large proteins on phage for functional selections. *J. Mol. Biol.* **296,** 487–495.

Simpson, R. J., Hammacher, A., Smith, D. K., Matthews, J. M., and Ward, L. D. (1997). Interleukin-6: Structure-function relationships. *Protein Sci.* **6,** 929–955.

Somers, W., Stahl, M., and Seehra, J. S. (1997). 1.9 A crystal structure of interleukin 6: Implications for a novel mode of receptor dimerization and signaling. *EMBO J.* **16,** 989–997.

Sprang, S. R., and Bazan, J. F. (1993). Cytokine structural taxonomy and mechanisms of receptor engagement. *Curr. Opin. Struct. Biol.* **3,** 815–827.

Sprecher, C. A., Grant, F. J., Baumgartner, J. W., Presnell, S. R., Schrader, S. K., Yamagiwa, T., Whitmore, T. E., O'Hara, P. J., and Foster, D. F. (1998). Cloning and characterization of a novel class I cytokine receptor. *Biochem. Biophys. Res. Commun.* **246,** 82–90.

Starr, R., Novak, U., Willson, T. A., Inglese, M., Murphy, V., Alexander, W. S., Metcalf, D., Nicola, N. A., Hilton, D. J., and Ernst, M. (1997). Distinct roles for leukemia inhibitory factor receptor alpha-chain and gp130 in cell type-specific signal transduction. *J. Biol. Chem.* **272,** 19982–19986.

Stewart, C. L., Kaspar, P., Brunet, L. J., Bhatt, H., Gadi, I., Kontgen, F., and Abbondanzo, S. J. (1992). Blastocyst implantation depends on maternal expression of leukemia inhibitory factor. *Nature* **359,** 76–79.

Strong, R. K. (2002). Asymmetric ligand recognition by the activating natural killer cell receptor NKG2D, a symmetric homodimer. *Mol. Immunol.* **38,** 1029–1037.

Sugamura, K., Asao, H., Kondo, M., Tanaka, N., Ishii, N., Nakamura, M., and Takeshita, T. (1995). The common gamma-chain for multiple cytokine receptors. *Adv. Immunol.* **59,** 225–277.

Sundberg, E. J., and Mariuzza, R. A. (2000). Luxury accommodations: The expanding role of structural plasticity in protein–protein interactions. *Structure Fold Des.* **8,** R137–R142.

Syed, R. S., Reid, S. W., Li, C., Cheetham, J. C., Aoki, K. H., Liu, B., Zhan, H., Osslund, T. D., Chirino, A. J., and Zhang, J. *et al.* (1998). Efficiency of signalling through cytokine receptors depends critically on receptor orientation. *Nature* **395,** 511–516.

Taga, T., and Kishimoto, T. (1992). Cytokine receptors and signal transduction. *FASEB J.* **6,** 3387–3396.

Taga, T., and Kishimoto, T. (1995). Signaling mechanisms through cytokine receptors that share signal transducing receptor components. *Curr. Opin. Immunol.* **7,** 17–23.

Taga, T., and Kishimoto, T. (1997). Gp130 and the interleukin-6 family of cytokines. *Annu. Rev. Immunol.* **15,** 797–819.

Tavernier, J., Devos, R., Cornelis, S., Tuypens, T., Van der Heyden, J., Fiers, W., and Plaetinck, G. (1991). A human high affinity interleukin-5 receptor (IL5R) is composed of an IL5-specific alpha chain and a beta chain shared with the receptor for GM-CSF. *Cell* **66,** 1175–1184.

Theze, J., Alzari, P. M., and Bertoglio, J. (1996). Interleukin 2 and its receptors: Recent advances and new immunological functions. *Immunol. Today* **17,** 481–486.

Timmermann, A., Kuster, A., Kurth, I., Heinrich, P. C., and Muller-Newen, G. (2002). A functional role of the membrane-proximal extracellular domains of the signal transducer gp130 in heterodimerization with the leukemia inhibitory factor receptor. *Eur. J. Biochem.* **269,** 2716–2726.

Trinchieri, G., Pflanz, S., and Kastelein, R. (2003). The IL-12 family of heterodimeric cytokines: New players in the regulation of T-cell responses. *Immunity* **19,** 641–644.

Vollmer, P., Oppmann, B., Voltz, N., Fischer, M., and Rose-John, S. (1999). A role for the immunoglobulin-like domain of the human IL-6 receptor. Intracellular protein transport and shedding. *Eur. J. Biochem.* **263,** 438–446.

Walter, M. R. (2002). Crystal structures of alpha-helical cytokine-receptor complexes: We've only scratched the surface. *Biotechniques* (Suppl. 46–48), 50–57.

Wells, J. A. (1996). Hormone mimicry. *Science* **273,** 449–450.

Wells, J. A., and de Vos, A. M. (1996). Hematopoietic receptor complexes. *Annu. Rev. Biochem.* **65,** 609–634.

Wilson, I. A., and Jolliffe, L. K. (1999). The structure, organization, activation, and plasticity of the erythropoietin receptor. *Curr. Opin. Struct. Biol.* **9,** 696–704.

Woodcock, J. M., McClure, B. J., Stomski, F. C., Elliott, M. J., Bagley, C. J., and Lopez, A. F. (1997). The human granulocyte-macrophage colony-stimulating factor (GM-CSF) receptor exists as a preformed receptor complex that can be activated by GM-CSF, interleukin-3, or interleukin-5. *Blood* **90,** 3005–3017.

Woodcock, J. M., Zacharakis, B., Plaetinck, G., Bagley, C. J., Qiyu, S., Hercus, T. R., Tavernier, J., and Lopez, A. F. (1994). Three residues in the common beta chain of the human GM-CSF, IL-3 and IL-5 receptors are essential for GM-CSF and IL-5 but not IL-3 high affinity binding and interact with Glu21 of GM-CSF. *EMBO J.* **13,** 5176–5185.

Yawata, H., Yasukawa, K., Natsuka, S., Murakami, M., Yamasaki, K., Hibi, M., Taga, T., and Kishimoto, T. (1993). Structure-function analysis of human IL-6 receptor: Dissociation of amino acid residues required for IL-6-binding and for IL-6 signal transduction through gp130. *EMBO J.* **12,** 1705–1712.

Zhang, J. L., Foster, D., and Sebald, W. (2003). Human IL-21 and IL-4 bind to partially overlapping epitopes of common gamma-chain. *Biochem. Biophys. Res. Commun.* **300,** 291–296.

Zurawski, S. M., Vega, F., Jr., Doyle, E. L., Huyghe, B., Flaherty, K., McKay, D. B., and Zurawski, G. (1993). Definition and spatial location of mouse interleukin-2 residues that interact with its heterotrimeric receptor. *EMBO J.* **12,** 5113–5119.

THE STRUCTURAL BASIS FOR BIOLOGICAL SIGNALING, REGULATION, AND SPECIFICITY IN THE GROWTH HORMONE–PROLACTIN SYSTEM OF HORMONES AND RECEPTORS

By ANTHONY A. KOSSIAKOFF

Department of Biochemistry and Molecular Biology
Institute for Biophysical Dynamics, University of Chicago
Cummings Life Sciences Center, Chicago, Illinois 60637

Abstract

The pituitary hormones growth hormone (GH), prolactin (PRL) and placental lactogen (PL), are members of an extensive cytokine superfamily of hormones and receptors that share many of the same general structure-function relationships in expressing their biological activities. The biology of the pituitary hormones involves a very sophisticated interplay of cross-reactivity and specificity. Biological activity is triggered via a hormone-induced receptor homodimerization process that is regulated by tertiary features of the hormone. These hormones have an asymmetric four α-helical bundle structure that gives rise to two receptor

ADVANCES IN PROTEIN CHEMISTRY, Vol. 68

0065-3233/04 $35.00

binding sites that have distinctly different topographies and electrostatic character. This feature plays an important role in the regulation of these systems by producing binding surfaces with dramatically different binding affinities to the receptor extracellular domains (ECD). As a consequence, the signaling complexes for systems that activate through receptor homodimerization are formed in a controlled sequential step-wise manner. Extensive biochemical and biophysical characterization of the two hormone-receptor interfaces indicate that the energetic properties of the two binding sites are fundamentally different and that the receptor shows extraordinary conformational plasticity to mate with the topographically dissimilar sites on the hormone. An unexpected finding has been that the two hormone binding sites are allosterically coupled; a certain set of mutations in the higher affinity site can produce both conformational and energetic effects in the lower affinity site. These effects are so large that at some level they must have played some role in the evolution of the molecule.

Prolactin (PRL), placental lactogen (PL) and growth hormone (GH) are pituitary hormones that regulate an extensive variety of important physiological functions. The biology of prolactin and growth hormone is integrated on many levels (Goffin *et al.*, 1996); however, over the 400 million years since they diverged from a common gene parent, evolution has built in different regulating components distinguishing them (Gertler *et al.*, 1996; Nicoll *et al.*, 1986). While growth hormone biology generally centers around the regulation and differentiation of muscle, cartilage, and bone cells, it is the prolactin hormones and receptors that display a much broader spectrum of activities ranging in diversity from their well known effects in mammalian reproductive biology to osmoregulation in fishes and nesting behavior in birds (DeVlaming, 1979).

The endocrine activities of prolactin and growth hormone are triggered by hormone-induced homodimerization of their cognate receptors. The receptors belong to the large hematopoietic receptor superfamily (Bazan, 1990; Cosman *et al.*, 1990). In primates, the growth hormone receptor (GHR) is activated solely by homodimerization through binding to growth hormone (De Vos *et al.*, 1992; Nicoll *et al.*, 1986). However, prolactin biology works through regulated cross-reactivity; most receptors are programmed to bind three hormones: prolactin (PRL), placental lactogen (PL), and growth hormone (GH) (Kelly *et al.*, 1991). Additionally, there is a set of activities that are induced by post-translational modified forms of prolactin that probably react through a non-cytokine type of receptor (Sinha, 1995).

The GHR and PRLR receptors have a three-domain organization: (1) an extracellular domain, which binds the activating hormone and is responsible for conferring specificity; (2) a short transmembrane α-helix segment (~25 residues); and (3) a cytoplasmic domain. While most of the receptors in the superfamily contain some number of FNIII domains in their extracellular portion and have a similar length transmembrane segment, their cytoplasmic portions generally have little similarity. Interestingly, it is the cytoplasmic portion that is the target for the JAK tyrosine kinases that trans-phosphorylate elements on themselves, the receptors and associated transcription factors, which leads to the first steps in signal transduction (Ihle *et al.*, 1994).

Within the cytokine super family, the growth hormone (GH)/prolactin (PRL) family of hormones and receptors are arguably the most extensively studied systems focused on structure-function issues and molecular recognition (Bazan, 1990; De Vos and Kossiakoff, 1992; Kossiakoff and De Vos, 1998; Kossiakoff *et al.*, 1994; Sprang and Bazan, 1993; Wells, 1991; Wells and De Vos, 1996). These studies and those of related cytokine systems have been instrumental in defining modes of hormone action and regulation (Banner *et al.*, 1996; De Vos *et al.*, 1992; Somers *et al.*, 1994; Syed *et al.*, 1998; Walter *et al.*, 1995; Wiesmann *et al.*, 1997). The structure-based mechanisms by which these systems activate are similar (Banner *et al.*, 1996; Sprang and Bazan, 1993; Wells and De Vos, 1996). However, although these mechanisms are conceptually simple—hormone induced receptor aggregation (Fig. 1)—the molecular strategies that are employed are complex and hardly predictable (Atwell *et al.*, 1997; De Vos *et al.*, 1992; Kossiakoff *et al.*, 1994; Somers *et al.*, 1994).

I. Structural Basis for Receptor Homodimerization

Tertiary structure plays a role in how the hormone regulates receptor activation. The hormones in this family are long-chain four α-helix bundle proteins (Kossiakoff and De Vos, 1998; Smith *et al.*, 1988; Sprang and Bazan, 1993). A notable feature of their tertiary structure is that it contains no symmetry that might support equivalent binding environments for the receptors. This feature distinguishes these hormones from the symmetric homodimer type like γ-interferon and IL-10, which use two identical binding sites to dimerize their cognate receptors (Randal and Kossiakoff, 2001).

The extracellular portion of the hGHR and PRLR receptors is composed of two canonical fibronectin type III (FNIII) β-sheet modules (Fig. 2). The N- and C-terminal domains each contain about 110 amino acids of very similar structures that are connected by a 5–6 residue linker (De Vos *et al.*, 1992; Somers *et al.*, 1994). However, the interdomain angles between the N and C-terminal domains differ substantially between the

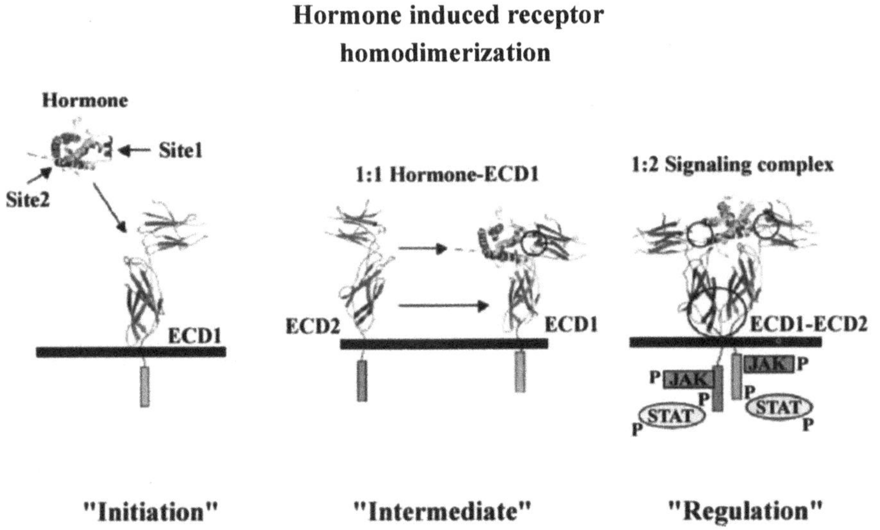

FIG. 1. Mechanism of hormone-induced receptor homodimerization. The GH/PRL receptors are 3-domain single-pass receptors containing an extracellular domain (ECD), a transmembrane section of about 25 amino acids, and a cytoplasmic domain that forms the binding site for the tyrosine kinase activities. The "Initiation" step involves the hormone binding event to the ECD (ECD1). The segments of the molecules that are involved in the contact (Site1) and colored red (hormone) and yellow (ECD1). This first step forms a stable 1:1 "Intermediate" which then recruits a second receptor in the "Regulation" step through two sets of contacts (Site2); one to the hormone and the other through forming receptor "stem-stem" contacts (ECD1-ECD2). This step forms the stable homodimer, which organizes the cytoplasmic components to initiate binding and phosphorylation. (See Color Insert.)

ECD domains of hGHR and PRLR and have consequential effects on the ternary structure of their respective complexes (see later) (Somers *et al.*, 1994).

This difference in inter-domain angles between the FNIII domains influences the global structure of the complex, which affects the binding footprint the receptors impart on the hormone, as well as the receptor-receptor interface between them (Kossiakoff *et al.*, 1994). These differences are primarily produced by subtle changes in the six residue domain linkers, which alters the disposition of each receptor domain relative to the other. The importance of such global changes to binding specificity and regulation of activity have been suggested from several structural studies (Kossiakoff *et al.*, 1994; Livnah *et al.*, 1996; Somers *et al.*, 1994; Wilson and Jolliffe, 1999). For instance, the ability of hGH to cross-react between hGHR and hPRLR has been proposed to be due in part to

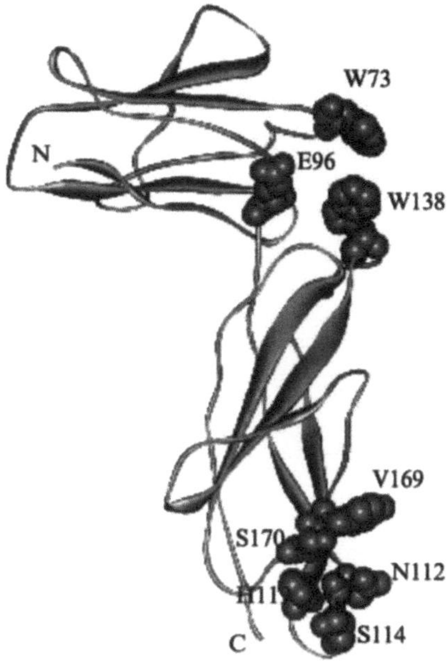

FIG. 2. Ribbon drawing of the hGH receptor ECD. The side chains of the ECD that are used by the receptor to bind at both Site1 and Site2 of the hormone are indicated. The side chains of the stem-stem contact are shown at the base of the C-terminal domain. (See Color Insert.)

concerted shifts in the orientations of the receptor binding loops through a global change affected by small changes in the inter-domain angles of the receptor domains (Kossiakoff *et al.*, 1994; Somers *et al.*, 1994).

The crystal structure of human growth hormone (hGH) bound to extracellular domain (ECD) of its receptor (hGHR) was the first structure to reveal how the two receptors bind to the asymmetric hormone (De Vos *et al.*, 1992). The structure showed that the two ECDs binding to Site1 and Site2, respectively, use essentially the same set of residues to bind to two sites on opposite faces of the hormone (De Vos *et al.*, 1992) (Fig. 1). Essentially the identical model is seen in a prolactin hormone-receptor complex (Elkins *et al.*, 2000). The two binding sites have distinctly different topographies and electrostatic character leading to different affinities for the receptor ECDs (Fig. 3). Thus this binding requires extraordinary local and global plasticity at the binding surfaces of both the hormone and receptor.

A consequence of the asymmetric nature of the hormone's binding sites is that the sites have very different inherent affinities for the receptor. The

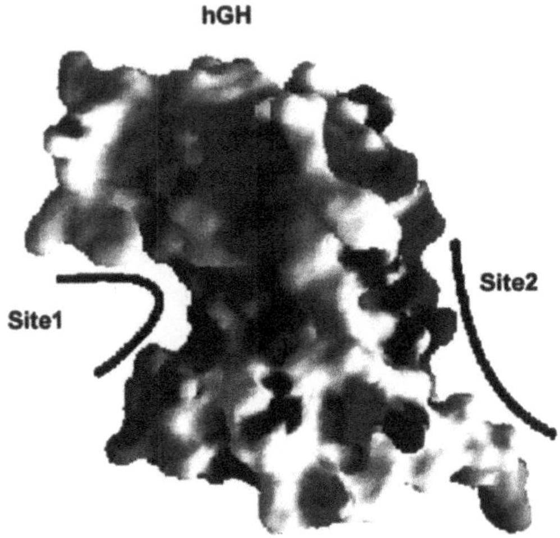

FIG. 3. Molecular surface of hGH showing the different topographies of the Site1 and Site2 binding sites. In addition, the two sites possess quite different electrostatic properties (red negative charge, blue positive charge; image rendered between ± 10 kT). (See Color Insert.)

high-affinity site, Site1, is always occupied first by ECD1 (Fuh *et al.*, 1990). This sequence of events is required because productive binding of ECD2 at Site2 of the hormone needs additional contacts to a patch of the C-terminal domain of ECD1. The binding of ECD2 is the programed regulatory step for triggering biological action and it involves a set of highly tuned interactions among binding interfaces in two spatially distinct binding sites (Bernat *et al.*, 2003; Walsh *et al.*, 2003). The energetic relationships between the ECD1-ECD2 contacts and the hormone-ECD2 Site2 interactions are important in producing the active ternary hormone-receptor complex (see later) (Bernat *et al.*, 2003).

II. Hormone Receptor Binding Sites

A. *Both ECD1 and ECD2 Use the Same Resides to Bind to the Hormone*

The surfaces of the receptors that bind to the topographically different Site1 and Site2 binding sites on the hormones are formed by six closely spaced surface loops (L1-L6) that extend from the β-sheet core in a manner somewhat similar to antigen binding loops in antibodies. Three loops reside in the N-terminal domain (L1-L3); two others in the C-terminal domain (L5-L6). Binding loop L4 serves as the five residue linker between the domains. The conformation of L4 plays a key role in orientation of the domains with respect to one another (Fig. 2). For instance, differences in the conformation of L3 in complexes of hGH binding to hGHR and hPRLR create significant changes in the global positioning of the N and C-terminal domains of the receptors bound at site (De Vos *et al.*, 1992; Elkins *et al.*, 2000; Somers *et al.*, 1994). Conformational changes in the receptor loops are also key for allowing for the stereochemical complementarity to Site1 and Site2 (De Vos *et al.*, 1992; Elkins *et al.*, 2000). Structurally, the C-terminal L5 loop (residues in the 160s loop) of the receptors undergoes ~8 Å backbone displacement going from its configuration in the Site1 to Site2 interface.

B. *Hormone Binding Site1, the High-Affinity Site*

Step 1 of hormone-induced receptor homodimerization involves the binding of ECD1 to the hormone using its high affinity Site1 (Fig. 1). Both hGH and hPRL have about 1–2 nM affinity for their cognate receptors (Cunningham *et al.*, 1990a). The hormone's Site1 is formed by residues that are exposed on helix 4 of the helix bundle, together with residues on the connecting loop between helix 1 and 2 to produce an extensive binding crevice (Fig. 3) (De Vos *et al.*, 1992). The GH and PRL complexes contain approximately the same number of inter-molecular H-bonds (8-9 H-bonds) (Elkins *et al.*, 2000; Somers *et al.*, 1994). The total surface area buried on the hormone in the hGH-hGHR and hGH-hPRLR complexes is about 1200 $Å^2$. In the ovine placental lactogen (oPL)-rat prolactin receptor (rPRLR) complex (a fully prolactin complex) about 850 $Å^2$ is buried on oPL (Elkins *et al.*, 2000). Similar surface areas are buried on the respective receptors.

The overall packing of the four helices of the hormones is very similar in all the complexes, indicating no global changes of the type seen in the analysis of the structure of an affinity mature hGH mutant (Ultsch *et al.*, 1994). The largest differences in the bound hormones are seen in a small "mini-helix" of two turns (residues 38–47) contained in the segment connecting helices 1 and 2. In the case of hGH binding to the ECD1 of hGHR and hPRLR, the mini-helix differs by about 3 Å between the respective complex (Somers *et al.*, 1994).

C. *Binding of ECD2 Involves Two Spatially Distinct Sites*

The binding of ECD2 to the 1:1 hGH:ECD1 intermediate complex is the regulatory step in the activation mechanism and involves a set of extensive interactions with both the hormone Site2 and the C-terminal portion of

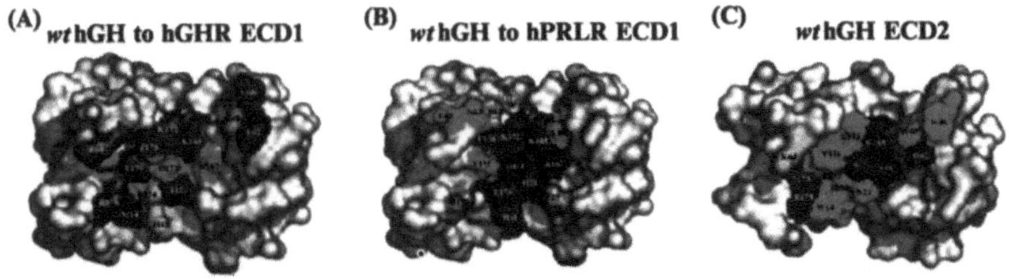

FIG. 4. The binding energy epitopes for (A) hGH binding to hGHR ECD1 (Site1) (B) hGH binding to hPRLR ECD1 (Site1), and (C) hGH binding to hGHR ECD2 (Site2). Color code as decrease in binding based on Ala-scan mutagenesis: Red, no binding >100-fold; magenta, 15-80-fold; pink, 5-15-fold; green, 2-5-fold; blue, no change; yellow, >2-fold increase in binding. (See Color Insert.)

ECD1 (Fig. 1). To form the active ternary complex, ECD2 binds to the preformed 1:1 hormone-ECD1 intermediate complex via two spatially distinct binding interactions (Fig. 4). Independently these sites are weak but together produce a tight ECD2 association. One point of contact for ECD2 is through its N-terminal domain to the Site2 interface of the hormone, while a portion of its C-terminal domain contacts an essentially identical region of the C-terminal domain of ECD1 (De Vos *et al.*, 1992; Kossiakoff *et al.*, 1994). To differentiate these two subsites the former contact is referred to here as the "Site2" contact and the C-terminal domain interaction is termed the "stem-stem" contact. The contacts to the hormone at Site2 are primarily hydrophobic in nature, involving the burial of W104 and W169 ECD2 residues. The so-called stem-stem contacts are significantly more hydrophilic in nature (Bernat *et al.*, 2003). Each of these ECD2 contacts points contribute significantly to the stability of the active ternary complex. The results of Bernat *et al.*, however, suggest that the stem-stem contact interface plays a somewhat more important role compared to the hormone Site2-ECD2 contact (Bernat *et al.*, 2003).

D. *The Hormone Site2 Interface*

For both hGH and oPL (a prolactin-like hormone), the Site2 binding epitope involves residues in helices 1 and 3. In contrast to the concave surface of the hormones at Site1, binding Site2 is a relatively flat surface. Upon binding, about 650 $Å^2$ of the oPL surface becomes buried in the interface with rPRLR-ECD2 (Elkins *et al.*, 2000). This compares to about 860 $Å^2$ that is buried in the equivalent hGHR-ECD2 interface (De Vos *et al.*, 1992). In contrast, a noteworthy difference between prolactin and growth hormone complexes is that the oPL-rPRLR ECD2 interface contains nine intermolecular H-bonds, while that of hGH-ECD2 contains only four. Thus, although it is somewhat smaller than its hGH counterpart, the oPL Site2 interaction contains over twice the number of H-bonds (Elkins *et al.*, 2000).

E. *The Receptor Stem Interaction*

The receptor-receptor interactions through the C-terminal domains of ECD1 and ECD2 are a conserved structural element of prolactin and growth hormone systems. The stem-stem ECD contact interface in hGHR is comprised of eight residues from ECD1 and six residues from ECD2 (Fig. 4). These residues are contained in the C-terminal FNIII domain (the stem region) of the hGHR ECD. In the complex, the two ECDs have a pseudo-symmetric arrangement that results from an approximately two-fold symmetry combined with an 8 Å translation along the pseudo two-fold axis of the complex (Fig. 1). As a result, the six contact residue positions on ECD2 are also used on ECD1 and represent a large subset of the eight ECD1 contact positions. Although the topology of the C-terminal domain of the rPRLR ECD is virtually identical to that of the C-terminal domain of hGHR, the stem-stem contact in these two complexes show a marked variation in their orientation and electrostatic character, and different portions of the receptor ECDs are involved in the interaction. The surface area buried in the stem-stem interaction between the rPRLRs is smaller than that buried between the hGHRs; the former being $\sim$370 Å^2 compared to $\sim$470 Å^2.

In contrast to the pseudo two-fold arrangement in the hGHR complexes, in the oPL:rPRLR complex, only about one third of the interface residues on ECD1 are also used in the interface on ECD2 (Elkins *et al.*, 2000). Eleven residues are involved in the contact interface on each rPRLR and in the receptor interface, there are four H-bonding interactions compared to six H-bonds or salt bridging interactions found between hGHRs (Table I). Although residue 201 in the ECD1 H-bonds across the interface in each complex, these interactions are not equivalent interactions because of the differences in orientations of the domains.

III. Hormone-Receptor Binding Energetics: The Binding "Hot-Spot" Concept

A. *Ala-Scan Mutagenesis of Site1*

The energetics of the high-affinity Site1 hGH-hGHR ECD1 binding that forms the intermediate 1:1 complex has led to a number of important insights into the relationships between binding and specificity

TABLE I
Receptor–Receptor H-Bonding Interactions (<3.2 Å) for PRL and hGHR Ternary Complexes

rPRL-R1	rPRL-R2	hGH-R1	hGH-R2
I188O	Q196Nϵ2	S145Oγ	D152Oδ2
Y200N	Q194Oϵ1	L146N	S201Oγ
D201Oδ1	K198N	T147Oγ	D152Oδ1
E187Oϵ2	K198Nz	H150Nϵ2	N143Oδ1
		D152Oδ2	Y200Oη
		S201Oγ	Y200Oη

R1 and R2 are ECD1 and ECD2, respectively.

determinants and protein-protein interactions in general (Clackson and Wells, 1995; Clackson *et al.*, 1998; Cunningham and Wells, 1989, 1993). Approximately 30 residues of hGH make contact in the Site1 interface with ECD1. However, it was shown by Ala-scanning mutagenesis that about 85% of the binding energy was developed through only eight of the residues, and there were no apparent distinguishing characteristics between the interaction that are energetically important from those of the other side chains that were energetically null (Fig. 4A) (Cunningham and Wells, 1993). Interestingly, it has been shown that it is possible to introduce multiple Ala mutations into the binding interface, as long as they are not the hot-spot residues, without producing an appreciable loss of binding (Clackson *et al.*, 1998). This type of energy epitope, characterized by the concentration of the binding determinants within a relatively small area of the contact surface, allows for separate adjacent areas to be used for specificity determinants without compromising the binding.

It has been determined that a similar type of hot-spot also exists for the hGH-hPRLR-ECD1 Site1 interaction and that the interface encompasses the same binding footprint as the hGHR ECD1 counterpart (Fig. 4B) (Cunningham and Wells, 1991; Somers *et al.*, 1994). The two systems differ, however, in how the energy is distributed within the footprint. For instance, residues like E174 and R167 make little energetic contribution to binding in the hGH-hGHR case but play extremely important roles in the hGH-hPRLR contact surface (>100-fold difference in binding between the two systems when alanine is substituted at each of these sites). Interestingly, R167 makes a salt-bridge to an acidic side chain in both receptors (to E127 in hGHR and D124 in hPRLR), yet Ala-scan mutagenesis indicates that only in the case of hGH-hPRLR does this salt-bridge have a positive effect on binding (Cunningham and Wells, 1991). This is an example of how the energetics of interactions is very context-dependent in protein-protein interfaces (Kossiakoff *et al.*, 1994).

B. *Ala-Scan Mutagenesis of hGH Site2*

The ECD2 binds to the 1:1 wt-hGH:ECD1 complex with an equilibrium dissociation constant (K_d) of 3.8 nM (Walsh *et al.*, 2003), which is similar to the binding affinity at the high-affinity Site1 (~1.2 nM) (Cunningham and Wells, 1993). This binding energy reflects the contributions from residues in both the Site2 interaction between the hormone and ECD2, as well as those in the ECD1-ECD2 stem-stem contact (Bernat *et al.*, 2003).

The Ala-scan data for hGH Site2 indicate that no single Site2 residue is absolutely critical for the binding of ECD2 to the hGH:ECD1 intermediate (Fig. 4C). Thus there is no equivalent binding "hot-spot" compared to the Site1 situation (see previously). Overall, most of the alanine substitutions produced 2 to 10-fold effects on binding affinity [K_d(Ala)/K_d(*wt*)] (Walsh *et al.*, 2003). Generally, the k_{on} rate constants show little fluctuation (<2.5-fold) over the 15-residue Ala-scan; the differences in binding result mainly from changes in the k_{off} rate constants. A similar trend of rate dependence due to off-rate kinetics has been seen in other cytokine systems: Site1 binding mutants of hGH and ECD1 (Clackson *et al.*, 1998), interleukins 2 (Liparoto and Ciardelli, 1999) and 4 (Shen *et al.*, 1996; Wang *et al.*, 1997), erythropoietin (Hensley *et al.*, 2000), and G-CSF (Nice *et al.*, 1993).

C. *The Energetics of the Stem-Stem Contact*

Although the stem-stem contact interfaces of the hGHR ECD1 and ECD2 are nearly identical in content and conformation, significant differences exists in the size and location of their respective binding "hot-spots" (Bernat *et al.*, 2003) (Fig. 5). In ECD1, Ala mutations at three sites produce decreases in binding of 20-fold or greater, with two others producing 5-fold effects. In ECD2, five of the eight scanned residues caused little change in binding affinity. However, Ala mutations at two positions (D152A, Y200A) completely eliminate ECD2 binding. Interestingly, a mutation at one of these positions, D152, causes idiopathic Type II short stature (Laron dwarfism) (Duquesnoy *et al.*, 1994). Based on this analysis and model building, it was suggested that the D152H mutation inhibits effective homodimerization in the context of its presence in ECD2, rather than ECD1 (Bernat *et al.*, 2003).

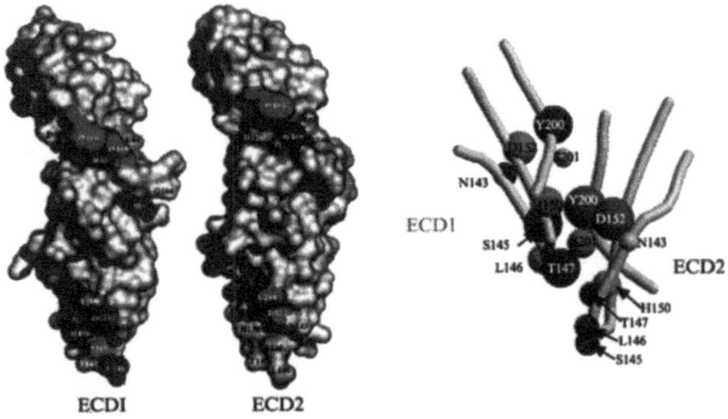

FIG. 5. (Right) Structural relationship of residues in the stem-stem interface. Decreases in binding affinity due to individual wild-type to Ala substitutions are represented by color-coding and the size of the representative spheres. (Center and Left) Space filling representation of Ala-scan data for ECD1 and ECD2 binding to hGH Site1 and Site2 and the stem-stem interaction. The residues in the N-terminal domains of the ECDs (top patch) represent only subsets of the corresponding contact interfaces including most, but not all the energetically important positions. The residues in the N-terminal domains of ECD1 and ECD2 bind to Site1 and Site2 of hGH, respectively. Color code as decrease in binding based on Ala-scan mutagenesis: Red, >50-fold; magenta, 15-50-fold; blue, 3–15 fold; light blue, 0-2-fold; yellow, >2-fold increase in binding. (See Color Insert.)

IV. Stereoselectivity of the Prolactin Receptor (hPRLR)

The cross-reactivity properties of the hPRLR is one of the most interesting and intriguing molecular recognition issues involved in the regulation of human GH and PRL biology. While in humans hGHR is only activated by its cognate hormone, growth hormone, hPRLR can be turned on and regulated by three hormones: hGH, hPL, and hPRL. As discussed previously, the binding hot spot in the hGH Site1 interface differs significantly in the context of hGHR or hPRLR (Cunningham and Wells, 1991) (Fig. 5). A central feature of this difference is that in the hGH-hPRLR interface, a bridging Zn ion located in hormone-receptor interface plays a pivotal role in binding (Cunningham *et al.*, 1990a). The presence of this Zn^{++} is also an essential element of hPL binding to hPRLR, but interestingly, not in the case of hPRL binding to its cognate receptor.

By and large, the respective binding hot-spots for hPL and hGH binding to hPRLR tracks well, where the side chains coordinating to the Zn^{++} serve as the principal binding determinants (Walsh and Kossiakoff, unpublished). One notable outlier is the R167A mutant, which causes a ~800-fold reduction in binding in hGH (Cunningham and Wells, 1991) but has only a 30-fold effect in the context of the hPL molecule.

On the receptor side, however, a distinctly different energy epitope is produced by each of the three hormones. The Ala-scan analysis of hPRLR binding its cognate hormone (hPRL) indicates there are five receptor residues, which when individually changed to Ala, eliminate binding (Walsh and Kossiakoff, unpublished): W104, D124, W169, D217, and H218 (hGHR numbering). These effects track well with the hGHR Ala-scan with its cognate hormone, hGH (Clackson and Wells, 1995). However, in the cross-reactivity situation (i.e., hPRLR-hGH), the energies contributed by the individual receptor residues are quite different. Surprisingly, W104A (a major player in hGH-hGHR binding) has little effect on binding, and W169A produces only a 40-fold decrease in binding (Walsh and Kossiakoff, unpublished). The major contributors to binding based on the Ala-scan data are G76, D124, F170, and D217. The fact that the G76 mutation has such a large effect on binding is presumably due to conformational factors.

The binding of hPL produces yet a third distinctive hot-spot pattern for the distribution of contributing hPRLR residues to the hormone's binding (Walsh and Kossiakoff, unpublished). As in the case for hGH, G76A essentially eliminates binding. However, W104 becomes more important and W169 less so compared to the effects seen for hGH. Interestingly, the Y122A mutant actually increases binding about 5-fold. Clearly, structures of the complexes of hPL and hPRL would be important complements to the structure of hGH-hPRLR (Somers *et al.*, 1994) to sort out the structural basis for the disparate effects.

V. Comparison of the hGH Site1 and Site2 Binding Energy Epitopes in the hGH-hGHR Ternary Complex

Site1 and Site2 differ in how the energetically important residues are organized at the contact interface. In Site1, the hot spots of the respective hormone and receptor surfaces are spatially matched; that is, the hot spot residues, W104 and W169 in ECD1, directly interact with their counterpart hot-spot residues of R64, K172, T175, and F176, which are located in helix 4 of the hGH (Fig. 4A). In contrast, the relationships between the most important residues for Site2 binding do not have a similar spatial complementarity. It is surprising, for instance, that while W104 is essential for binding at Site2, removing hormone residues that contact it have a rather small or modest influence on binding (Walsh *et al.*, 2003). The opposite situation is observed for mutations involving W169 and its packing with its hormone neighbors. The W169A mutant has a modest effect on binding, but removing the contacting residue in the hormone (R16A) has a significantly larger effect on ECD2 binding. This suggests that the spatial complementarity of the hot-spot residues might be an important attribute for providing very tight binding interactions, which is the case for Site1. Conversely, such an arrangement is not necessary or perhaps desirable for weaker protein-protein interactions like the hGH Site2 interface, whose function is to fine-tune the transient protein association processes, which regulate the system's activity. In this regard it was determined that it is the receptor-receptor contact that contributes most energy to formation of the final ternary complex, and that while the Site2 hGHR interaction is required for receptor signaling, the energy derived from it is somewhat less (Bernat *et al.*, 2003).

VI. Protein Engineering to Create New Properties

A. *Hormone Specificity*

The pattern of specificity and cross-reactivity involves some rather significant molecular recognition challenges since GHs and PRLs have little (~25%) sequence conservation even among the residues involved in receptor binding (Elkins *et al.*, 2000; Somers *et al.*, 1994). The structure of hGH bound to the prolactin receptor (hPRLR) showed that in these systems, local conformational flexibility of the receptor binding loops, together with rigid-body movements of the receptor domains, facilitates the creation of specific but different interactions with the same binding site. The effects of conformational change on altering specificity were also observed in protein engineering studies that "converted" the binding Site1 of two PRLR specific hormones, hPRL and hPL, into hGH (Cunningham and Wells, 1991; Cunningham *et al.*, 1990b; Lowman *et al.*, 1991a). This could be accomplished by substituting the hGH sequence at as few as 5–6 places in their sequence. Surprisingly, several of these positions map outside the Site1 hormone-receptor interface. Presumably they must act as indirect specificity determinants by inducing conformational changes that subtly reorganize the contact residues into productive binding interactions. The implications of this finding are considerable and may open up totally new ways to look at how specificity and cross-reactivity are developed in cytokine systems.

B. *Binding Affinity*

Using a phage display mutagenesis approach, Wells and colleagues produced a number of novel hGH variant molecules having properties that differed significantly from the wild-type (*wt*) hormone (Atwell *et al.*, 1997; Lowman and Wells, 1993; Lowman *et al.*, 1991b). One of particular interest was a variant (hGHv) that was affinity matured to bind much more tightly to ECD1 (Lowman and Wells, 1993). The variant (hGHv) had 15 mutations in its Site1 binding site. This high-affinity molecule retained its biological activity as a growth hormone: however, its increased binding does not translate into high biological activity (Pearce *et al.*, 1999). Additionally, hGHv has lost the ability to cross-react with the prolactin receptor (Lowman *et al.*, 1991).

An interesting and unanticipated finding was that a surprisingly large number of mutations that appeared to be stereochemically improbable based on the *wt* complex. Since many of the mutations involved altering the stereochemical character of the *wt* hGH residues, the energy landscape of the hormone was necessarily altered. For instance, several *wt* hGH residues that were involved in salt bridges or strong H-bonds to the receptor were inexplicably converted to hydrophobic side chains in the phage display selections. There appeared to be a systematic conversion of hydrophilic to hydrophobic side chains. The dominance of hydrophobic over hydrophilic in developing binding energy in itself is not unusual; however, the partners of the hydrophilic groups on the receptor side remain hydrophilic.

It is noteworthy that the production of this hGHv variant was not a case of fine-tuning the initial binding interface. Schiffer *et al.*, have shown that the new contact interface, the so-called *structural epitope*, has significantly more hydrophobic character, a rearranged H-bonding network, and involved large conformational changes in both the hormone and receptor interfaces (Schiffer *et al.*, 2002). Because the stereochemical character of these interfaces is so different, it is not surprising that the functional epitope, which describes the distribution of the binding energy in Site1 of the high-affinity hGHv, has been shown to be fundamentally different from that of the *wt* hormone in several notable ways (Pal *et al.*, 2003). In *wt* hGH, the functional epitope of Site1 is dominated by a small number of key residues that are spatially clustered into a so-called binding "hot-spot" (Clackson and Wells, 1995; Cunningham and Wells, 1989). In contrast, even though the binding of hGHv is much tighter, the binding contributions of the original hot-spot residues are highly attenuated. This attenuation appears to be more than compensated for by the addition of a new set of contributing residues located at the periphery of the binding epitope, in effect resulting in an expanded and more diffuse hot-spot (Pal *et al.*, 2003).

VII. Allosteric Effects, A Big Surprise

The phage display affinity maturation process that focused solely on Site1 produced some unexpected and important consequences in Site2. Site1 and Site2 are on opposite faces of the hormone and are separated by over 15 Å. While it was anticipated that Site1 interactions would be altered as a result of the mutations at this site in the hormone, it was surprising that the largest changes in structural conformation were found in the Site2 interface, where no mutations were made (Schiffer *et al.*, 2002). This interface in hGHv has only limited structural relationship to its *wt* counterpart. The same sets of hormone and receptor residues are used, but their stereochemical relationships are completely different, several groups differ by more than 10 Å. Interestingly, this new reconfigured hGHv-ECD2 interface has a comparable binding association as that found in the *wt*-complex (~4 nM). This structure is an excellent example of the structural cooperativity that exists between binding sites in these systems. Another important aspect of this structure is that the distribution of binding energy among the residues' energy at Site1 and Site2 are different than their counterparts in the *wt* ternary complex (Walsh and Kossiakoff, unpublished results).

A. *Site1 and Site2 Are Structurally and Functionally Coupled*

It is likely that this new binding solution for hGHv-hGH-ECD2 is triggered by a structural mechanism linking Site2 to a subset of the mutations in Site1 introduced in the phage display experiments. It is noteworthy that the structurally distinct conformation of hGHv at Site2 was under no selection pressure and supports binding of the second receptor as tightly as in the *wt* complex. A specific example of the structural coupling is observed from the altered roles of Asp116 in Site2 of the hormone in the two complexes (Schiffer *et al.*, 2002). Asp116 is located near the center of helix3, and thus the side chain extends off a fairly rigid scaffold. Although it is adjacent to several important receptor side chains, in the *wt* complex it appears to play a bystander role, making no H-bond to the receptor. It is probable that the small movements of Asp116 that are a consequence of the repacking of the four-helix bundle in hGHv effectively trigger the new H-bonding scheme where by the carboxylate side chain makes new H-bonds to the receptor through the indole nitrogens of the side chains of Trp104 and Trp169, as well as to the side chain of Arg43. As seen in Fig. 6, the Trp side chains undergo a significant reorganization to facilitate the formation of these H-bonds. It is noteworthy that these structural changes involving D116 are also reflected in the energetics of binding. Whereas the Ala-scan of Site2 in the *wt* hormone indicate that D116 plays the bystander role in that interaction (Walsh *et al.*, 2003), in the context of the hGHv complex it plays a very central role presumably through the formation of the H-bonds to the Trp side chains (Fig. 6B). These

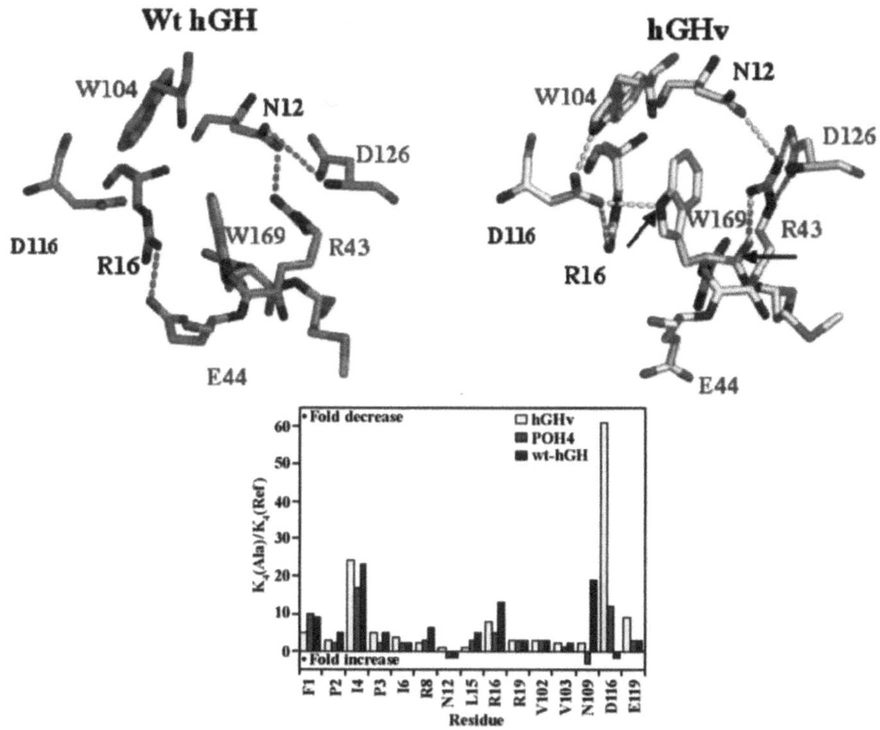

FIG. 6. Site2 interfaces of hGHv and wt-hGH with their respective ECD2. (A) The *wt* hGH (green) and the hGHv (yellow) interface around D116. The hormone and receptor residues are labeled in black and red, respectively. The intermolecular H-bonds and salt bridges between the hormone and the receptor are depicted as broken lines. The arrows point to the peptide bond flip observed for W169$_{E2}$ between the two crystal structures. (B) Bar graphs of the fold decreases/increases for the Site2 Ala-scan of the R2 to all three hormones. The K_d is the reference Site2 dissociation constant K_d for each hormone. The asterisks indicate residues that when mutated to alanine results in no detectable Site2 binding affinity. (See Color Insert.)

observations clearly have fundamental import to understanding the inherent efficiencies of these cytokine hormones and receptors as binding entities even outside of evolutionary control.

The concept that the two spatially distinct binding sites on cytokine hormones are structurally and functionally coupled as displayed in the hGHv complex is novel, and the process whereby new binding surfaces are synthesized by indirect molecule effects has been termed ''functional cooperativity'' (Schiffer *et al.*, 2002). In this mechanism it is not the mutation(s) in one site that alone affect the other site. A set of concerted changes is also involved among the hormone and the domains of the receptor ECDs. The finding of strong cross-molecular interaction induced during receptor dimerization establishes a new molecular recognition paradigm and opens up fundamental new areas of investigation relating to the mechanisms of biological regulation by protein-protein associations. However, it remains an open question as to how general this is and whether evolution actually uses this strategy to influence the receptor signaling of GH/PRL systems in biologically important ways.

VIII. Biological Implications of Transient Receptor Dimerization

The role of receptor homodimerization is more complicated than simply bringing the cytoplasmic elements of the receptors together. For example, structural studies of EPO and EPO-R indicate that a function of the hormone is to establish a fairly exact receptor alignment, as well as to induce dimerization (Livnah *et al.*, 1998, 1999; Remy *et al.*, 1999; Wilson and Jolliffe, 1999). Based on patterns of cross-hormone and cross-species activities and the known structural differences in the active complexes, exact receptor orientation is probably not as crucial for prolactin and growth hormone systems, although this has not been directly established.

Although strict orientation effects may not be crucial, it appears that the dynamics governing the stability of the aggregated signaling complex are an important regulatory element for the prolactins. Consequently, the "inefficient" Site2 binding is likely an evolved characteristic of homologous prolactin systems, distinguishing their homodimerization process from those of GH and EPO. To explain mutagenesis data influencing Site2 binding, Herman *et al.* (1999) have suggested a "minimal time" mechanism. This mechanism is based on the assumption that signal transduction requires a minimal persistence life-time for the homodimer to facilitate effective transphosphorylation of the associated JAK2 kinases. Once this goal is achieved, the existence of the dimerized receptors is no longer obligatory. It is proposed that this minimal time is generally shorter for PRLRs than for GHRs, perhaps because the JAK2 kinase is preassociated in the case of PRLRs (Lebrun *et al.*, 1994) but not GHRs (Argetsinger *et al.*, 1993). The minimal time hypothesis is also supported by a study by Pearce *et al.*, who engineered tighter and weaker binding interactions between hGH and hGHR (Pearce *et al.*, 1996). They found that increasing affinities of the hGH-hGHR associations at both Site1 and Site2 produced no measurable increases in biological activity. However, reducing affinity at Site1 by 30-fold marked a point that appeared to correspond to a threshold where activity was affected suggesting that wild-type hGH-hGHR affinity is higher at Site1 than it needs to be to sustain full biological activity.

IX. Mechanism of ECD2 Binding, the Regulation Step in Biological Signaling

The binding of ECD2 to the 1:1 hormone-ECD1 intermediate complexes proceeds through a pathway having two general characteristics. First, the binding kinetics show this to be a relatively slow on/slow off process (Bernat *et al.*, 2003; Walsh *et al.*, 2003). Second, the binding strengths of the different mutants correlate exclusively with their relative k_{off} values, the k_{on} rate constants are remarkably constant over the range of Site2 mutants analyzed. Such a situation indicates that the binding transition state (TS) energy is not affected by factors at the level of the structural changes introduced by the mutations—even the very big ones. When taken with the slow on/off kinetics, this suggests that the TS energy reflects some large global conformational change that is required to occur prior (or concurrently) to forming a highly organized protein-protein interface. Further, it is noteworthy that the k_{on} rates for receptor binding are remarkably similar whether the binding occurs at Site1 or Site2, which are topographically very dissimilar sites (Walsh *et al.*, 2003). This also holds true for essentially all the Site1 and Site2 mutants. Taken together, these observations can be interpreted to suggest that there is a common high-energy element in the binding, which is independent of the topography of the binding surface itself. Since the binding at both Site1 and Site2 requires coordination of binding elements from the two loosely linked receptor domains, the productive alignment of these domains is a plausible candidate for the high-energy step in binding.

X. Future Directions

Although the growth hormone-prolactin system is arguably the most extensively studied hormone-receptor system with regard to investigating the structure-function basis for binding, specificity, and signaling, we have figuratively just "scratched the surface." New surprises such as the allosteric coupling between the binding sites keep emerging. There is a tendency in research to strip-mine the easily accessible information then move on. While this might be an efficient strategy, it leaves many of the most important features of the system buried and underappreciated. Perhaps the greatest next challenge will be to try to establish the molecular details that drive the functional processes of the cytoplasmic portion of the system. There are many inherent technical difficulties in doing this, but it is here that holds the promise to provide fundamental new insights into how the signaling systems operate at the molecular level.

Acknowledgment

I would like to thank Dr. Scott Walsh for providing data prior to publication.

References

Argetsinger, L., Campbell, G., Yang, X., Witthuhn, B., Silvennoinen, O., Ihle, J., and Carter-Su, C. (1993). Identification of JAK2 as a growth hormone receptor-associated tyrosine kinase. *Cell* **74,** 237–244.

Atwell, S., Ultsch, M. H., De Vos, A. M., and Wells, J. A. (1997). Structural plasticity in a remodeled protein-protein interface. *Science* **278,** 1125–1128.

Banner, D. W., D'Arcy, A. D., Chene, C., Winkler, F. K., Guha, A., Konigsberg, W. H., Nermerson, Y., and Kirchhofer, D. (1996). The crystal structure of the complex of blood coagulation factor VIIa with soluble tissue factor. *Nature* **380,** 41–46.

Bazan, J. F. (1990). Structural design and molecular evolution of a cytokine receptor superfamily. *Proc. Natl. Acad. Sci. USA* **87,** 6934–6938.

Bernat, B., Pal, G., Sun, M., and Kossiakoff, A. A. (2003). Determination of the energetics governing the regulatory step in growth hormone-induced receptor homodimerization. *Proc. Natl. Acad. Sci. USA* **100**(3), 952–957.

Clackson, T., Ultsch, M. H., Wells, J. A., and De Vos, A. M. (1998). Structural and functional analysis of the 1:1 growth hormone:receptor complex reveals the molecular basis for receptor affinity. *J. Mol. Biol.* **277**(5), 1111–1128.

Clackson, T., and Wells, J. A. (1995). A hot spot of binding energy in a hormone-receptor interface. *Science* **267**(5196), 383–386.

Cosman, D., Lyman, S. D., Idzerda, R. L., Reckmann, M. P., Park, L. S., Goodwin, R. G., and March, C. J. (1990). A new cytokine receptor superfamily. *Trends Biochem. Sci.* **15,** 265–270.

Cunningham, B. C., Bass, S., Fuh, G., and Wells, J. A. (1990a). Zinc mediation of the binding of human growth hormone to the human prolactin receptor. *Science* **250,** 1709–1712.

Cunningham, B. C., Henner, D. J., and Wells, J. A. (1990b). Engineering human prolactin to bind to the human growth hormone receptor. *Science* **247**(4949 Pt 1), 1461–1465.

Cunningham, B. C., and Wells, J. A. (1989). High-resolution epitope mapping of hGH-receptor interactions by alanine-scanning mutagenesis. *Science* **244,** 1081–1085.

Cunningham, B. C., and Wells, J. A. (1991). Rational design of receptor-specific variants of human growth hormone. *Proc. Natl. Acad. Sci. USA* **88,** 3407–3411.

Cunningham, B. C., and Wells, J. A. (1993). Comparison of a structural and a functional epitope. *J. Mol. Biol.* **234,** 554–563.

De Vos, A. M., and Kossiakoff, A. A. (1992). Receptor action and interaction. *Curr. Opin. Struct. Biol.* **2,** 852–858.

De Vos, A. M., Ultsch, M., and Kossiakoff, A. A. (1992). Human Growth Hormone and Extracellular Domain of its Receptor: Crystal Structure of the Complex. *Science* **255,** 306–312.

DeVlaming, V. (1979). Actions of prolactin among the vertebrates. *In* "Hormones and Evolution" (E. J. W. Barrington, Ed.), pp. 561–642. Academic Press, New York.

Duquesnoy, P., Sobrier, M. L., Duriez, B., Dastot, F., Buchanan, C. R., Savage, M. O., Preece, M. A., Craescu, C. T., Blouquit, Y., and Goossens, M. *et al.* (1994). A single amino acid substitution in the exoplasmic domain of the human growth hormone (GH) receptor confers familial GH resistance (Laron syndrome) with positive GH-binding activity by abolishing receptor homodimerization. *EMBO J.* **13**(6), 1386–1395.

Elkins, P. A., Christinger, H. W., Sandowski, Y., Sakal, E., Gertler, A., De Vos, A. M., and Kossiakoff, A. A. (2000). Ternary complex between placental lactogen and the extracellular domain of the prolactin receptor. *Nat. Struct. Biol.* **7**(9), 808–815.

Fuh, G., Mulkerrin, M. G., Bass, S., McFarland, N., Brochier, M., Bourell, J. H., Light, D. R., and Wells, J. A. (1990). The human growth hormone receptor. Secretion from *Escherichia coli* and disulfide bonding pattern of the extracellular binding domain. *J. Biol. Chem.* **265**(6), 3111–3115.

Gertler, A., Grosclaude, J., Strasburger, C. J., Nir, S., and Djiane, J. (1996). Real-time kinetic measurements of the interactions between lactogenic hormones and prolactin-receptor extracellular domains from several species support the model of hormone-induced transient receptor dimerization. *J. Biol. Chem.* **271**(40), 24482–24491.

Goffin, V., Shiverick, K. T., Kelly, P. A., and Martial, J. A. (1996). Sequence-function relationships within the expanding family of prolactin, growth hormone, placental lactogen, and related proteins in mammals. *Endocrine Reviews* **17,** 385–410.

Hensley, P., Doyle, M. L., Myszka, D. G., Woody, R. W., Brigham-Burke, M. R., Erickson-Miller, C. L., Griffin, C. A., Jones, C. S., McNulty, D. E., O'Brien, S. P., Amegadzie, B. Y., MacKenzie, L., Ryan, M. D., and Young, P. R. (2000). Evaluating energetics of erythropoietin ligand binding to homodimerized receptor extracellular domains. *Methods Enzymol.* **323,** 177–207.

Herman, A., Helman, D., Livnah, O., and Gertler, A. (1999). Ruminant placental lactogens act as antagonists to homologous growth hormone receptors and as agonists to human or rabbit growth hormone receptors. *J. Biol. Chem.* **274,** 7631–7639.

Ihle, J. N., Witthuhn, B. A., Quelle, F. W., Yamamoto, K., Thierfelder, W. E., Kreider, B., and Silvennoinen, O. (1994). Signaling by the cytokine receptor superfamily: JAKs and STATs. *Trends Biochem. Sci.* **19,** 222–227.

Kelly, P. A., Djiane, J., Banville, D., Ali, S., Edery, M., and Rozakis, M. (1991). The growth hormone/prolactin receptor gene family. *Oxford Surveys On Eukaryotic Genes* **7,** 29–50.

Kossiakoff, A. A., and De Vos, A. M. (1998). Structural basis for cytokine hormone-receptor recognition and receptor activation. *Advances In Protein Chemistry* **52,** 67–108.

Kossiakoff, A. A., Somers, W., Ultsch, M., Andow, K., Muller, Y., and De Vos, A. M. (1994). Comparison of the intermediate complexes of human growth hormone bound to the human growth hormone and prolactin receptors. *Protein Science* **3,** 1697–1705.

Lebrun, J. J., Ali, S., Sofer, L., Ullrich, A., and Kelly, P. A. (1994). Prolactin-induced proliferation of Nb2 cells involves tyrosine phosphorylation of the prolactin receptor and its associated tyrosine kinase JAK2. *J. Biol. Chem.* **269,** 14021–14026.

Liparoto, S. F., and Ciardelli, T. L. (1999). Biosensor analysis of the interleukin-2 receptor complex. *J. Mol. Recognit.* **12**(5), 316–321.

Livnah, O., Johnson, D. L., Stura, E. A., Farrell, F. X., Barbone, F. P., You, Y., Liu, K. D., Goldsmith, M. A., He, W., Krause, C. D., Pestka, S., Jolliffe, L. K., and Wilson, I. A. (1998). An antagonist peptide-EPO receptor complex suggests that receptor dimerization is not sufficient for activation. *Nat. Struct. Biol.* **5**(11), 993–1004.

Livnah, O., Stura, E. A., Johnson, D. L., Middleton, S. A., Mulcahy, L. S., Wrighton, N. C., Dower, W. J., Jolliffe, L. K., and Wilson, I. A. (1996). Functional mimicry of a protein hormone by a peptide agonist: The EPO receptor complex at 2.8 Å. *Science* **273**(5274), 464–471.

Livnah, O., Stura, E. A., Middleton, S. A., Johnson, D. L., Jolliffe, L. K., and Wilson, I. A. (1999). Crystallographic evidence for preformed dimers of erythropoietin receptor before ligand activation. *Science* **283**(5404), 987–990.

Lowman, H. B., Bass, S. H., Simpson, N., and Wells, J. A. (1991a). Selecting high-affinity binding proteins by monovalent phage display. *Biochemistry* **30,** 10832–10838.

Lowman, H. B., Cunningham, B. C., and Wells, J. A. (1991b). Mutational analysis and protein engineering of receptor-binding determinants in human placental lactogen. *J. Biol. Chem.* **266,** 10982–10988.

Lowman, H. B., and Wells, J. A. (1993). Affinity maturation of human growth hormone: Monovalent phage display. *J. Mol. Biol.* **234,** 564–578.

Nice, E., Layton, J., Fabri, L., Hellman, U., Engstrom, A., Persson, B., and Burgess, A. W. (1993). Mapping of the antibody- and receptor-binding domains of granulocyte colony-stimulating factor using an optical biosensor. Comparison with enzyme-linked immunosorbent assay competition studies. *J. Chromatogr.* **646**(1), 159–168.

Nicoll, C. S., Mayer, G. L., and Russel, S. M. (1986). Structural features of prolactins and growth hormones that can be related to their biological properties. *Endocrine Rev.* **7,** 169–203.

Pal, G., Kossiakoff, A. A., and Sidhu, S. S. (2003). The functional binding epitope of a high affinity variant of human growth hormone mapped by shotgun alanine-scanning mutagenesis: Insights into the mechanisms responsible for improved affinity. *J. Mol. Biol.* **332**(1), 195–204.

Pearce, K. H., Jr., Cunningham, B. C., Fuh, G., Teeri, T., and Wells, J. A. (1999). Growth hormone binding affinity for its receptor surpasses the requirements for cellular activity. *Biochemistry* **38**(1), 81–89.

Pearce, K. H. J., Ultsch, M. H., Kelley, R. F., De Vos, A. M., and Wells, J. A. (1996). Structural and mutational analysis of affinity-inert contact residues at the growth hormone-receptor interface. *Biochemistry* **35**(32), 10300–10307.

Randal, M., and Kossiakoff, A. A. (2001). The structure and activity of a monomeric interferon-gamma:alpha-chain receptor signaling complex. *Structure (Camb)* **9**(2), 155–163.

Remy, I., Wilson, I. A., and Michnick, S. W. (1999). Erythropoietin receptor activation by ligand-induced conformation change science. *Science* **283,** 990–993.

Schiffer, C., Ultsch, M., Walsh, S., Somers, W., De Vos, A. M., and Kossiakoff, A. (2002). Structure of a phage display-derived variant of human growth hormone complexed to two copies of the extracellular domain of its receptor: Evidence for strong structural coupling between receptor binding sites. *J. Mol. Biol.* **316**(2), 277–289.

Shen, B. J., Hage, T., and Sebald, W. (1996). Global and local determinants for the kinetics of interleukin-4/interleukin-4 receptor alpha chain interaction. A biosensor study employing recombinant interleukin-4-binding protein. *Eur. J. Biochem.* **240**(1), 252–261.

Sinha, Y. N. (1995). Structural Variants of Prolactin: Occurrence and Physiological Significance. *Endocrine Reviews* **16**(3), 354–369.

Smith, J. L., Corfield, P. W. R., Hendrickson, W. A., and Low, B. W. (1988). Refinement at 1.4 Å resolution of a model of erabutoxin b: Treatment of ordered solvent and discrete disorder. *Acta. Crystallogr. Sect. A* **44,** 357–368.

Somers, W., Ultsch, M., De Vos, A. M., and Kossiakoff, A. A. (1994). The X-ray structure of the growth hormone-prolactin receptor complex. *Nature* **372,** 478–481.

Sprang, S. R., and Bazan, J. F. (1993). Cytokine structural taxonomy and mechanisms of receptor engagement. *Curr. Opin. Struct. Biol.* **3,** 815–827.

Syed, R. S., Reid, S. W., Li, C., Cheetham, J. C., Aoki, K. H., Liu, B., Zhan, H., Osslund, T. D., Chirino, A. J., Zhang, J., Finer-Moore, J., Elliott, S., Sitney, K., Katz, B. A., Matthews, D. J., Wendoloski, J. J., Egrie, J., and Stroud, R. M. (1998). Efficiency of signalling through cytokine receptors depends critically on receptor orientation. *Nature* **395**(6701), 511–516.

Ultsch, M., Somers, W., Kossiakoff, A. A., and De Vos, A. M. (1994). The crystal structure of affinity-matured human growth hormone at 2 Å resolution. *J. Mol. Biol.* **236,** 286–299.

Walsh, S. T., Jevitts, L. M., Sylvester, J. E., and Kossiakoff, A. A. (2003). Site2 binding energetics of the regulatory step of growth hormone-induced receptor homodimerization. *Protein Sci.* **12**(9), 1960–1970.

Walter, M. R., Windsor, W. T., Nagabhushan, T. L., Lundell, D. J., Lunn, C. A., Zauodny, P. J., and Narula, S. K. (1995). Crystal structure of a complex between interferon-gamma and its soluble high-affinity receptor. *Nature* **376,** 230–235.

Wang, Y., Shen, B. J., and Sebald, W. (1997). A mixed-charge pair in human interleukin 4 dominates high-affinity interaction with the receptor alpha chain. *Proc. Natl. Acad. Sci. USA* **94**(5), 1657–1662.

Wells, J. A. (1991). Systematic mutational analyses of protein-protein interfaces. *Methods Enzymol.* **202,** 390–411.

Wells, J. A., and De Vos, A. M. (1996). Hematopoietic receptor complexes. *Annu. Rev. Biochem.* **65,** 609–634.

Wiesmann, C., Fuh, G., Christinger, H. W., Eigenbrot, C., Wells, J. A., and De Vos, A. M. (1997). Crystal structure at 1.7 Å resolution of VEGF in complex with domain 2 of the FLt-1 receptor. *Cell* **5,** 695–704.

Wilson, I. A., and Jolliffe, L. K. (1999). The structure, organization, activation and plasticity of the erythropoietin receptor. *Curr. Opin. Struct. Biol.* **9**(6), 696–704.

STRUCTURAL ANALYSIS OF IL-10 AND TYPE I INTERFERON FAMILY MEMBERS AND THEIR COMPLEXES WITH RECEPTOR

By MARK R. WALTER

University of Alabama at Birmingham,
Department of Microbiology and Center for Biophysical Sciences and Engineering,
Birmingham, Alabama 35294

I. Introduction

Helical cytokines are secreted proteins that function as intracellular messengers. The information content contained on their α-helical scaffolds is decoded by extracellular high-fidelity interactions with cell surface receptor chains that initiate intracellular signaling cascades leading to diverse cellular functions. Amino acid sequence analysis of the extracellular cytokine binding domains revealed two distinct receptor subclasses, class-1 and class-2 (Bazan, 1990a,b). The α-helical cytokines were subsequently defined based on the class of receptor (class 1 or class 2) they

ADVANCES IN PROTEIN CHEMISTRY, Vol. 68

0065-3233/04 $35.00

interact with. This review focuses on the structural and functional analysis of α-helical class 2 cytokines (HCII) and their receptors (HCRII).

A. *Biological Activity*

The HCII family is comprised of the type I interferons (IFN-α, IFN-β, IFN-κ, IFN-ω), IFN-γ, and the interleukins IL-10, IL-19, IL-20, IL-24, IL-26, IL-28, and IL-29. The biological activities of the family are diverse, often redundant, and sometimes paradoxical. Together, the HCIIs play a very important, but not exclusive, role in mounting and controlling innate and adaptive host immune responses against invading pathogens. For example, the type I IFNs, IL-28, IL-29 and IFN-γ provide a first line of defense against viral infection by inducing the expression of numerous cellular genes resulting in an antiviral state (Kotenko *et al.*, 2003; Pestka, 1997; Sheppard *et al.*, 2003). Although IFN-γ exhibits anti-viral properties, transgenic mice missing their high affinity IFN-γ receptor chain (IFN-γR1) respond normally to viral challenge but are greatly impaired in their ability to resist infection by microbial pathogens (Bach *et al.*, 1997; Huang *et al.*, 1993). Thus IFN-γ plays an especially important role in innate host resistance to microbial pathogens and more recently has been shown to induce immune responses to tumor cells (Ikeda *et al.*, 2002).

In contrast to the macrophage activating activities of IFN-γ, IL-10 is a potent immunosuppressive cytokine that protects the host from an over exuberant immune response (Moore *et al.*, 2001). This is largely due to IL-10s ability to suppress the production of pro-inflammatory cytokines (CSIF activity) such as IFN-γ, IL-1, TNF-α, IL-6, and IL-2 and down-regulate MHC class I and II molecules. The functional importance of IL-10 was first demonstrated by using IL-10 knock-out mice infected with *H. hepaticus* that spontaneously develop enterocolitis (Kuhn *et al.*, 1993; Kullberg *et al.*, 1998, 2001). Infection of these mice with other pathogens generally leads to an increased ability to clear infections, but it comes at the cost of a greatly increased risk of the death from a toxic shock-like syndrome compared to the wildtype animals (Moore *et al.*, 2001). Administration of exogenous IL-10 in almost all studies limits the anti-pathogen responses. Thus IL-10 levels must be delicately balanced to ensure an immune response sufficient to clear the pathogen without damaging the host. Despite its largely immunosuppressive activities, IL-10 also functions as a potent growth factor for mast cells, a growth and differentiation factor for B cells and can proliferate CD8+ T cells (MacNeil *et al.*, 1990; Rousset *et al.*, 1992).

The critical role type I IFNs, IFN-γ, and IL-10 play in host defense is further substantiated by identification of soluble IFN-α and IFN-γ binding proteins in the genomes of several poxviruses (Smith, 1996). These molecules bind with high affinity to the IFNs, preventing interactions with their receptors and subsequently blocking their biological activities. Similarly, viral IL-10 molecules have been located in the genomes of several viruses including the Epstein Barr virus (EBV) (Vieira *et al.*, 1991), cytomegalovirus (CMV) (Kotenko *et al.*, 2000; Lockridge *et al.*, 2000), ORF virus (ORF) (Fleming *et al.*, 2000), and the yabba-like disease virus (YLDV) (Lee *et al.*, 2001). Studies to date show the viral IL-10s induce only a subset of the activities observed for the cellular IL-10s but signal through the same cellular IL-10 receptor complex. The viral protein mimics share between 27–85% sequence identity with their cellular homologs. The functional advantage gained by the pox and herpes viruses that contain IFN antagonists or IL-10 mimics is currently not understood. However, consistent with the function of IL-10 in immunosuppression, the viruses that encode viral IL-10s are able to establish lifelong persistent infections.

Many of the cellular HCIIs (IL-19, IL-22, IL-24, IL-26) were only recently discovered, and their detailed functional properties are just now being determined (Dumoutier *et al.*, 2000a; Gallagher *et al.*, 2000; Jiang *et al.*, 1995; Knappe *et al.*, 2000; Kotenko *et al.*, 2003; Sheppard *et al.*, 2003; Xie *et al.*, 2000). Current data suggest that all of these molecules are involved in host inflammatory responses. IL-19 has been shown to up-regulate the expression of the pro-inflammatory cytokines IL-6 and TNF-α, but its *in vitro* function remains unknown (Liao *et al.*, 2002). Basal expression of IL-19 is observed in monocytes but expression can be induced by exposure to lipopolysaccharide (LPS). Over-expression of IL-20 in transgenic mice results in an abnormal skin morphology similar to psoratic skin (Blumberg *et al.*, 2001). Treatment of human keratinocytes with IL-20 resulted in up-regulation of TNF-α, MCP-1, and IL-1, suggesting it is an inflammatory mediator and proliferation factor for keratinocytes. IL-22 does not induce pro-inflammatory cytokines. Rather, IL-22 up-regulates the production of early systemically circulated defense proteins (acute phase proteins) such as serum amyloid A, α1-antichymotrypsin, and haptoglobin in hepatocytes (Dumoutier *et al.*, 2000c) as well as PAP1 protein, which is observed at high levels in pancreatitis (Aggarwal *et al.*, 2001). IL-22 also induces the production of reactive oxygen species in resting B cells. These studies suggest IL-22 may induce inflammatory responses initiated from the liver or panacreas. IL-22 activity is regulated by a soluble binding protein, IL-22BP, which is not a soluble form of the membrane bound IL-22 receptor but a unique receptor chain (Dumoutier *et al.*, 2001; Kotenko *et al.*, 2001; Xu *et al.*, 2001).

IL-24 exhibits two intriguing functional properties depending on how it is expressed. When secreted at physiological concentrations, IL-24 exhibits pro-inflammatory cytokine functions similar to IL-19 such as inducing IL-6, IFN-γ,

TNF-α, IL-12, and GMCSF (Caudell *et al.*, 2002). However, when IL-24 is over expressed via an adenovirus gene delivery system, it selectively induces apoptosis of cancer cells but not normal cells (Madireddi *et al.*, 2000; Mhashilkar *et al.*, 2001). The mechanisms that distinguish the different biological activities of IL-24 are currently under intense investigation. Finally, IL-26 was identified as a molecule secreted by T-cells infected with herpesvirus saimiri (Knappe *et al.*, 2000). IL-26 has been shown to be up-regulated in NK and T cells by IL-2/IL-12 and anti-CD3 antibody. However, the specific functional properties of IL-26 have not been reported.

B. HCRII Organization

The HCRIIs are single membrane spanning proteins composed of an extracellular cytokine binding module called the class 2 homology region (C2HR), a membrane spanning helix, and an intracellular domain (ID) (Fig. 1). Except for IFNAR1, all C2HRs consist of approximately 200 amino acids that fold into two fibronectin type III β-sandwhich domains of ~100 amino acids each. The extracellular region of IFNAR1 consists of approximately 400 amino acids, which corresponds to a duplication of the 200 amino acid C2HR (Bazan, 1990a). The function of the C2HR is not restricted to cytokine binding since tissue factor (TF), the membrane bound activator of factor VIIa, is a member of the family (Morrissey, 2001). Further genome mining studies may reveal more diverse functions for the C2HR fold.

The intracellular domains (IDs) of the receptors share essentially no amino acid sequence identity. However, they can be placed into two groups based on the size of their domains. The IDs of the six human HCRIIs identified thus far are composed of 200–300 amino acids (ID_L), while a second group of four receptors have short C-terminal tails (ID_S) of ~60–100 amino acids. Both ID_L and ID_S domains provide recruitment sites for JAK kinases, while only the ID_L chains provide STAT recruitment sites. Together they activate intracellular pathways required to complete the signal transduction cascade originated by the cytokine-receptor interaction. Several informative reviews on JAK/STAT interactions have recently been published (Kerr *et al.*, 2003; Kisseleva *et al.*, 2002; O'Shea *et al.*, 2002).

C. Organization of the Receptor Complexes

The functional HCII receptor complexes are composed of one $HCRII_{IDL}$ and one $HCRII_{IDS}$ (Fig. 1). For example, IL-10R1 and IL-10R2 with intracellular domains of 318 and 82 amino acids, respectively, are

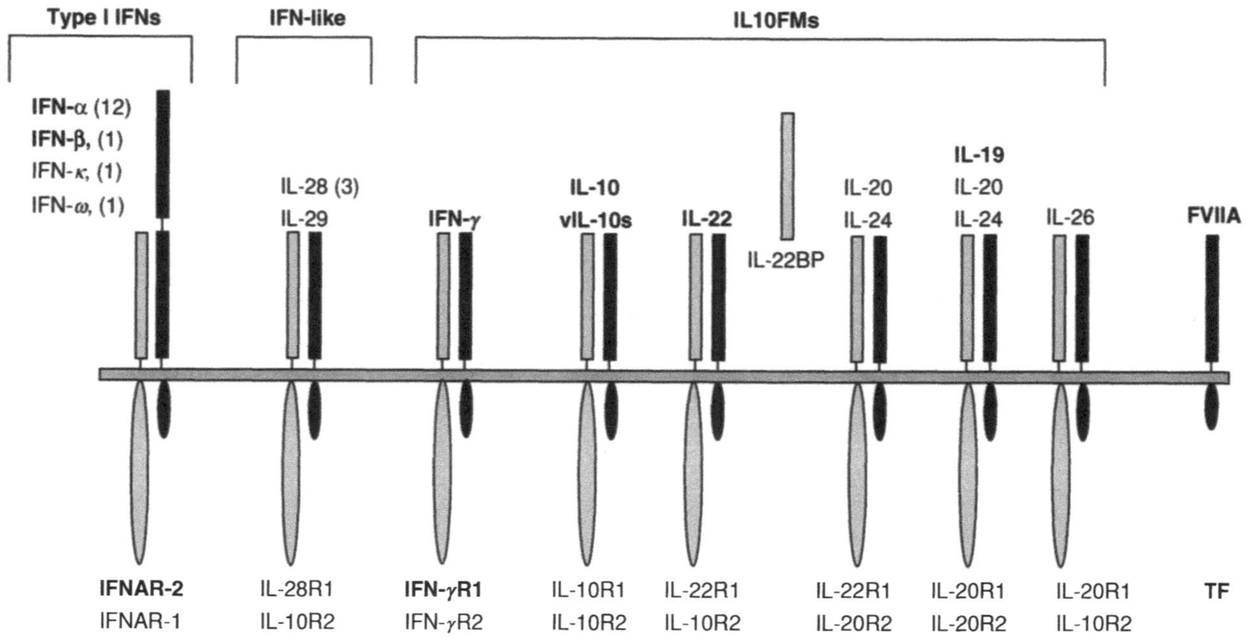

FIG. 1. Schematic diagram of the functional human HCII/HCRII signaling complexes. C2HRs are shown as rectangles and the intracellular domains as ellipses. The names of HCIIs are written above the receptor complex with which they interact. HCII and HCRII labels are in bold if the crystal or NMR structure of the molecule has been completed.

required for IL-10 activity (Kotenko *et al.*, 1997). Receptors for IFN-γ, IL-10, IL-22, and the type I IFNs are distinguished by large differences in affinity for cytokine. For example, IL-10 binds to a soluble fragment of IL-10R1 with a dissociation constant of ~1 nM while the affinity of soluble IFN-10R2 is approximately 1 mM (Logsdon *et al.*, 2002). Similar results have been obtained for IFN-γ. Because of the low affinity of IFN-γR2 for IFN-γ and the IFN-γ/IFN-γR1 complex, it was unclear how IFN-γR2 could be rapidly recruited into the ternary complex to initiate signal transduction. However, recent FRET studies have confirmed that high-affinity IFN-γR1 and low affinity IFN-γR2 chains are associated with one another in the absence of cytokine on the cell surface (Krause *et al.*, 2002). This is hypothesized to be a general organizational feature of the HCRII heterodimeric receptor complexes. For IFN-γ, IL-10, IL-22, and the type I IFNs, the higher-affinity receptor chains contain the larger intracellular domains ($C2HR_L$) while the low-affinity receptors have short intracellular domains. However, qualitative gel filtration experiments suggest IL-20R2, which contains only a small intracellular domain ($C2HR_S$), has higher affinity for IL-19 and IL-20 than IL-20R1 (Pletnev *et al.*, 2003).

D. *Sequence Diversity and Gene Structure of the HCIIs*

The constant battle between host and pathogen is hypothesized to be the driving force in the rapid expansion of the type I IFN family to include 15 different subtypes comprised of 12 IFN-αs, one IFN-β, one IFN-κ, and one IFN-ω (Murphy, 1993). The twelve IFN-α subtypes share approximately 80% sequence identity with each other and 60%, 40%, and 30% with IFN-ω, IFN-β, and IFN-ω, respectively (Derynck *et al.*, 1980; Hauptmann and Swetly, 1985; LaFleur *et al.*, 2001; Pestka, 1997). Despite this sequence diversity, all 15 molecules bind and signal through the IFNAR1/IFNAR2 heterodimer complex. In addition to anti-viral activities, the type I IFNs display cell growth inhibitory functions and enhance the activity of natural killer cells. Despite high sequence identities, activity profiles of the purified type I IFN subtypes reveal differences in antiviral and NK activities of up to 100 and 10,000 fold, respectively (Pestka, 1997). Thus, subtle amino acid changes in the scaffolds of the type I IFNs can drastically after their biological activity profiles.

In contrast to the high sequence identity of the type I IFNs, IL-10, IL-19, IL-20, IL-22, IL-24, and IL-26 share only about 25% amino acid sequence identity with each other. This group of molecules is referred to as the *IL-10 family* based on sequence identity and a common gene structure consisting of five exons (Kotenko, 2002). Despite having a different gene organization (4 exons) and almost no sequence identity with IL-10, structural studies of IFN-γ revealed it is also a distant member of the IL-10 family (Walter and Nagabhushan, 1995; Zdanov *et al.*, 1995). In contrast to the type I IFNs, the IL-10 family cytokines interact with at least six different receptor complexes (Fig. 1).

The most recently discovered HCIIs are IL-28 and IL-29, which form a family of four molecules (IL-28a, b, c, and IL29) that share ~70% sequence identity with one another but only about 14% with the type I IFNs and IL-10 family members (IL10FMs). Amino acid sequence analysis and their anti-viral activity properties suggest IL-28 and IL-29 are most similar to the type I IFNs but yet they signal through a receptor complex that contains IL-28R1 and the IL-10R2 chain. Thus in some ways they may be considered hybrids of the type I IFNs and IL-10 families. This classification is consistent with studies on the origin of the HCII family, which suggests the type I IFN and IL-10 families originated from a primordial IL-28/IL-29 gene (Lutfalla *et al.*, 2003).

At present, the HCII family consists of 29 molecules that interact with eight different heterodimeric receptor complexes. The biological activities of the HCII family members are dependent on vast array of high-fidelity molecular recognition events. This review will provide a summary of the progress made in defining the structural mechanisms of HCII cytokine-receptor recognition. A list of X-ray and NMR structural analyses of the HCII family of molecules is presented in Table I. There is insufficient space to review all of the structural results. Rather the goal of this review is to present a unified structural analysis of the HCIIs, the HC2Rs, and their complexes. As a result, much of the work reported is original, synthesized from the structures determined by the groups referenced in Table I. Additional comparisons are made between the class 1 (HCI) and class 2 (HCII) cytokine families by using the prototypical HCI receptor complex between growth hormone (GH) and the growth hormone receptor (GHR) to highlight key differences between the families. A complete review of the GH/GHR system is made elsewhere in this volume.

II. Structures of the HCIIs

A. *The HCII Fold*

The predominant feature of each α-helical cytokine structure is a left-handed anti-parallel four-helix bundle (Fig. 2) (Presnell and Cohen, 1989; Sprang and Bazan, 1993). The four helices of the bundle are connected by two long overhand connections and one short segment to form a distinct up-up-down-down topology first described for porcine growth hormone

TABLE I
Summary of Class 2 Cytokine, Receptor, and Receptor Complex Structures

Class/Molecule	Species	Method	Res. (Å)	PDB	Reference
Type I IFNs					
1. IFN-β	Murine	X-ray	2.2	1RMI	Senda *et al.*, 1995
2. IFN-α2b	Human	X-ray	2.9	1RH2	Radhakrishnan *et al.*, 1996
3. IFN-α2a	Human	NMR	—	1ITF	Klaus *et al.*, 1997
4. IFN-β	Human	X-ray	2.2	1AU1	Karpusas *et al.*, 1997
5. IFN-τ	Ovine	X-ray	2.1	1B5L	Radhakrishnan *et al.*, 1999
C2HRs					
1. TF	Human	X-ray	2.2	1BOY	Harlos *et al.*, 1994
2. TF	Human	X-ray	1.7	2HFT	Muller *et al.*, 1996
3. TF	Rabbit	X-ray	2.4	1A21	Muller *et al.*, 1998
4. IFNγR1(D1): A6Fab	Human: Mouse	X-ray	2.8	1997	Sogabe *et al.*, 1997
5. IFNAR2	Human	NMR	—	1N6V	Chill *et al.*, 2003
IL-10FMs					
1. IFN-γ	Human	X-ray	3.5	1HIG	Ealick *et al.*, 1991
2. IFN-γ (backbone)	Human	NMR	—	—	Grzesiek *et al.*, 1992
3. IL-10	Human	X-ray	1.8	1ILK	Zdanov *et al.*, 1995
4. IL-10	Human		2.0	1INR	Walter and Nagabhushan, 1995
5. IL-10	Human	X-ray	1.6	2ILK	Zdanov *et al.*, 1996
6. IL-10	ebv	X-ray	1.9	1VLK	Zdanov *et al.*, 1995
7. IFN-γ	Bovine	X-ray	2.0	1D9C	Randal and Kossiakoff, 2000
8. IFN-γ	Bovine	X-ray	2.9	1D9G	Randal and Kossiakoff, 2000
9. IFN-γSC	Human	X-ray	2.9	1EKU	Landar *et al.*, 2000
10. IL-10M1: 9D7Fab	Human:Rat	X-ray	2.0	1LK3	Josephson *et al.*, 2002b
11. IL-22	Human	X-ray	2.0	1M4R	Nagem *et al.*, 2002
12. IL-19	Human	X-ray	2.0	1N1F	Chang *et al.*, 2003
HCII/C2HR and FVIIa/TF Complexes					
1. IFN-γ/IFN-γR1	Human	X-ray	2.9	—	Walter *et al.*, 1995
2. FVIIa/TF	Human	X-ray	2.0	1DAN	Banner *et al.*, 1996
3. IFN-γ/IFN-γR1	Human	X-ray	2.9	1FG9	Thiel *et al.*, 2000
4. IL-10/IL-10R1	Human	X-ray	2.9	1J7V	Josephson *et al.*, 2001
5. IFN-γSC/ IFNγR1	Human	X-ray	2.0	1FYH	Randal and Kossiakoff, 2001
6. cmvIL-10/ IL-10R1	cmv/Human	X-ray	2.7	1LQS	Jones *et al.*, 2002a

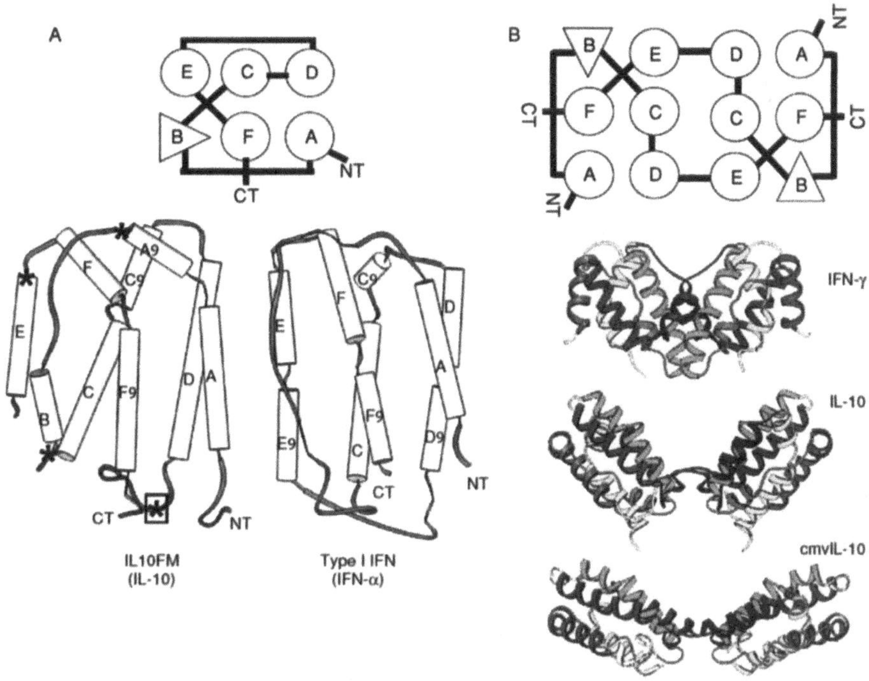

FIG. 2. Topology and structure of the HCIIs. (A) Domain structure of the HCIIs. Schematic diagram of the six secondary structural elements (A-F) are shown along with cylinder diagrams of the IL10FM, IL-10 and the type I IFN, IFN-α. The location of intron-exon boundaries of IL-10 are denoted by asterisks. The boxed asterisk denotes the missing intron-exon junction in IFN-γ. (B) HCII intercalated dimer structures. View of the schematic diagram is perpendicular to the view of the ribbon diagrams.

(Abdel-Meguid *et al.*, 1987). Despite the tremendous explosion in the number of X-ray and NMR structures in the protein data bank in the past 17 years, this simple topology has only been observed for α-helical cytokines.

In the past, α-helices for each structure have been labeled sequentially starting with the N-terminal helix as helix A. However, because different HCIIs have different numbers of α-helices, structurally equivalent α-helices may be labeled differently. To avoid this problem, in this review the cytokines are described by six secondary structural elements that pack into two layers (A, F, B and D, C, F) to form the globular folding unit or core structure (Fig. 2). For the HCIIs, five of the structural elements (A, and C-F) are α-helices, while structural element B adopts a number of different secondary structures. In fact, this is one structural feature of the HCIIs that distinguishes it from the HCIs. The six roughly parallel structural elements are connected by four loops (AB, BC, CD, and EF loops), which reverse the direction of the chain. Further dissection of the HCIIs into the IL-10 and type I IFN structural families are presented later.

B. The IL-10 Family

Crystal structures have been determined for six IL-10 family members (IL10FMs) IFN-γ, IL-10, ebvIL-10, human cmvIL-10, IL-19, and IL-22 (Table I). IL-10, ebvIL-10, cmvIL-10, and IFN-γ exist as dimers while IL-19 and IL-22 are monomers. In this section the structural domains of these six cytokines are compared (Fig. 3). The six secondary structural elements of the IL10FMs can be enclosed in a rectangular prism 32Å wide by 45Å long, and 25Å thick. The domain of IFN-γ is the most compact, allowing it to be encompassed in a prism 10Å shorter than the other molecules. However, most helices in IFN-γ are the same length as the other IL-10FMs (Table II). The main exception is helix F′ which is only 4 residues long in IFN-γ but 8-16 residues long in the other IL10FMs. The longer F′

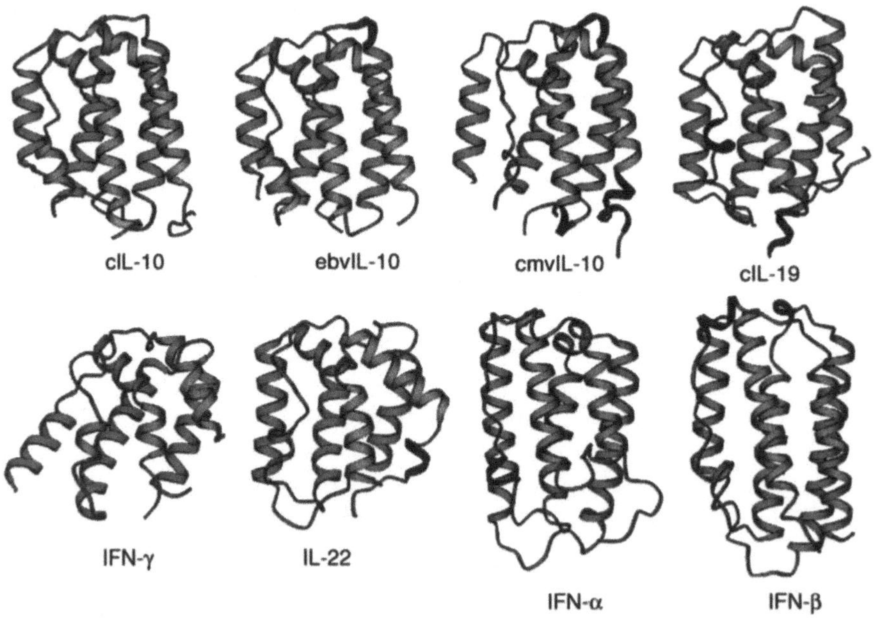

FIG. 3. Structures of IL-10FM and type I IFN domains. Each domain is viewed looking into the high affinity receptor binding site as determined experimentally for IL-10, ebvIL-10, cmvIL-10, IFN-γ, IFN-α, and IFN-β.

TABLE II
Helix Angles in the HCIIs

	IL-10	cmv10	IL-19	IL-22	IFN-γ	IFN-α	IFN-β	IFN-τ
Helix A	174.0 (13)	163.1 (13)	164.2 (9)	139.9 (15)	153.0 (11)	167.2 (13)	151.2 (10)	166.8 (20)
Helix A′	121.3 (7)	108.0 (4)	124.2 (6)	—	—	—	161.6 (10)	—
Helix B	24.5 (7)	—	—	38.8 (4)	21.6 (6)	—	40.7 (4)	32.5 (5)
Helix C	153.2 (14)	142.9 (14)	151.9 (17)	149.0 (16)	139.1 (20)	159.2 (15)	152.5 (20)	146.1 (13)
Helix C′	160.0 (8)	159.7 (8)	159.2 (4)	—	—	111.5 (4)	—	—
Helix D	8.4 (18)	13.5 (18)	26.2 (24)	26.9 (15)	24.5 (16)	17.9 (9)	33.8 (9)	32.5 (20)
Helix D′	—	—	—	—	—	23.1 (8)	41.9 (14)	—
Helix E	174.8 (11)	172.4 (12)	164.8 (14)	171.6 (17)	131.4 (11)	160.8 (9)	148.8 (13)	156.7 (9)
Helix E′	—	—	—	—	—	171.4 (10)	155.3 (9)	163.9 (9)
Helix F	47.1 (8)	57.0 (6)	47.1 (10)	51.2 (9)	49.5 (10)	11.7 (11)	18.4 (13)	22.7 (10)
Helix F′	— (16)	— (14)	— (11)	— (12)	— (4)	— (9)	— (8)	— (14)

Helix angles, in degrees, relative to helix F′. Number of residues in each helix is shown in parentheses.

helix in the other IL-10FMs is replaced by a 17 residue non-structured C-terminal tail in IFN-γ. The shorter length of helix F′ together with larger skew angles of helices C and E result in its smaller domain size.

1. Helix F and Conserved Fingerprint Residues

The central feature of the IL-10 family is a pronounced bend (47°–57°) in the middle of helix F (Figs. 2 and 3, Table II). Comparison of IFN-γ and IL-10 structures when they were first solved revealed five conserved residues that cluster at the bend in helix F (Leu-47, Phe-71, Tyr-72, Ala-139, and Glu-142) (Walter and Nagabhushan, 1995). These residues were predicted to be critical for maintaining the IL-10 fold and useful as a "fingerprint" for recognizing other cytokines that adopt IL-10-like structures. A portion of this fingerprint is most easily identified in the conserved linear amino acid sequence KAMSEFDIFI, which in the three dimensional structure of IL-10 and other IL-10FMs corresponds to the bend in helix F where it crosses under the AB loop.

With structure determinations completed for several new IL-10FM family members of viral and human origin, it is now possible to extend the analysis of the initial fingerprint region (Table III, Fig. 4). Comparison of the structures of IL-10, cmvIL-10, IL-19, and IL-22 suggests that the IL-10 fingerprint consists of 11 residues, of which nine are mostly buried and contribute to the structural integrity of the molecules. Surprisingly, over half of these residues are charged or polar. Residue pairs Lys-34/Asp-144, Asp-41/Lys-138 form conserved salt bridge interactions while the Oδ1 and Oδ2 atoms of Glu-142 form hydrogen bonds with the backbone nitrogens of Leu-47 in the AB-loop.

Of the charged residues, Glu-142 is probably the most important residue for the structural integrity of the AB loop since it is conserved in all class 2 cytokines. However, Asp-41 and Asp-144 are not conserved in IFN-γ or the Type I IFNs. In IFN-γ, the Asp-41/Lys-138 salt-bridge appears to be replaced by a hydrogen bond between the backbone carbonyl of Gly-18 in the AB loop with His-111 on helix F. Mutation of His-111 results in drastic loss of IFN-γR1 binding and biological activity (Lunn *et al.*, 1992b). The extreme phenotype of His-111 mutations, especially to an aspartate, may be at least partly due to the role His-111 plays in stabilizing the conformation

TABLE III
IL-10FM Fingerprint Residues

	AB loop				Helix C			Helix F			
	K34	D41	**L47**	L48	**F71**	**Y72**	V76	**K138**	**A139**	**E142**	D144
IL-10	—	—	—	—	—	—	—	—	—	—	—
cmvIL-10	R	—	W	—	R	—	—	—	G	—	—
rcmvIL-10	R	—	—	F	H	—	—	—	—	—	—
IFN-γ	—	V	—	F	—	—	L	—	—	—	I
IL-19	—	—	I	—	—	—	—	—	S	—	—
IL-20	R	—	I	—	L	—	—	—	—	—	—
IL-22	—	—	—	I	—	T	—	—	—	—	—
IL-24	—	—	—	—	—	—	—	—	—	—	—
IL-26	—	—	—	—	—	F	—	—	—	—	—
IFN-α	R_{22}	C_{29}	F_{36}	F_{38}	F_{123}	V_{63}	L_{66}	C_{138}	V_{143}	E_{146}	M_{148}
IFN-β	W_{22}	C_{31}	F_{38}	I_{40}	Y_{123}	I_{66}	I_{69}	C_{141}	V_{146}	E_{149}	L_{151}
IL-28	L	C	F	P	L	T	L	S	V	N	F
IL29	L	C	F	P	L	T	L	S	V	N	F

Bolded residues correspond to the original IL-10 fingerprint described in Walter and Nagabhushan, 1995. rcmv = rhesus cmvIL-10.

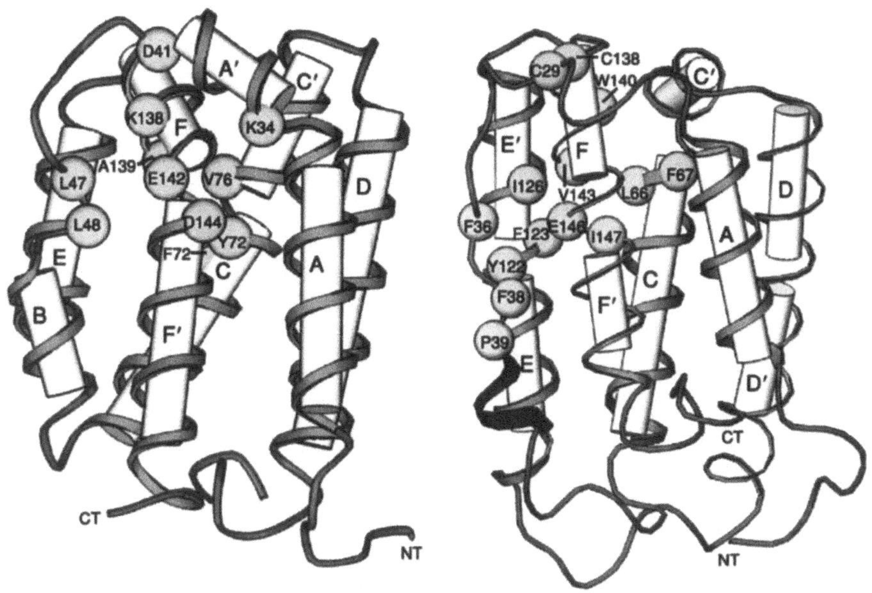

FIG. 4. Fingerprint residues in the IL10FMs and type I IFNs. Ribbon diagram of IL-10 and IFN-α2b are shown. The location of structurally important and conserved residues on the IL-10 and IFN-α scaffolds are shown as spheres and labeled. See Tables IV and V for equivalent residues in other IL-10FMs or type I IFNs, respectively.

of the AB loop. Regardless, the ionic nature of the conserved residues in the IL-10FMs is striking and may explain, in part, the sensitivity of IL-10 and IFN-γ to unfolding when exposed to low pH solutions (Syto *et al.*, 1998). It has also been proposed that the high conservation of the charged residues in the fingerprint is due to their dual functional roles in stabilizing the fold as well as participating in receptor binding (Jones *et al.*, 2002a).

In addition to the charged interactions, several mostly conserved hydrophobic residues cluster at the bend in helix F (Leu-47, Leu-48, Phe-71, Tyr-72, Val-76, Ala-139). The hydrophobic methylene groups of Lys-138 and Glu-142 also participate in the tight packing of the core. Amino acids Leu-47 and Leu-48 play an especially important role in the attachment of the C-terminal end of the AB loop to the helical core. All together, the conserved cluster ensures the precise orientation of the AB loop relative to helix F. Mutagenesis and crystallographic studies have demonstrated the critical role the AB loop plays in specific high affinity receptor binding (Jones *et al.*, 2002a; Lundell *et al.*, 1994).

NMR and crystallographic studies show the AB loop region which connects the secondary structural elements A and B is largely flexible in solution but adopts a highly ordered structure upon receptor binding (Grzesiek *et al.*, 1992; Randal and Kossiakoff, 2001; Walter *et al.*, 1995; Zdanov *et al.*, 1997). In the case of IFN-γ, the AB loop region adopts a short 3_{10} helix upon binding IFN-γR1, while IL-10s AB-loop exists in a more extended conformation. The AB loop region of IL-10 has an inherit propensity to adopt its "receptor bound conformation" since the conformation may be induced by either crystal contacts or the receptor itself (Josephson *et al.*, 2001; Zdanov *et al.*, 1996).

2. *Helix Angles*

Helix-helix angles were calculated for each IL-10FM by using helix F′ as the reference helix (Table II). Inspection of this table shows the IL-10 family members may be split into molecules that have a bend (~40–50 degrees) in helix A (IL-10, cmvIL-10, IL-19) and those that do not (IL-22 and IFN-γ). Interestingly, molecules with a bend in helix A also have a corresponding bend in helix C (C′) while IL-22 and IFN-γ, which have straight A helices, also have straight C helices. Interestingly, the separation of the IL-10s and IL-19 from IL-22 and IFN-γ based on their structures is consistent with their different chromosomal locations. For example, IL-10, IL-19, IL-20, and IL-24 are located in the gene locus 1q31-1q32 junction, while IFN-γ, IL-22, and IL-26 are located on 12q15 (Dumoutier *et al.*, 2000b; Eskdale *et al.*, 1997; Huang *et al.*, 2001).

Overall, F′/′C, F′/D, and F′/E angles differ from one another by 14°, 19°, and 44°, respectively. The large deviation of the F′/E angle in IFN-γ from other IL-10FMs is directly related to its involvement in the extensive IFN-γ dimer interface. If the IFN-γ F′/E angle is excluded from the analysis, the F′/E angles for other IL-10 family members differ from one another by only 10°. As a result, the F′/A helix angle shows the greatest variation (34°) among IL-10FMs, with helices from IL-22 and IL-10 exhibiting the smallest (140°) and largest (174°) values, respectively.

3. *Superpositions*

Structural superposition of IL10FMs domains were performed by using Dali (Holm and Sander, 1995, 1999) and refined by the method of Rao and Rossman as implemented in SUPPOS (Fig. 5). Using IL-10 as a reference, the viral IL-10s are most similar to IL-10 followed by IL-19, IL-22, and IFN-γ, respectively (Table IV). The comparisons reveal a common helical core centered on the bend in helix F (Fig. 5). Residues with

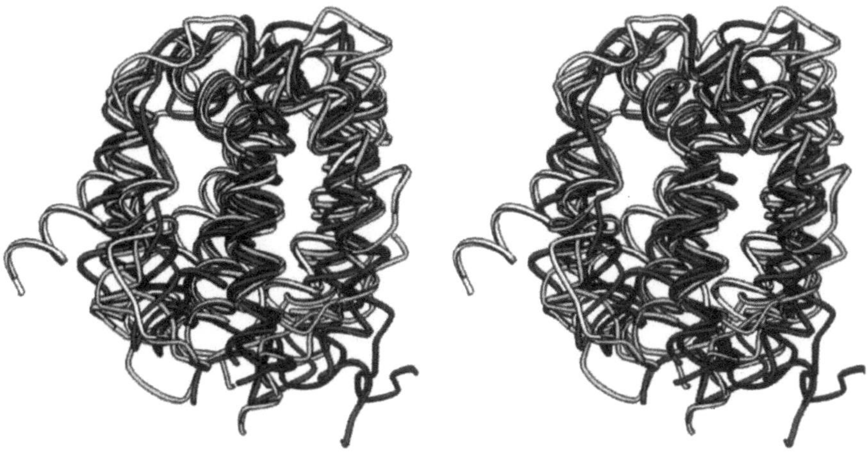

FIG. 5. Superposition of the IL-10FM domains. The domains are oriented as shown in Fig. 2A. IFN-γ is colored yellow, IL-10 is blue, cmvIL-10 is magenta, IL-19 is cyan, and IL-22 is green. (See Color Insert.)

TABLE IV
Alpha Carbon Atom Superpositions of IL-10FM Domains

	cmvIL-10	IFN-γ	IL-19	IL-22
IL-10	1.27, 113 23%	2.6, 101 15%	1.07, 92 19%	1.40, 90 19%
IL-22	1.40, 87 13%	2.8, 104 15%	1.34, 102 14%	—
IL-19	1.37, 104 19%	2.56, 100 13%	—	—
IFN-γ	2.26, 94 11%	—	—	—

Entries show rms deviation in Å, number of residue pairs and the % sequence identity of the structurally equivalent residue pairs.

the largest deviations between the structures are located on either end of the bundles and correspond to the N-terminus, the BC loop, and helix D. The data are consistent with the helical angle data that cluster IL-10, ebvIL-10, cmvIL-10, and IL-19 into a distinct group from IL-22 and IFN-γ. However, the distinctions between the groups are much less obvious by this general comparative method. As observed for all comparative studies,

IFN-γ exhibits the lowest structural similarity with the other IL-10FMs largely because of the very different orientations of helices C and E that contribute to its unique and extensive dimer interface.

4. *Disulfide Bonds*

IFN-γ is the only IL-10FM that contains no cysteine residues in its amino acid sequence. Other IL-10FMs contain 2–6 cysteine residues positioned at five different locations on the IL-10 scaffold corresponding to the N-terminus, the N-terminal end of helix C, the C-terminal end of helix C, the DE loop, and the C-terminus (Fig. 6). All disulfide bonds are located on either end of the helix bundle in regions that exhibit considerable structural diversity among the different family members. The lack of extensive structural constraints at the ends of the helices appears to enable the IL-10FMs to adopt different disulfide pairings. Later the observed and predicted pairings for IL-10FMs are discussed.

IL-10 forms two disulfide bonds. The first occurs between Cys-12 and Cys-108 and links the N-terminus to the DE loop while the second between Cys-62 and Cys-114 links the N-terminal end of helix C with the DE loop. These two disulfide bonds are conserved in the structures of ebvIL-10, cmvIL-10, and IL-19, and based on conserved cysteine sequence positions are predicted to be present in the structures of all viral and cellular IL-10s, IL-20, and IL-26. In addition to the two disulfide bonds described previously, IL-19 and IL-20 form a third disulfide bond that creates a second link between helix C and the DE loop. IL-22 conserves only the first linkage (Cys-12/Cys-108) between the N-terminus and the DE loop, while Cys-62 now pairs with a cysteine located on the C-terminal end of helix F rather than the DE loop. Yet another type of disulfide linkage is predicted for the structure of IL-24. IL-24 contains only two cysteine residues corresponding to IL-10 residues Cys-12 and Cys-62. Assuming the N-terminus of IL-24 is similar to IL-19, the location of these cysteines would be near enough to one another to form a disulfide bond when IL-24 is secreted from the cell.

Most of the IL-10s, as well as IL-19, IL-20, IL-22, and IL-24, have an even number of cysteine residues that all participate in disulfide bonds. However, human, simian, and baboon cmvIL-10 as well as IL-26 contain one additional unpaired cysteine residue (Fig. 6). The crystal structure of human cmvIL-10 revealed the unpaired cysteine forms an inter-chain disulfide bond resulting in a covalently linked dimer structure that differs from the non covalent dimers of ebv and cellular IL-10 and IFN-γ (Jones *et al.*, 2002b). The unpaired cysteine is located in the BC loop of cmvIL-10. However, the position of the unpaired cysteine in simian and baboon

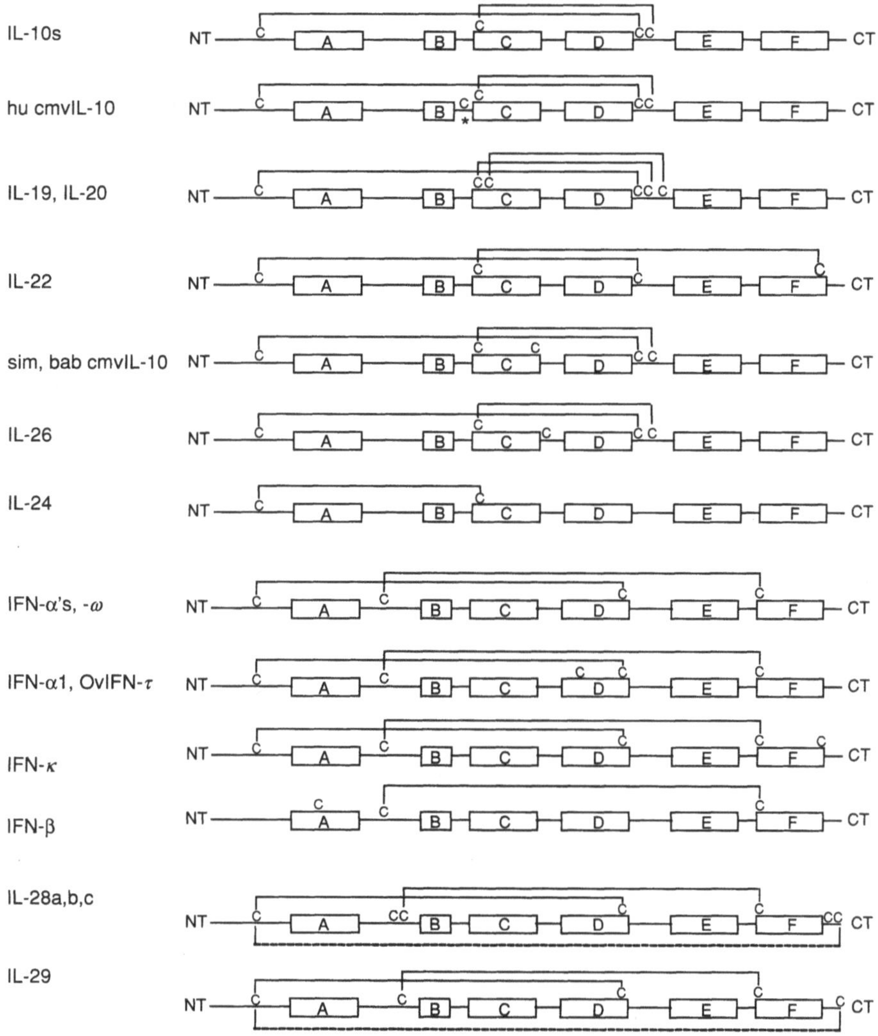

FIG. 6. Cysteine positions in the HCIIs. The cysteine positions of HCIIs are placed on a schematic diagram of HCII secondary structure. Helices are shown as rectangles and loops as lines. Experimentally determined and predicted disulfide bond linkages are shown.

cmvIL-10s and IL-26 is located at the C-terminal end of helix C rather than the BC loop. Based on the IL-10 and IFN-γ structural scaffolds this cysteine is not predicted to form a cysteine linked dimer.

5. Intercalated Dimers

Cellular and viral IL-10s and IFN-γ form intercalated dimers (Fig. 2). The intercalated dimer is formed from the first four secondary structural elements from one chain (A-D) and the final two (E and F) from the other. The intertwined dimer fold observed for IFN-γ, IL-10, and many other protein structures, is proposed to be an evolutionary mechanism of protein oligomerization referred to as 3D domain swapping (Bennett *et al.*, 1995). This theory suggests IFN-γ and IL-10 evolved from a monomeric protein by exchanging structural domains (α-helices E and F) with another monomer to create the dimer. This hypothesis is strengthened by the discovery of the monomeric IL-10FMs, IL-19, IL-20, and IL-22. In fact, the close relationship between the intertwined dimeric IFN-γ and monomeric type I IFNs was clearly identified when the first structure of IFN-γ was completed in 1991 (Ealick *et al.*, 1991).

Although both IFN-γ and IL-10 swap the same secondary structural elements (helices E and F) to form the dimers, their quaternary structures are very different. IFN-γ and IL-10 adopt inter-domain angles of approximately 60° and 90°, respectively. More recently, the structure of cmvIL-10 revealed a domain angle of approximately 150° (Jones *et al.*, 2002b). In addition, the two cmvIL-10 peptide chains form an interchain disulfide bond while IFN-γ and cellular IL-10 are non-covalent dimers. The domain orientations of each dimer are essentially fixed at one unique inter domain angle which alters the orientation of the cell surface receptors and may ultimately modulate cellular signal transduction events.

The orientation of the domains greatly influences the nature of the interdomain interface and the accessibility of the helices for possible interactions with their cell surface receptors. For example, the 60° interdomain angle of IFN-γ results in an extensive dimer interface (1742Å^2 of buried surface area) comprised of numerous hydrophobic residues on helices C and E. In contrast, the IL-10s exhibit larger interdomain angles and bury smaller amounts of surface area into the interfaces (700–1191Å^2). There are two reasons for the large variation in the amount of surface area buried in the dimer interfaces of different IL-10 structures. First, the conformations of the loops in the interface (BC, CD, DE) are different in different IL-10 structures. Second, the model of IL-10 refined at 1.6Å resolution (2ilk) has 4 additional N-terminal residues that contribute buried surface area to the interface. Despite these differences, all α-helices in each domain of cellular IL-10 or cmvIL-10 are solvent exposed and accessible for interaction with receptor molecules. From analysis of IL-10 and IFN-γ crystal structures determined in multiple crystal forms, the domain angles for the dimers are essentially fixed.

6. *The Role of the Intercalated Dimers in Signal Transduction*

What is the functional role of dimeric IFN-γ and IL-10 To address this question, several "functionally monomeric IFN-γs and a monomeric IL-10 (IL-10M1) were designed and their biological activities assayed (Josephson *et al.*, 2000a; Krause *et al.*, 2000; Landar *et al.*, 2000; Randal and Kossiakoff, 1998, 2001). Because of the extensive hydrophobic dimer interface in IFN-γ, a true monomeric IFN-γ was not created. Rather, the two chains of IFN-γ were coupled together by using various linker peptides to form a "linked" or "single chain" IFN-γ (IFN-γSC) (Lunn *et al.*, 1992a). The IFN-γSC mutant allows asymmetric mutagenesis to destroy one IFN-γR1 binding site while the other remains intact creating a "functionally monomeric" IFN-γ. IFN-γSC constructs are functionally monomeric since they bind only one IFN-γR1 chain despite maintaining the intertwined dimer structure.

Studies by Randal and Kossiakoff and Landar *et al.* used a structure-based approach to design short peptide linkers that connected the C-terminus of one IFN-γ chain to the N-terminus of the other. The resulting IFN-γSC mutants contained short linkers of 7 or 8 amino acids that replaced one of the 18 residue flexible tails found in wild-type IFN-γ. The structural integrity of each IFN-γSC construct was confirmed by crystal structure determination (Landar *et al.*, 2000; Randal and Kossiakoff, 2001). In a third study by Krause *et al.*, the original construct of Lunn *et al.* was used where two full length copies of IFN-γ (residues 1–137) were linked by an 18 amino acid linker from IgA1 (Krause *et al.*, 2000; Lunn *et al.*, 1992a). The structural integrity of this IFN-γSC variant was confirmed to be wild-type by using 1D NMR (Lunn *et al.*, 1992a). In addition to differences in the peptide linkers, the IFN-γSC designed by Krause *et al.* (2000) and Randal and Kossiakoff (2001) used IFN-γ1–137 while Landar *et al.* (2000) used the IFN-γD′ (residues 1–141) construct first reported by Ealick *et al.* (1991).

Overall, the functional results of each study are in good agreement. The anti-viral activity of each IFN-γSC construct is 0.1–3% of wild-type IFN-γ. Variations in activity may be due to true functional differences between the constructs as well as the use of different cell types and virus in the assays. Nonetheless, the data clearly show a single IFN-γ binding site is sufficient to induce IFN-γ biological activity. Landar *et al.* and Krause *et al.* also looked at MHC class I regulation by the mutants (Krause *et al.*, 2000; Landar *et al.*, 2000). As seen in the anti-viral assays, the activity of the IFN-γSC molecules was 1–3% of the wild-type dimers. Additional studies by Krause *et al.* showed IFN-γSC could up-regulate STAT1a levels (measured at a 15 minute time point) to about half the levels obtained with the IFN-γ dimer. These data suggest that signal transduction is not linear but complex, consisting of several yet-to-be-fully elucidated pathways. In this regard, it is interesting that Landar *et al.* reported IFN-γSC induced proliferation of TF-1 cells at 40% the activity of wild-type IFN-γ compared with the ~1% activity level of IFN-γSC in MHC class I up-regulation and induction of the antiviral state. The reasons for this difference remain to be determined.

The smaller domain interface of IL-10 enabled the design of a monomeric IL-10 (IL-10M1) by increasing the length of the DE loop by six amino acids (Josephson *et al.*, 2000b). Helices D and E are connected by a twelve residue linker of which the final four residues (Glu-115, Asn-116, K-117, and Ser-118) extend across the dimer interface. Molecular modeling studies show the final four residues of the linker are too short to allow helices E and F to insert into the hydrophobic cleft formed by helices A through D (see Fig. 2B). However, increasing the length of the linker by inserting 6 amino acids between linker residues Asn-116 and Lys-117 resulted in the formation of a monomeric IL-10. The structure of IL-10M1 was initially confirmed by CD and later by crystal structure analysis (Josephson *et al.*, 2002a). In proliferation assays with a murine cell line transfected with either human or murine IL-10R1, IL-10M1 exhibited between 6 and 10% of dimeric IL-10 activity.

Monomeric forms of IFN-γ and IL-10 are able to activate their respective signal transduction machinery and induce the expected biological activity. Thus there is not a fundamental difference in receptor activation by monomeric and dimeric HCIIs. Furthermore, the dimeric receptor complexes do not appear to recruit different molecules into the complex than the monomeric mutants since the monomers are able to activate the same biological activities (proliferation, MHC up-regulation, and antiviral activity) as the dimers. The results of these studies suggest the function of the dimers is to optimize the stability of the receptor complexes to allow maximal activation at specific receptor levels found on normal cells. In this regard it is interesting that cell surface levels of the low-affinity IFN-γR2 chain level are regulated on T-cells (Bach *et al.*, 1995; Pernis *et al.*, 1995) while the high-affinity IL-10R1 chain is regulated in human neutrophils (Crepaldi *et al.*, 2001; Ding *et al.*, 2001). By modulating receptor levels, the dimers may allow signaling to occur at receptor levels that are lower than required for signaling for a monomeric counterpart. This may provide a mechanism to enable two cytokines who share common receptors to selectively activate different cell types. It appears that cellular and ebvIL-10 use this mechanism to produce drastically different functional effects (Liu *et al.*, 1997).

C. *The Type I IFNs*

Knowledge of type I IFN fold is based on structures of two human IFN-α subtypes (IFN-α2b, IFN-α2c), murine and human IFN-β, and Ovine IFN-τ (Karpusas *et al.*, 1997; Klaus *et al.*, 1997; Radhakrishnan *et al.*, 1996; Radhakrishnan *et al.*, 1999). The type I IFNs are essentially the same size (45Å × 35Å × 25Å) and contain the same six secondary structural elements connected in the same manner as the IL-10FMs. Although they are similar in overall structure to the IL-10s, there are a number of subtle differences between the IL-10 and type I IFN families that give rise to their different biological and biochemical properties.

1. *Helix F and Conserved Fingerprint Residues*

The type I IFNs contains a much smaller bend in helix F (12°–23° versus 47°–57°) than observed in the IL-10FMs. However, they also conserve a cluster of amino acids centered at the bend in helix F and properly position the AB loop. Compared with the IL-10 family, the conserved cluster is larger consisting of 14 residues that are almost entirely hydrophobic (Table V, Fig. 4). Despite this difference, the most conserved linear amino acid sequence in the type I IFNs (SPCAWEVVRAEI, residues 136–147) corresponds to the N-terminal end of helix F, where it intersects with the AB loop as observed for the IL-10FMs. Although the sidechain chemistry of the fingerprint residues differ, their locations are generally conserved. For example, the salt bridge between Asp-41 in the AB loop and Lys-138 on helix F in the IL-10FMs is replaced with a disulfide bond between Cys-29 and Cys-138 in the type I IFN structures. Glu-142 in IL-10 is conserved in the type I IFNs (Glu-146 in IFN-α) at the center of the bend in helix F. However, the Oϵ1 atom of Glu-146 makes a hydrogen bond with

TABLE V
Type I IFN Fingerprint Residues

	AB loop				Helix C		Helix D			Helix F				
	C29	F36	F38	P39	L66	F67	Y122	F123	I126	C138	W140	V143	E146	I147
IFN-β	—	—	—	—	I	—	—	Y	—	—	—	—	—	—
IFN-τ	—	—	L	—	—	—	—	—	—	—	—	—	—	—
IFN-ω	—	—	—	—	—	—	—	—	—	—	—	—	—	—
IFN-κ	—	—	L	—	I	—	—	—	—	—	—	I	—	—
IL-28	—	—	R	T	—	E	W	L	L	—	E	—	N	L
IL-29	—	—	G	N	—	E	W	L	L	—	E	—	N	L

the OH group of Tyr-122 located on helix E rather than the mainchain nitrogens of Leu-47 in the AB loop as osbserved in the IL-10FMs. Phe-36 and Phe-38 in the type I IFNs corresponds to hydrophobic residues Leu-47 and Leu-48 in the IL-10FM fingerprint, which attaches the C-terminal end of the AB loop to the core of the molecule. Finally, several hydrophobic residues on helix C (Leu-66 and Phe-67) and helix F (C138, W140, V143, and I147) complete the conserved core. It should be noted there are other conserved residues in the type I IFN sequences, but these 14 cluster around helix F in a manner analogous to the IL-10 fingerprint described previously.

In addition to helix F, helices D and E also contain small bends in IFN-α2b and IFN-β. In each molecule, the bend in helix E is centered on Tyr-122, which hydrogen bonds to Glu-146 on helix F. Thus the bend in helix E occurs at the same place as helix F in the center of the conserved structural core. The bend in helix D occurs at the position of the buried polar residue Gln-91 in IFN-α2b which forms a hydrogen bond with Ser-14 on helix A. A similar interaction is observed in IFN-β between the structurally identical Gln-94 and Gln-10 on helix A. It is notable that Gln-91 is conserved in all type I IFNs.

2. Superpositions

Almost all of the type I IFNs may be superimposed with rms deviations of approximately 1 Å for ~100 Ca atom pairs (Table VI). Structurally conserved residues form the central portion of each cylindrically shaped molecule and include secondary structural elements B and E. Each molecule exhibits large deviations at the ends of the helix bundle corresponding to the AB, CD, and DE loops as well as the N and C terminus. In general, the cores of the type I IFNs are structurally similar

TABLE VI
Alpha Carbon Atom Superpositions of the Type I IFNs

	IFN-β	mIFN-β	OvIFN-τ
IFN-α2	1.15, 106 36%	1.07, 98 23%	1.17, 109 50%
IFN-β	—	0.40, 112 48%	1.04, 97 33%
mIFN-β	—	—	1.04, 104 28%

Entries show r.m.s. deviation in Å, number of residue pairs and the % sequence identity of the structurally equivalent residue pairs.

to one another than IL-10FMs which is consistent with their higher sequence identity.

3. Differences Between IFN-α and IFN-β

Superposition of IFN-α2 and IFN-β highlight several differences between the subtypes as well as general features of the type I IFN fold (Fig. 7). First, helix A is seven residues longer in IFN-β than IFN-α. Helix A in IFN-β contains a small bend that occurs approximately where helix A ends in IFN-α2b. This allows the longer helix A of IFN-β to pack efficiently against the helical bundle. In contrast to the helical N-terminus of IFN-β, the N-terminus of IFN-α forms a flexible loop that is connected to the DE loop by disulfide bond.

The most flexible regions of the type I IFNs are located on the same end of the helix bundle near the C-terminus of the molecule. They include the N-terminus (1–8), the BC (residues 44–50) and DE loops (100–112), and

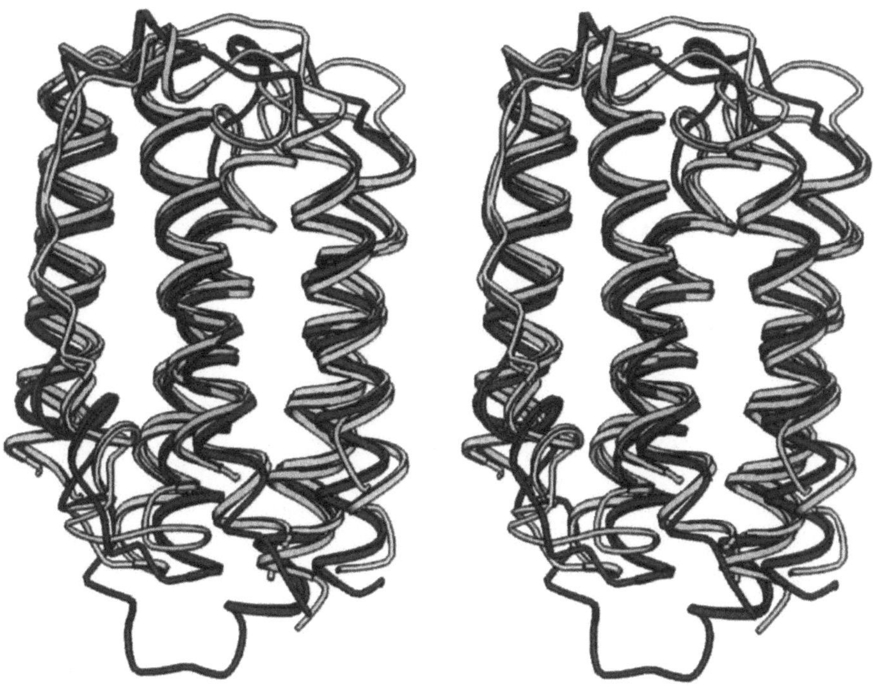

FIG. 7. Stereoview comparison of the type I IFNs. IFN-α2b (green), human IFN-β (magenta), and ovine IFN-τ (cyan) are shown. The view is looking into the IFNAR2 binding site. (See Color Insert.)

the C-terminus (160–165). Even helices D and E themselves appear to be quite mobile. For example, the N-terminal end of helix E in IFN-β differs by six amino acids in length between the non-crystallographically related molecules (Karpusas *et al.*, 1997). Similarly, helix D in IFN-α differed in length by 11 amino acids between the six non-crystallographically related molecules (Radhakrishnan *et al.*, 1996). One possible reason for this observation is each helix contains buried polar residues that allow the helices to adopt multiple conformations.

In contrast to the flexible C-terminal region, the highly conserved fingerprint residues are located at the N-terminus of helix F and include a disulfide bond that locks the conformation of AB loop and helix F. Residues on this end of the helix bundle are more flexible than the central core but adopt well-defined structures in the crystal and in solution. Thus, structural differences on this end of the bundle are thought to influence the different receptor binding and biological properties of IFN-α and IFN-β.

In particular, the position of the 3_{10} helix in the AB loop is shifted four residues toward structural element B in IFN-β relative to IFN-α2. This difference certainly influences IFNAR2 binding (Piehler and Schreiber, 1999). A second major difference between IFN-α2 and IFN-β occurs in the CD loop, which is located directly behind the AB loop (Figs. 2A and 7). The CD loop in IFN-α2 contains a short α-helix (helix C′), which is bent at an angle of $\sim 50^\circ$ relative to helix C, while the loop adopts a more extended conformation in IFN-β. This structural difference may be important for IFNARI binding (Radhakrishnan *et al.*, 1996). Interestingly, the largest structural changes in ovine IFN-τ also occur in the AB and CD loop, suggesting these two regions play a major role in modulating the activity of the different IFN subtypes (Radhakrishnan *et al.*, 1999).

4. Disulfide Bonds

The type I IFNs are distinguished from the IL-10FMs by a conserved disulfide bond (Cys-29:Cys-138) that links the AB loop to the N-terminus of helix F. IFN-β forms just this one disulfide bond, while IFN-α subtypes and ovine IFN-τ contain a second disulfide, analogous to the IL-10FMs, which links the N-terminus and the DE loop. IFN-β and IFN-α1 each contain one additional unpaired cysteine residue located in the center of helix A, and D, respectively, which are both partially buried and do not participate in disulfide bond formation. IFN-κ also contains one unpaired cysteine located at the C-terminal end of helix F.

In contrast to the three cysteines in IFN-β and 4 in IFN-α, the IL-28's and IL-29 contain 7 and 5 cysteine residues, respectively (Fig. 6). Four of the residues found in both molecules are predicted to form disulfide bonds similar to the IFN-αs. Of the three additional cysteines in IL-28, one is found in the AB loop adjacent to Cys 29 (IFN-α2b numbering), and the other two are located at the C-terminus of the molecule near the position of the unpaired cysteine residue in IFN-κ. The most C-terminal cysteine in IL-28 corresponds to the fifth cysteine residue in IL-29. The structural role of these cysteines has not been defined.

III. Structures of the Class 2 Homology Region (C2HR)

A. *Overview of the C2HR Module*

The structures of three cytokine receptor C2HRs and tissue factor (TF) have been determined (see Table I and Chill *et al.*, 2003; Harlos *et al.*, 1994; Josephson *et al.*, 2001; Randal and Kossiakoff, 2001; Thiel *et al.*, 2000; Walter *et al.*, 1995). The general features of the C2HR are shown in Fig. 8. It consists of two β-sandwich domains, D1 and D2, connected by a short linker containing 1 turn of α- or 3_{10} helix. The cytokine binding site is comprised predominantly of the loops located at the D1 and D2 interface. The N-terminal D1 domain is most distal from the cell membrane, while the C-terminal D2 domain is followed by a short tether of 5–12 amino acids in length before the beginning of the transmembrane helix.

B. *D1 and D2 Domain Comparisons*

D1 and D2 are comprised of seven β-strands, each of which forms a three (β-strands A, B, and E) and four-strand (β-strands C, C′, F, and G) anti-parallel β-sheet that pack against one another in a face-to-face arrangement. Each domain may be considered a "stripped down" immunoglobulin (Ig) domain. The topology of the domains are closely related to the fibronectin type III (FBN-III) fold that has been described as a c2 (Chothia and Jones, 1997) or s-type set (Bork *et al.*, 1994) by different groups. The β-strands are connected to one another by structurally diverse loops that are defined by the β-strands they connect followed by a superscript denoting D1 or D2. Thus, the loop connecting β-strands A and B in D1 is designated the AB^{D1} loop.

Sequence comparisons have identified 15 residues that are highly conserved in the domains of the C2HRs. A subset of these residues were used originally to define the HCRII family (Bazan, 1990b). The location of the

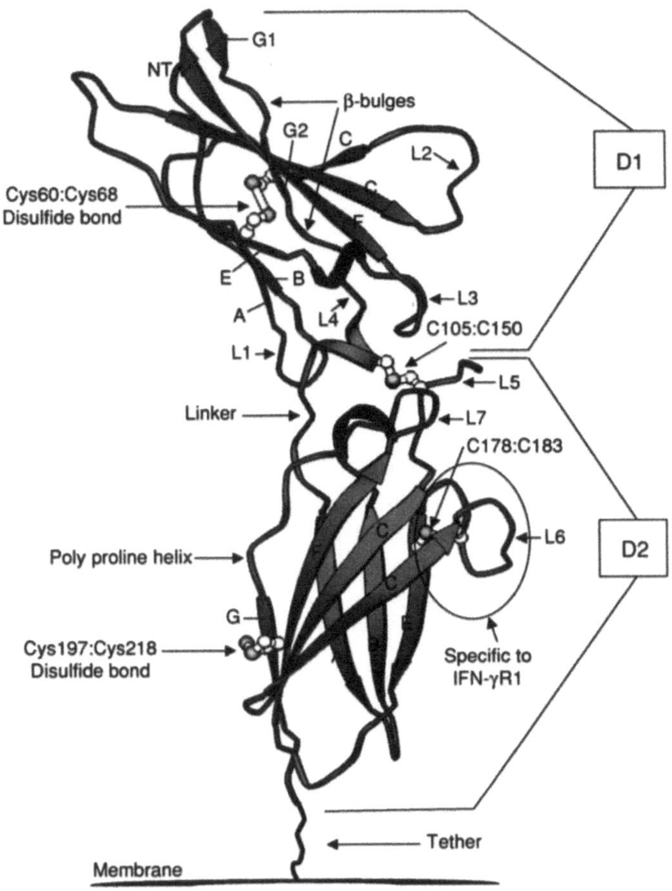

FIG. 8. Ribbon diagram of the IFN-γR1. Key features in common to all C2HRs or regions that differ as described in the text are shown highlighted.

residues is shown on the structure of IFN-γR1 in Fig. 9. Nine are located in D1 (IFN-γR1 residues Pro-15, Asn-24, Val-29, Trp-31, Cys-66, Cys-68, Arg-85, Val-85, and Ala-87) and six (Pro-112, Pro-113, Tyr-155, Cys-197, Ser-214, and Cys-218) are found in D2. Most of the residues cluster at the N-terminal ends of the domains near the start of β-strand A^{D1} and A^{D2}. Besides the cysteine residues, which form one disulfide bond in each domain, Asn-24 is the only residue that is not located in one of the two clusters. Asn-24 is located on AB^{D1} loop, where it hydrogen bonds to the domain linker.

The D1 and D2 domains consist of approximately 100 residues each. D1/D1 and D2/D2 comparisons reveal ~70 Cα atom pairs may be superimposed

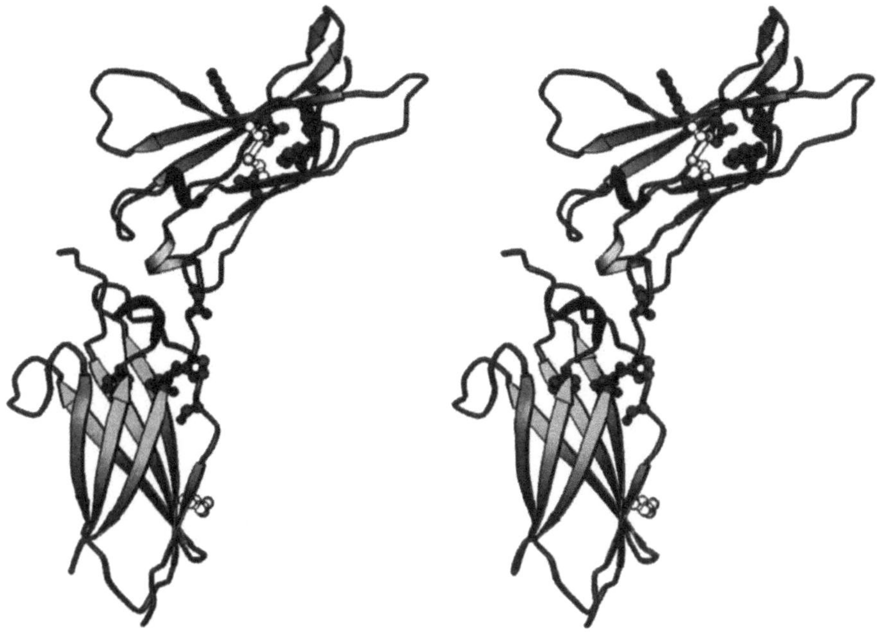

FIG. 9. Stereoview of the location of conserved residues in the C2HRs. Conserved residues are shown on the C2HR of IFN-γR1. The sidechains of the 15 residues conserved in at least 10 of the 12 C2HR sequences compared are shown in ball and stick representation. Conserved cysteines are white and other residues are grey.

with root mean square (r.m.s) deviations of 0.9–1.8 Å (Tables VII and VIII). Residues excluded from the comparisons are mostly located on the loop regions. The greatest deviations occur for the BC^{D1} and EF^{D1} (L3) loops in D1 and the BC^{D2} (L5), $C'E^{D2}$ (L6), and FG^{D2} (L7) loops in D2 (Fig. 10). The distinct loop conformations of the domains in different C2HRs along with sequence variation and domain orientation play key roles in forming specific high-affinity interactions with the cytokines. A surprising finding of the comparisons is the high structural similarity observed between TF and the cytokine receptor domains. The domain structures of TF are most similar to IFN-γR1 and IFNAR2. In IL-10R1, D1 is most similar to D1 of IFN-γR1, but its D2 is once again most similar to D2 of TF. The excellent structural similarity between the D1 domains of IL-10R1 and IFN-γR1 reflects, in part, the conserved nature of the cytokine binding surface (site Ia corresponding to the bend in helix F) that D1 interacts with.

Another surprise from the domain comparisons first noted by Walter *et al.* was the high degree of structural similarity between D2 of the growth

TABLE VII
Comparison of C2HR D1s

	IL-10R1 (D1)	IFN-γR1 (D1)	IFNAR2 (D1)	TF (D1)	GHR (D1)	GHR (D2)
IL-10R1 (D1)	—	0.9/75	1.8/79	1.2/73	1.8/65	1.1/67
IFN-γR1 (D1)	1.6/95 19%	—	1.7/75	0.7/71	1.5/64	1.0 /68
IFNAR2 (D1)	2.5/94 14%	2.6/93 16%	—	1.4/69	1.8/69	1.8/77
TF (D1)	2.2/92 21%	1.6/88 24%	2.2/85 28%	—	1.7/66	1.2/65
GHR (D1)	2.6/80 13%	2.1/80 10%	2.7/84 13%	2.3/78 10%	—	
GHR (D2)	2.1/90 12%	1.9/89 13%	2.3/84 13%	2.2/85 20%	2.2/75 13%	—

Entries correspond to r.m.s. deviation in Å followed by the number of Ca atom pairs. The percent sequence identity of residues compared is also listed.

TABLE VIII
Comparison of C2HR D2s

	IL-10R1 (D2)	IFN-γR1 (D2)	IFNAR2 (D2)	TF (D2)	GHR (D2)
IL-10R1 (D2)	—	1.2/66	1.8/64	1.4/77	1.3/73
IFN-γR1 (D2)	2.4/87 14%	—	1.5/68	1.0/60	1.5/73
IFNAR2 (D2)	3.5/84 13%	2.6 /87 18%	—	1.8/64	2.0/71
TF (D2)	2.5/96 17%	2.4/81 17%	3.1/80 23%	—	1.2/70
GHR (D2)	2.2/91 18%	2.3/88 13%	3.2/85 14%	2.5/89 16%	—

Entries correspond to r.m.s. deviation in Å followed by the number of Ca atom pairs. The percent sequence identity of residues compared is also listed.

hormone receptor (GHR) with D1 from IFN-γR1 (Walter *et al.*, 1995). Extending this analysis to include the more recent structures of IL-10R1 and IFNAR2 shows the D1 and D2 domains of IFN-γR1, IL-10R1, and TF are very similar to GHR D2 (Tables VII and VIII). However, r.m.s deviations calculated between GHR D2 and the domains IFNAR2 are significantly higher. These comparisons certainly provide some initial clues about the origins of the C2HRs.

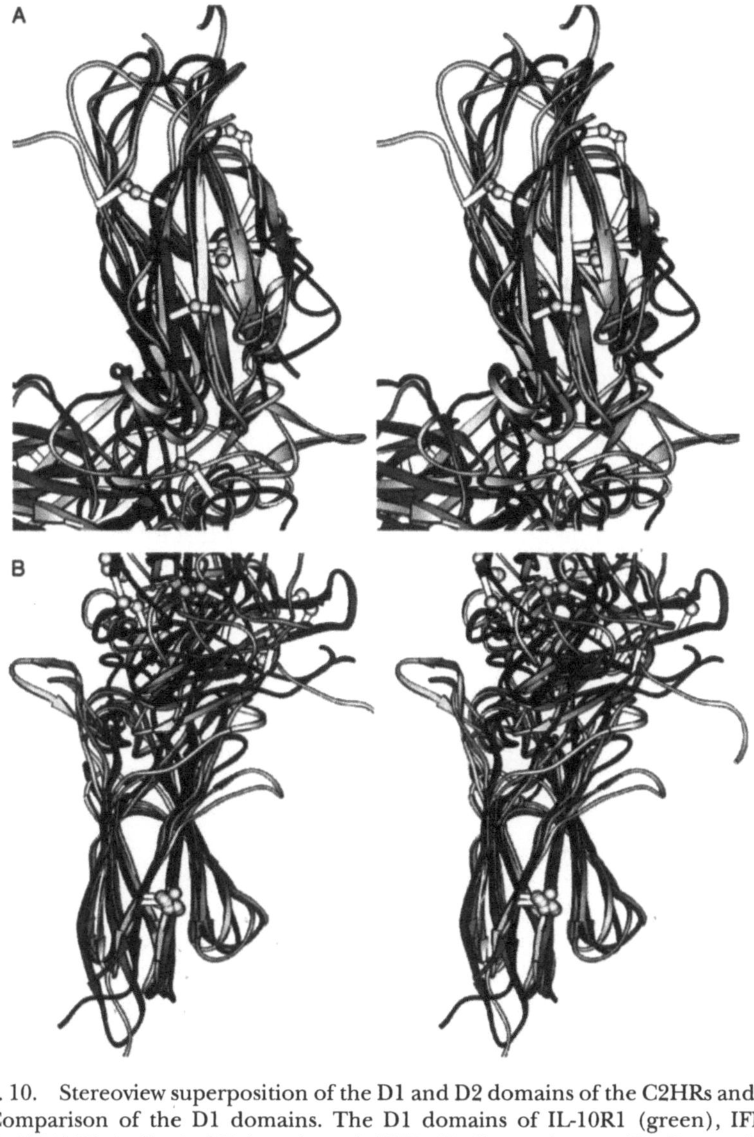

FIG. 10. Stereoview superposition of the D1 and D2 domains of the C2HRs and GHR. (A) Comparison of the D1 domains. The D1 domains of IL-10R1 (green), IFN-γR1 (blue), IFNAR2 (yellow), TF (cyan), and GHR (Red) are shown. (B) Comparison of D2 domains. Coloring is as described for the D1 domains.

C. Disulfide Bonds

Almost all C2HRs contain four cysteine residues that form two conserved disulfide bonds. The first occurs in D1 between Cys-60 and Cys-68 (IFN-γR1 numbering) and links β-strands C'^{D1} and E^{D1} (Fig. 8). This disulfide bond is conserved in all D1 domains of the C2HRs except IL-10R1, where the Cys-54 on C'^{D1} forms a disulfide bond with Cys-35 located on β-strand C^{D1} rather than E^{D1}. The second conserved disulfide bond is located in D2 where Cys-197 and Cys-218 in IFN-γR1 links β-strands F^{D2} and G^{D2}. The one known exception to this disulfide bond occurs in IL-22R1. In IL-22R1, a cysteine on β-strand G^{D2} corresponding to Cys-218 in IFN-γR1 is conserved but the disulfide is formed with a cysteine residue located on β-strand A^{D2} rather than F^{D2} (Logsdon *et al.*, 2002).

The C2HR of IFN-γR1 contains a total of 8 cysteine residues. In addition to the two conserved disulfides described previously, Cys-105 and Cys-150 connect the α-helical linker and the C-terminal end of the BC^{D2} loop, while Cys-178 and Cys-183 link β-strand E^{D2} with the $C'E^{D2}$ loop (Fig. 8). Both of these disulfide bonds are unique to IFN-γR1. A total of six cysteine residues are found in the C2HRs of IFNAR2 and IL-28R1. The additional unique disulfide bond in IFNAR2 links β-strand A^{D1} and G^{D1} and may influence the secondary structure of β-strand G^{D1}. Amino acid sequence alignments position the two additional cysteine residues of IL-28R1 on the EF^{D1} and BC^{D2} loops. Simple modeling experiments suggest these cysteines will form a disulfide that links D1 and D2 in an analogous manner to the Cys-105:Cys-150 disulfide in IFN-γR1.

D. Structural Diversity in β-Strand G

Excluding the loops, the greatest structural differences in D1 occur in the edge β-strand G^{D1} that contains one to two β-bulges that separate β-strand G^{D1} into β-strand $G1^{D1}$ and $G2^{D1}$ (Fig. 11). β-strand G^{D1} of IFNAR2 contains a single one residue β-bulge (bulge 1) consisting of Leu-92, while IFN-γR1 and IL-10R1 contain a two-residue bulge 1. Residues that form β-bulge 1 in IFN-γR1 and IL-10R1 are Ala-$95^{IFN\text{-}\gamma R1}$/Tyr-$96^{IFN\text{-}\gamma R1}$ and Asn-$89^{IL\text{-}10R1}$/Trp-$90^{IL\text{-}10R1}$, respectively. The mainchain O atoms of Ala-$95^{IFN\text{-}\gamma R1}$ and Asn-$89^{IL\text{-}10R1}$ in the center of the bulge hydrogen bond with the Oγ atom of the preceding serine residues (Ser-94 in IFN-γR1 and Ser-88 in IL-10R1) while Tyr-$96^{IFN\text{-}\gamma R1}$ and Trp-$90^{IL\text{-}10R1}$ participate in an extended π-cation stacking motif first observed in D2 of the growth hormone and prolactin receptor C1HRs (de Vos *et al.*, 1992; Livnah *et al.*, 1996; Somers *et al.*, 1994). β-bulge 1 in D1 of IFN-γR1 and IL-10R1 is followed by two

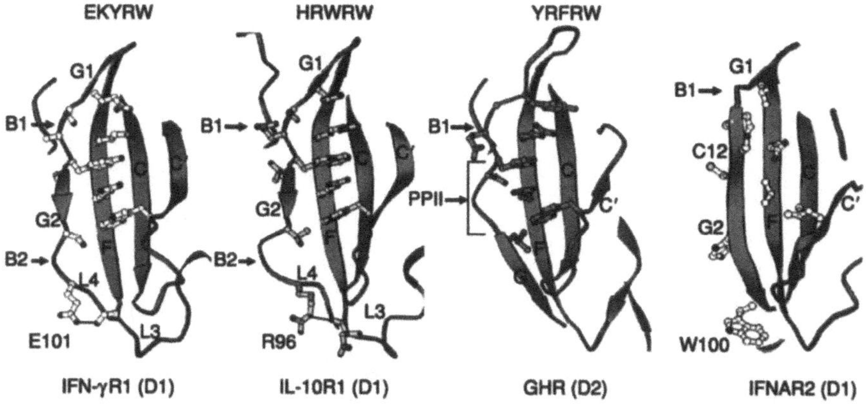

FIG. 11. Ribbon diagram of the GFCC' β-sheet face in D1. The π cation stacking motif and structural differences observed in the edge β-strand G^{D1} are shown. β-bulges (B1 or B2) and the left handed poly proline helix (PPII) in GHR D2 are denoted with arrows. Residues at the end of the β-strand GD1 that participated in cytokine contacts are labeled on the figure. IFNAR2 Cys-12 which forms a disulfide bond with cysteine on β-strand A^{D1} is also labeled.

residues of regular anti-parallel hydrogen bonded β-structure (β-strand $G2^{D1}$) before a second β-bulge (bulge-2) of two residues in IFN-γR1 and three residues in IL-10R1. The same hydrogen bonding scheme used in β-bulge 1 is used in β-bulge 2 of IFN-γR1 and IL-10R1 where the main-chain O atom hydrogen bonds to the proceding serine residue. However, the final residues in bulge 2 (Glu-$100^{IFN\gamma R1}$ and Arg-$96^{IL\text{-}10R1}$) do not participate in cation stacking motifs but form hydrogen bonds to the N-terminal end of β-strand F^{D1}. β-bulge 2 is especially important since it properly orients the sidechains of Glu-101 in IFN-γR1 and Arg-96 in IL-10R1 for contact with their respective ligands (Josephson *et al.*, 2001).

In contrast to IFN-γR1 and IL-10R1, IFNAR2 does not contain a second β-bulge. Rather, β-strand $G2^{D1}$ extends all the way to domain linker. Also unique to IFNAR2 is the presence of a cysteine residue located in the middle of the β-strand $G2^{D1}$, which forms a disulfide bond with β-strand A^{D1}. Along with the loss of the 2 residue β-bulge 1 described for IFN-γR1 and IL-10R1, D1 of IFNAR2 does not contain a π-cation stacking motif. Although the hydrogen bonding and conformation of β-strand $G2^{D1}$ is different in IFNAR2, Trp-100, which is important in IFN-α and IFN-β binding, is positioned at the end of β-strand $G2^{D1}$ near the position of Arg-96 and Asp-101 in IL-10R1 and IFN-γR1. Like IFNAR2, TF does not form bulge-1 nor the π-cation system. However, D1 from TF does form a bulge-2 structure essentially identical to IFN-γR1. It is interesting that D1 from GHR is much more like IFNAR2 where β-strand G^{D1} forms β-bulge-1 but β-bulge-2 is replaced by a disulfide bond between Cys-108 and Cys-122 that constrains β-strand G to adopt regular anti-parallel hydrogen bonded β-structure. The structural variations observed for β-strand G^{D1} are consistent with its importance in properly positioning residues for ligand recognition and high affinity binding.

1. The Poly-Proline Helix

As observed in D2 of GHR (Fig. 11), residues preceding β-strand G^{D2} in IL-10R1, IFN-γR1, IFNAR2, and TF all form 4- or 5-residue left-handed poly proline II (PPII) helices (Fig. 8; Adzhubei and Sternberg, 1993). This secondary structural element was first recognized in the tandem FNBIII structure of neuroglian (Huber *et al.*, 1994). The PPII helices in D2 of the C2HRs are comprised of very similar bulge motifs described for β-strand G^{D1}. Besides a conserved serine residue and common phi, psi angles (−75, 145), the amino acid sequences that form the PPII helices in each receptor are quite different (IFN-γR1, residues 212–216, EKSKE; IL-10R1, 197–200, WSKE; IFNAR2, 195–198, KSPL; TF, 204–207, ESPV, GHR, 225–228, FSET). The PPII helix is located between the start of β-strand G^{D2} on its C-terminal end and the conformationally diverse residues following L7 on its N-terminal end. Although the PPII helix is conserved in D2 of all the class 2 cytokine receptor structures determined at this time, it has no known functional role.

E. C2HR Linker and the Inter-Domain Angle

D1 and D2 are connected by a linker of 9–11 amino acids that form one turn of α or 3_{10} helix followed by residues in an extended conformation that make a variety of mainchain and sidechain interactions with β-strand A^{D1} and residues preceding the PPII helix in D2 (Fig. 8). To date, the helical linker between D1 and D2 distinguishes the C1HR and C2HRs from the repeating FBNIII domains found on other molecules such as neuroglian (Huber *et al.*, 1994) and fibronectin (Leahy *et al.*, 1992, 1996). The domain linker along with the AB^{D1}, EF^{D1}, BC^{D2}, and FG^{D2} loops form the domain interface that gives rise to distinct interdomain angles observed for each C1HR and C2HR.

To compare the orientations of the C2HRs as well as GHR in a quantitative manner, interdomain angles were calculated by using the procedure developed by Deivanayagam *et al.*, 2000. As shown in Fig. 12, the C2HRs

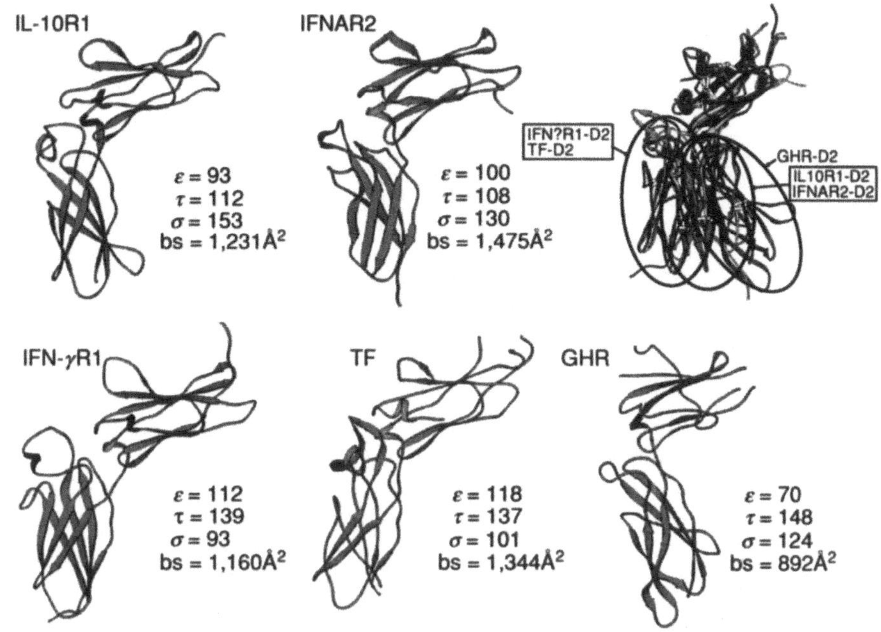

FIG. 12. C2HR domain orientations. Each C2HR is labeled with its characteristic tilt (ϵ), roll (τ), and swivel (σ), parameters (Deivanayagam *et al.*, 2000) along with the amount of amino acid surface area buried in its domain interface (bs). The C2HRs adopt three distinct D1/D2 orientations. Similar orientations are observed for IL-10R1 and IFNAR2 (group 1), IFN-γR1 and TF (group 2), and GHR (group 3). The difference between each group is most obvious if shown by superimposing the D1 domains of the receptors. (See Color Insert.)

have larger tilt angles (ϵ) of 90–118° as compared with GHR, the prototypical C1HR, with a e value of 70°. Based on the analysis of tilt (ϵ), roll (τ), and swivel (σ), the C2HRs of IL-10R1 and IFNAR2 can be separated into a separate subclass from IFN-γR1 and tissue factor (TF). In addition to the rotations, the domains also undergo translations yielding yet another mechanism to generate structure diversity in the ligand binding site for modulating ligand specificity and affinity. In addition to different domain tilt angles and translations, the C2HRs have more extensive domain interfaces that bury 1231Å^2–1474Å^2 of surface area, while the C1HR of GHR buries 892Å^2. Based on the size of the interfaces and comparisons of the IL-10R1 and IFN-γR1 C2HRs in several different crystal forms, their inter-domain angles are fixed in one orientation for ligand recognition.

IV. Structure of the High-Affinity HCII/C2HR Interfaces

A. *Site Ia and Ib Contact Regions*

Crystal structures have been determined for the dimeric HCIIs IFN-γ, IL-10, and cmvIL-10 bound to their respective high-affinity C2HRs. In each structure, two C2HRs bind the two-fold related surfaces of the dimers comprised of helix A, the AB loop, and helix F to form a 1:2 dimer:C2HR complex. The complementary receptor interactions are formed from the L2-L5 and L7 loops. The contact site can be divided into two distinct surfaces, site Ia and site Ib (Fig. 13). Site Ia is centered in the bend in helix F and includes the AB loop, while site Ib is located near the C-terminus of helix F and the N-terminus of helix A. Receptor binding loops L2-L4 contributed from D1 interact exclusively with site Ia while loops L5-L7 from D2 interact with Site Ib. As shown in Fig. 5, the site Ia contact site is the most structurally conserved region in the IL-10FMs, while site Ib is structurally diverse. Similarly, the D1 domains of the C2HRs, including their ligand binding loops (L2-L4), are more conserved than the D2 domains that present loops L5-L7 to the binding interface.

Although the crystal structure of the IFN-α2/IFNAR2 complex has not been reported, a model has been derived by docking the structures

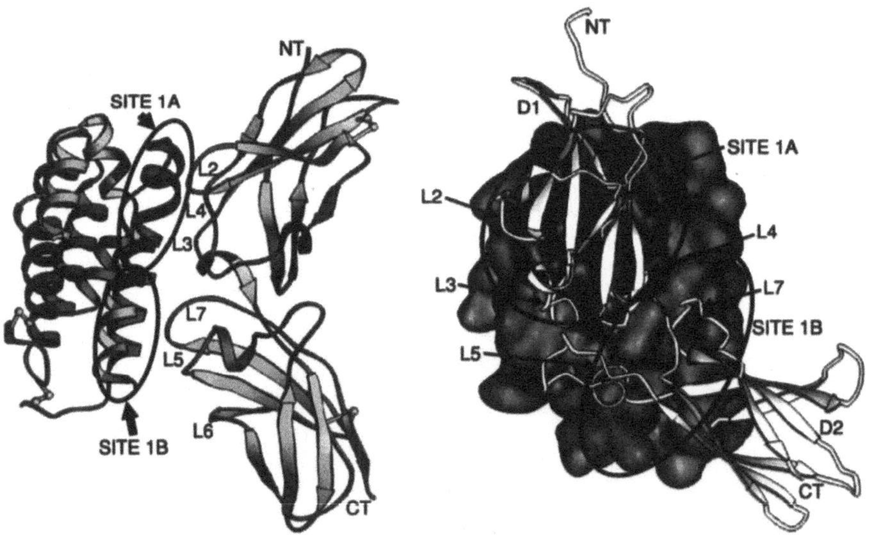

Fig. 13. The HCII/C2HR high-affinity interaction site. The IL-10/IL-10R1 complex structure is used as the prototype complex. (See Color Insert.)

of IFN-α2 and IFNAR2 by using constraints obtained by double mutant cycle analysis (Chill *et al.*, 2002, 2003; Roisman *et al.*, 2001). Although a higher resolution X-ray or NMR structure must still be determined, the model confirms IFN-α2 uses the same structural regions as the IL-10FMs, comprised of helix A, the AB-loop, and helix F to contact IFNAR2. However, in contrast to IL-10/IL-10R1 and IFN-γ/IFN-γR1 complexes, only IFNAR2 binding loops L2–L4 participate in IFN-α2 binding. Thus, site Ia contacts contributes close to 100% of the buried surface to the IFN-α2/IFNAR2 interface compared to ~60% for the IL-10, IFN-γ and the GH/GHR complexes.

The characteristics of each HCII/C2HR interface are summarized in Table IX. The interfaces are comprised of 17–25 residues donated from the HCII and an additional 19–23 residues from the C2HR. Residues that bury surface area into the IL-10/IL-10R1, IFN-γ/IFN-γR1, and IFN-α2/IFNAR2 interfaces are shown in Figs. 14–16. The change in accessible surface area on complex formation is between 800Å^2–969Å^2 for the cytokines and 800Å^2–1131Å^2 for the C2HRs. These interaction surfaces are slightly smaller than the GH/GHR complex, which consists of 29 GH and 33 GHR residues that each bury ~1300Å^2 of surface area. Residues in each interface are predominantly polar. The greatest number of polar and charged residues are found in the IL-10/IL-10R1 and cmvIL-10/IL-10R1 complexes (75%), while IFN-α and IFN-γ have slightly lower number of polar residues (~60%).

B. *Characteristics of the HCII/C2HR Interfaces*

Each interface contains 11 or 12 hydrogen bonds or salt bridge interactions. Although the GH/GHR complex buries more surface area into its interface, it still contains only 12 specific interactions. Three of the IFN-γ/IFN-γR1 interactions occur between mainchain atoms in the AB loop of IFN-γ and the L2 loop of IFN-γR1. In addition to the 12 receptor-cytokine interactions, seven water molecules bridge the donor and acceptor groups of IFN-γ and IFN-γR1 via hydrogen bonds (Randal and Kossiakoff, 2001). Since the B-factors for the water and protein atoms in the complex are essentially the same, the water molecules are considered to be an integral part of the interface. The presence of ordered water molecules in the IFN-γ/IFN-γR1 interface is not likely unique, since the IL-10/IL-10R1 interface is even more polar. However, the poorer diffraction quality of IL-10 complex crystals (~2.9Å versus 2.0Å) obtained at this time prevents a detailed description of the interfacial waters. The water molecules in the

TABLE IX
Statistics of Several Cytokine Receptor Interfaces

Cytokine	#Res	%Polar	B.S.(Å^2)	Receptor	#Res	%Polar	B.S.(Å^2)	#MC	#MC/SC	T
IL-10	24	75	969	IL-10R1	23	74	1132	1	6	11
cmvIL-10	25	76	945	IL-10R1	21	71	1007	1	6	12
IFN-γ	23	61	863	IFN-γR1	21	76	866	3	3	12
IFN-α	17	53	~800	IFNAR2	19	58	~800	—	—	—
GH	29	69	1,278	GHR	33	58	1294	1	3	12

Only residues contributing more than 5Å^2 of surface area to the interface are included in the summaries. B.S., buried surface; MC, number of mainchain hydrogen bonds; MC/SC, number of mainchain/sidechain hydrogen bonds, T, total number of interactions.

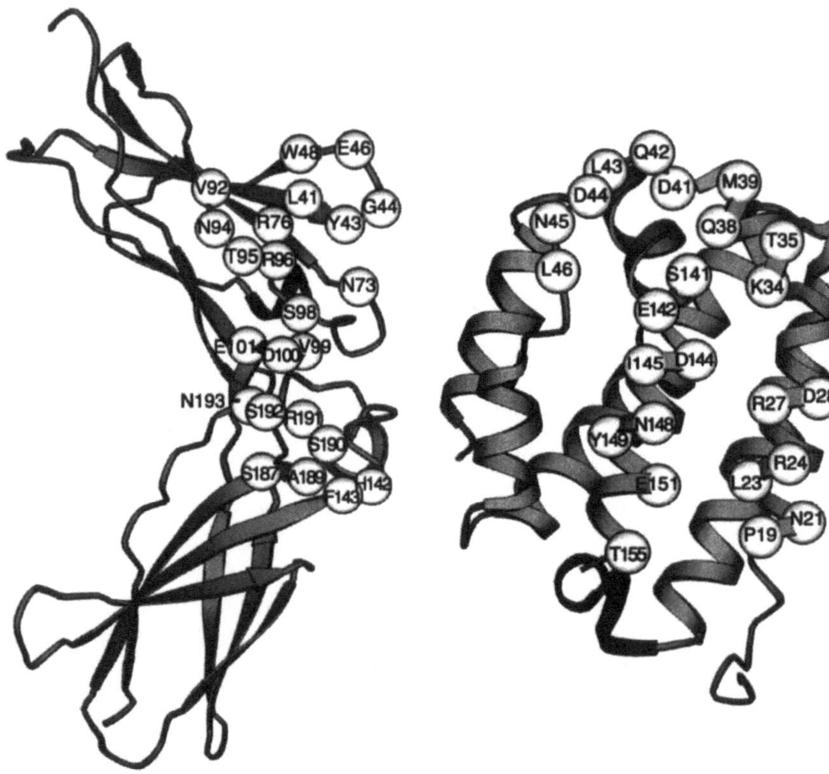

FIG. 14. Residues that form the IL-10/IL-10R1 interface. IL-10R1 and IL-10 are shown as ribbon diagrams. The position of each IL-10R1 and IL-10 residue that buries surface area into the interface is shown as a sphere corresponding to the Ca atom for each residue.

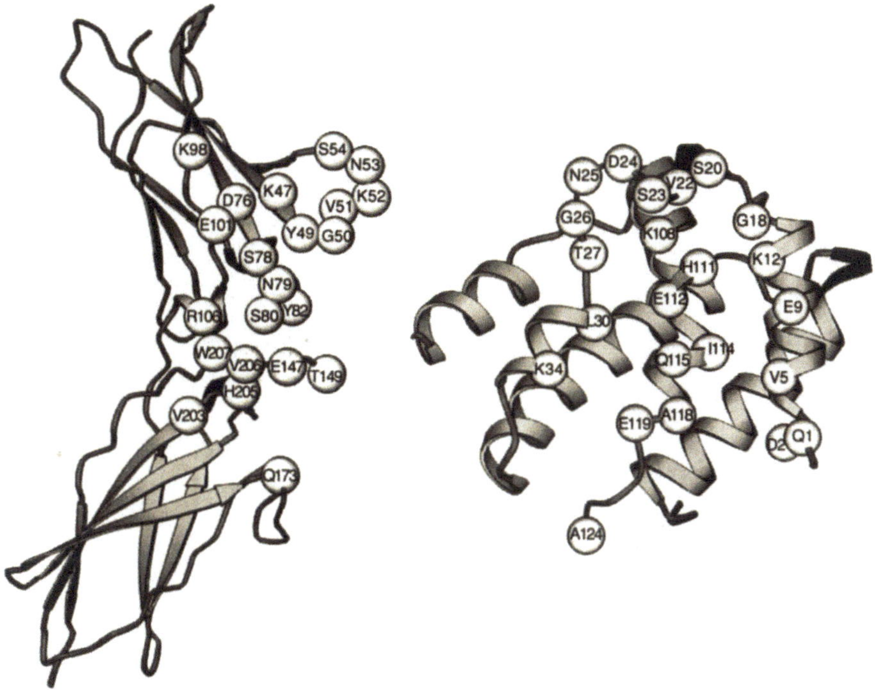

FIG. 15. Residues that form the IFN-γ/IFN-γR1 interface. IFN-γR1 and IFN-γ are shown as ribbon diagrams. The position of each IFN-γR1 and IFN-γ residue that buries surface area into the interface is shown as a sphere corresponding to the Ca atom for each residue.

interface of the IFN-γ/IFN-γR1 complex have been hypothesized to be important for IFNγ/IFN-γR1 interactions (Randal and Kossiakoff, 2001). However, detailed experiments to test this hypothesis have not been reported.

Six of the 12 interactions in the IL-10/IL-10R1 complex occur between a sidechain atom on one side of the interface and a mainchain atom on the other. These mainchain-mainchain interactions are important because they are highly dependent on the proper three-dimensional structure of the cytokine scaffold. Thus, subtle amino acid changes to either binding partner can dramatically change their interaction properties. Although specific hydrogen bonds have not been determined for the IFN-α2/IFNAR2 complex, it is assumed that several mainchain-mainchain and mainchain-sidechain interactions similar to those described for IFN-γ/IFNγR1 occur in the IFN-α2/IFNAR2 interface.

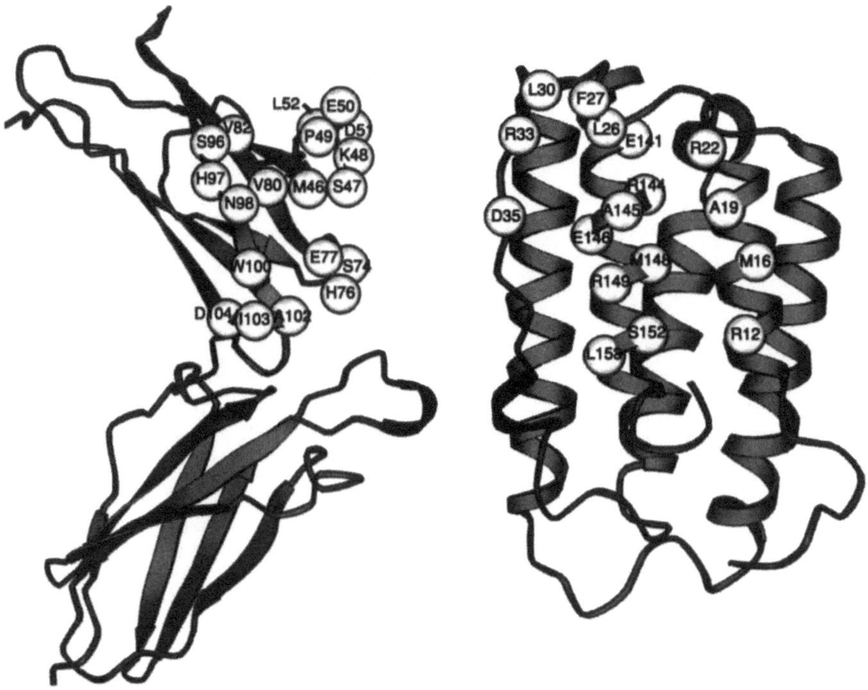

FIG. 16. Residues that form the IFN-α2/IFNAR2 interface. IFNAR2 and IFN-α2 are shown as ribbon diagrams. The position of each IFNAR2 and IFN-α2 residue that buries surface area into the interface is shown as a sphere corresponding to the Ca atom for each residue.

C. Receptor Loop Contributions to Cytokine Binding

Ligand recognition and high-affinity binding are modulated by changes in the orientation of the ligand that changes the extent of contacts made with the C2HR binding loops. To further define the contribution of each binding loop in HCII binding, the surface area buried by each receptor loop (L2-L7) in each interface was calculated (Table X). The results show considerable variability in the amount of surface area buried by each loop in the different HCII/C2HR complexes. For example, the L2 loop of IFN-γR1 and IFNAR2 contribute ~38% and 50% of the total surface area of the IFN-γ/IFN-γR1 and IFN-α2/IFNAR2 complexes, respectively. However, only 18% of the total buried surface is contributed by the L2 loop of IL-10R1 in the IL-10 and cmvIL-10 complexes. The large amounts of surface area contributed by L2 in the IFN-α2 and IFN-γ complexes correlates with the 3_{10} helical conformations of their AB loops. The helical AB loop regions of IFN-γ and IFN-α2 allow more extensive interactions with their receptors

TABLE X
Surface Area Buried by L2-L7 C2HR Loops

	IL-10 IL-10R1	cmvIL-10 IL-10R1	IFN-γ IFN-γR1	IFN-α2 IFNAR2	GH GHR
L1	—	—	—	—	66, 2 (5.1)
L2	206, 5 (18.2)	181, 5 (17.8)	328, 7 (**37.9**)	~400, 7 (**50**)	95, 3 (7.3)
L3	147, 2 (13.0)	139, 3 (13.7)	154, 5 (17.8)	~200, 5 (25)	289, 8 (22.3)
L4	350, 8 (**30.9**)	286,6 (**28.2**)	82, 2 (9.4)	~160, 7 (20)	264, 7 (20.4)
L5	146, 2 (12.9)	134, 2 (13.2)	67, 2 (7.7)	—	363, 8 (**28.1**)
L6	—	—	17, 1 (1.9)	—	—
L7	283, 6 (25)	275, 5 (27.0)	219, 4 (25.3)	—	217, 5 (16.8)
Total b.s. (Å^2)	1132	1007	866	~800	1294
Total res.	23	21	21	~19	33
Å^2/res.	49	48	41	~42	39

The three entries correspond to: buried surface in Å^2, number of residues in the loop, % of the total amount of buried surface contributed by the loop.

than the extended AB loop conformations found in IL-10 and cmvIL-10. The smaller contribution of the IL-10R1 L2 loop in the IL-10/IL-10R1 and cmvIL-10/IL-10R1 complexes is offset by the increased surface area buried by the L4 loop which accounts for 30% of the total buried surface area in both complexes. In addition to the extensive surface area contributions made by receptor loops in the site Ia interface (L2-L4), the L7 loop of IL-10R1 and IFNγR1 (Site Ib) contribute the second largest amount of buried surface area (~25% of the total) to their respective cytokine interfaces.

Although the amount of surface area contributed by L2-L7 loops is different in the Il-10/IL-10R1 and IFN-γ/IFN-γR1 complexes, five structurally conserved IL-10 and IFN-γR1 amino acid positions form most of the interactions with IL-10 and IFN-γ, respectively. The structurally conserved residues Tyr-43$^{\text{IL-10R1}}$ and Tyr-49$^{\text{IFN-}\gamma\text{R1}}$ are located on L2, Arg-76$^{\text{IL-10R1}}$ and Trp-82$^{\text{IFN-}\gamma\text{R1}}$ are located on L3, Arg-96$^{\text{IL-10R1}}$ and Glu-101$^{\text{IFN-}\gamma\text{R1}}$ are located on L4 and two residues (Ser-190$^{\text{IL-10R1}}$ and Arg-191$^{\text{IL-10R1}}$, Val-206$^{\text{IFN-}\gamma\text{R1}}$, Trp-207$^{\text{IFN-}\gamma\text{R1}}$) are located on L7. When D1 of IL-10R1 and IFN-γR1 are superimposed, residue pairs Tyr-43$^{\text{IL-10R1}}$ /Tyr-49$^{\text{IFN-}\gamma\text{R1}}$, Arg-76$^{\text{IL-10R1}}$/Trp-82$^{\text{IFN-}\gamma\text{R1}}$, and Arg-96$^{\text{IL-10R1}}$/Glu-101$^{\text{IFN-}\gamma\text{R1}}$ differ by 1.7Å, 0.1Å, and 0.9Å, respectively. In contrast, superposition of IL10R1 and IFN-γR1 D2 domains positions the residue pairs on L7 4.5Å-8Å apart from one another. Analysis of buried surface area and specific contacts in the IFN-γ and IL-10 interfaces suggests all five residues are important for ligand binding with Tyr-43$^{\text{IFN-}\gamma\text{R1}}$ and Arg-96$^{\text{IL-10R1}}$ being essential to their respective high-affinity interactions. Of course this prediction will have to be tested by detailed energetic analyses of the interfaces. Alanine scanning data on the IFN-γ interface has been collected but not reported at this time (Randal and Kossiakoff, 2001).

Superposition of D1 of IFNAR2 onto the IL-10R1 and IFN-γR1 D1 domains reveals IFNAR2 residues Met-46, Val-80, and Trp-100 correspond to Tyr-43, Arg-76, and Arg-96 in IL-10R1 and Tyr-43, Trp-82, and Glu-101 in IFN-γR1. As predicted from the structural comparison with IL-10R1 and IFN-γR1, mutation of IFNAR2 Met-46 to an alanine causes a 500-fold decrease in affinity for IFN-α2, while mutation of Trp-100 to an alanine causes only a 4-fold decrease in ligand affinity (Piehler and Schreiber, 1999). Interestingly, Trp-100 is much more important in the IFN-β/IFNAR2 interaction, where its mutation to an alanine results in at least a 100-fold drop in IFN-β affinity (Piehler and Schreiber, 1999). Furthermore, mutation of Arg-35 in IFN-β has a much smaller impact on the IFN-β/IFNAR2 interaction than the IFN-α2/IFNAR2 complex (Runkel *et al.*, 1998). Thus sequence and structure differences between IFN-α2 and IFN-β lead to different orientations of the ligands on IFNAR2. How these differences alter biological activity remains to be determined.

D. *Structural Features that Modulate Cytokine-Receptor Orientation*

A comparison of the HCII IL-10/IL-10R1 and HCI GH/GHR receptor complexes provides several insights into the structural mechanisms that modulate cytokine-receptor orientation. As shown in Fig. 17, when D1 of IL-10R1 and GHR are superimposed, IL-10 and GH differ in orientation by about 50° about an axis perpendicular to the plane of the figure. As a result, L5 in GHR buries the most surface area into GH, while IL-10R1 L4 buries the most surface area into IL-10 (Table X). Energetic analysis of the GH/GHR interaction confirms that Trp-169 located on L5 is the most critical residue for high-affinity binding (Clackson and Wells, 1995). Other energetically important residues in the GH/GHR contact are found on GHR L1 (Arg-43, Glu-44), which is located adjacent to L5 and Trp-104, Ile-105, and Pro-106 located on GHR L3. As a result of its orientation on GHR, the GH hot spot is located at the C-terminal end of helix F (called *helix D* in the growth hormone literature and equivalent to siteIb in Fig. 13), essentially opposite the hot spot of IFN-α2 and the predicted hot spots for IL-10 and IFN-γ (Fig. 18).

What causes the difference in orientation of GH and IL-10? Two structural features that play a major role are the bend in helix F and the interdomain angles of the C2HRs. Interestingly, the N-terminal ends of helix F in IL-10 and GH are oriented in approximately the same direction (Fig. 17). However, helix F in IL-10 bends by 47° away from the D1/D2

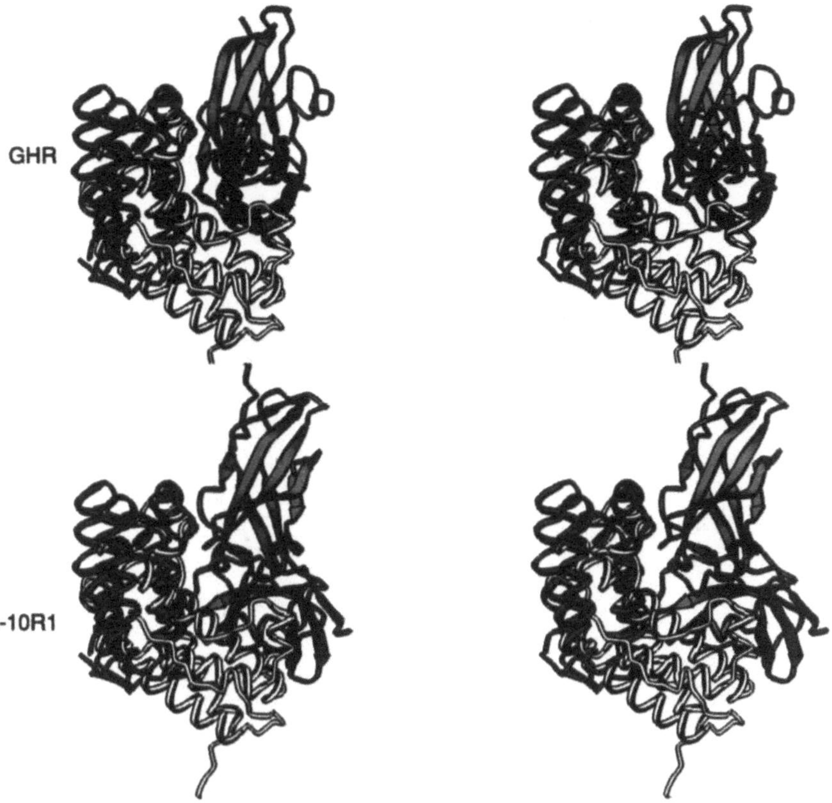

FIG. 17. Stereoview comparison of the orientation of IL-10 and GH on their C2HRs. D1s of IL-10R1 and GHR were superimposed as in Fig. 10A. The resulting orientations of IL-10 (magenta) and GH (white) are shown with GHR (top stereoview) or IL-10R1 (bottom stereoview). (See Color Insert.)

interface, while helix F in GH is essentially straight. The bend in helix F allows the AB loop region of IL-10 (the site Ia interface) to make extensive interactions with D1. In contrast, the C-terminal end of the straight helix F in GH makes the greatest interaction with D2 of GHR. For IL-10 and GH, the interactions are critically dependent on the domain orientation of their respective receptors. As shown in Fig. 17, GHR D2 is essentially perpendicular to the page, which allows the straight helix F of GH to make extensive contacts with its L5 loop. However, because of the bend in IL-10 helix F, GHR D2 does make any contacts with IL-10. The situation is different when GH and IL-10 are analyzed on the IL-10R1. In this case, the interdomain angle of IL-10R1 rotates IL-10R1 D2 up towards the bent

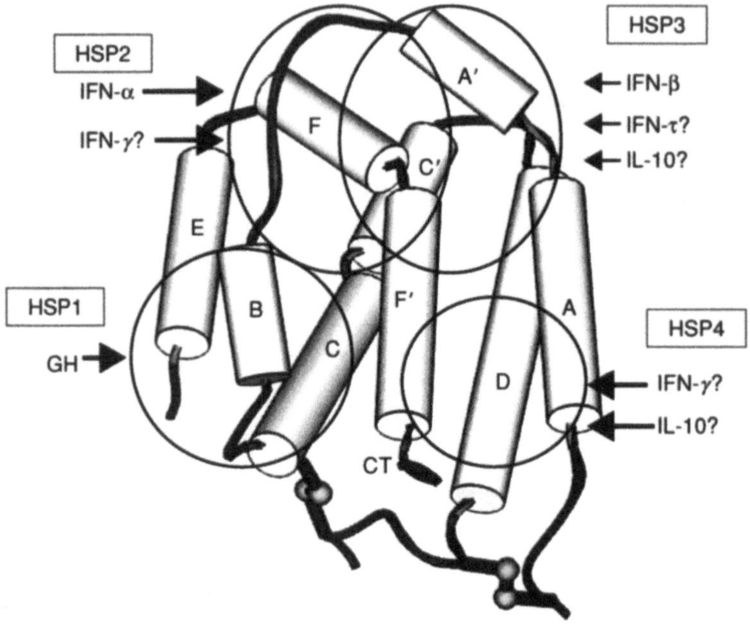

FIG. 18. Known and predicted hot spot (HSP) locations on the HCIIs and GH. HSP regions for GH, IFN-α2, IFN-β, IL-10, and IFN-γ are roughly shown in the circled regions on the cylinder representation of the IL-10 scaffold. Predicted HSPs are denoted with question marks.

IL-10 helix F to compliment the different orientation and resulting in significant contacts between IL-10 the IL-10R1 L7 loop. In contrast, the straight helix F of GH makes numerous steric clashes with IL-10R1 L5 which would preclude any possibility of GH interacting with IL-10R1.

The structural differences in IL-10 and GH helix F and the complimentary changes in the inter-domain angles of IL-10R1 and GHR to optimize their binding interfaces provides an excellent example of the co-evolution of the cytokine-receptor binding pairs. Another example of this mechanism is observed in the structure of the IFN-γ/IFN-γR1 complex. The IFN-γR1 L6 loop is unique among the C2HRs. It contains an additional 8 residues, compared to the other C2HR sequences, which are linked to β-strand E^{D2} by a disulfide bond (Fig. 8). The orientation of IFN-γ observed on D1 is only possible because IFN-γ's F′ helix is only 4 residues long (Table II), which prevents a steric clash with the IFN-γR1 L6. IFN-γR1 L6 is composed predominantly of glutamate and aspartate residues positioned to interact with the unstructured basic tail of IFN-γ (Walter *et al.*,

1995). Thus the structural as well as chemical evolution of IFN-γR1 L6 perfectly complements the short helix F′ and basic tail of IFN-γ.

E. Putative Membrane Complexes

Two C2HRs bind the dimeric HCIIs IFN-γ, IL-10, and cmvIL-10 to form 1:2 dimer:C2HR complexes (Fig. 19). In contrast to the GH/GHR complex, the two-fold related C2HRs do not interact with one another. The C-terminal ends of the C2HRs are separated by 88Å (Thiel *et al.*, 2000) or 102Å (Walter *et al.*, 1995) in the IFN-γ/IFN-γR1 complex, 110Å in the IL-10/IL-10R1 complex (Josephson *et al.*, 2001), and 105Å in the cmvIL-10/IL-10R1 complex (Jones *et al.*, 2002a) at the point where they are predicted to enter the membrane. Thus each dimer complex maintains similar two-fold relationships and C-terminal spacing of its high-affinity receptor chains. However, because of the different domain arrangements of IFN-γ, IL-10, and cmvIL-10, the orientation of the two-fold related receptor chains is very different. As shown in Fig. 20A, IFN-γR1 assumes an angle of approximately 60° with respect to the cell membrane, while IL-10R1s bound to IL-10 and cmvIL-10 adopt angles of 32° and 18°,

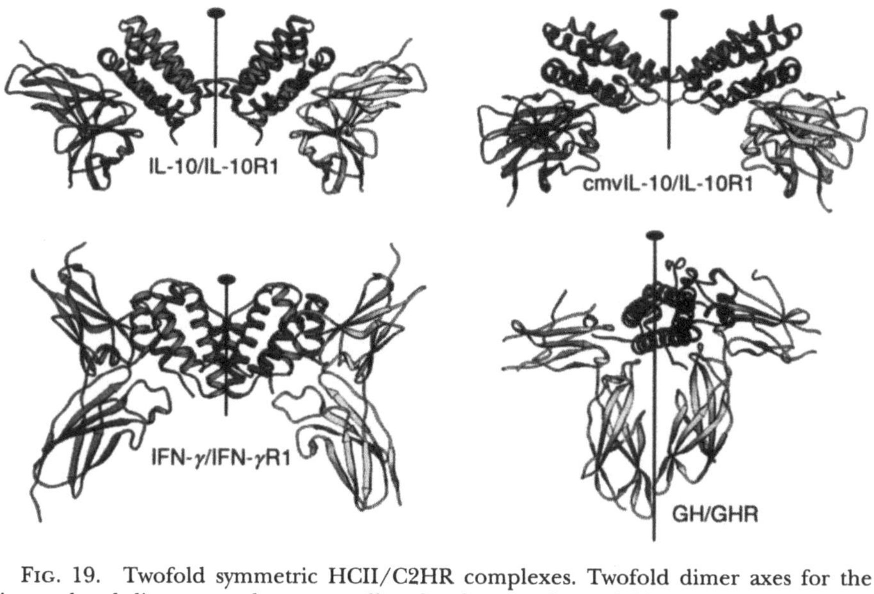

FIG. 19. Twofold symmetric HCII/C2HR complexes. Twofold dimer axes for the intercalated dimers are shown as well as for the pseudo twofold axis in the GH/GHR complex (de Vos *et al.*, 1992).

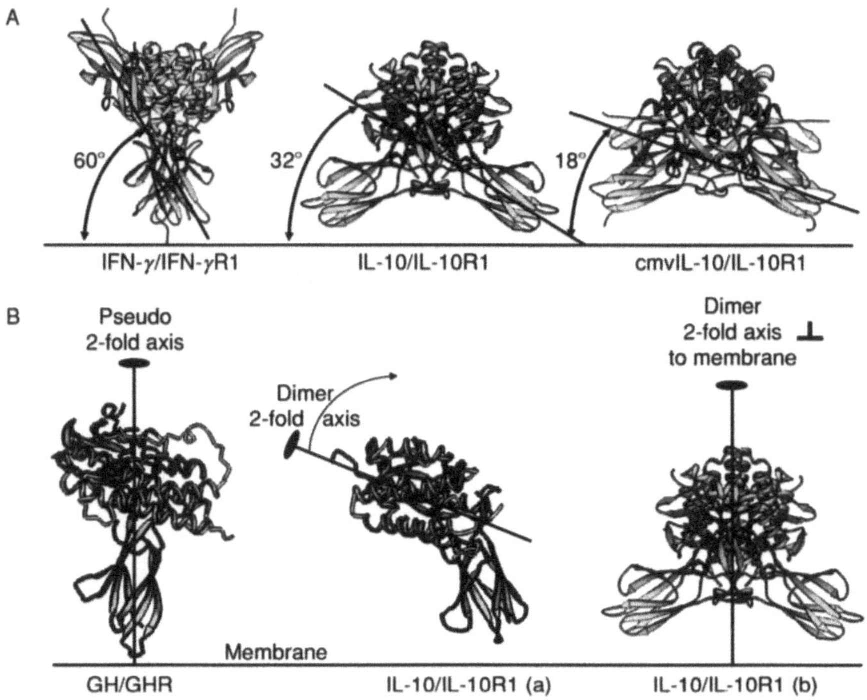

FIG. 20. Receptor orientation in the IL-10 and IFN-γ receptor complexes. (A) Receptor complexes are shown rotated $\sim 90^\circ$ about the twofold axes shown in Fig. 19. (B) Constraints on ligand orientation in the HCI and HCII receptor complexes. As shown on the figure, the pseudo twofold axis of the GHR binding sites is perpendicular to the putative position of the cell membrane. The 1:1 IL-10/IL-10R1 complex (a) was oriented by aligning IL-10R1 D1 to the D1 of GHR. The orientation of the IL-10/IL-10R1 complex in (b) is rotated to position the twofold axis of the dimer perpendicular to the membrane. This allows the twofold related IL-10 binding sites to simultaneously contact the membrane tethered IL-10R1s. This positions the α-helices of IL-10 approximately parallel to the cell membrane rather than horizontal as observed for the GHR complex.

respectively. The differences in the orientations of the high-affinity 1:2 receptor complexes likely influence the binding properties of the lower-affinity R2 chains.

Each C2HR is tethered to the membrane by a 5–12 residue linker. This general feature of α-helical cytokine receptors requires the cytokine binding sites be positioned essentially parallel to the cell membrane. For this to occur in the HCII dimers IFN-γ and IL-10, the two-fold symmetry axis of the molecules must be oriented perpendicular to the cell membrane (Fig. 19). In contrast to the two-fold symmetry axes of IFN-γ and IL-10, GH contains a pseudo two-fold axis that runs perpendicular to the helical bundle axis (Figs. 19–20). As a result, the GHRs recognize GH α-helices oriented parallel to the membrane in an upright orientation. In contrast, IL-10R1 and IFN-γR1 recognize IL-10 and IFN-γ helices that are more perpendicular to the membrane requiring the receptors to adopt angles that significantly deviate from the upright orientation of the GHRs (Fig. 20B). How the lower-affinity IL-10R2 and IFN-γR2 chain recognize the upright helices of IL-10 and IFN-γ in their respective ternary complexes remains to be determined.

V. Concluding Remarks

This chapter provides a series of structural comparisons of the HCIIs, C2HRs, and their complexes. The goal of the comparisons is to highlight structural features of the molecules that are important for their functional properties. It is hoped that this work will provide a common reference point as new structures in the family are determined. Noticeably missing from the review is a structure of a ternary complex between a HCII and the heterodimeric C2HRs. This is exclusively a technical issue because of the extremely low affinity of the many of the R2 chains characterized thus far. In this respect, recent experiments showing the ability to purify a ternary complex consisting of IL-19, IL-20R2, and IL-20R1 holds the most promise for obtaining this structural information (Pletnev *et al.*, 2003). However, it is unlikely that one ternary complex structure will be sufficient to understand mechanistic differences between family members that ultimately lead to their unique signaling properties. The main challenge for the future is to define the functional correspondence between the structural features of the extracellular receptor-cytokine complexes and the initiation of intracellular signaling cascades and the resulting biological activity.

Acknowledgments

The author would like to thank past and present members of the Walter lab for their hard work to elucidate many of the concepts presented in this chapter, Mike Carson for use of his Ribbons package, and the support from the National Institutes of Health (AI47300).

References

Abdel-Meguid, S. S., Shieh, H. S., Smith, W. W., Dayringer, H. E., Violand, B. N., and Bentle, L. A. (1987). Three-dimensional structure of a genetically engineered variant of porcine growth hormone. *Proc. Natl. Acad. Sci. USA* **84,** 6434–6437.

Adzhubei, A. A., and Sternberg, M. J. (1993). Left-handed polyproline II helices commonly occur in globular proteins. *J. Mol. Biol.* **229,** 472–493.

Aggarwal, S., Xie, M. H., Maruoka, M., Foster, J., and Gurney, A. L. (2001). Acinar cells of the pancreas are a target of interleukin-22. *J. Interferon. Cytokine Res.* **21,** 1047–1053.

Bach, E. A., Aguet, M., and Schreiber, R. D. (1997). The IFN gamma receptor: A paradigm for cytokine receptor signaling. *Annu. Rev. Immunol.* **15,** 563–591.

Bach, E. A., Szabo, S. J., Dighe, A. S., Ashkenazi, A., Aguet, M., Murphy, K. M., and Schreiber, R. D. (1995). Ligand-induced autoregulation of IFN-gamma receptor beta chain expression in T helper cell subsets. *Science* **270,** 1215–1218.

Banner, D. W., D'Arcy, A., Chene, C., Winkler, F. K., Guha, A., Konigsberg, W. H., Nemerson, Y., and Kirchhofer, D. (1996). The crystal structure of the complex of blood coagulation factor VIIa with soluble tissue factor. *Nature* **380,** 41–46.

Bazan, J. F. (1990a). Shared architecture of hormone binding in type I and type II interferon receptors. *Cell* **61,** 753–754.

Bazan, J. F. (1990b). Structural design and molecular evolution of a cytokine receptor superfamily. *Proc. Natl. Acad. Sci. USA* **87,** 6934–6938.

Bennett, M. J., Schlunegger, M. P., and Eisenberg, D. (1995). 3D domain swapping: A mechanism for oligomer assembly. *Protein Sci.* **4,** 2455–2468.

Blumberg, H., Conklin, D., Xu, W. F., Grossmann, A., Brender, T., Carollo, S., Eagan, M., Foster, D., Haldeman, B. A., Hammond, A., Haugen, H., Jelinek, L., Kelly, J. D., Madden, K., Maurer, M. F., Parrish-Novak, J., Prunkard, D., Sexson, S., Sprecher, C., Waggie, K., West, J., Whitmore, T. E., Yao, L., Kuechle, M. K., Dale, B. A., and Chandrasekher, Y. A. (2001). Interleukin 20: Discovery, receptor identification, and role in epidermal function. *Cell* **104,** 9–19.

Bork, P., Holm, L., and Sander, C. (1994). The immunoglobulin fold. Structural classification, sequence patterns and common core. *J. Mol. Biol.* **242,** 309–320.

Caudell, E. G., Mumm, J. B., Poindexter, N., Ekmekcioglu, S., Mhashilkar, A. M., Yang, X. H., Retter, M. W., Hill, P., Chada, S., and Grimm, E. A. (2002). The protein product of the tumor suppressor gene, melanoma differentiation-associated gene 7, exhibits immunostimulatory activity and is designated IL-24. *J. Immunol.* **168,** 6041–6046.

Chang, C., Magracheva, E., Kozlov, S., Fong, S., Tobin, G., Kotenko, S., Wlodawer, A., and Zdanov, A. (2003). Crystal Structure of Interleukin-19 Defines a New Subfamily of Helical Cytokines. *J. Biol. Chem.* **278,** 3308–3313.

Chill, J. H., Nivasch, R., Levy, R., Albeck, S., Schreiber, G., and Anglister, J. (2002). The human interferon receptor: NMR-based modeling, mapping of the IFN-alpha 2 binding site, and observed ligand-induced tightening. *Biochemistry* **41,** 3575–3585.

Chill, J. H., Quadt, S. R., Levy, R., Schreiber, G., and Anglister, J. (2003). The human type I interferon receptor: NMR structure reveals the molecular basis of ligand binding. *Structure (Camb)* **11,** 791–802.

Chothia, C., and Jones, E. Y. (1997). The molecular structure of cell adhesion molecules. *Annu. Rev. Biochem.* **66,** 823–862.

Clackson, T., and Wells, J. A. (1995). A hot spot of binding energy in a hormone-receptor interface. *Science* **267,** 383–386.

Crepaldi, L., Gasperini, S., Lapinet, J. A., Calzetti, F., Pinardi, C., Liu, Y., Zurawski, S., de Waal Malefyt, R., Moore, K. W., and Cassatella, M. A. (2001). Up-regulation of IL-10R1 expression is required to render human neutrophils fully responsive to IL-10. *J. Immunol.* **167,** 2312–2322.

de Vos, A. M., Ultsch, M., and Kossiakoff, A. A. (1992). Human growth hormone and extracellular domain of its receptor: Crystal structure of the complex. *Science* **255,** 306–312.

Deivanayagam, C. C., Rich, R. L., Carson, M., Owens, R. T., Danthuluri, S., Bice, T., Hook, M., and Narayana, S. V. (2000). Novel fold and assembly of the repetitive B region of the *Staphylococcus aureus* collagen-binding surface protein. *Structure Fold Des.* **8,** 67–78.

Derynck, R., Content, J., DeClercq, E., Volckaert, G., Tavernier, J., Devos, R., and Fiers, W. (1980). Isolation and structure of a human fibroblast interferon gene. *Nature* **285,** 542–547.

Ding, Y., Qin, L., Zamarin, D., Kotenko, S. V., Pestka, S., Moore, K. W., and Bromberg, J. S. (2001). Differential IL-10R1 expression plays a critical role in IL-10-mediated immune regulation. *J. Immunol.* **167,** 6884–6892.

Dumoutier, L., Lejeune, D., Colau, D., and Renauld, J. C. (2001). Cloning and characterization of IL-22 binding protein, a natural antagonist of IL-10-related T cell-derived inducible factor/IL-22. *J. Immunol.* **166,** 7090–7095.

Dumoutier, L., Louahed, J., and Renauld, J. C. (2000a). Cloning and characterization of IL-10-related T cell-derived inducible factor (IL-TIF), a novel cytokine structurally related to IL-10 and inducible by IL-9. *J. Immunol.* **164,** 1814–1819.

Dumoutier, L., Van Roost, E., Ameye, G., Michaux, L., and Renauld, J. C. (2000b). IL-TIF/IL-22: Genomic organization and mapping of the human and mouse genes. *Genes Immun.* **1,** 488–494.

Dumoutier, L., Van Roost, E., Colau, D., and Renauld, J. C. (2000c). Human interleukin-10-related T cell-derived inducible factor: Molecular cloning and functional characterization as an hepatocyte-stimulating factor. *Proc. Natl. Acad. Sci. USA* **97,** 10144–10149.

Ealick, S. E., Cook, W. J., Vijay-Kumar, S., Carson, M., Nagabhushan, T. L., Trotta, P. P., and Bugg, C. E. (1991). Three-dimensional structure of recombinant human interferon-gamma. *Science* **252,** 698–702.

Eskdale, J., Kube, D., Tesch, H., and Gallagher, G. (1997). Mapping of the human IL10 gene and further characterization of the 5′ flanking sequence. *Immunogenetics* **46,** 120–128.

Fleming, S. B., Haig, D. M., Nettleton, P., Reid, H. W., McCaughan, C. A., Wise, L. M., and Mercer, A. (2000). Sequence and functional analysis of a homolog of interleukin-10 encoded by the parapoxvirus orf virus. *Virus Genes* **21,** 85–95.

Gallagher, G., Dickensheets, H., Eskdale, J., Izotova, L. S., Mirochnitchenko, O. V., Peat, J. D., Vazquez, N., Pestka, S., Donnelly, R. P., and Kotenko, S. V. (2000). Cloning, expression and initial characterization of interleukin-19 (IL-19), a novel homologue of human interleukin-10 (IL-10). *Genes Immun.* **1,** 442–450.

Grzesiek, S., Dobeli, H., Gentz, R., Garotta, G., Labhardt, A. M., and Bax, A. (1992). 1H, 13C, and 15N NMR backbone assignments and secondary structure of human interferon-gamma. *Biochemistry* **31,** 8180–8190.

Harlos, K., Martin, D. M., O'Brien, D. P., Jones, E. Y., Stuart, D. I., Polikarpov, I., Miller, A., Tuddenham, E. G., and Boys, C. W. (1994). Crystal structure of the extracellular region of human tissue factor. *Nature* **370,** 662–666.

Hauptmann, R., and Swetly, P. (1985). A novel class of human type I interferons. *Nucleic Acids Res.* **13,** 4739–4749.

Holm, L., and Sander, C. (1995). Dali: A network tool for protein structure comparison. *Trends Biochem. Sci.* **20,** 478–480.

Holm, L., and Sander, C. (1999). Protein folds and families: Sequence and structure alignments. *Nucleic Acids Res.* **27,** 244–247.

Huang, E. Y., Madireddi, M. T., Gopalkrishnan, R. V., Leszczyniecka, M., Su, Z., Lebedeva, I. V., Kang, D., Jiang, H., Lin, J. J., Alexandre, D., Chen, Y., Vozhilla, N., Mei, M. X., Christiansen, K. A., Sivo, F., Goldstein, N. I., Mhashilkar, A. B., Chada, S., Huberman, E., Pestka, S., and Fisher, P. B. (2001). Genomic structure, chromosomal localization and expression profile of a novel melanoma differentiation associated (mda-7) gene with cancer specific growth suppressing and apoptosis inducing properties. *Oncogene* **20,** 7051–7063.

Huang, S., Hendriks, W., Althage, A., Hemmi, S., Bluethmann, H., Kamijo, R., Vilcek, J., Zinkernagel, R. M., and Aguet, M. (1993). Immune response in mice that lack the interferon-gamma receptor. *Science* **259,** 1742–1745.

Huber, A. H., Wang, Y. M., Bieber, A. J., and Bjorkman, P. J. (1994). Crystal structure of tandem type III fibronectin domains from Drosophila neuroglian at 2.0 A. *Neuron* **12,** 717–731.

Ikeda, H., Old, L. J., and Schreiber, R. D. (2002). The roles of IFN gamma in protection against tumor development and cancer immunoediting. *Cytokine Growth Factor Rev.* **13,** 95–109.

Jiang, H., Lin, J. J., Su, Z. Z., Goldstein, N. I., and Fisher, P. B. (1995). Subtraction hybridization identifies a novel melanoma differentiation associated gene, mda-7, modulated during human melanoma differentiation, growth and progression. *Oncogene* **11,** 2477–2486.

Jones, B. C., Logsdon, N. J., Josephson, K., Cook, J., Barry, P. A., and Walter, M. R. (2002a). Crystal structure of human cytomegalovirus IL-10 bound to soluble human IL-10R1. *Proc. Natl. Acad. Sci. USA* **99,** 9404–9409.

Jones, B. C., Logsdon, N. J., Josephson, K., Cook, J., Barry, P. A., and Walter, M. R. (2002b). Crystal structure of human cytomegalovirus IL-10 bound to soluble human IL-10R1. *Proc. Natl. Acad. Sci. USA* **99,** 9404–9409.

Josephson, K., DiGiacomo, R., Indelicato, S. R., Iyo, A. H., Nagabhushan, T. L., Parker, M. H., and Walter, M. R. (2000a). Design and analysis of an engineered human interleukin-10 monomer. *J. Biol. Chem.* **275,** 25054.

Josephson, K., DiGiacomo, R., Indelicato, S. R., Iyo, A. H., Nagabhushan, T. L., Parker, M. H., Walter, M. R., and Ayo, A. H. (2000b). Design and analysis of an engineered human interleukin-10 monomer. *J. Biol. Chem.* **275,** 13552–13557.

Josephson, K., Jones, B. C., Walter, L. J., DiGiacomo, R., Indelicato, S. R., and Walter, M. R. (2002a). Noncompetitive antibody neutralization of IL-10 revealed by protein engineering and x-ray crystallography. *Structure (Camb)* **10,** 981–987.

Josephson, K., Jones, B. C., Walter, L. J., DiGiacomo, R., Indelicato, S. R., and Walter, M. R. (2002b). Non-competitive antibody neutralization of IL-10 revealed by protein engineering and x-ray crystallography. *Structure* **10,** 981–987.

Josephson, K., Logsdon, N. J., and Walter, M. R. (2001). Crystal structure of the IL-10/IL-10R1 complex reveals a shared receptor binding site. *Immunity* **15,** 35–46.

Karpusas, M., Nolte, M., Benton, C. B., Meier, W., Lipscomb, W. N., and Goelz, S. (1997). The crystal structure of human interferon beta at 2.2-A resolution. *Proc. Natl. Acad. Sci. USA* **94,** 11813–11818.

Kerr, I. M., Costa-Pereira, A. P., Lillemeier, B. F., and Strobl, B. (2003). Of JAKs, STATs, blind watchmakers, jeeps and trains. *FEBS Lett.* **546,** 1–5.

Kisseleva, T., Bhattacharya, S., Braunstein, J., and Schindler, C. W. (2002). Signaling through the JAK/STAT pathway, recent advances and future challenges. *Gene* **285,** 1–24.

Klaus, W., Gsell, B., Labhardt, A. M., Wipf, B., and Senn, H. (1997). The three-dimensional high resolution structure of human interferon alpha-2a determined by heteronuclear NMR spectroscopy in solution. *J. Mol. Biol.* **274,** 661–675.

Knappe, A., Hor, S., Wittmann, S., and Fickenscher, H. (2000). Induction of a novel cellular homolog of interleukin-10, AK155, by transformation of T lymphocytes with herpesvirus saimiri. *J. Virol.* **74,** 3881–3887.

Kotenko, S. V. (2002). The family of IL-10-related cytokines and their receptors: Related, but to what extent? *Cytokine Growth Factor Rev.* **13,** 223–240.

Kotenko, S. V., Gallagher, G., Baurin, V. V., Lewis-Antes, A., Shen, M., Shah, N. K., Langer, J. A., Sheikh, F., Dickensheets, H., and Donnelly, R. P. (2003). IFN-lambdas mediate antiviral protection through a distinct class II cytokine receptor complex. *Nat. Immunol.* **4,** 69–77.

Kotenko, S. V., Izotova, L. S., Mirochnitchenko, O. V., Esterova, E., Dickensheets, H., Donnelly, R. P., and Pestka, S. (2001). Identification, cloning, and characterization of a novel soluble receptor that binds IL-22 and neutralizes its activity. *J. Immunol.* **166,** 7096–7103.

Kotenko, S. V., Krause, C. D., Izotova, L. S., Pollack, B. P., Wu, W., and Pestka, S. (1997). Identification and functional characterization of a second chain of the interleukin-10 receptor complex. *EMBO J.* **16,** 5894–5903.

Kotenko, S. V., Saccani, S., Izotova, L. S., Mirochnitchenko, O. V., and Pestka, S. (2000). Human cytomegalovirus harbors its own unique IL-10 homolog (cmvIL-10). *Proc. Natl. Acad. Sci. USA* **97,** 1695–1700.

Krause, C. D., Lunn, C. A., Izotova, L. S., Mirochnitchenko, O., Kotenko, S. V., Lundell, D. J., Narula, S. K., and Pestka, S. (2000). Signaling by covalent heterodimers of interferon-gamma. Evidence for one-sided signaling in the active tetrameric receptor complex. *J. Biol. Chem.* **275,** 22995–23004.

Krause, C. D., Mei, E., Xie, J., Jia, Y., Bopp, M. A., Hochstrasser, R. M., and Pestka, S. (2002). Seeing the light: Preassembly and ligand-induced changes of the interferon gamma receptor complex in cells. *Mol. Cell. Proteomics* **1,** 805–815.

Kuhn, R., Lohler, J., Rennick, D., Rajewsky, K., and Muller, W. (1993). Interleukin-10-deficient mice develop chronic enterocolitis. *Cell* **75,** 263–274.

Kullberg, M. C., Rothfuchs, A. G., Jankovic, D., Caspar, P., Wynn, T. A., Gorelick, P. L., Cheever, A. W., and Sher, A. (2001). Helicobacter hepaticus-induced colitis in interleukin-10-deficient mice: Cytokine requirements for the induction and maintenance of intestinal inflammation. *Infect. Immun.* **69,** 4232–4241.

Kullberg, M. C., Ward, J. M., Gorelick, P. L., Caspar, P., Hieny, S., Cheever, A., Jankovic, D., and Sher, A. (1998). Helicobacter hepaticus triggers colitis in specific-pathogen-free interleukin-10 (IL-10)-deficient mice through an IL-12- and gamma interferon-dependent mechanism. *Infect. Immun.* **66,** 5157–5166.

LaFleur, D. W., Nardelli, B., Tsareva, T., Mather, D., Feng, P., Semenuk, M., Taylor, K., Buergin, M., Chinchilla, D., Roshke, V., Chen, G., Ruben, S. M., Pitha, P. M., Coleman, T. A., and Moore, P. A. (2001). Interferon-kappa, a novel type I interferon expressed in human keratinocytes. *J. Biol. Chem.* **276,** 39765–39771.

Landar, A., Curry, B., Parker, M. H., DiGiacomo, R., Indelicato, S. R., Nagabhushan, T. L., Rizzi, G., and Walter, M. R. (2000). Design, characterization, and structure of a biologically active single-chain mutant of human IFN-gamma. *J. Mol. Biol.* **299,** 169–179.

Leahy, D. J., Aukhil, I., and Erickson, H. P. (1996). 2.0 A crystal structure of a four-domain segment of human fibronectin encompassing the RGD loop and synergy region. *Cell* **84,** 155–164.

Leahy, D. J., Hendrickson, W. A., Aukhil, I., and Erickson, H. P. (1992). Structure of a fibronectin type III domain from tenascin phased by MAD analysis of the selenomethionyl protein. *Science* **258,** 987–991.

Lee, H. J., Essani, K., and Smith, G. L. (2001). The genome sequence of Yaba-like disease virus, a yatapoxvirus. *Virology* **281,** 170–192.

Liao, Y. C., Liang, W. G., Chen, F. W., Hsu, J. H., Yang, J. J., and Chang, M. S. (2002). IL-19 induces production of IL-6 and TNF-alpha and results in cell apoptosis through TNF-alpha. *J. Immunol.* **169,** 4288–4297.

Liu, Y., de Waal Malefyt, R., Briere, F., Parham, C., Bridon, J. M., Banchereau, J., Moore, K. W., and Xu, J. (1997). The EBV IL-10 homologue is a selective agonist with impaired binding to the IL-10 receptor. *J. Immunol.* **158,** 604–613.

Livnah, O., Stura, E. A., Johnson, D. L., Middleton, S. A., Mulcahy, L. S., Wrighton, N. C., Dower, W. J., Jolliffe, L. K., and Wilson, I. A. (1996). Functional mimicry of a protein hormone by a peptide agonist: The EPO receptor complex at 2.8 A. *Science* **273,** 464–471.

Lockridge, K. M., Zhou, S. S., Kravitz, R. H., Johnson, J. L., Sawai, E. T., Blewett, E. L., and Barry, P. A. (2000). Primate cytomegaloviruses encode and express an IL-10-like protein. *Virology* **268,** 272–280.

Logsdon, N. J., Jones, B. C., Josephson, K., Cook, J., and Walter, M. R. (2002). Comparison of interleukin-22 and interleukin-10 soluble receptor complexes. *J. Interferon Cytokine Res.* **22,** 1099–1112.

Lundell, D., Lunn, C. A., Senior, M. M., Zavodny, P. J., and Narula, S. K. (1994). Importance of the loop connecting A and B helices of human interferon-gamma in recognition by interferon-gamma receptor. *J. Biol. Chem.* **269,** 16159–16162.

Lunn, C. A., Davies, L., Dalgarno, D., Narula, S. K., Zavodny, P. J., and Lundell, D. (1992a). An active covalently linked dimer of human interferon-gamma. Subunit orientation in the native protein. *J. Biol. Chem.* **267,** 17920–17924.

Lunn, C. A., Fossetta, J., Dalgarno, D., Murgolo, N., Windsor, W., Zavodny, P. J., Narula, S. K., and Lundell, D. (1992b). A point mutation of human interferon gamma abolishes receptor recognition. *Protein Eng.* **5,** 253–257.

Lutfalla, G., Crollius, H. R., Stange-Thomann, N., Jaillon, O., Mogensen, K., and Monneron, D. (2003). Comparative genomic analysis reveals independent expansion of a lineage-specific gene family in vertebrates: The class II cytokine receptors and their ligands in mammals and fish. *BMC Genomics* **4,** 29.

MacNeil, I. A., Suda, T., Moore, K. W., Mosmann, T. R., and Zlotnik, A. (1990). IL-10, a novel growth cofactor for mature and immature T cells. *J. Immunol.* **145,** 4167–4173.

Madireddi, M. T., Su, Z. Z., Young, C. S., Goldstein, N. I., and Fisher, P. B. (2000). Mda-7, a novel melanoma differentiation associated gene with promise for cancer gene therapy. *Adv. Exp. Med. Biol.* **465,** 239–261.

Mhashilkar, A. M., Schrock, R. D., Hindi, M., Liao, J., Sieger, K., Kourouma, F., Zou-Yang, X. H., Onishi, E., Takh, O., Vedvick, T. S., Fanger, G., Stewart, L., Watson, G. J., Snary, D., Fisher, P. B., Saeki, T., Roth, J. A., Ramesh, R., and Chada, S. (2001). Melanoma differentiation associated gene-7 (mda-7): A novel anti-tumor gene for cancer gene therapy. *Mol. Med.* **7,** 271–282.

Moore, K. W., de Waal Malefyt, R., Coffman, R. L., and O'Garra, A. (2001). Interleukin-10 and the interleukin-10 receptor. *Annu. Rev. Immunol.* **19,** 683–765.

Morrissey, J. H. (2001). Tissue factor: An enzyme cofactor and a true receptor. *Thromb. Haemost.* **86,** 66–74.

Muller, Y. A., Kelley, R. F., and de Vos, A. M. (1998). Hinge bending within the cytokine receptor superfamily revealed by the 2.4 A crystal structure of the extracellular domain of rabbit tissue factor. *Protein Sci.* **7,** 1106–1115.

Muller, Y. A., Ultsch, M. H., and de Vos, A. M. (1996). The crystal structure of the extracellular domain of human tissue factor refined to 1.7 A resolution. *J. Mol. Biol.* **256,** 144–159.

Murphy, P. M. (1993). Molecular mimicry and the generation of host defense protein diversity. *Cell* **72,** 823–826.

Nagem, R. A., Colau, D., Dumoutier, L., Renauld, J. C., Ogata, C., and Polikarpov, I. (2002). Crystal structure of recombinant human interleukin-22. *Structure (Camb)* **10,** 1051–1062.

O'Shea, J. J., Gadina, M., and Schreiber, R. D. (2002). Cytokine signaling in 2002: New surprises in the Jak/Stat pathway. *Cell* **109**(Suppl), S121–S131.

Pernis, A., Gupta, S., Gollob, K. J., Garfein, E., Coffman, R. L., Schindler, C., and Rothman, P. (1995). Lack of interferon gamma receptor beta chain and the prevention of interferon gamma signaling in TH1 cells. *Science* **269,** 245–247.

Pestka, S. (1997). The human interferon-alpha species and hybrid proteins. *Semin. Oncol.* **24,** S9-4–S9-17.

Piehler, J., and Schreiber, G. (1999). Mutational and structural analysis of the binding interface between type I interferons and their receptor Ifnar2. *J. Mol. Biol.* **294,** 223–237.

Pletnev, S., Magracheva, E., Kozlov, S., Tobin, G., Kotenko, S. V., Wlodawer, A., and Zdanov, A. (2003). Characterization of the recombinant extracellular domains of human interleukin-20 receptors and their complexes with interleukin-19 and interleukin-20. *Biochemistry* **42,** 12617–12624.

Presnell, S. R., and Cohen, F. E. (1989). Topological distribution of four-alpha-helix bundles. *Proc. Natl. Acad. Sci. USA* **86,** 6592–6596.

Radhakrishnan, R., Walter, L. J., Hruza, A., Reichert, P., Trotta, P. P., Nagabhushan, T. L., and Walter, M. R. (1996). Zinc mediated dimer of human interferon-alpha 2b revealed by X-ray crystallography. *Structure* **4,** 1453–1463.

Radhakrishnan, R., Walter, L. J., Subramaniam, P. S., Johnson, H. M., and Walter, M. R. (1999). Crystal structure of ovine interferon-tau at 2.1 A resolution. *J. Mol. Biol.* **286,** 151–162.

Randal, M., and Kossiakoff, A. A. (1998). Crystallization and preliminary X-ray analysis of a 1:1 complex between a designed monomeric interferon-gamma and its soluble receptor. *Protein Sci.* **7,** 1057–1060.

Randal, M., and Kossiakoff, A. A. (2000). The 2.0 A structure of bovine interferon-gamma; assessment of the structural differences between species. *Acta Crystallogr. D Biol. Crystallogr.* **56**(Pt 1), 14–24.

Randal, M., and Kossiakoff, A. A. (2001). The structure and activity of a monomeric interferon-gamma: Alpha-chain receptor signaling complex. *Structure (Camb)* **9,** 155–163.

Roisman, L. C., Piehler, J., Trosset, J. Y., Scheraga, H. A., and Schreiber, G. (2001). Structure of the interferon-receptor complex determined by distance constraints from double-mutant cycles and flexible docking. *Proc. Natl. Acad. Sci. USA* **98,** 13231–13236.

Rousset, F., Garcia, E., Defrance, T., Peronne, C., Vezzio, N., Hsu, D. H., Kastelein, R., Moore, K. W., and Banchereau, J. (1992). Interleukin 10 is a potent growth and differentiation factor for activated human B lymphocytes. *Proc. Natl. Acad. Sci. USA* **89,** 1890–1893.

Runkel, L., Pfeffer, L., Lewerenz, M., Monneron, D., Yang, C. H., Murti, A., Pellegrini, S., Goelz, S., Uze, G., and Mogensen, K. (1998). Differences in activity between alpha and beta type I interferons explored by mutational analysis. *J. Biol. Chem.* **273,** 8003–8008.

Senda, T., Saitoh, S., and Mitsui, Y. (1995). Refined crystal structure of recombinant murine interferon-beta at 2.15 A resolution. *J. Mol. Biol.* **253,** 187–207.

Sheppard, P., Kindsvogel, W., Xu, W., Henderson, K., Schlutsmeyer, S., Whitmore, T. E., Kuestner, R., Garrigues, U., Birks, C., Roraback, J., Ostrander, C., Dong, D., Shin, J., Presnell, S., Fox, B., Haldeman, B., Cooper, E., Taft, D., Gilbert, T., Grant, F. J., Tackett, M., Krivan, W., McKnight, G., Clegg, C., Foster, D., and Klucher, K. M. (2003). IL-28, IL-29 and their class II cytokine receptor IL-28R. *Nat. Immunol.* **4,** 63–68.

Smith, G. L. (1996). Virus proteins that bind cytokines, chemokines or interferons. *Curr. Opin. Immunol.* **8,** 467–471.

Sogabe, S., Stuart, F., Henke, C., Bridges, A., Williams, G., Birch, A., Winkler, F. K., and Robinson, J. A. (1997). Neutralizing epitopes on the extracellular interferon gamma receptor (IFNgammaR) alpha-chain characterized by homolog scanning mutagenesis and X-ray crystal structure of the A6 fab-IFNgammaR1-108 complex. *J. Mol. Biol.* **273,** 882–897.

Somers, W., Ultsch, M., De Vos, A. M., and Kossiakoff, A. A. (1994). The X-ray structure of a growth hormone-prolactin receptor complex. *Nature* **372,** 478–481.

Sprang, S. R., and Bazan, J. F. (1993). Cytokine structural taxonomy and mechanisms of receptor engagement. *Curr. Opin. Struct. Biol.* **3,** 815–821.

Syto, R., Murgolo, N. J., Braswell, E. H., Mui, P., Huang, E., and Windsor, W. T. (1998). Structural and biological stability of the human interleukin 10 homodimer. *Biochemistry* **37,** 16943–16951.

Thiel, D. J., le Du, M. H., Walter, R. L., D'Arcy, A., Chene, C., Fountoulakis, M., Garotta, G., Winkler, F. K., and Ealick, S. E. (2000). Observation of an unexpected third receptor molecule in the crystal structure of human interferon-gamma receptor complex. *Structure Fold Des.* **8,** 927–936.

Vieira, P., de Waal-Malefyt, R., Dang, M. N., Johnson, K. E., Kastelein, R., Fiorentino, D. F., de Vries, J. E., Roncarolo, M. G., Mosmann, T. R., and Moore, K. W. (1991). Isolation and expression of human cytokine synthesis inhibitory factor cDNA clones: Homology to Epstein-Barr virus open reading frame BCRFI. *Proc. Natl. Acad. Sci. USA* **88,** 1172–1176.

Walter, M. R., and Nagabhushan, T. L. (1995). Crystal structure of interleukin 10 reveals an interferon gamma-like fold. *Biochemistry* **34,** 12118–12125.

Walter, M. R., Windsor, W. T., Nagabhushan, T. L., Lundell, D. J., Lunn, C. A., Zauodny, P. J., and Narula, S. K. (1995). Crystal structure of a complex between interferon-gamma and its soluble high-affinity receptor. *Nature* **376,** 230–235.

Xie, M. H., Aggarwal, S., Ho, W. H., Foster, J., Zhang, Z., Stinson, J., Wood, W. I., Goddard, A. D., and Gurney, A. L. (2000). Interleukin (IL)-22, a novel human cytokine that signals through the interferon receptor-related proteins CRF2-4 and IL-22R. *J. Biol. Chem.* **275,** 31335–31339.

Xu, W., Presnell, S. R., Parrish-Novak, J., Kindsvogel, W., Jaspers, S., Chen, Z., Dillon, S. R., Gao, Z., Gilbert, T., Madden, K., Schlutsmeyer, S., Yao, L., Whitmore, T. E., Chandrasekher, Y., Grant, F. J., Maurer, M., Jelinek, L., Storey, H., Brender, T., Hammond, A., Topouzis, S., Clegg, C. H., and Foster, D. C. (2001). A soluble class II cytokine receptor, IL-22RA2, is a naturally occurring IL-22 antagonist. *Proc. Natl. Acad. Sci. USA* **98,** 9511–9516.

Zdanov, A., Schalk-Hihi, C., Gustchina, A., Tsang, M., Weatherbee, J., and Wlodawer, A. (1995). Crystal structure of interleukin-10 reveals the functional dimer with an unexpected topological similarity to interferon gamma. *Structure* **3,** 591–601.

Zdanov, A., Schalk-Hihi, C., Menon, S., Moore, K. W., and Wlodawer, A. (1997). Crystal structure of Epstein-Barr virus protein BCRF1, a homolog of cellular interleukin-10. *J. Mol. Biol.* **268,** 460–467.

Zdanov, A., Schalk-Hihi, C., and Wlodawer, A. (1996). Crystal structure of human interleukin-10 at 1.6 A resolution and a model of a complex with its soluble receptor. *Protein Sci.* **5,** 1955–1962.

ASSEMBLY OF POST-RECEPTOR SIGNALING COMPLEXES FOR THE TUMOR NECROSIS FACTOR RECEPTOR SUPERFAMILY

By HAO WU

Department of Biochemistry, Weill Medical College of Cornell University, New York, New York 10021

ADVANCES IN PROTEIN CHEMISTRY, Vol. 68

Abstract

The tumor necrosis factor (TNF) receptor (TNFR) superfamily comprises more than 20 type-I transmembrane proteins that are structurally related in their extracellular domains and specifically activated by the corresponding superfamily of TNF-like ligands. Members of this receptor superfamily are widely distributed and play important roles in many crucial biological processes such as lymphoid and neuronal development, innate and adaptive immunity, and maintenance of cellular homeostasis. A remarkable dichotomy of the TNFR superfamily is the ability of these receptors to induce the opposing effects of gene transcription for cell survival, proliferation, and differentiation and of apoptotic cell death. The intracellular signaling proteins known as *TNF receptor associated factors* (TRAFs) are the major signal transducers for the cell survival effects, while the death-domain-containing proteins mediate cell death induction. This review summarizes recent structural, biochemical, and functional studies of these signal transducers and proposes the molecular mechanisms of the intracellular signal transduction.

I. Introduction

A. *Remarkable Dichotomy: Survival and Death*

The tumor necrosis factor (TNF) receptor (TNFR) superfamily comprises more than 20 type-I transmembrane proteins that are structurally related in their extracellular domains and specifically activated by the corresponding superfamily of TNF-like ligands (Locksley *et al.*, 2001). Members of this receptor superfamily are widely distributed and play important roles in many crucial biological processes such as lymphoid and neuronal development, innate and adaptive immunity, and maintenance of cellular homeostasis. Agents that manipulate the signaling of these receptors are being used or showing promise towards the treatment and prevention of many human diseases (Ashkenazi and Dixit, 1998; Leonen, 1998; Newton and Decicco, 1999).

A remarkable dichotomy of the TNFR superfamily is the ability of these receptors to induce the opposing effects of gene transcription for cell survival, proliferation, and differentiation and of apoptotic cell death (Gravestein and Borst, 1998; Locksley *et al.*, 2001; Smith *et al.*, 1994). Some members of the superfamily—such as TNF-R2, CD40, CD30, Ox40, 4-1BB, LTβR, and TRANCE-R (also known as RANK)—induce mostly survival effects, while others—such as Fas, DR4 and DR5—are mostly pro-apoptotic. In addition, receptors such as TNF-R1 and DR3 induce either cell survival or cell death in different cellular contexts. This functional divergence within the receptor superfamily is a consequence of the varied intracellular domains, leading to the assembly of different intracellular signaling complexes (Fig. 1).

B. *Identification of TNF: Historical Perspective*

Anecdotal but convincing associations of tumor necrosis or regression with bacterial infections have been noted throughout history and all over the world. In particular, pioneering clinicians in the late 19th century began adopting the idea of provoking acute skin infections, such as erysipelas, for the treatment of various kinds of tumors including sarcomas, cancers of the bone and connective tissues, breast cancer, ovarian cancer, Hodgkin's disease, and melanoma (Coley, 1893).

Our understanding on the underlying mechanism of this novel cancer treatment was significantly advanced by the discovery in 1975 that bacterial endotoxin induced the production and release of an anti-tumor activity from host cells like macrophages. This activity caused hemorrhagic necrosis of transplanted tumors in mice and killed transformed cell lines (Carswell *et al.*, 1975). The promise of TNF as a cancer cure prompted many laboratories to search the molecular identity of TNF, which eventually led to the purification, characterization and cloning of TNF (Beutler and Cerami, 1986; Pennica *et al.*, 1984; Shirai *et al.*, 1985; Wang *et al.*, 1985).

However, it was soon discovered that TNF is a pleiotropic cytokine important in host defense against pathogens and capable of inducing cell survival, proliferation, and differentiation as well as cell death (Fiers, 1991; Goeddel *et al.*, 1986). These collections of effects are mediated by the two receptors of TNF, TNF-R1, and TNF-R2 (Lewis *et al.*, 1991). In fact, TNF does not generally provoke cell killing as in its anti-tumor activity but more often promotes gene transcription and cell activation. The opposing cell death and cell survival functions of TNF have since become the major functional characteristics of the TNF and the TNFR superfamily.

C. *Intracellular Signaling Pathways: TRAFs and DD Proteins*

Upon receptor activation, different intracellular signaling complexes are assembled for different members of the TNFR superfamily, depending on their intracellular domains and sequences (Fig. 1). Receptors that do not contain a structural module known as the death domain (DD) in their intracellular domains are survival receptors, which directly recruit adapter

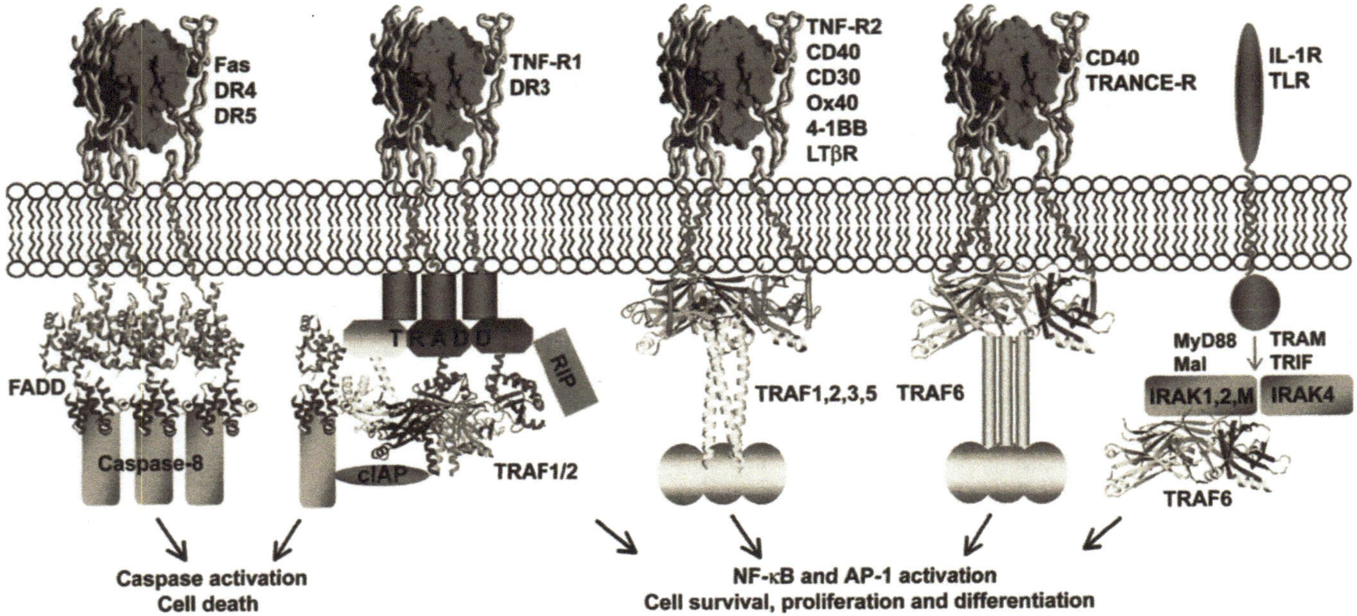

FIG. 1. Intracellular signaling pathways for the TNFR superfamily and the IL-1R/TLR superfamily. Proteins with known structures are shown as ribbon drawings. Hypothetical transmembrane helices are built to connect extracellular rand intracellular domains of these receptors (shown in orange). The amino terminal domains of TRAFs are shown as yellow spheres. (See Color Insert.)

proteins known as the *TNF receptor associated factors* (TRAFs) (Arch *et al.*, 1998; Chung *et al.*, 2002; Rothe *et al.*, 1994). Six mammalian TRAFs (TRAF1-6) have been identified so far, out of which, TRAF1, 2, 3, 5, and 6 participate in the signal transduction of the TNFR superfamily (Cao *et al.*, 1996b; Cheng *et al.*, 1995; Ishida *et al.*, 1996a,b; Mizushima *et al.*, 1998; Mosialos *et al.*, 1995; Nakano *et al.*, 1996; Regnier *et al.*, 1995; Rothe *et al.*, 1994; Sato *et al.*, 1995). The Epstein-Barr virus oncoprotein LMP1 also constitutively recruits TRAFs to promote growth and transformation (Mosialos *et al.*, 1995). Within the TRAF family, TRAF1, 2, 3, and 5 are considered TRAF2-like because they are recruited to a shared set of receptor family members by recognizing the same sequences on these receptors (Arch *et al.*, 1998).

TRAF6, on the other hand, has a unique sequence specificity that does not overlap with that of other TRAFs (Darnay *et al.*, 1999; Pullen *et al.*, 1998). It directly interacts with a subset of the TNFR superfamily such as CD40 and TRANCE-R. In addition to signal transduction for the TNFR superfamily, TRAF6 is also a major signal transducer for the interleukin-1 receptor (IL-1R)/Toll-like receptor (TLR) superfamily. The intracellular domains of IL-1Rs and TLRs contain a structural module known as the *TIR domain,* which recruit a family of TIR-domain containing intracellular signaling proteins including MyD88, Mal/TIRAP, TRIF, and TRAM (O'Neill *et al.*, 2003) and several other adapter proteins such as Tollip and SARM (Burns *et al.*, 2000; O'Neill *et al.*, 2003). In turn, these signaling complexes recruit Ser/Thr kinases, IRAK1, IRAK2, IRAK-M, and IRAK4 (Cao *et al.*, 1996a; Muzio *et al.*, 1997; Suzuki *et al.*, 2002; Wesche *et al.*, 1999), which directly interact with TRAF6 to activate downstream signal transduction.

The downstream effectors of TRAF signaling are transcription factors in the nuclear factor κ-B (NF-κB) and activator protein-1 (AP-1) family (Ghosh and Karin, 2002; Shaulian and Karin, 2002), which can turn on numerous genes involved in many aspects of cellular and immune functions. While the carboxyl terminal TRAF domain, containing a coiled-coil TRAF-N domain and a conserved TRAF-C domain, is both necessary and sufficient for TRAF self-association and receptor interaction, the amino terminal domain, containing RING and zinc-finger motifs, is important for downstream functions (Rothe *et al.*, 1994).

TRAF2 and TRAF6 apparently utilize different molecular pathways for NF-κB and AP-1 activation (Chung *et al.*, 2002; Wu and Arron, 2003). However, in both cases, it has been shown that a unique form of non-degradative polyubiquitination plays an important role in TRAF downstream signaling. *In vitro* biochemical reconstitution showed that TRAF6, as a RING domain protein, directly supported the synthesis of these unique polyubiquitin chains, together with a ubiquitin conjugating enzyme system (Deng *et al.*, 2000; Wang *et al.*, 2001a). Similarly, negative regulation of NF-κB activity by a TRAF2-interacting deubiquitination enzyme specific for non-degradative polyubiquitin chains implicated the role of ubiquitination in TRAF2-mediated NF-κB activation (Brummelkamp *et al.*, 2003; Kovalenko *et al.*, 2003; Trompouki *et al.*, 2003).

Receptors that contain an intracellular DD are known as death receptors, which are exemplified by Fas and TNF-R1 (Ashkenazi and Dixit, 1998; Nagata, 1997). Fas is an effective prototypical cell killing receptor. The intracellular DD of Fas directly recruits a DD-containing protein known as Fas-associated DD (FADD) via DD-DD interactions (Chinnaiyan *et al.*, 1995). FADD also contains a death effector domain (DED), which further recruits the DED-containing pro-caspase-8 or pro-caspase-10 to elicit caspase activation and apoptosis (Boldin *et al.*, 1996; Muzio *et al.*, 1996; Wang *et al.*, 2001b).

TNF-R1-like death receptors, on the other hand, possess the intrinsic capability of both cell death and cell survival induction. The underlying mechanism for this duality lies on the recruitment of a multifunctional protein, TNF receptor-associated DD (TRADD), via DD-DD domain interactions, by TNF-R1 (Hsu *et al.*, 1996b). The amino terminal domain of TRADD (TRADD-N) recruits TRAF2 (Hsu *et al.*, 1996b), while the carboxyl terminal DD of TRADD can recruit FADD and a DD-containing Ser/Thr kinase, receptor-interacting protein (RIP), via DD-DD interactions (Hsu *et al.*, 1996a,b; Stanger *et al.*, 1995). Both TRAF2 and RIP contribute to survival signaling (Kelliher *et al.*, 1998; Yeh *et al.*, 1997), while FADD recruits and activates caspases to induce apoptosis. TNF-R1-induced apoptosis appears to involve the switch from an initial plasma membrane bound complex consisting of TNF-R1, TRADD, RIP1, and TRAF2 to a cytoplasmic complex consisting of TRADD, RIP1, FADD, and caspase-8 (Micheau and Tschopp, 2003). The control between the survival and death pathways from TNF-R1 is likely to be rather complex and may involve cellular inhibitors of apoptosis (cIAPs), FLICE-inhibitory proteins (FLIPs), and c-Jun N-terminal kinase (JNK) (Deng *et al.*, 2003; Irmler *et al.*, 1997; Micheau and Tschopp, 2003; Wang *et al.*, 1998).

D. *Structural and Functional Studies of Intracellular Signaling Pathways*

During the past few years, a large number of crystal and NMR structures (Table I), structure-based mutations (Table II) and thermodynamic data (Table III) have become available. These studies have led to a much more advanced understanding towards the molecular basis of signaling transduction of the TNFR superfamily.

TABLE I
Experimental Structures of Intracellular Signaling Proteins

Protein	Binding partner and sequence[a]	Method	Resolution	Protein, partner[b]	PDB code, reference
TRAF2 (327–501)			2.2 Å	6	1CA4 (Park *et al.*, 1999)
TRAF2 (310–501)	TNF-R2 (420–428) QVPFSKEEC	Co-crystallization	2.3 Å	6, 2	1CA9 (Park *et al.*, 1999)
TRAF2 (310–501)	CD40 (250–266) PVQETLHGCQPV TQEDG	Co-crystallization	2.7 Å	3, 2	1CZZ (Ye *et al.*, 1999)
TRAF2 (327–501)	CD40 (250–254) PVQET	Co-crystallization	2.0 Å	8, 8	1D00 (Ye *et al.*, 1999)
TRAF2 (327–501)	CD40 V251I mutant (249–254) YPIQET	Co-crystallization	2.0 Å	8, 8	1QSC (McWhirter *et al.*, 1999)
TRAF2 (327–501)	CD30 (576–583) MLSVEEEG	Co-crystallization	2.0 Å	6, 3	1D01 (Ye *et al.*, 1999)
TRAF2 (327–501)	Ox40 (262–266) PIQEE	Co-crystallization	2.0 Å	6, 6	1D0A (Ye *et al.*, 1999)
TRAF2 (327–501)	m4-1BB (231–236) GAAQEE	Co-crystallization	2.5 Å	6, 5	1D0J (Ye *et al.*, 1999)
TRAF2 (327–501)	LMP1 (204–210) PQQATDD	Co-crystallization	2.0 Å	3, 2	1CZY (Ye *et al.*, 1999)
TRAF3 (341–568)			2.8 Å	2	1FLK (Ni *et al.*, 2000)
TRAF3 (341–568)	CD40 (247–266) TAAPVQETLHGC QPVTQEDG	Soaking	3.5 Å	2, 2	1FLL (Ni *et al.*, 2000)
TRAF3 (377–568)	TANK (178–195) SVPIQCTDKTDK QEALFK	Soaking	2.9 Å	1, 1	1L0A (Li *et al.*, 2002)

(*continued*)

TABLE I (*continued*)

Protein	Binding partner and sequence[a]	Method	Resolution	Protein, partner[b]	PDB code, reference
TRAF3 (377–568)	TANK (171–191) IATDTQCSVPIQCT DKTDKQE	Soaking	3.5 Å	1, 1	1KZZ (Li *et al.*, 2002)
TRAF3 (377–568)	LTβR (385–408) PYPIPEEGDPGPPG LSTPHQEDGK	Soaking	3.5 Å	1, 1	1RF3 (Li *et al.*, 2003)
TRAF6 (346–504)			2.4 Å	1	1LB4 (Ye *et al.*, 2002)
TRAF6 (346–504)	CD40 (230–238) KQEPQEIDF	Co-crystallization	1.8 Å	1, 1	1LB6 (Ye *et al.*, 2002)
TRAF6 (346–504)	TRANCE-R (342–349) QMPTEDEY	Co-crystallization	2.0 Å	1, 1	1LB5 (Ye *et al.*, 2002)
TRAF2 (327–501)	TRADD-N (7–163)	Co-crystallization	2.0 Å	1, 1	1F3V (Park *et al.*, 2000)
TRADD-N (1–169)		NMR			1F2H (Tsao *et al.*, 2000)
Fas DD (202–319)		NMR			1DDF (Huang *et al.*, 1996)
FADD DD (89–183)		NMR			1FAD (Jeong *et al.*, 1999)
TNFR-1 DD (316–426) R347A		NMR			1ICH (Telliez *et al.*, 2000)
FADD DED (1–83) F25Y		NMR			1A1W (Eberstadt *et al.*, 1998)

[a]m: mouse; otherwise from human.
[b]Number of protein and partner molecules per crystallographic asymmetric unit.

TABLE II
Structure-Based Mutational Studies

TRAF	Receptor/ adapter, motif position	K_d[a]	Method[b]	Reference
TRAF3	CD40 (Q263A) P_{11}	–	SPR	(Ni *et al.*, 2000)
	CD40 (T254A) P_2	– –		
TRAF2, 3, 5	LTβR (P387A) P_{-3}	+, – –	GST-pulldown	(Li *et al.*, 2003)
	LTβR (P389A) P_{-1}	+		
	LTβR (D393A) P_3	+		
	LTβR (E390A/ E391A) $P_{0/1}$	–, – – –		
	LTβR (E390A/ E391A/D393A) $P_{0/1/3}$	– – –		
TRAF2, 3	TANK (Q182A) P_0	– –	GST-pulldown	(Li *et al.*, 2002)
	TANK (T184A) P_2	– –		
	TANK (D185A) P_3	– –		
	TANK (D188A) P_6	–		
	TANK (F194A) P_{12}	–		
TRAF6	CD40 (P237A) P_{-2}	+	GST-pulldown and NF-κB activation	(Ye *et al.*, 2002)
	CD40 (P237Q) P_{-2}	– –		
	CD40 (E239Q) P_0	– –		
	CD40 (D242A) P_3	–		
	CD40 (Q235A) P_{-4}	+		
	TRANCE-R (E342A, E375A, E449A) P_0	– –	NF-κB activation	
	TRANCE-R (E342A, E375A) P_0/P_0	+		
	TRANCE-R (E342A, E449A) P_0/P_0	+		
	TRANCE-R (E375A, E449A) P_0/P_0	+		
	IRAK (E706A) P_0	–	NF-κB activation	
	IRAK (E587A, E706A) P_0/P_0	– –		
	IRAK (E544A, E587A, E706A) $P_0/P_0/P_0$	– – –		
TRAF6 (R392A)	IRAK	– –	TRAF6 dominant negative effect on NF-κB activation	
TRAF6 (F471A)		– – –		
TRAF6 (Y473A)		– – –		
TRAF2	TRADD (Y16A)	32	SPR	(Park *et al.*, 2000)
	TRADD (Y16A, F18A)	>641		

(*continued*)

TABLE II (*continued*)

TRAF	Receptor/ adapter, motif position	K_d[a]	Method[b]	Reference
	TRADD (H65A)	9.1		
	TRADD (S67A)	9.2		
	TRADD (Q143A)	1.3		
	TRADD (D145K)	2.2		
	TRADD (R146A)	17		
	TRADD (L147A)	3.2		
TRAF2 ($S454A^{TRAF1}$)	TRADD	0.6	SPR	(Park *et al.*, 2000)
TRAF2 ($T401M^{TRAF3}$)		18		
TRAF2 ($L471K^{TRAF4}$)		U.D.		
TRAF2 ($L471R^{TRAF5}$)		U.D.		
TRAF2 ($D450K^{TRAF6}$)		5.8		
TRAF2 ($S467F^{TRAF4, 6}$)		5.4		

[a]+: no effect; −: decreased; − −: greatly decreased; − − −: drastically decreased; numbers: relative K_d to wild type interactions; U. D.: undetectable.
[b]SPR: surface plasmon resonance.

II. DOMAIN AND OLIGOMERIC STRUCTURES OF TRAFs

A. *TRAF-C Domains: Anti-Parallel β-Sandwiches with a Unique Topology*

The structure of a TRAF-C domain was first revealed from the crystal structure of the TRAF domain of human TRAF2 (Fig. 2A, E), alone and in complex with a receptor peptide from TNF-R2 (Park *et al.*, 1999). The main structural architecture of the TRAF-C domain comprises an eight-stranded anti-parallel β-sandwich, with strands $\beta1$, $\beta8$, $\beta5$, and $\beta6$ in one sheet and $\beta2$, $\beta3$, $\beta4$, and $\beta7$ in the other. Visual inspection of the SCOP structure database (Murzin *et al.*, 1995) and an automatic structural similarity search with the Dali program (Holm and Sander, 1995) showed that the TRAF-C domain represent a novel fold for an eight-stranded anti-parallel β-sandwich. However, the topology observed

TABLE III
Thermodynamic Characterizations of TRAF-Receptor Interactions

TRAF	Receptor/ adapter and sequence[a]	K_d[b]	Method[c]	Reference
TRAF2 (310–501)	CD30 (573–583) SDVMLSVEEEG $\Delta H = -14.0 \pm 0.8$ kcal/mol; $-T\Delta S = 8.03$ kcal/mol; $\Delta C_p = -245$ cal/mol·K	40 μM	ITC	(Ye and Wu, 2000)
	CD40 (250–266) PVQETLHGCQPVTQEDG $\Delta H = -9.5 \pm 1.0$ kcal/mol; $-T\Delta S = 3.87$ kcal/mol	60 μM		
	Ox40 (262–266) PIQEE $\Delta H = -13.0 \pm 0.9$ kcal/mol; $-T\Delta S = 7.22$ kcal/mol	50 μM		
	TNF-R2 (420–428) QVPFSKEEC	0.5 mM		
	m4-1BB (231–236) GAAQEE	1.0 mM		
	LMP1 (204–210) PQQATDD	1.9 mM		
TRAF2 (327-501)	TRADD (7–163)	7.8 μM	SPR	(Park *et al.*, 2000)
TRAF3 (341-568)	TANK (178–195) SVPIQCTDKTDKQEALFK	23.9 μM	ITC	(Li *et al.*, 2002)
TRAF6 (333-508)	CD40 (216–245) KKVAKKPTNK APHPKQEPQEINFPDDLPGS	59.9 μM	ITC	(Ye *et al.*, 2002)
	CD40 (230–238) KQEPQEIDF	84.0 μM		
	mTRANCE-R (337–345) RKIPTEDEY	78.0 μM		
	mTRANCE-R (370–378) FQEPLEVGE	770.0 μM		
	mTRANCE-R (444–452) GNTPGEDHE	763.0 μM		
	IRAK (539–548) PPSPQENSYV	518.1 μM		
	IRAK (582–590) PNQPVESDE	79.0 μM		
	IRAK (701–710) RQGPEESDEF	54.3 μM		
	IRAK-2 (523–532) SNTPEETDDV	66.2 μM		
	IRAK-M (475–483) PSIPVEDDE	142.2 μM		

[a]m: mouse; otherwise from human.
[b]K_d: dissociation constant; ΔC_p: heat capacity change with temperature.
[c]ITC: isothermal titration calorimetry; SPR: surface plasmon resonance.

for this domain may be reached by circular permutations of the β strands in Cu, Zn superoxide dismutase (PDB entry 2SOD) (Tainer *et al.*, 1982) and the C2 domain from synaptotagmin I (PDB entry 1rsy) (Sutton *et al.*, 1995).

The TRAF-C domain of TRAF2 contains several additional structural features. In particular, strands $\beta2$ and $\beta7$ are highly twisted with a β-bulge in each strand. Preceding the $\beta1$ strand, residues 348-350 form a parallel β structure ($\beta0$) with strand $\beta2$, immediately after the β-bulge in this strand. The side chains of residues in $\beta0$ partly cover one edge of the β-sandwich. Therefore the twisting of $\beta2$ appears to play a structural role in the TRAF-C domain. Since the $\beta7$ strand contains the primary receptor peptide interaction site, the β-bulge and the twist in this strand may also play important structural and biological roles. A three-turn helix is present in the crossover connection between strands $\beta1$ and $\beta2$. Comparison among the 48 independent copies of the TRAF-C domain of TRAF2 (McWhirter *et al.*, 1999; Park *et al.*, 1999; Ye *et al.*, 1999) showed that with the exception of the flexible $\beta7$-$\beta8$ loop (up to 3–4 Å in Cα distance), the structures are highly conserved in different crystal packing environment and they superimpose with r.m.s.d. of around 0.3–0.6 Å. Structural comparison of TRAF-C domain structures in the absence and presence of receptor peptide interactions has revealed that the TRAF-C domain is fairly rigid in its overall architecture and does not undergo large conformational changes upon receptor peptide binding.

Subsequently, structures of the TRAF-C domains of TRAF3 (Li *et al.*, 2002; Ni *et al.*, 2000) and TRAF6 (Ye *et al.*, 2002) were determined and shown to exhibit a similar structural architecture. TRAF3 structures are extremely similar to TRAF2 (Fig. 2B), with r.m.s.d. in Cα positions of around 0.5–0.8 Å, similar to the variations observed within the different TRAF3 structures. This structural conservation is consistent with the shared receptor binding specificity of TRAF2 and TRAF3. In comparison to TRAF2, TRAF3 contains an insertion at the $\beta5$-$\beta6$ loop that makes this loop somewhat more flexible and a deletion at the $\beta7$-$\beta8$ loop that makes the loop more ordered and defined.

In contrast to TRAF3, structural differences between TRAF2 and TRAF6 are much more pronounced (Fig. 2C), resulting in r.m.s.d. of 1.1–1.2 Å for 127 aligned Cα positions within 3.0 Å. While the central β-sheet superimposes well, significant differences are observed in almost all loop regions, including the $\alpha1$ helix within the $\beta1$-$\beta2$ loop, the $\beta3$-$\beta4$ loop (one residue insertion), the $\beta5$-$\beta6$ loop (three residue insertion), the $\beta7$ strand (one residue insertion) and the $\beta7$-$\beta8$ loop (one residue deletion). Relative to TRAF2, the $\beta3$-$\beta4$ loop of the TRAF-C domain of TRAF6 exhibits an

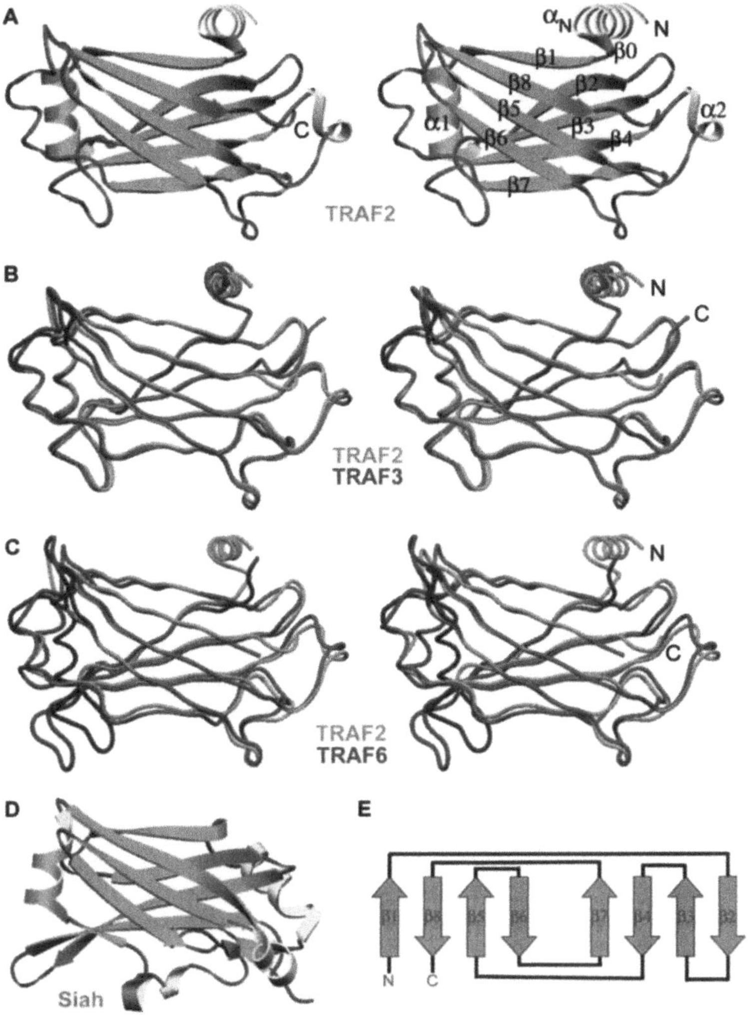

FIG. 2. TRAF domain structures. (A) Stereo drawing of the TRAF domain of TRAF2 with labeled secondary structures. (B) Superposition of the TRAF domain of TRAF2 (cyan) and TRAF3 (magenta). Regions with large differences between the two structures are shown in blue for TRAF2 and red for TRAF3. (C) Superposition of the TRAF domain of TRAF2 (cyan) and TRAF6 (magenta). Regions with large differences between the two structures are shown in blue for TRAF2 and red for TRAF6. (D) Ribbon drawing of Siah. (E) Topology of TRAF-C domains. (See Color Insert.)

up to 12 Å movement in Cα positions, so that it no longer interacts with the peptides in the TRAF6 complexes (see Section V). The remaining loops exhibit 2–5 Å movement in Cα positions. Due to the deletion in β7-β8 loop, this loop is more similar to TRAF3 than to TRAF2. The β6-β7 loop is disordered in the absence of receptor binding. Interestingly, the β2-β3 and the β4-β5 loops involved in TRAF trimerization are conserved between TRAF6 and TRAF2, demonstrating that on the structural level TRAF6 can form similar trimers.

Sequence analysis showed that a diverse set of proteins with unrelated functions to TRAFs appear to contain the TRAF-C domain. These include meprins, a family of extracellular metalloproteases (Uren and Vaux, 1996), MUL, the product of the causative gene in Mulibrey Nanism syndrome, USP7 (HAUSP), an ubiquitin protease, and SPOP, a POZ (poxvirus and zinc finger) domain-containing protein (Zapata *et al.*, 2001). Because of its similarities with meprins, TRAF-C domain was also dubbed meprin- and TRAF-homology (MATH) domain (Uren and Vaux, 1996).

Although sharing no significant sequence homology, a recent crystal structure of seven in absentia homolog (Siah), a member of the E3 ubiquitin ligase RING domain proteins, surprisingly revealed that its substrate-binding domain (SBD) is dimeric and adopts an eight-stranded anti-parallel β-sandwich fold that is highly similar to the TRAF-C domain (Polekhina *et al.*, 2002) (Fig. 2D, E). The Siah structure also reveals two novel zinc fingers in a region with sequence similarity to TRAFs. In addition, it appears that the SBD of Siah potentiates TNF-mediated NF-κB activation, suggesting potential functional similarities as well between Siah and TRAFs.

B. *Conserved Trimeric Structures of TRAF Domains: Energetics and Specificity*

A most striking structural feature of the TRAF domain, comprising a coiled coil TRAF-N domain followed by the TRAF-C domain, is the formation of a mushroom-shaped trimer with the TRAF-C domain as the cap and the coiled-coil domain as the stalk (McWhirter *et al.*, 1999; Park *et al.*, 1999; Ye *et al.*, 1999) (Fig. 3A, B). The trimer obeys perfect or near perfect threefold symmetry. The diameter of the mushroom cap ranges between 50 to 80 Å while the stalk is approximately 50 Å long for 5 heptad repeats (residues 311–347).

Both the coiled coil domain and the TRAF-C domain mediate TRAF domain trimerization. The three-stranded parallel coiled coil structure is stabilized by hydrophobic residues at positions A and D of the

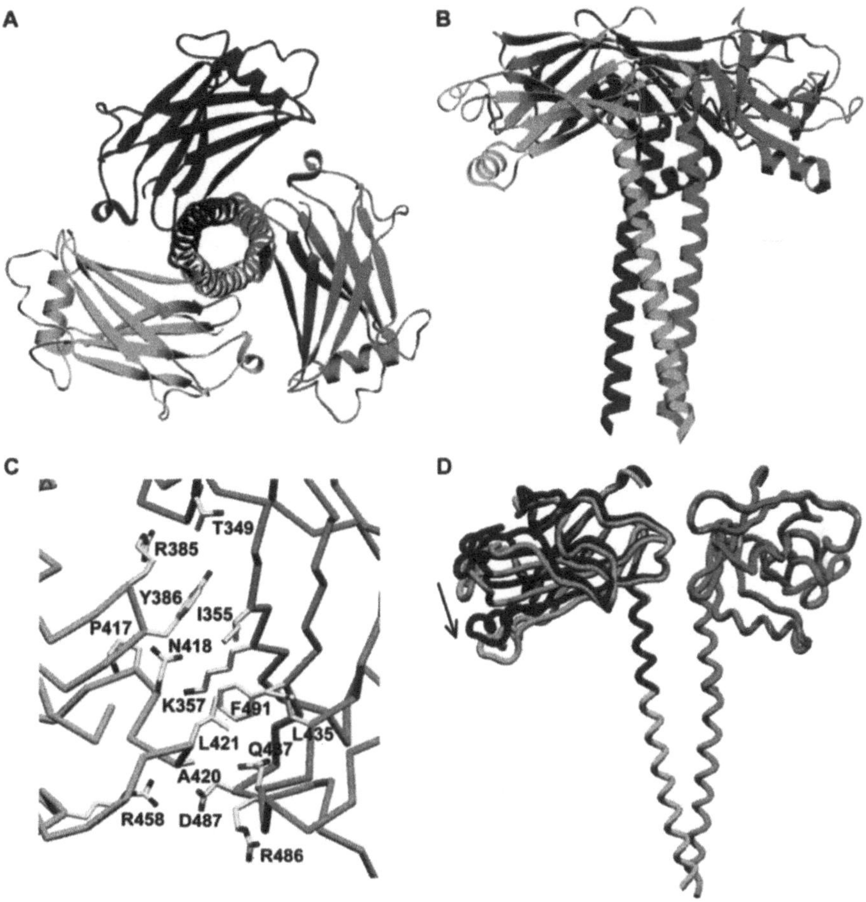

FIG. 3. TRAF trimerization. (A, B) Trimeric structure of the TRAF domain of TRAF2, shown with the three-fold axis into the page and vertical, respectively. (C) Detailed interaction between the TRAF-C domains in the trimer. (D) Observed structural variation among TRAF domain trimers. When superimposed onto one of the protomers in the trimer, a neighboring protomer may exhibit a different tilt of the mushroom cap-shaped TRAF-C domain. (See Color Insert.)

heptad repeats of the coiled coil. The trimeric interface of the TRAF-C domain is formed by packing one end of the β-sandwich (the β2-β3, β4-β5 and β6-β7 connections) against an edge and a face of the β-sandwich (β0, β1, and β8 strands, β5-β6 and β7-β8 connections) of the neighboring protomer (Fig. 3C). Most residues that participate in the formation of this interface are rather hydrophobic (such as I355, Y386, A420, L421 and F491). Hydrophilic interactions are also observed at this trimer interface, involving the side chains of K357, R385, R458, and D487.

The coiled coil domain plays an important role in stabilizing TRAF domain trimerization. Each TRAF-C domain buries roughly 640 Å^2 surface area upon trimerization (Park *et al.*, 1999), which is rather small compared to other stable protein-protein interactions, suggesting that the TRAF-C domain alone is not sufficient for trimerization (Janin *et al.*, 1988). Consistent with this analysis, solution studies on several TRAF domain constructs of TRAF2 showed that minimally three heptad repeats (residues 327–347), which increase the surface area burial to 1060 Å^2, are required for trimer formation (Park *et al.*, 1999). Interestingly, the coiled-coil domain of TRAF4 contains only three heptad repeats, the shortest among TRAFs. In comparison, the coiled-coil domain of TRAF2 appears to contain up to 14 heptad repeats, which could stretch to 140 Å long.

Structural and computational analyses suggest that the TRAF-C domain is the major specificity determinant of TRAF domain trimerization. Amino acid residues contributing to trimerization of the TRAF domain of TRAF2 are largely conserved among the TRAF family members (Park *et al.*, 1999), suggesting that all TRAFs may be able to form similar homotrimers as well. This hypothesis has been experimentally verified for the TRAF domains of TRAF1, TRAF2, TRAF3, and TRAF6 (Ni *et al.*, 2000; Park *et al.*, 1999; Pullen *et al.*, 1999b; Ye *et al.*, 2002). Formation of heterotrimers, as has been shown for TRAF1 and TRAF2, may also be possible (Rothe *et al.*, 1994). On the other hand, the coiled coil domains, which may be important as an energetic determinant of trimerization, do not contain conserved signature sequences characteristic of trimeric coiled coils (Harbury *et al.*, 1993). Prediction of coiled coil structures using the Multicoil program (Wolf *et al.*, 1997) showed the preference of TRAF1, TRAF2, and TRAF6 for dimeric, rather than trimeric, coiled coils. Therefore it appears that the TRAF-C domain, rather than the coiled coil domain, determines the observed specificity of TRAF trimerization.

Structural comparison among 18 different TRAF2 trimers and 6 TRAF3 trimers revealed that the trimeric structure is highly conserved. However, slight variations in the relative disposition of the protomers in the TRAF domain trimer are observed. These structural differences are exemplified by a flexing of the head of the mushroom relative to the stalk, on the order of 2–6° (Fig. 3D) and are unlikely to have functional implications. The coiled coil domains are more flexible, especially near the ends remote to the TRAF-C domains.

III. TRAF2-Receptor Interactions: Establishment of the Paradigm

A. *Conserved Recognition of Diverse Receptors*

The first glimpse of a TRAF2-receptor interaction was provided by the crystal structure of the TRAF domain of TRAF2 in complex with a receptor peptide from TNF-R2 (Park *et al.*, 1999). Each peptide is bound symmetrically to a shallow surface depression on the side of the mushroom-shaped trimer, extending from the top to the bottom rim of the mushroom cap (Fig. 4A, B). The peptide contacts one TRAF domain exclusively, with no contacts to the other two molecules of the trimer. This mode of interaction is distinct from the interaction between TNF and the extracellular domain of its receptor, where each receptor binds at the interface between neighboring protomers in the TNF trimer. Therefore, TRAF2-receptor interactions do not rely structurally on TRAF2 trimerization, but rely energetically on avidity-mediated affinity enhancement afforded by TRAF2 and receptor trimerization.

Because TRAF2 interacts with many different receptors, a major structural question is the molecular basis of this diversity. Towards understanding this question, a total of eight crystal structures of the TRAF domain of TRAF2 in complex with diverse receptor peptides have been determined (McWhirter *et al.*, 1999; Park *et al.*, 1999; Ye *et al.*, 1999) (three structures are with CD40) (Fig. 4C). These different structures encompass the three TRAF2 binding motifs proposed previously from biochemical and functional studies, the PxQx(T/S/D) (x = any amino acid) motif in LMP1, CD30, CD40 and CD27 (Aizawa *et al.*, 1997; Akiba *et al.*, 1998; Boucher *et al.*, 1997; Brodeur *et al.*, 1997; Devergne *et al.*, 1996; Franken *et al.*, 1996; Gedrich *et al.*, 1996; Sandberg *et al.*, 1997), the FSxEE (F = large hydrophobe) sequence in TNF-R2 and CD30 (Boucher *et al.*, 1997; Rothe *et al.*, 1994) and the QEE motif in 4-1BB and Ox40 (Arch and Thompson, 1998). The structures provide multiple observations, in different crystal packing environments, for the binding modes of receptor peptides from each of the three proposed TRAF binding motifs, four structures for the PxQxT motif (LMP1 and CD40), two for the FSxEE motif (CD30 and TNF-R2), and two for the QEE motif (Ox40 and 4-1BB). This collection of the different structures allowed a unified understanding and detailed comparison of the binding modes of the various receptor peptides.

Despite the high degree of sequence variation in the receptor peptides the structures surprisingly revealed that the peptides have a conserved binding mode and share a common binding site on the TRAF domain.

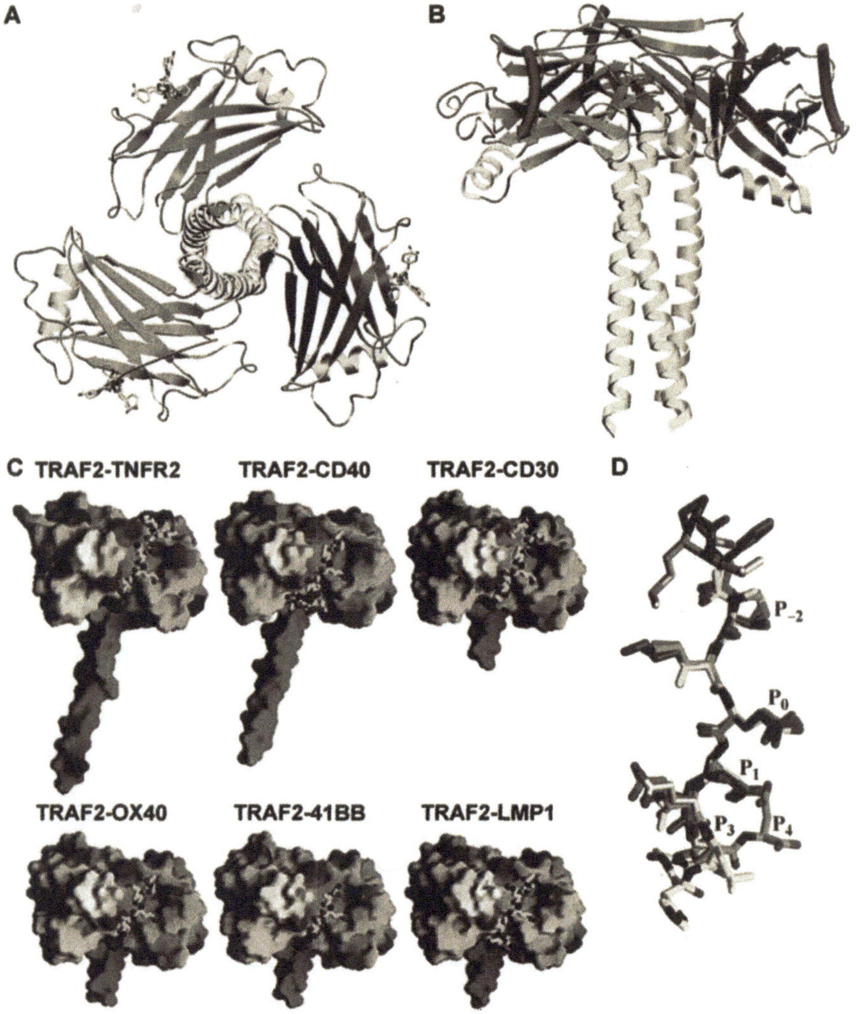

FIG. 4. TRAF2-receptor interactions. (A) Ribbon diagram of a TRAF2-receptor complex, looking down the three-fold axis. The bound receptor chains are shown as stick models. (B) Ribbon diagram of a TRAF2-receptor complex with the threefold axis vertical. The bound receptor chains are shown as arrows. (C) Surface electrostatic representation of TRAF2-peptide complexes. (D) Superposition of bound receptor peptides, showing the structural conservation of the main chain conformations and the side chain conformations at P_2, P_0, and P_1 positions. Parts of this figure were modified from Ye *et al.* (1999). (See Color Insert.)

Structural superposition of the 7 different structures of receptor peptide complexes indicated a most highly conserved central core of four residues, giving rise to an r.m.s.d of less than 0.1 Å among the main chain atoms of these residues (Fig. 4D). The structure-based sequence alignment of the receptor peptides showed that the third position of this four-residue core is invariably occupied by either a Gln or Glu residue and possesses the highest degree of sequence conservation. This residue was denoted as the zero position (P_0) of the TRAF binding motif. The conserved structural core of the receptor peptides, and therefore the TRAF binding motif, then extends from the P_{-2} to the P_1 positions.

The receptor peptides assume an essentially extended main chain conformation in the complex and cut across four β-strands ($\beta6$ in the first sheet, $\beta7$, $\beta4$, and $\beta3$ in the second sheet) on one side of the β-sandwich structure of the TRAF-C domain. Although the direction of the peptide chain is essentially perpendicular to these β strands, a portion of the peptide chain (P_{-1} to P_1) runs anti-parallel and adjacent to the latter half of strand $\beta7$ (residues 466–468), immediately after the β-bulge in this strand. This leads to three anti-parallel β-edge main chain hydrogen bonds between the peptide and the $\beta7$ of TRAF2, extending the four-stranded second β-sheet by one strand (Fig. 5A). In addition, the main chain amide group of the following residue is within hydrogen-bonding distance to the carboxylate group of D399 in TRAF2.

The formation of a β-sheet has been frequently observed in protein-peptide interactions, such as substrate recognition by certain serine proteases (Tong *et al.*, 1998) and peptide recognition by the PTB and PDZ domains (Kuriyan and Cowburn, 1997). Detailed analysis further revealed that the central portion of the receptor peptide (P_{-2}, P_0, and P_1 positions) is more twisted than a regular β-strand to possess the polyproline II (PPII) helix conformation. The PPII conformation is also frequently used in protein-peptide interactions such as those seen in the peptide recognition by SH3 domains (Lim *et al.*, 1994) and class II MHC molecules (Stern *et al.*, 1994). This conformation allows the peptide chain to twist in order to maximize the interaction of its side chains with a protein surface. As a consequence, large proportions of the side chains at the P_{-2}, P_0, and P_1 positions of the receptor peptides are buried at the TRAF2 interface. Therefore, in the case of TRAF2-receptor interactions, the main chain hydrogen bonds and the PPII conformation maximize both main chain and side chain interactions with the TRAF2 surface.

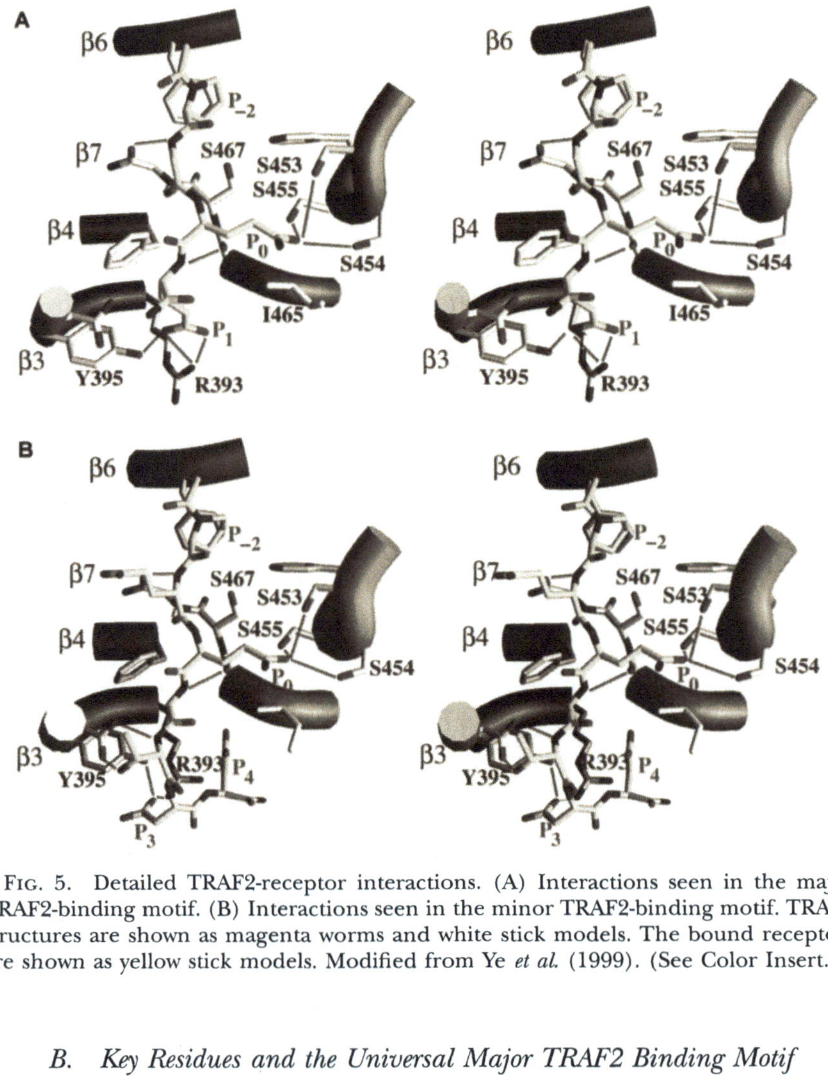

FIG. 5. Detailed TRAF2-receptor interactions. (A) Interactions seen in the major TRAF2-binding motif. (B) Interactions seen in the minor TRAF2-binding motif. TRAF2 structures are shown as magenta worms and white stick models. The bound receptors are shown as yellow stick models. Modified from Ye *et al.* (1999). (See Color Insert.)

B. Key Residues and the Universal Major TRAF2 Binding Motif

The major structural determinants of TRAF2-receptor peptide interactions appear to reside on the side chains at the P_{-2}, P_0, and P_1 positions within the TRAF2 binding sequences (Fig. 5A). They each engage into distinct pockets on the surface of TRAF2. The P_{-2} side chains are completely buried by the TRAF domain surface. The most frequently occupied residues at this position are Pro and Ser, both of which make extensive van der Waals contacts with TRAF2. In addition, Ser forms a hydrogen bond between its hydroxyl and the side chain of S467 in TRAF2. Model building of other amino acids at this position suggests that the size and the enclosure of the pocket may allow other non-charged medium-sized side chains such as Thr, Cys and Ile but may restrict the accommodation of larger side chains such as Gln. In 4-1BB, this position is an Ala, which would be expected to fit less well, consistent with the weaker TRAF2 binding affinity and the weaker electron density in this structure.

Shape complementarity and hydrogen bonding interactions are the major determinants for the selectivity of Glu and Gln residues at the P_0 position. The aliphatic portion of the side chains pack against I465 while the hydrophilic tip is surrounded by the three hydroxyls of S453, S454 and S455 in TRAF2. The Gln side chain is within hydrogen bonding distances to all three hydroxyl groups, perhaps making this one of the strongest anchoring points in the interaction. However, when this position is a Glu (as in CD30 and TNF-R2), only one hydrogen bond may be possible as the carboxylate side chain is positioned further away from the TRAF2 surface. Since there are no charged residues near the vicinity of the P_0 site, this difference may arise from the need for the negative charge in Glu to be more heavily solvated than its Gln counterpart.

The P_1 position in most TRAF2 binding sequences is occupied by a Glu residue, although LMP1 has an Ala at this position (see Subsection C). The carboxylate moiety of the Glu residue makes bi-dentate ion-pair interactions with the side chain guanidinium group of R393, and an additional hydrogen bond with the hydroxyl of Y395. These hydrogen-bonding interactions appear to require Glu specifically, as an Asp residue is too short to reach R393 and Y395 in TRAF2.

The sequence and structural conservations at the P_{-2}, P_0, and P_1 positions define a major TRAF2 binding motif that bears the consensus sequence of px(Q/E)E, in which Pro is shown in lower case because it can be substituted for other medium sized non-charged residues. Most of the binding sequences identified so far for TRAF1, 2, 3, and 5 are consistent with the motif, thereby explaining the recognition of diverse receptor sequences by TRAF2 (Fig. 6).

C. The Minor TRAF2 Binding Motif

The major TRAF binding motif has a conserved Glu residue at the P_1 position, which is involved in important ion-pair and hydrogen-bonding interactions with TRAF2. However, the receptor peptide from LMP1 has an Ala at the P_1 position (Fig. 6), which cannot participate in the ion-pair and hydrogen-bonding interactions. The structure of the LMP1 peptide in complex with TRAF2 (Ye *et al.*, 1999) showed that the interactions with the

	P_{-4}	P_{-3}	P_{-2}	P_{-1}	P_0	P_1	P_2	P_3	P_4	P_5	P_6
hTNF-R2 (422-432)	P	F	S	K	E	E	C	A	F	R	S
hCD40 (248-258)	A	A	P	V	Q	E	T	L	H	G	C
hCD30 (576-586)	M	L	S	V	E	E	E	G	K	E	D
hCD30 (559-569)	H	Y	P	E	Q	E	T	E	P	P	L
hCD27 (244-254)	T	I	P	I	Q	E	D	Y	R	K	P
hLTβR (386-410)	Y	P	I	P	E	E	G	D	P	G	P
hLTβR (400-410)	S	T	P	H	Q	E	D	G	K	A	W
hATAR (266-276)	T	V	A	V	E	E	T	I	P	S	F
hOx40 (260-270)	R	T	P	I	Q	E	E	Q	A	D	A
m4-1BB (230-240)	T	G	A	A	Q	E	E	D	A	C	S
m4-1BB (242-252)	R	C	P	Q	E	E	E	G	G	G	G
h4-1BB (232-242)	V	Q	T	T	Q	E	E	D	G	C	S
h4-1BB (244-254)	R	F	P	E	E	E	E	G	G	C	E
bLMP1 (204-214)	R	T	P	V	Q	E	S	G	Y	P	D
bLMP1 (219-229)	R	P	P	V	Q	E	T	G	G	G	G
bLMP1 (243-253)*	H	P	P	V	Q	E	T	G	G	G	G
bLMP1 (315-325)	H	P	P	V	Q	E	T	G	E	G	G
bLMP1 (359-369)	H	P	P	I	Q	E	T	G	N	G	G
hTANK (178-188)	S	V	P	I	Q	C	T	D	K	T	D
hLMP1 (202-212)	P	H	P	Q	Q	A	T	D	D	S	S
rLMP1 (315-325)#	P	Y	P	I	Q	A	T	D	G	G	N
rLMP1 (377-387)	P	H	P	I	Q	A	T	D	G	A	N
rLMP1 (425-435)	P	H	P	V	Q	A	S	D	G	G	D
Major Motif			p	x	Q/E	E					
Minor Motif			p	x	Q/E	x	x	D			

FIG. 6. Sequence alignment of TRAF2 binding sequences, illustrating the two TRAF2-binding motifs. h: human; m: mouse; b: bovine; r: rat. Modified from Ye *et al.* (1999). (See Color Insert.)

R393 and Y395 residues are mediated by an Asp at the P_3 position of this peptide (Fig. 5B). The side chain of R393 undergoes a small conformational change to accommodate this new interaction.

This structural information, together with sequence analysis, suggests that there is another TRAF2 binding consensus sequence motif, px(Q/E)xxD, in which the P_1 Glu is replaced by a P_3 Asp (Fig. 6). The distinguishing factor between the two motifs resides on the last residues: the Glu at the P_1 position for the major consensus and the Asp at the P_3 position for the minor consensus sequence. Structurally, similar interactions are seen between the acidic side chains of Glu at P_1 or Asp at P_3 and the conserved TRAF2 residues R393 and Y395.

In addition to LMP1, the intracellular protein, TANK (also known as I-TRAF) (Cheng and Baltimore, 1996; Rothe *et al.*, 1996), possesses the minor TRAF2 binding consensus motif (Fig. 6) and may interact with TRAFs similarly as seen in the TRAF2-LMP1 complex. This hypothesis is consistent with the crystal structure of the TRAF3-TANK peptide complex (see Section IV) (Li *et al.*, 2002).

D. *Extent and Variations*

Outside the conserved core of P_{-2} to P_1 residues, residues at P_2 show some degree of conservation of the main chain conformation, whereas additional C-terminal residues (P_3 and beyond) have large conformational differences among the various peptides (Fig. 4D). At the N-terminal side, the P_{-3} residue appears to have reasonable conservation of the main chain conformation as well, although several of the peptides only have an acetyl group at this position. Therefore a more relaxed definition of the TRAF binding core sequence would include the P_{-3} to the P_2 residue, covering 6 residues. In addition, both amino and carboxyl terminal extensions of the core make further contacts with the TRAF domain of TRAF2. At the amino terminal end, the TRAF domain complex with TNF-R2 contains ordered residues starting from the P_{-4} position, even though additional residues exist at the amino terminus in the peptide. At the carboxyl terminal end, the TRAF2-CD40 complex contains ordered residues up to the P_6 position. These results suggest that a complete TRAF2 binding sequence may contain eleven residues (from P_{-4} to P_6), which covers the entire span of one face of the TRAF domain surface. Residues at the variable positions within or outside the length of the TRAF2 binding motifs are generally exposed on the surface of complex, explaining the tolerance to substitutions. It should be kept in mind, however, that the conformations of end residues appear highly dependent on their side chain chemistry and thus the actual lengths of the TRAF binding regions may vary from receptor to receptor.

While it appears that all TRAF2 binding sequences identified so far bear either the major or minor TRAF2 binding motif, the presence of the motifs may not be sufficient for the interactions and that residues at other positions may be important as well in specific cases. For example, the side chain of the P_2 residue is situated close to that of D399 in TRAF2. When P_2 is a Thr, as in CD40, potential hydrogen bonding interactions are observed with the side chain of D399. Mutational studies have shown that this residue is important for TRAF2 interaction (Pullen *et al.*, 1999a). In addition, given the low affinity of these TRAF2-receptor interactions, it will not be too surprising if other variations of the TRAF2-interaction motifs are present.

The natural lengths of the intracellular domains of the TNF receptors may range from 36 residues as in Ox40 to several hundred residues as in TRANCE-R. Amino acid analyses using the PHD program (Rost and

Sander, 1994) suggest that most of these receptors may exist in near random coil conformations with low secondary structure components. This observation further supports the hypothesis that the TRAF2 binding sites of these receptors are primarily composed of linear sequences rather than three-dimensional composites. However, secondary interactions may ensue after the central TRAF2 binding sequences are docked onto the TRAF domain surface. The flanking regions may also modulate the exposure and the dynamic behavior of the TRAF2 binding determinants, thereby exerting influences on their functions.

E. *Conservation of Receptor Interaction in TRAF1, 2, 3, and 5*

All residues of TRAF2 that recognize the TRAF2 binding motifs are conserved among TRAF1, 2, 3, and 5, explaining the overlapping receptor-binding specificity among these TRAFs. In contrast, these residues are not conserved in TRAF4 and TRAF6. In TRAF4, three important TRAF2 residues, R393, Y395, and S467, are changed to Ser, Phe, and Phe, respectively. Curiously, TRAF4 has never been shown to interact with any receptor. In TRAF6, Y395, and S467 of TRAF2 are changed to His and Phe, respectively. The S467F change introduces the bulky phenyl ring on the surface, which may disallow the binding of receptor peptides in the same manner as observed here. This analysis is consistent with the unique receptor-binding specificity of TRAF6.

IV. TRAF3-Receptor Interactions: Similarities and Differences with TRAF2

The surface of TRAF2 used for interacting with the core regions of the receptor peptides is highly conserved among TRAF1, 2, 3, and 5, suggesting that these TRAFs would recognize receptors in a similar fashion. Crystal structures of TRAF3 in complex with peptides from CD40, TANK, and LTβR (Li *et al.*, 2002; Ni *et al.*, 2000) (Fig. 7A, B, C) have mostly confirmed this hypothesis, as seen in the crystal structures of TRAF3 in complex with TANK and LTβR. However, added insights and sometimes surprising observations have also been revealed by these structures.

One important insight derived from these structures is that, residues beyond, especially carboxyl terminal to, the core motifs, may also directly contact TRAF3 and play an additional energetic and functional role in TRAF3 interaction (Fig. 7A, B). In the TRAF3 complex with CD40 and LTβR, the peptides assume a reverse turn conformation that folds back onto the side of the TRAF domain. In the TRAF3 complex with TANK, the peptide chain continues to the bottom of the TRAF domain, making a boomerang-like shape on the surface of the TRAF domain.

Interestingly, both TANK and LTβR interacted with TRAF3 in a similar manner as seen in the TRAF2-LMP1 complex (Fig. 7D), which belong to the minor TRAF binding motif that we classified based on structural observations. TANK, bearing a sequence motif of PIQCTD, was predicted to interact similarly as in TRAF2-LMP1 complex. In the two independent structures of the TRAF3-TANK complexes, both peptides superimpose well with the receptor peptide conformations in the core regions of the TRAF2-LMP1 complexes and both showed the same hydrogen-bonding interactions between the P_3 Asp and R393 and Y395 of TRAF3. However, the side chain conformations of the P_0 residues are different between the two observations with one in a similar position to interact with TRAF3 via hydrogen bonding interactions and the other not. Mutational studies of TANK confirmed the critical role of the P_0, P_2, and P_3 residues and supported additional roles from the more remote residues, P_6 and P_{12} in TRAF3 interaction (Table II).

On the other hand, it was not apparent at all that the LTβR sequence (IPEEGD) could fall into the TRAF2 binding motif because the P_{-2} position is occupied by the Ile, not the neighboring Pro in this sequence. In fact, this LTβR sequence is a composite of the major and minor motifs since P_1 is a Glu and P_3 is an Asp. In the structure, both P_1 Glu and P_3 Asp interact with the conserved R393 and Y395 residues. Mutational studies showed that triple mutations at P_0, P_1, and P_3 of the LTβR sequence completely abolish the interactions with both TRAF2 and TRAF3.

The biggest surprise is from the structure of TRAF3 in complex with a CD40 peptide (Fig. 7C). A very similar peptide was also used in the crystallization of TRAF2 in complex with CD40 (Ye *et al.*, 1999). What is surprising was not that the CD40 peptide forms a hairpin loop on the surface of TRAF3 to make additional contacts with the TRAF domain, but the lack of conserved interactions of the P_0 and P_1 positions of the peptide with TRAF3 (Fig. 7E), despite high degree of structural conservation of TRAF3 with TRAF2, due to a significant shift in the main chain of the CD40 peptide. Instead, T254 at the P_2 position makes intra-chain hydrogen bonds with the main chain of E264 and D265 at the adjacent strand and Q263 of CD40 interacts with Y395 and D399 of TRAF2. Competitive binding studies using surface plasmon resonance (SPR) confirmed the importance of T254 and Q263 in TRAF3 binding.

It is likely that the weak affinity between TRAFs and receptor peptides made it possible for these interactions to be influenced by flanking residues in the receptor peptides and even by crystal packing. The lack of peptide interaction due to packing constraint has been seen in several

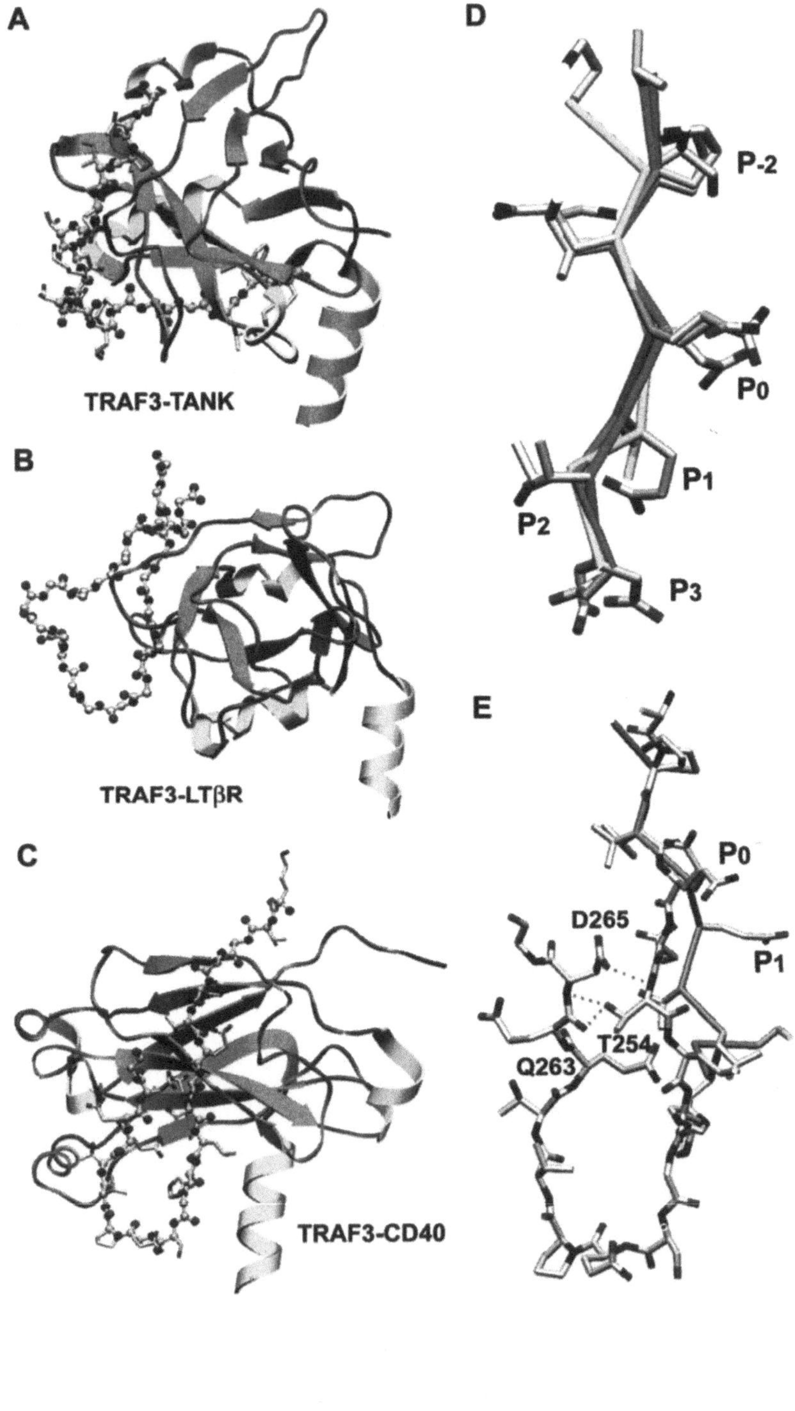
A
TRAF3-TANK
B
TRAF3-LTβR
C
TRAF3-CD40
D
P-2
P0
P1
P2
P3
E
P0
D265
P1
T254
Q263

of the crystal forms of TRAF2. For TRAF3, both the CD40 and TANK peptides also have a number of packing interactions in their crystals. This adds another layer of complexity in dissecting and extrapolating the modes of interactions in physiological settings.

V. TRAF6-Receptor Interactions: Distinct Specificity

In contrast to the shared receptor binding specificity among TRAF1, 2, 3, and 5, TRAF6 has a unique receptor interaction specificity that is not shared by other members of the TRAF family. In addition, TRAF6 is the only TRAF family member that also participates in the signal transduction of the IL-1R/TLR superfamily by its specificity for IRAKs.

In keeping with the distinct specificity, crystal structures of TRAF6 in complex with peptides from CD40 and TRANCE-R revealed striking differences in the mode of peptide binding (Ye *et al.*, 2002). Although a similar region of the TRAF domain is used for receptor interaction, the receptor chain cuts across the TRAF domain in a dramatically different trajectory that is 40° away from the TRAF2 direction (Fig. 8A, B). As a result, side chains of TRAF6 binding peptides interact with surface pockets on TRAF6 that are completely different from those on TRAF2 (Fig. 8C).

Among the numerous structural differences between TRAF6 and TRAF2, the insertion of P468 in the β-bulge of the $\beta 7$ strand makes possible for TRAF6 binding peptides to form more extensive main chain hydrogen bonds with this strand of the TRAF-C domain (residues 234–238 of CD40 and 344–349 of TRANCE-R with residues P468-G472 of TRAF6) (Fig. 8D). Rather than the more twisted PPII helix conformation for the core region of TRAF2 binding peptides, the TRAF6 binding peptides assume a typical β conformation. The peptides no longer interact with the $\beta 3$-$\beta 4$ loop in the TRAF6 complexes, due to a 12 Å movement in the position of this loop (Fig. 8B).

Similar to the nomenclature used for TRAF2 binding peptides, residues E235 of CD40 and E346 of TRANCE-R were designated as the P_0 position of TRAF6 binding peptides, because they occupy a similar, although not identical location to the P_0 residue (Q/E) in the TRAF2 binding motif

Fig. 7. TRAF3-receptor interactions. (A, B, C) Ribbon diagrams of TRAF3-TANK, TRAF3-LtβR and TRAF3-CD40 complexes. (D) Superposition of the TANK peptide (green) and the LTβR peptide (yellow) with the LMP1 peptide (pink) in the TRAF2-LMP1 complex. Only the core residues are shown. (E) Details of the CD40 conformation in the TRAF3-CD40 complex. Shown in pink is the superimposed CD40 conformation seen in the TRAF2-CD40 complex. (See Color Insert.)

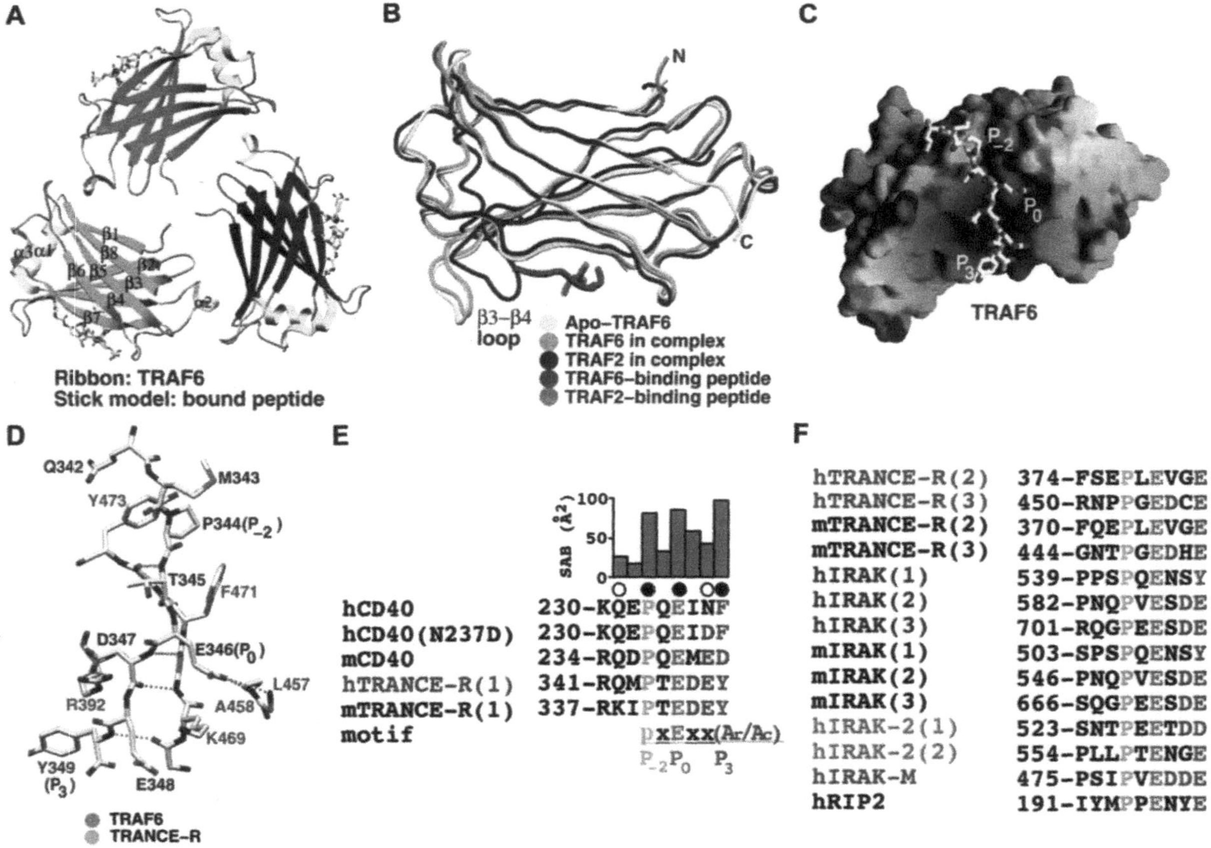
A
Ribbon: TRAF6
Stick model: bound peptide
B
N
C
β3–β4 loop
Apo–TRAF6
TRAF6 in complex
TRAF2 in complex
TRAF6–binding peptide
TRAF2–binding peptide
C
P-2
P0
P3
TRAF6
D
Q342
M343
Y473
P344(P-2)
T345
F471
D347
E346(P0)
L457
R392
A458
K469
Y349 (P3)
E348
TRAF6
TRANCE–R
E
SAB (Å2)
100
50
0
hCD40 230-KQEPQEINF
hCD40(N237D) 230-KQEPQEIDF
mCD40 234-RQDPQEMED
hTRANCE-R(1) 341-RQMPTEDEY
mTRANCE-R(1) 337-RKIPTEDEY
motif pxExx(Ar/Ac)
P-2 P0 P3
F
hTRANCE-R(2) 374-FSEPLEVGE
hTRANCE-R(3) 450-RNPPGEDCE
mTRANCE-R(2) 370-FQEPLEVGE
mTRANCE-R(3) 444-GNTPGEDHE
hIRAK(1) 539-PPSPQENSY
hIRAK(2) 582-PNQPVESDE
hIRAK(3) 701-RQGPEESDE
mIRAK(1) 503-SPSPQENSY
mIRAK(2) 546-PNQPVESDE
mIRAK(3) 666-SQGPEESDE
hIRAK-2(1) 523-SNTPEETDD
hIRAK-2(2) 554-PLLPTENGE
hIRAK-M 475-PSIPVEDDE
hRIP2 191-IYMPPENYE

(corresponding Cα distance 2.3 Å). The P_0 position is near the point of intersection between the two classes of peptides. Based on this naming system, peptide residues from P_{-4} to P_3 are in direct contact with TRAF6. Analyses of surface area burial and specific side chain interactions suggest that the P_{-2}, P_0, and P_3 residues contribute most to the structural interaction, each of which interacts with a specific pocket on the surface of TRAF6 (Fig. 8D).

The P_{-2} Pro residue interacts with the TRAF6 surface pocket formed by hydrophobic residues F471 and Y473 of TRAF6. The carboxylate of the P_0 Glu residue is recognized by hydrogen bonding with main chain amide nitrogen atoms of L457 and A458 and the aliphatic portion of the side chain exhibits a close fit with the TRAF6 surface. In addition, the carboxylate of the P_0 residue may form a favorable charge-charge interaction with the side chain of K469, although the interaction is not within hydrogen bonding distance. The P_3 residue, F238 of CD40 or Y349 of TRANCE-R, is adjacent to several aromatic and basic residues of TRAF6, including H376, R392, H394, and R466. An amino-aromatic interaction is observed between Y349 of TRANCE-R and R392 of TRAF6. Structurally, a similar interaction should be possible for F238 of CD40, or for an acidic residue, which is present in mouse CD40.

Interestingly, the peptide interaction sites of TRAF6 are quite analogous to those of TRAF2. The P_{-2} pocket forming residues are replacement of S467 and C469, respectively, which form the P_{-2} pocket in TRAF2 that is about 3 Å away from the corresponding TRAF6 pocket. The P_0 pocket residues L457 and A458 correspond to S454 and S455 of TRAF2, the side chains of which provide the hydroxyls for hydrogen bonding with the Glu/Gln side chains at the P_0 position. Similarly, residues R392 and H394 are structural correspondent of R393 and Y395 of TRAF2, two critical residues of the TRAF2 P_1 pocket. Evolutionarily speaking, TRAF6 is one of the oldest TRAF family members. The previous observations suggest an evolutionary mechanism in which the same mutations that abolish the interaction with TRAF6 also create new specificity for TRAF2.

FIG. 8. TRAF6-receptor interactions. (A) TRAF6 trimeric model. (B) Superposition of TRAF6 with TRAF2, with bound receptor peptides. (C) Electrostatic surface of TRAF6 in complex with the TRANCE-R peptide. (D) Detailed interactions between TRAF6 and TRANCE-R. (E) Sequence alignment of CD40 and TRANCE-R and the corresponding surface area burial (SAB) and the TRAF6-binding motif. (F) TRAF6-binding sequences identified based on the structurally defined motif. Modified from Ye *et al.* (2002). (See Color Insert.)

Structure-based sequence alignment of TRAF6 binding sites in human and mouse CD40 and TRANCE-R led to the definition of a TRAF6 binding consensus sequence pxExx(Ar/Ac) (p shown in lowercase because of tolerance for other small to medium sized residues, Ar for aromatic and Ac for acidic residues) (P_{-2} to P_3) (Fig. 8E, F). Similarly as observed for TRAF2 binding peptides, the P_{-2} position is likely able to accommodate conserved changes to small and medium-sized hydrophobic residues. Consistent with this analysis, mutational studies on CD40 showed that the Pro residue can be changed to Ala without showing qualitative differences in TRAF6 interaction and signaling, while a change to Gln abolishes interaction and signaling (Table II). The P_0 position could similarly accommodate Gln as well as Glu, but a change to Ala drastically reduced interaction and signaling. Similarly, removing the side chain of the P_3 residues drastically reduced TRAF6 interaction and signaling. In addition, it appears that the P_1 and P_2 positions may have a preference for acidic residues due to their complementarity to the basic TRAF6 surface formed in particular by the side chains of R392 and K469 at this region. Isothermal titration calorimetry (ITC) measurements showed that peptides with acidic residues at these positions possess higher affinity to TRAF6 (Table III).

VI. Thermodynamics of TRAF-Receptor Interactions

A. *Weak Affinity and Avidity*

TRAF recruitment requires ligand-induced changes in receptors that allow simultaneous interactions of each TRAF trimer with three receptor intracellular domains. This observation implicates that monomeric TRAF-receptor interactions are of low affinity so that the interactions do not occur in the absence of receptor activation. A number of quantitative biophysical characterizations with isothermal titration calorimetry (ITC) and surface plasma resonance (SPR) have provided solid support to this view (Table III).

Measurements of TRAF2-receptor peptide interactions using ITC showed that TRAF2 interacts with CD40, CD30, and Ox40 peptides with dissociation constants in the range of 40–60 μM and with TNF-R2, 4-1BB, and LMP1 with dissociation constants in the 0.5–1.9 mM range (Ye and Wu, 2000). Similarly, TRAF3 interacts with TANK with a dissociation constant of 24 μM (Li *et al.*, 2002). In the case of TRAF6, a range of dissociation constant between 50–770 μM was measured by ITC (Ye *et al.*, 2002). In all cases, the heat release during the titration of receptor peptides into a

TRAF solution exhibited excellent agreement with ideal binding, indicating the presence of a single type of binding site and the lack of cooperativity in the interaction.

The quantitative measurements between TRAFs and receptor peptides likely represent TRAF-receptor interactions in the context of the full-length intracellular domains. The ability of the short five-residue peptide of Ox40 to confer an affinity to TRAF2 as high as the longer peptides is consistent with the structural observation that a core of a few residues appears to dominate the interaction with TRAF2 (Park *et al.*, 1999; Ye *et al.*, 1999), although it is likely that residues beyond the core contribute further to the interactions. In the case of CD40, characterization by SPR of the interaction between TRAF2 and monomeric full-length cytoplasmic domain of CD40 gave rise to a dissociation constant of 30 μM (Pullen *et al.*, 1999b), similar to the 60 μM dissociation constant derived from the ITC measurement on a CD40 peptide.

The measured TRAF-receptor interaction affinities are much lower than most protein-protein or protein-peptide interactions involved in signal transduction (Kuriyan and Cowburn, 1997), suggesting that TRAF recruitment is entirely dependent on affinity enhancement through avidity. However, the exact magnitudes of this enhancement by trimerization are not really clear and may depend on the exact separation, geometry and conformational state of the oligomerized receptors. In an artificial experiment using trimeric coiled coil to trimerize the intracellular domain of CD40, a 12-fold higher affinity was observed using SPR experiments. However, it may be expected that in optimal cases, the avidity effect should be exponential.

B. Favorable Enthalpy, Unfavorable Entropy, and Induced Fit

ITC measurements on TRAF2-receptor interactions have revealed invariably favorable enthalpies and unfavorable entropies, indicating that all these interactions are energetically driven by exothermic enthalpy. The enthalpy of the interaction showed a relative large negative linear dependence with temperature, as measured for the TRAF2-CD30 interaction at 10, 20 and 30 °C (Table III).

Structurally, TRAF2-receptor interactions possess both hydrophilic and hydrophobic components. For example, in the TRAF2-CD30 interaction, approximately 750 Å^2 of hydrophobic surface area and 400 Å^2 polar surface area are buried at the interface. Therefore the favorable enthalpy may arise from the significant amount of polar interactions, including main chain hydrogen bonds, side chain hydrogen bonds and salt bridges in the complex. On the other hand, the relatively large negative heat capacity change is consistent with the presence of significant hydrophobic component in the interactions.

The large dependence of enthalpy with temperature is indicative of a specific interaction, even though that the affinities of these monomeric interactions between TRAF2 and receptors are rather low. As suggested from thermodynamic studies of protein-DNA interactions, a non-specific weak complex held together by electrostatic forces often exhibits little temperature dependence of enthalpy (Ladbury, 1995).

The observed unfavorable entropy appears to contradict the presumably favorable solvation entropy from the burial of significant hydrophobic areas at the TRAF2-receptor interfaces. It is likely that the unfavorable entropy may be largely due to conformational restraints on the receptor peptides upon TRAF2 interaction. Secondary structure predictions of cytoplasmic tails of most TRAF-interacting TNFRs suggest that these receptor tails do not have a pre-formed well-ordered three-dimensional structure. Rather, linear peptides from localized regions of the receptors are responsible for TRAF2 interaction. Therefore it is likely that the peptides, or full-length receptors, are flexible before docking onto the protein and penalized by conformational entropy. This suggests the involvement of conformational changes and induced fit in the interaction between TRAF2 and receptor peptides.

VII. TRAF2-TRADD Interaction: A Novel Mode of TRAF Signaling

A. *The TRADD-N Domain*

The TRAF2-TRADD interaction is mediated by the TRAF domain of TRAF2 and the N-terminal domain of TRADD (TRADD-N). The TRADD-N domain has so far only been found in mammalian TRADD proteins. It folds into an α-β sandwich with a four-stranded β-sheet and six α-helices, each forming one layer of the structure (Park *et al.*, 2000; Tsao *et al.*, 2000) (Fig. 9A). The β-sheet is entirely anti-parallel and slightly twisted with a strand order of $\beta 2$, $\beta 3$, $\beta 1$, and $\beta 4$. There are two helices each in the $\beta 1$-$\beta 2$ and $\beta 3$-$\beta 4$ crossover connections while the $\beta 2$-$\beta 3$ connection is hairpin-like. The remaining two helices (E and F) are near the carboxy-terminus of the domain; the loop in between (EF loop) partly covers one end of the exposed face of the β-sheet.

The basic topology of the TRADD-N domain resembles the family of ferredoxin-like α-β sandwiches (Orengo and Thornton, 1993), which are often present as domains in larger structures such as the palm domain of

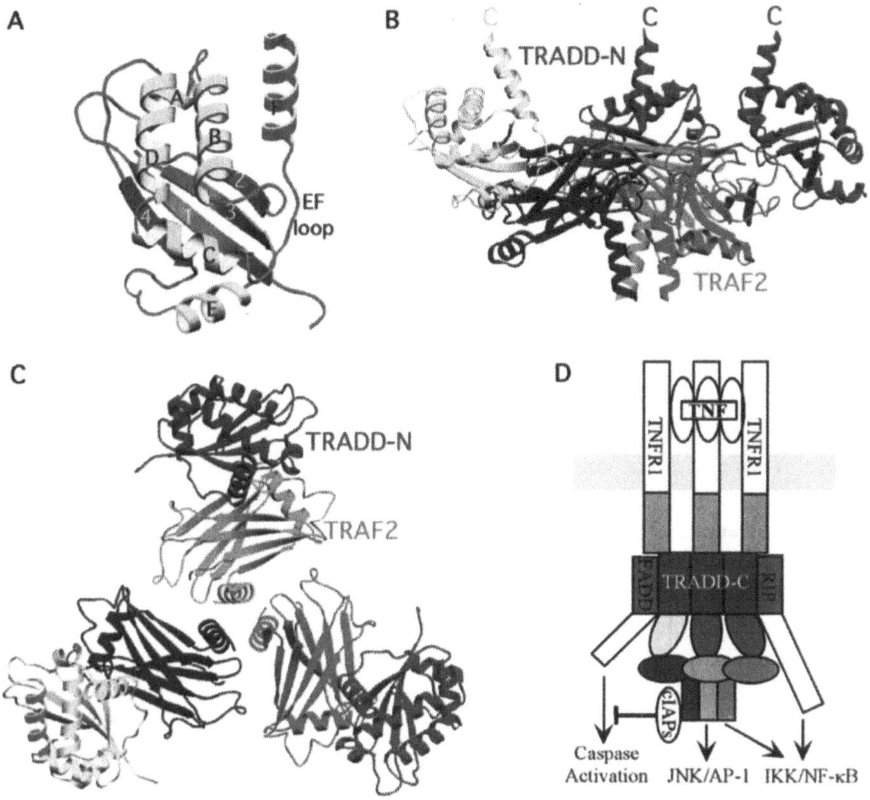

FIG. 9. TRAF2-TRADD interaction. (A) Ribbon drawing of the TRADD-N domain. (B, C) Ribbon diagrams of the TRAF2-TRADD complex. (D) Schematic representation of the TNF-R1 signaling complex. Reproduced from Park *et al.* (2000). (See Color Insert.)

many polymerases and the dimerization domain of carboxypeptidases. However, the TRADD-N structure is much more elaborate than a typical ferredoxin-like fold, as represented by extra helices in $\beta1$-$\beta2$ and $\beta3$-$\beta4$ connections and two additional helices near the carboxy-terminus.

B. *TRADD-TRAF2 Interface: Interactions and Energetics*

The trimeric TRAF domain of TRAF2 imposes the threefold symmetry to the stoichiometrically bound TRADD-N (Fig. 9B, C). When viewed from the side of the mushroom-shaped trimeric structure of TRAF2,

TRADD-N interacts with TRAF2 at the upper rim of the mushroom cap and adds a wing-like structure to the mushroom. The carboxyl terminus of TRADD-N projects up to the membrane-proximal direction of the complex, indicating the possible location of the carboxyl terminal death domain of TRADD. In this orientation, TRADD can be recruited to TNF-R1 via its death domain and forms the central platform for recruiting other signaling molecules such as FADD, RIP and TRAF2 (Fig. 9D).

The TRAF2-TRADD interaction occupies a similar surface at the edge of the β-sandwich of TRAF2, indicating the competitive nature of TRAF2-TRADD and TRAF2-receptor interactions. Each TRADD-N contacts one protomer of the TRAF domain. The binding is at the β-sandwich domain exclusively, away from the coiled-coil domain. The interaction between TRAF2 and TRADD buries a total of 1500 Å^2 surface area, in contrast with the smaller protein-peptide contacts in TRAF2-receptor interactions. There are small local conformational adjustments in the Cα positions of TRAF2 (0.5–1.0 Å) within or immediately adjacent to the TRADD binding site. We suspect that TRADD-N (especially the EF loop) may undergo larger conformational rearrangement upon TRAF2 binding based on the relative instability of TRADD-N in its isolated state.

The actual molecular contacts of TRAF2 with TRADD and with receptors are entirely different, and TRADD does not possess TRAF2 binding motifs. The interface between TRADD-N and the TRAF domain of TRAF2 possesses dual "ridge into groove" contacts, in which both TRADD-N and TRAF2 contain reciprocal elevations and depressions (Fig. 10A). Two distinct and adjacent regions may be defined (Fig. 10B). Region I is mediated by the exposed shallow face of the β-sheet of TRADD and a surface protrusion of TRAF2 formed by β7, the following loop and the connection between β3 and β4. Many residues, such as Y16, F18, H65, and S67 and I72 of TRADD and T401, H406, L471 and P474 of TRAF2, collectively contribute to this interaction (Fig. 10C). Region II is mediated by a highly charged prominent ridge formed by TRADD residues 143–149 in the EF loop and a surface depression of TRAF2 presented by strand β6 and the following loop. Residues 145–147 of TRADD form anti-parallel main chain hydrogen bonds with residues 448–450 of TRAF2 in the connection between β6 and β7. Many side chain hydrogen bonds and salt bridges exit at this interface including the hydrogen bonds between R146 of TRADD and D445 of TRAF2, between R76 of TRADD and D450 of TRAF2, between Q143 of TRADD and S454 of TRAF2, and between D145 of TRADD and main chain of G468 of TRAF2 (Fig. 10D). Water molecules abound at Region II of the interaction and at the boundary between the two regions (Fig. 10E).

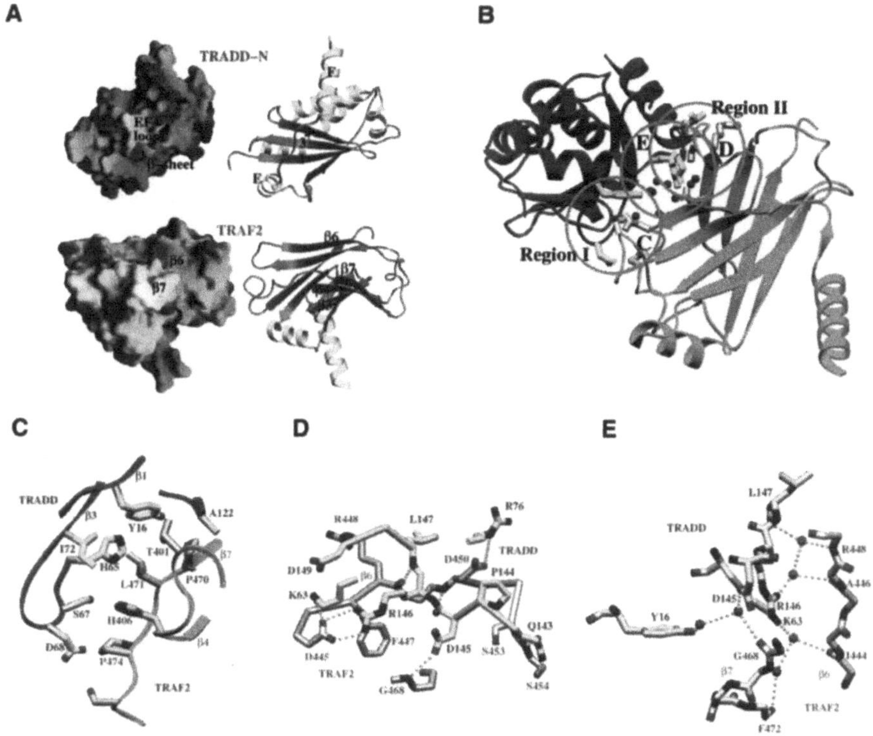

FIG. 10. Detailed TRAF2-TRADD interaction. (A) Interaction surfaces and their locations on the individual structures (in red). (B) Molecular interactions at the two regions of the interactions. (C, D, E) Details of region I, region II and water-mediated interactions, respectively. (See Color Insert.)

Residues in the two regions showed differential effects on the binding affinity, in a manner unrelated to their surface area burial at the interface (Table II). In general, alanine substitutions of residues in Region I (Y16, F18, H65, and S67) had much more drastic effects, despite their smaller surface area contribution. This may be explained by their complete solvent inaccessibility in the complex and the largely hydrophobic nature of the contact. Residues in Region II were selected for their large surface area burial and their abilities to form direct and water-mediated hydrogen bonds or salt bridges at the interface (Q143, D145, R146 and L147). Rather unexpectedly, mutations on Q143, D145 and L147 had minimal effects on the binding energy ($\leq$3-fold). The only residue that exhibits a large effect is R146, which forms buried bi-dentate hydrogen bonds and a salt bridge to residue D445 of TRAF2. These structure-based mutational studies reveal that Region I is the primary energetic determinant of the interaction, while Region II contributes to specificity and perhaps long range attraction forces to facilitate association.

C. Higher Affinity and Distinct Specificity: More Efficient Signaling and Suppression of Apoptosis

SPR measurements revealed that the TRAF2-TRADD interaction possesses a dissociation constant of 7.8 μM, which is significantly stronger than direct TRAF2-receptor interactions ($K_d = 40$ μM–1.9 mM) (Ye and Wu, 2000). The higher affinity of the TRAF2-TRADD interaction suggests that TRADD might be a stronger inducer of TRAF2 signaling. Comparison of signal transduction from the two TNF receptors, TNF-R1, a TRADD-mediated TRAF2 signaling receptor, and TNF-R2, a direct receptor-mediated TRAF2 signaling receptor, substantiated this hypothesis. TNF-induced JNK activation was used as the readout for TRAF2-mediated signal transduction because TRAF2 is the major activator of JNK (Yeh *et al.*, 1997). Indeed, in response to TNF, the TNF-R1-expressing fibroblasts had significantly higher level of JNK activation than the TNF-R2-expressing fibroblasts when normalized to receptor and cell numbers.

In addition to higher affinity, swapping mutagenesis showed a specificity of TRADD for TRAF1 and TRAF2, but not TRAF3, 4, 5, and 6 (Table II). This is in contrast to the direct TRAF recruitment by the subgroup of TNF receptors that do not contain a death domain, which exhibits a more promiscuous specificity for TRAF1, 2, 3, and 5 (Ni *et al.*, 2000; Park *et al.*, 1999; Ye *et al.*, 1999). Interestingly, TRAF1 and TRAF2 interact constitutively with the cellular caspase inhibitors cIAP1 and cIAP2 (Fig. 9D), which were originally isolated from the TNF-R2 signaling complex (Rothe *et al.*, 1995). In addition, this interaction with cIAPs requires both TRAF1 and TRAF2 (Rothe *et al.*, 1995), suggesting that TRAF1/2 hetero-oligomer interacts with cIAP1 and cIAP2. The specificity of TRADD for TRAF1 and TRAF2 ensures the recruitment of cIAPs to the signaling complex, which may be important for direct caspase-8 inhibition (Wang *et al.*, 1998) and the immediate suppression of apoptosis at the apical point of the cascade.

The high affinity and restricted specificity of TRADD for TRAF1/2 explain that although TNF-R1 and related death receptors are capable of both cell survival promotion and cell death induction, under most circumstances, they rarely induce cell death. TRAF2 signaling plays an important role in protection from apoptosis because TRAF2-deficient mice are overly sensitive to TNF-induced cell death (Yeh *et al.*, 1997). This TRAF1/2-mediated protection from apoptosis is dependent on the high affinity of

TRADD-TRAF2 interaction because TRADD mutants with even moderately decreased affinity to TRAF2, such as H65A, Y16A and S67A, showed significantly increased cell death induction. The much higher affinity of TRAF2 recruitment by TRADD versus the direct TRAF2 recruitment by TNF receptors without a death domain suggests the importance of preserving this high affinity in the signal transduction by TNF-R1 and related receptors. A modest decrease in the monomeric affinity, which may translate into a larger difference in the multimeric interaction, could lead to an imbalance in the regulation between cell survival and cell death.

There may be two ways that TRAF2 can protect cells from apoptosis and the cooperation of the two mechanisms may be necessary for efficient apoptosis suppression. The first mechanism has to do with recruitment of the cellular caspase inhibitors, cIAP1 and cIAP2, to the TNF-R1 signaling complex by TRAF1 and TRAF2 to inhibit caspase-8 activation (Rothe *et al.*, 1995; Wang *et al.*, 1998). This mechanism acts at the entry point of TNF-mediated apoptosis, is independent of gene transcription and explains the specificity of TRADD for TRAF1 and TRAF2 but not other members of the TRAF family. The second mechanism may be related to the ability of TRAF2 to activate NF-κB, which may induce the expression of anti-apoptotic genes to suppress cell death (Beg and Baltimore, 1996). Since apoptosis induction can be fast and does not require gene transcription and protein synthesis, the first mechanism is likely to be crucial in placing apoptosis under check while the second mechanism may further strengthen the anti-apoptotic function of TRAF2.

However, under certain circumstance, the TNF-R1 signaling complex can switch to its apoptotic mode. One possible mechanism may have to do with the mitochondrial release of Smac in a JNK-dependent manner (Deng *et al.*, 2003). As an IAP-interacting protein, Smac may compete with TRAF1/2 for cIAP interaction to remove cIAPs from the TNF-R1 signaling complex. Another possible mechanism is the involvement of NF-κB-inducible gene product FLIP. TNF-R1-induced apoptosis appears to involve the switch from an initial plasma membrane bound complex consisting of TNF-R1, TRADD, RIP1 and TRAF2 to a cytoplasmic complex consisting of TRADD, RIP1, FADD, and caspase-8 (Micheau and Tschopp, 2003).

VIII. TRAF Signaling Inhibitors

Because of the role of TRAFs in inflammation and tumorigenesis, downregulation of TRAF signaling may serve potential therapeutic benefits to many diseases. A way to inhibit TRAF signaling is to identify reagents that inhibit TRAF-receptor interactions. The ability of short peptides to interact with TRAFs provides templates for peptidomimetic drug identification approaches. In addition, small molecules may be identified that fit into the specific surface pockets of TRAFs, such as the hydrophobic P_{-2} site.

As a proof of principle that inhibition of TRAF-receptor interactions could act as potential therapeutic means, cell permeable TRAF6-interacting decoy peptides were constructed by fusing the TRAF6 binding sequences from TRANCE-R with a hydrophobic sequence of the Kaposi fibroblast growth factor signal peptide (Ye *et al.*, 2002). They were tested for their inhibitory effects on TRANCE-R-mediated signal transduction. Pre-treatment with the decoy peptides inhibited endogenous TRANCE-R mediated NF-κB activation upon TRANCE stimulation in a dose-dependent manner. Further, the decoy peptides inhibited TRANCE-induced osteoclast differentiation in a cell line model and in primary cells, without affecting cell viability. These results demonstrate that peptides containing the TRAF6 binding motif can act as decoys to inhibit TRAF6 signaling and associated biological functions.

The effectiveness of these decoy peptides to compete with endogenous oligomeric interactions raises the optimism that it is possible for monomeric interactions to compete with the endogenous oligomeric forms of the interactions. There may be two possible mechanisms by which the cell permeable decoy peptides could work. First, endogenous concentrations of receptors may be significantly low to allow competition by the decoy peptides due to concentration advantage. Second, it is possible that the hydrophobic leaders of the decoy peptides may render the peptides to associate with cellular membranes to achieve higher local concentrations of the peptides.

Several aspects of the TRAF-receptor interactions may assist the design of high affinity TRAF binding inhibitors. The low affinity interaction indicates a non-ideal steric or chemical complementarity between TRAFs and these receptor peptides, increasing the possibility for affinity improvement. Surface pockets, especially the hydrophobic P_{-2} pocket, may be targets for small molecules. In addition, as reduction of conformational entropy may contribute negatively to the interaction, an increase in affinity may be achieved by rigidifying potential TRAF2 binding moieties.

IX. DD and DD-DD Interactions

Death domains (DDs) are found in the intracellular portion of death receptors such as Fas and TNF-R1 and death receptor-interacting adapter proteins such as FADD and TRADD. They are protein-protein interaction domains (Fesik, 2000). The structure of DDs, as first revealed by the NMR structure of the death domain of Fas, consists of six amphipathic antiparallel α-helices arranged in a unusual Greek Key topology (Huang *et al.*, 1996) (Fig. 11A). The overall fold of other death domains, such as the NGFR p75 (Liepinsh *et al.*, 1997), FADD (Jeong *et al.*, 1999), and TNFR-1 (Sukits *et al.*, 2001; Telliez *et al.*, 2000) were found to be similar to the Fas death domain with only minor differences in the length and orientation for some of the α-helices (Fesik, 2000) (Fig. 11B).

Structural studies have revealed that several other domains involved in cell death and inflammatory signaling transduction, including the death effector domain (DED) (Fig. 11C), the caspase recruitment domain (CARD) and the Pyrin domain (PYD), also possess the same six helix bundle structures of DDs (Chou *et al.*, 1998; Eberstadt *et al.*, 1998; Hiller *et al.*, 2003), forming the death domain superfamily. Interestingly, interactions have only been observed among proteins within the same subfamilies with no cross interactions between proteins from different subfamilies.

Despite the availability of a large number of structures of isolated domains in the death domain superfamily, only two complex structures have been determined, the Pelle/Tube DD/DD complex (Xiao *et al.*, 1999) and the Apaf-1/Procaspase-9 CARD/CARD complex (Qin *et al.*, 1999), revealing two samplings of modes of interactions in this superfamily (Fig. 11D, E). Regardless of the details of the interactions, the biggest surprise, in both cases, is perhaps the asymmetry of the interactions, considering what might have been expected for homophilic interactions. Pelle and Tube are DD-containing proteins involved in Toll signaling in *Drosophila* with Pelle as a Ser/Thr kinase and Tube as an adapter protein. The Pelle/Tube interaction is mediated by the insertion of the α4 helix and the following loop of Pelle into a groove of Tube formed by the α1-α2 corner, α6 and the preceding loop and by the insertion of the C-terminal tail of Tube into a cavity formed by the α4-α5 and α2-α3 hairpins of Pelle (Qin *et al.*, 1999). Apaf-1 and Procaspase-9, on the other hand, are essential components of the mitochondrial apoptotic pathway. The Apaf-1/Procaspase-9 interaction is mediated by the mutual recognition of the slightly concave surface of procaspase-9 formed by the positively charged α-1a, α-1b and α4 helices and the convex surface of Apaf-1 formed by the negatively charged α2 and α3 helices.

The structural basis of DD/DD interactions involved in death receptor signaling, such as those in the Fas-FADD complex, remains largely unresolved. Although no solid experimental evidence is available, the Fas-FADD and TNF-R1/TRADD complexes are likely to be trimeric, most likely with Fas and TNF-R1 possessing oligomerization surfaces. Therefore such complexes might comprise at least two interfaces, a self-oligomerization surface

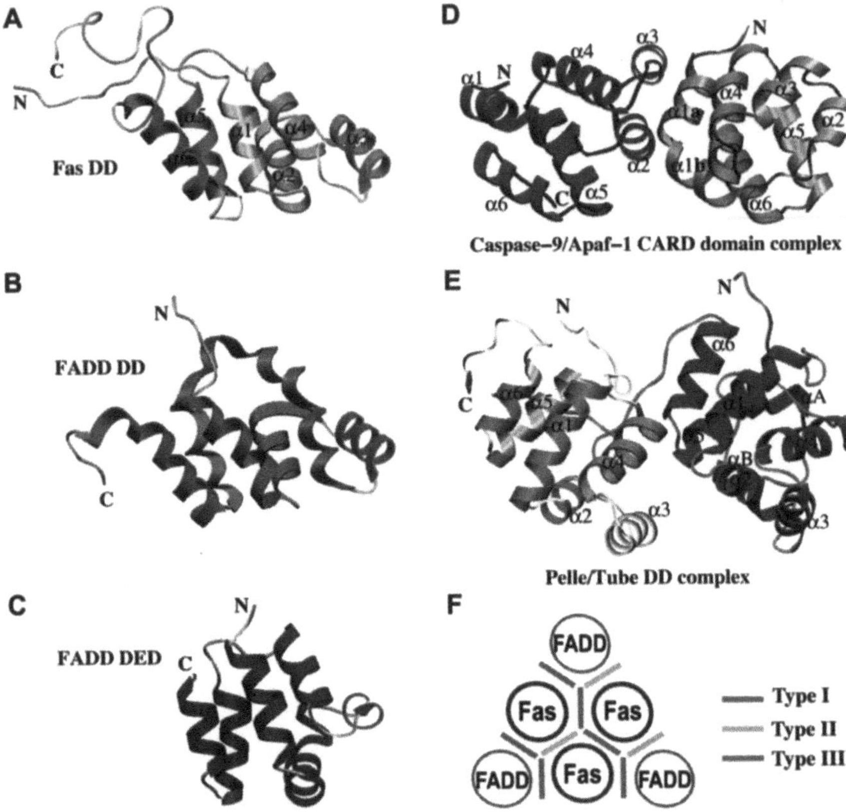

FIG. 11. Structures of DDs and DEDs. (A) DD of Fas. (B) DD of FADD. (C) DED of FADD. (D) CARD domain complex of caspase-9 (red) and Apaf-1 (orange). (E) DD complex of Pelle (pink) and Tube (purple). (F) A hypothetical model of the oligomeric Fas-FADD complex, showing the potential involvement of three types of interfaces. (See Color Insert.)

and a Fas-FADD interaction surface. This conjecture is supported by mutational data that showed that residues important for binding and/or function spread throughout the surfaces of Fas (Martin *et al.*, 1999), FADD (Hill *et al.*, 2004), TRADD (Park and Baichwal, 1996) and TNF-R1 (Telliez *et al.*, 2000).

An interesting model of the Fas-FADD complex has been proposed based on the Pelle/Tube and Apaf-1/Procaspase-9 structures (Weber and Vincenz, 2001) (Fig. 11F). Although there is no particularly convincing rational for using these structures as models, assuming that Fas has both a surface like Procaspase-9 for interaction with Apaf-1 (type I) and a surface like Pelle for interaction with Tube (type II), superposition of Pelle and Procaspase-9 in their respective complexes brought Tube and Apaf-1 into interaction vicinity to create a third interface. In addition, further propagation of alignment created a hetero-hexameric complex of Fas-FADD. Although details of the interactions are unlikely to be correct, the opposed surfaces in the complex appear to be chemically compatible and exhibit no serious steric clashes. In the model, FADD binds at the oligomerization interface of Fas and therefore only interacts with oligomerized Fas. One potential problem with this model is the lack of three-fold symmetry for Fas oligomerization. It nonetheless provided an interesting conceptual model to perceive DD-DD complexes. It remains to be shown by oligomeric DD-DD complex structures whether these interactions and models are true and whether they have any general implications for the DD superfamily.

X. Conclusion: Emerging Principles of Post-Receptor Signal Transduction

A. *Ligand-Induced Receptor Activation: Re-orientation of Intracellular Domains into Closer Proximity for Signaling*

TNF and related cytokine ligands form trimers or higher order oligomers in solution and on cell surface (Eck and Sprang, 1989). Upon receptor interaction, three receptor molecules interact with a TNF trimer to form 3:3 ligand-receptor complexes (Banner *et al.*, 1993). In the conventional view, this ligand-mediated receptor trimerization is the induction of receptor signal transduction across the cell membrane. However, more recent studies have demonstrated that TNFRs exist in pre-formed non-signaling trimers before ligand binding through a region of the extracellular domain named pre-ligand-binding assembly domain (PLAD), which is physically separate from the ligand binding site (Chan *et al.*, 2000; Siegel *et al.*, 2000). PLAD appears to be required for all aspects of receptor signaling, including ligand interaction, receptor activation and dominant interference of mutant receptors. The presorting of receptor chains into homotypic complexes on the cell surface makes biological sense in that it could promote rapidity and specificity of ligand interaction and prevent interference among different receptors of a same ligand.

Receptor pre-assembly has also been described for other receptor families and appears to be a fundamental principle for the signal transduction of oligomeric receptors. For example, the insulin receptor is a constitutive disulfide-linked dimer, while both epidermal growth factor receptor (EGFR) and erythropoietin receptor (EpoR) may exist as dimers independent of ligand interaction (Constantinescu *et al.*, 2001; Grotzinger, 2002; Moriki *et al.*, 2001; Yu *et al.*, 2002). The erythropoietin receptor dimers undergo a scissors-type movement to accommodate the ligand (Livnah *et al.*, 1999; Remy *et al.*, 1999) and the extracellular domain orientations are apparently tightly coupled to transmembrane helix rotations to activate the receptors (Moriki *et al.*, 2001; Seubert *et al.*, 2003).

Whether or not transmembrane helix rotation is a crucial event that leads to receptor activation in the case of TNFR superfamily, a net effect of ligand binding and receptor activation appears to be the induced closer proximity of the intracellular domains (Fig. 12A), as shown by fluorescence energy transfer experiments (Chan *et al.*, 2000). In the structures of TRAF-receptor complexes, the distance between bound receptor peptides is approximately 50 Å. This suggests that a distance of separation between the intracellular domains of receptor chains on the order of 50 Å may be optimal for TRAF recruitment and signaling transduction.

B. Geometry of TRAF-Receptor Interactions

Structural, biochemical, biophysical and cell biological analyses of the TRAF-receptor interactions have eluted to the following structural mechanisms of the interactions (Fig. 12A). First, the intracellular domains of members of the TNFR superfamily do not appear to have well-defined tertiary structures. Rather, they may be fairly flexible and disordered in the absence of TRAF interaction. Second, short linear sequences within the intracellular domains are responsible for TRAF recruitments and the interactions are driven by favorable enthalpy. The interactions between core TRAF binding sequences and TRAFs and possibly secondary interactions mediated by residues flanking the core sequences provide conformational restraint to the intracellular domains upon TRAF interactions. This is implicated from the entropic loss in TRAF-receptor interactions. Third, the directionality of receptor peptides in TRAF-receptor complexes places the mushroom-shaped cap against the membrane, allowing the amino-terminal domains in full-length TRAFs to be exposed to the cytosol for interaction with down-stream signaling molecules (Fig. 1). This geometry of the TRAF-receptor interaction requires minimal linker residues between the membrane and the TRAF binding site. In keeping with this analysis, the 17-residue TRAF2 binding CD40 peptide has been shown capable of both TRAF interaction and wild type like NF-κB activation when linked immediately after the transmembrane region of CD40 (Cheng and Baltimore, 1996).

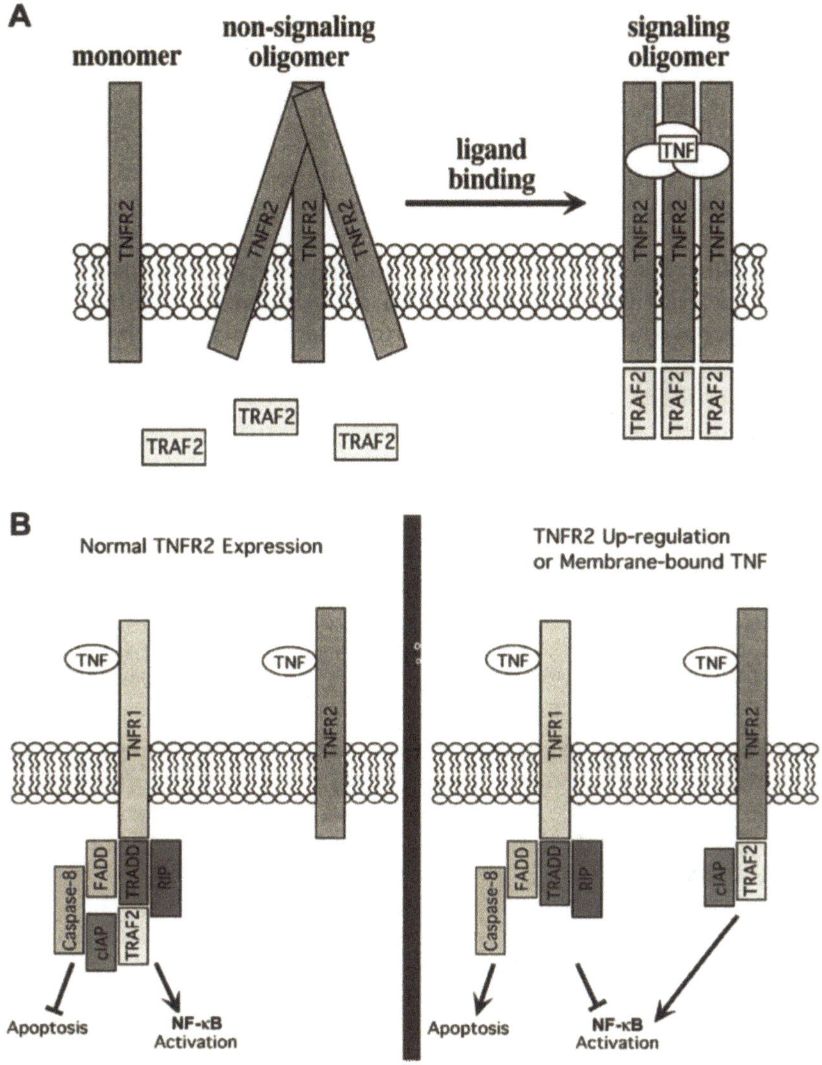

FIG. 12. Principles of post-receptor signal transduction. (A) Receptor activation and TRAF recruitment. (B) Competitive TRAF recruitments and regulation of cell survival and death. (See Color Insert.)

This oligomeric TRAF-receptor interaction geometry provides avidity to increase the affinity of TRAF recruitment upon receptor activation. This is necessary because interactions between TRAFs and monomeric receptors are relatively weak (Table III), which ensures that TRAFs do not interact with non-activated receptors. The dependence of TRAF recruitment on ligand-induced receptor activation has been shown for several TNFRs including TNF-R2, CD40 and LTβR (Kuhne *et al.*, 1997; Shu *et al.*, 1996; VanArsdale *et al.*, 1997) and is considered a common feature of the TNF receptor superfamily. Both affinity and specificity of the interaction will be greatly amplified by the avidity contribution from this oligomeric association, transforming a low affinity interaction into a tight and highly specific one.

For TRAF recruitment, since each receptor peptide contacts one TRAF domain only, avidity is the only factor that contributes to enhanced affinity of TRAFs for activated receptors. Due to the lack of structural information on DD interactions, it is not known whether ligand-induced receptor recruitment of FADD and TRADD (Hsu *et al.*, 1996b) is also purely driven by avidity or the intracellular DD of the receptors also form composite binding sites for FADD and TRADD upon receptor activation. In keeping with the latter scenario, a protein known as silencer of death domains (SODD) (Jiang *et al.*, 1999) has been shown to associate with the intracellular DD of TNF-R1 and get released upon ligand stimulation to activate the receptor.

C. Energetics: Affinity Differences of TRAF Recruitment and Different Avidity Requirements for Efficient Signaling

Quantitative characterization of monomeric TRAF-receptor and TRAF-adapter protein interactions showed a wide range of affinities from 7.8 μM to 1.9 mM (Table III), raising the question whether these receptors require different avidity contribution for efficient TRAF signaling. Many of the TNF-like cytokine ligands are membrane-bound and therefore may be capable of inducing membrane patching and higher order of receptor aggregation. Even though the minimal aggregation state of TRAFs appears to be trimeric, higher orders of aggregation may be possible as well in response to the higher order of receptor aggregation. This would increase the avidity in the TRAF-receptor interaction and the strength of the signal transduction. In keeping with this hypothesis, soluble trimeric CD40L can be fairly inefficient in inducing CD40 signaling under certain circumstances, compared to cell-bound or cross-linked hexameric CD40L (Pullen *et al.*, 1999b). In addition, TNF-R2 is mostly activated by membrane-bound form of TNF (Grell *et al.*, 1995). In some instances, multiple TRAF binding sequences in a single protein may also allow additional avidity from the interaction of TRAFs with neighboring TRAF binding sites. Therefore, TRAF-mediated signal transduction may be modulated at several levels including affinity and avidity.

The consideration on the combined effects of affinity and avidity explains why TNF-R2 is rather non-responsive to soluble TNF but mostly respond to cell-bound TNF (Grell *et al.*, 1995), while TNF-R1 can be stimulated efficiently with soluble TNF. TNF is synthesized as a transmembrane ligand and efficiently converted to soluble TNF under most but not all circumstances. The affinity of TRAF2 for TNF-R2 (K_d = 0.5 mM) is remarkably lower than the TRADD-TRAF2 interaction ($K_d = 7.8\ \mu M$). The cell-bound TNF may be capable of creating higher order of receptor clustering than soluble TNF and may therefore provide enhanced avidity to TRAF2 recruitment by TNF-R2 to potentiate this signal transduction. TNF-R2 signaling can also be induced by the up-regulation of expression, which again increases the avidity contribution of TRAF2 recruitment (Fig. 12B).

The generally weaker affinity between TRAF2 and many members of the TNF receptor superfamily such as CD40, CD30, Ox40 and 4-1BB ($K = 40\ \mu M$–1.0 mM) implies that higher order of aggregation may be required for the optimal signal transduction of these receptors. In keeping with this observation, the corresponding ligands for these receptors are membrane bound, which could induce more aggregation and provide higher avidity for the TRAF2-receptor interactions. In these cases, soluble ligands are often inefficient in eliciting signal transduction and may even act as decoys to down-regulate the signal (Hodgkin *et al.*, 1997; Kehry and Castle, 1994).

D. Specificity and Diverse Recognition: Conserved Interaction with Key Residues

A fundamental principle derived from the structural studies of TRAF-receptor complexes is that although the interactions are of low affinity, they are highly specific as shown by the identical interactions in different crystal packing environments and with different lengths of receptor peptides. This is in contrast to low affinity non-specific protein-DNA interactions that are mediated solely by charge attractions.

In addition to the specific recognition of a particular receptor peptide by TRAFs, a diverse set of receptor sequences may be recognized in a conserved mode of interaction by TRAFs. This ability of TRAFs to recognize diverse receptor sequences forms the basis for the wide spectrum of biological effects that TRAFs mediate. The molecular basis of this recognition resides on the conserved interaction of TRAFs with a few conserved key residues of receptor sequences rather than structural plasticity.

E. Biological Interplay: Competitive TRAF Recruitments and Context-Dependent Regulation of Survival and Death

Because TRAFs can be recruited to many different receptors, including survival receptors and receptors with potential for both cell death and cell survival induction, the competitive TRAF recruitment may modulate the signaling components of these receptors to induce differential biological effects. This would relate the outcome of particular receptor activation to the repertoire of TRAFs and TRAF binding partners in a particular type of cell and at a certain stage of differentiation.

This competitive recruitment hypothesis explains how TNF-R2, a survival receptor, could induce apoptosis (Fig. 12B). It has been observed that mice deficient in TNF-R2 showed decreased cell death following TNF treatment (Erickson *et al.*, 1994), while overexpression of TNF-R2 could lead to increased sensitivity to TNF-induced apoptosis (Chan and Lenardo, 2000; Haridas *et al.*, 1998; Heller *et al.*, 1992; Vandenabeele *et al.*, 1995; Weiss *et al.*, 1997). Moreover, TNF-induced apoptosis of activated primary T lymphocytes has also been shown to require TNF-R2 (Sarin *et al.*, 1995; Zheng *et al.*, 1995). Functionally, this may be important for Fas-independent peripheral deletion of T lymphocytes and the regulation of mature T-cell homeostasis.

The observed cooperation of TNF-R2 in TNF-induced apoptosis could be explained by a decreased recruitment of TRAF1, TRAF2 and cIAPs in the TNF-R1 signaling complex. Overexpression or up-regulation of expression of TNF-R2 upon T-cell activation could lead to an increased association of TRAF1 and TRAF2 with TNF-R2, sequestering intracellular TRAF1, TRAF2 and the constitutively associated cIAPs away from the signaling complex of TNF-R1 and related receptors. In keeping with this proposal, the TRAF2 binding region of TNF-R2 is required for this potentiation of TNF-R1-mediated apoptosis (Weiss *et al.*, 1998).

This interplay between TNF-R2 and TNF-R1 may be extended to other-members of the TNF receptor superfamily such as CD40, CD30, LTβR and CD27. These receptors have been shown to induce cell death under certain circumstances (Grell *et al.*, 1999). Similar to TNF-R2, activation of any of these receptors could lead to a sequestration, and possibly degradation (Duckett and Thompson, 1997), of TRAF1, TRAF2 and cIAPs. As these receptors have been shown to induce the expression of TNF (Grell *et al.*, 1999), the thus activated TNF-R1 would be tipped to apoptosis induction under the lack of recruitment of TRAF1, TRAF2 and cIAPs.

F. Remaining Questions

While the current structural and functional studies have shed lights on the molecular mechanisms of post-receptor signal transduction by the TNFR superfamily, many important questions remain. One such question is the structural basis for the formation of death receptor signaling complexes, involving DD-DD and DED-DED interactions. Another question is the molecular basis of TRAF downstream signaling. Does it involve oligomerization and proximity induced activation of down-stream effectors, or conformational modulations? Because of the importance of the TNFR superfamily in human disease, an ultimate question lies on the translation of structural and functional studies into therapeutic applications.

Acknowledgments

This work was funded by the National Institute of Health (AI45937 and AI47831), the Pew Charitable Trust and the Rita Allen Foundation. I apologize to all whose work has not been appropriately reviewed or cited due to space limitations.

References

Aizawa, S., Nakano, H., Ishida, T., Horie, R., Nagai, M., Ito, K., Yagita, H., Okumura, K., Inoue, J., and Watanabe, T. (1997). Tumor necrosis factor receptor-associated factor (TRAF) 5 and TRAF2 are involved in CD30-mediated NFkappaB activation. *J. Biol. Chem.* **272,** 2042–2045.

Akiba, H., Nakano, H., Nishinaka, S., Shindo, M., Kobata, T., Atsuta, M., Morimoto, C., Ware, C. F., Malinin, N. L., Wallach, D. *et al.* (1998). CD27, a member of the tumor necrosis factor receptor superfamily, activates NF-kappaB and stress-activated protein kinase/c-Jun N-terminal kinase via TRAF2, TRAF5, and NF-kappaB-inducing kinase. *J. Biol. Chem.* **273,** 13353–13358.

Arch, R. H., Gedrich, R. W., and Thompson, C. B. (1998). Tumor necrosis factor receptor-associated factors (TRAFs)-a family of adapter proteins that regulates life and death. *Genes Dev.* **12,** 2821–2830.

Arch, R. H., and Thompson, C. B. (1998). 4-1BB and Ox40 are members of a tumor necrosis factor (TNF)-nerve growth factor receptor subfamily that bind TNF receptor-associated factors and activate nuclear factor kappaB. *Mol. Cell. Biol.* **18,** 558–565.

Ashkenazi, A., and Dixit, V. M. (1998). Death receptors: Signaling and modulation. *Science* **281,** 1305–1308.

Banner, D. W., D'Arcy, A., Janes, W., Gentz, R., Schoenfeld, H. J., Broger, C., Loetscher, H., and Lesslauer, W. (1993). Crystal structure of the soluble human 55 kd TNF receptor-human TNF beta complex: Implications for TNF receptor activation. *Cell* **73,** 431–445.

Beg, A. A., and Baltimore, D. (1996). An essential role for NF-kappaB in preventing TNF-alpha-induced cell death. *Science* **274,** 782–784.

Beutler, B., and Cerami, A. (1986). Cachectin and tumour necrosis factor as two sides of the same biological coin. *Nature* **320,** 584–588.
Boldin, M. P., Goncharov, T. M., Goltsev, Y. V., and Wallach, D. (1996). Involvement of MACH, a novel MORT1/FADD-interacting protease, in Fas/APO-1-and TNF receptor-induced cell death. *Cell* **85,** 803–815.
Boucher, L. M., Marengere, L. E., Lu, Y., Thukral, S., and Mak, T. W. (1997). Binding sites of cytoplasmic effectors TRAF1, 2, and 3 on CD30 and other members of the TNF receptor superfamily. *Biochem. Biophys. Res. Commun.* **233,** 592–600.
Brodeur, S. R., Cheng, G., Baltimore, D., and Thorley-Lawson, D. A. (1997). Localization of the major NF-kappaB-activating site and the sole TRAF3 binding site of LMP-1 defines two distinct signaling motifs. *J. Biol. Chem.* **272,** 19777–19784.
Brummelkamp, T. R., Nijman, S. M., Dirac, A. M., and Bernards, R. (2003). Loss of the cylindromatosis tumour suppressor inhibits apoptosis by activating NF-kappaB. *Nature* **424,** 797–801.
Burns, K., Clatworthy, J., Martin, L., Martinon, F., Plumpton, C., Maschera, B., Lewis, A., Ray, K., Tschopp, J., and Volpe, F. (2000). Tollip, a new component of the IL-1RI pathway, links IRAK to the IL-1 receptor. *Nat. Cell Biol.* **2,** 346–351.
Cao, Z., Henzel, W. J., and Gao, X. (1996a). IRAK: A kinase associated with the interleukin-1 receptor. *Science* **271,** 1128–1131.
Cao, Z., Xiong, J., Takeuchi, M., Kurama, T., and Goeddel, D. V. (1996b). TRAF6 is a signal transducer for interleukin-1. *Nature* **383,** 443–446.
Carswell, E. A., Old, L. J., Kassel, R. L., Green, S., Fiore, N., and Williamson, B. (1975). An endotoxin-induced serum factor that causes necrosis of tumors. *Proc. Natl. Acad. Sci. USA* **72,** 3666–3670.
Chan, F. K., Chun, H. J., Zheng, L., Siegel, R. M., Bui, K. L., and Lenardo, M. J. (2000). A domain in TNF receptors that mediates ligand-independent receptor assembly and signaling. *Science* **288,** 2351–2354.
Chan, F. K., and Lenardo, M. J. (2000). A crucial role for p80 TNF-R2 in amplifying p60 TNF-R1 apoptosis signals in T lymphocytes. *Eur. J. Immunol.* **30,** 652–660.
Cheng, G., and Baltimore, D. (1996). TANK, a co-inducer with TRAF2 of TNF- and CD 40L-mediated NF-kappaB activation. *Genes Dev.* **10,** 963–973.
Cheng, G., Cleary, A. M., Ye, Z. S., Hong, D. I., Lederman, S., and Baltimore, D. (1995). Involvement of CRAF1, a relative of TRAF, in CD40 signaling. *Science* **267,** 1494–1498.
Chinnaiyan, A. M., O'Rourke, K., Tewari, M., and Dixit, V. M. (1995). FADD, a novel death domain-containing protein, interacts with the death domain of Fas and initiates apoptosis. *Cell* **81,** 505–512.
Chou, J. J., Matsuo, H., Duan, H., and Wagner, G. (1998). Solution structure of the RAIDD CARD and model for CARD/CARD interaction in caspase-2 and caspase-9 recruitment. *Cell* **94,** 171–180.
Chung, J. Y., Park, Y. C., Ye, H., and Wu, H. (2002). All TRAFs are not created equal: Common and distinct molecular mechanisms of TRAF-mediated signal transduction. *J. Cell. Sci.* **115,** 679–688.
Coley, W. B. (1893). The treatment of malignant tumors by repeated inoculations of erysipelas: With a report of ten original cases. *Am. J. Med. Sci.* **105,** 487–511.
Constantinescu, S. N., Keren, T., Socolovsky, M., Nam, H., Henis, Y. I., and Lodish, H. F. (2001). Ligand-independent oligomerization of cell-surface erythropoietin receptor is mediated by the transmembrane domain. *Proc. Natl. Acad. Sci. USA* **98,** 4379–4384.
Darnay, B. G., Ni, J., Moore, P. A., and Aggarwal, B. B. (1999). Activation of NF-kappaB by RANK requires tumor necrosis factor receptor-associated factor (TRAF) 6 and

NF-kappaB-inducing kinase. Identification of a novel TRAF6 interaction motif. *J. Biol. Chem.* **274,** 7724–7731.

Deng, L., Wang, C., Spencer, E., Yang, L., Braun, A., You, J., Slaughter, C., Pickart, C., and Chen, Z. J. (2000). Activation of the IkappaB kinase complex by TRAF6 requires a dimeric ubiquitin-conjugating enzyme complex and a unique polyubiquitin chain. *Cell* **103,** 351–361.

Deng, Y., Ren, X., Yang, L., Lin, Y., and Wu, X. (2003). A JNK-dependent pathway is required for TNFalpha-induced apoptosis. *Cell* **115,** 61–70.

Devergne, O., Hatzivassiliou, E., Izumi, K. M., Kaye, K. M., Kleijnen, M. F., Kieff, E., and Mosialos, G. (1996). Association of TRAF1, TRAF2, and TRAF3 with an Epstein-Barr virus LMP1 domain important for B-lymphocyte transformation: Role in NF-kappaB activation. *Mol. Cell. Biol.* **16,** 7098–7108.

Duckett, C. S., and Thompson, C. B. (1997). CD30-dependent degradation of TRAF2: Implications for negative regulation of TRAF signaling and the control of cell survival. *Genes Dev.* **11,** 2810–2821.

Eberstadt, M., Huang, B., Chen, Z., Meadows, R. P., Ng, S.-C., Zheng, L., J., L. M., and Fesik, S. W. (1998). NMR structure and mutagenesis of the FADD (Mort1) death-effector domain. *Nature* **392,** 941–945.

Eck, M. J., and Sprang, S. R. (1989). The structure of tumor necrosis factor-alpha at 2.6 A resolution. Implications for receptor binding. *J. Biol. Chem.* **264,** 17595–17605.

Erickson, S. L., de Sauvage, F. J., Kikly, K., Carver-Moore, K., Pitts-Meek, S., Gillett, N., Sheehan, K. C., Schreiber, R. D., Goeddel, D. V., and Moore, M. W. (1994). Decreased sensitivity to tumour-necrosis factor but normal T-cell development in TNF receptor-2-deficient mice. *Nature* **372,** 560–563.

Fesik, S. W. (2000). Insights into programmed cell death through structural biology. *Cell* **103,** 273–282.

Fiers, W. (1991). Tumor necrosis factor. Characterization at the molecular, cellular and *in vivo* level. *FEBS Lett.* **285,** 199–212.

Franken, M., Devergne, O., Rosenzweig, M., Annis, B., Kieff, E., and Wang, F. (1996). Comparative analysis identifies conserved tumor necrosis factor receptor-associated factor 3 binding sites in the human and simian Epstein-Barr virus oncogene LMP1. *J. Virol.* **70,** 7819–7826.

Gedrich, R. W., Gilfillan, M. C., Duckett, C. S., Van Dongen, J. L., and Thompson, C. B. (1996). CD30 contains two binding sites with different specificities for members of the tumor necrosis factor receptor-associated factor family of signal transducing proteins. *J. Biol. Chem.* **271,** 12852–12858.

Ghosh, S., and Karin, M. (2002). Missing pieces in the NF-kappaB puzzle. *Cell* **109**(Suppl.), S81–96.

Goeddel, D. V., Aggarwal, B. B., Gray, P. W., Leung, D. W., Nedwin, G. E., Palladino, M. A., Patton, J. S., Pennica, D., Shepard, H. M., Sugarman, B. J. *et al.* (1986). Tumor necrosis factors: Gene structure and biological activities. *Cold Spring Harb. Symp. Quant. Biol.* **51** (Pt. 1), 597–609.

Gravestein, L. A., and Borst, J. (1998). Tumor necrosis factor receptor family members in the immune system. *Sem. Immunol.* **10,** 423–434.

Grell, M., Douni, E., Wajant, H., Lohden, M., Clauss, M., Maxeiner, B., Georgopoulos, S., Lesslauer, W., Kollias, G., Pfizenmaier, K. *et al.* (1995). The transmembrane form of tumor necrosis factor is the prime activating ligand of the 80 kDa tumor necrosis factor receptor. *Cell* **83,** 793–802.

Grell, M., Zimmermann, G., Gottfried, E., Chen, C. M., Grunwald, U., Huang, D. C., Wu Lee, Y. H., Durkop, H., Engelmann, H., Scheurich, P. *et al.* (1999). Induction of cell death by tumour necrosis factor (TNF) receptor 2, CD40 and CD30: A role for TNF-R1 activation by endogenous membrane-anchored TNF. *EMBO J.* **18,** 3034–3043.

Grotzinger, J. (2002). Molecular mechanisms of cytokine receptor activation. *Biochim. Biophys. Acta* **1592,** 215–223.

Harbury, P. B., Zhang, T., Kim, P. S., and Alber, T. (1993). A switch between two-, three-, and four-stranded coiled coil in GCN4 leucine zipper mutants. *Science* **262,** 1401–1407.

Haridas, V., Darnay, B. G., Natarajan, K., Heller, R., and Aggarwal, B. B. (1998). Overexpression of the p80 TNF receptor leads to TNF-dependent apoptosis, nuclear factor-kappa B activation, and c-Jun kinase activation. *J. Immunol.* **160,** 3152–3162.

Heller, R. A., Song, K., Fan, N., and Chang, D. J. (1992). The p70 tumor necrosis factor receptor mediates cytotoxicity. *Cell* **70,** 47–56.

Hill, J. M., Morisawa, G., Kim, T., Huang, T., Wei, Y., and Werner, M. H. (2004). Identification of an expanded binding surface on the FADD death domain responsible for interaction with CD95/Fas. *J. Biol. Chem.* **279,** 1474–1481.

Hiller, S., Kohl, A., Fiorito, F., Herrmann, T., Wider, G., Tschopp, J., Grutter, M. G., and Wuthrich, K. (2003). NMR structure of the apoptosis- and inflammation-related NALP1 pyrin domain. *Structure (Camb)* **11,** 1199–1205.

Hodgkin, P. D., Chin, S. H., Bartell, G., Mamchak, A., Doherty, K., Lyons, A. B., and Hasbold, J. (1997). The importance of efficacy and partial agonism in evaluating models of B lymphocyte activation. *Int. Rev. Immunol.* **15,** 101–127.

Holm, L., and Sander, C. (1995). Dali: A network tool for protein structure comparison. *Trends Biochem. Sci.* **20,** 478–480.

Hsu, H., Huang, J., Shu, H. B., Baichwal, V., and Goeddel, D. V. (1996a). TNF-dependent recruitment of the protein kinase RIP to the TNF receptor-1 signaling complex. *Immunity* **4,** 387–396.

Hsu, H., Shu, H.-B., Pan, M.-G., and Goeddel, D. V. (1996b). TRADD-TRAF2 and TRADD-FADD interactions define two distinct TNF receptor 1 signal transduction pathways. *Cell* **84,** 299–308.

Huang, B., Eberstadt, M., Olejniczak, E. T., Meadows, R. P., and Fesik, S. W. (1996). NMR structure and mutagenesis of the Fas (APO-1/CD95) death domain. *Nature* **384,** 638–641.

Irmler, M., Thome, M., Hahne, M., Schneider, P., Hofmann, K., Steiner, V., Bodmer, J. L., Schroter, M., Burns, K., Mattmann, C. *et al.* (1997). Inhibition of death receptor signals by cellular FLIP [see comments]. *Nature* **388,** 190–195.

Ishida, T., Mizushima, S., Azuma, S., Kobayashi, N., Tojo, T., Suzuki, K., Aizawa, S., Watanabe, T., Mosialos, G., Kieff, E. *et al.* (1996a). Identification of TRAF6, a novel tumor necrosis factor receptor-associated factor protein that mediates signaling from an amino-terminal domain of the CD40 cytoplasmic region. *JBC* **271,** 28745–28748.

Ishida, T. K., Tojo, T., Aoki, T., Kobayashi, N., Ohishi, T., Watanabe, T., Yamamoto, T., and Inoue, J. (1996b). TRAF5, a novel tumor necrosis factor receptor-associated factor family protein, mediates CD40 signaling. *Proc. Natl. Acad. Sci. USA* **93,** 9437–9442.

Janin, J., Miller, S., and Chothia, C. (1988). Surface, subunit interfaces and interior of oligomeric proteins. *J. Mol. Biol.* **204,** 155–164.

Jeong, E. J., Bang, S., Lee, T. H., Park, Y. I., Sim, W. S., and Kim, K. S. (1999). The solution structure of FADD death domain. Structural basis of death domain interactions of Fas and FADD. *J. Biol. Chem.* **274,** 16337–16342.

Jiang, Y., Woronicz, J. D., Liu, W., and Goeddel, D. V. (1999). Prevention of constitutive TNF receptor 1 signaling by silencer of death domains. *Science* **283,** 543–546.

Kehry, M. R., and Castle, B. E. (1994). Regulation of CD40 ligand expression and use of recombinant CD40 ligand for studying B cell growth and differentiation. *Semin. Immunol.* **6,** 287–294.

Kelliher, M. A., Grimm, S., Ishida, Y., Kuo, F., Stanger, B. Z., and Leder, P. (1998). The death-domain kinase RIP mediates the TNF-induced NF-kB signal. *Immunity* **8,** 297–303.

Kovalenko, A., Chable-Bessia, C., Cantarella, G., Israel, A., Wallach, D., and Courtois, G. (2003). The tumour suppressor CYLD negatively regulates NF-kappaB signaling by deubiquitination. *Nature* **424,** 801–805.

Kuhne, M. R., Robbins, M., Hambor, J. E., Mackey, M. F., Kosaka, Y., Nishimura, T., Gigley, J. P., Noelle, R. J., and Calderhead, D. M. (1997). Assembly and regulation of the CD40 receptor complex in human B cells. *J. Exp. Med.* **186,** 337–342.

Kuriyan, J., and Cowburn, D. (1997). Modular peptide recognition domains in eukaryotic signaling. *Annu. Rev. Biophys. Biomol. Struct.* **26,** 259–288.

Ladbury, J. E. (1995). Counting the calories to stay in the groove. *Structure* **3,** 635–639.

Leonen, W. A. M. (1998). Editorial overview: CD27 and (TNFR) relatives in the immune system: Their role in health and disease. *Sem. Immunol.* **10,** 417–422.

Lewis, M., Tartaglia, L. A., Lee, A., Bennett, G. L., Rice, G. C., Wong, G. H., Chen, E. Y., and Goeddel, D. V. (1991). Cloning and expression of cDNAs for two distinct murine tumor necrosis factor receptors demonstrate one receptor is species specific. *Proc. Natl. Acad. Sci. USA* **88,** 2830–2834.

Li, C., Ni, C. Z., Havert, M. L., Cabezas, E., He, J., Kaiser, D., Reed, J. C., Satterthwait, A. C., Cheng, G., and Ely, K. R. (2002). Downstream regulator TANK binds to the CD40 recognition site on TRAF3. *Structure (Camb)* **10,** 403–411.

Li, C., Norris, P. S., Ni, C. Z., Havert, M. L., Chiong, E. M., Tran, B. R., Cabezas, E., Reed, J. C., Satterthwait, A. C., Ware, C. F., and Ely, K. R. (2003). Structurally distinct recognition motifs in lymphotoxin-beta receptor and CD40 for TRAF-mediated signaling. *J. Biol. Chem.* **278,** 50523–50529.

Liepinsh, E., Ilag, L. L., Otting, G., and Ibanez, C. F. (1997). NMR structure of the death domain of the p75 neurotrophin receptor. *EMBO J.* **16,** 4999–5005.

Lim, W. A., Richards, F. M., and Fox, R. O. (1994). Structural determinants of peptide-binding orientation and of sequence specificity in SH3 domains. *Nature* **372,** 375–379.

Livnah, O., Stura, E. A., Middleton, S. A., Johnson, D. L., Jolliffe, L. K., and Wilson, I. A. (1999). Crystallographic evidence for preformed dimers of erythropoietin receptor before ligand activation. *Science* **283,** 987–990.

Locksley, R. M., Killeen, N., and Lenardo, M. J. (2001). The TNF and TNF receptor superfamilies: Integrating mammalian biology. *Cell* **104,** 487–501.

Martin, D. A., Zheng, L., Siegel, R. M., Huang, B., Fisher, G. H., Wang, J., Jackson, C. E., Puck, J. M., Dale, J., Straus, S. E. *et al.* (1999). Defective CD95/APO-1/Fas signal complex formation in the human autoimmune lymphoproliferative syndrome, type Ia. *Proc. Natl. Acad. Sci. USA* **96,** 4552–4557.

McWhirter, S. M., Pullen, S. S., Holton, J. M., Crute, J. J., Kehry, M. R., and Alber, T. (1999). Crystallographic analysis of CD40 recognition and signaling by human TRAF2. *Proc. Natl. Acad. Sci. USA* **96,** 8408–8413.

Micheau, O., and Tschopp, J. (2003). Induction of TNF receptor I-mediated apoptosis via two sequential signaling complexes. *Cell* **114,** 181–190.

Mizushima, S., Fujita, M., Ishida, T., Azuma, S., Kato, K., Hirai, M., Otsuka, M., Yamamoto, T., and Inoue, J. (1998). Cloning and characterization of a cDNA encoding the human homolog of tumor necrosis factor receptor-associated factor 5 (TRAF5). *Gene* **207,** 135–140.

Moriki, T., Maruyama, H., and Maruyama, I. N. (2001). Activation of preformed EGF receptor dimers by ligand-induced rotation of the transmembrane domain. *J. Mol. Biol.* **311,** 1011–1026.

Mosialos, G., Birkenbach, M., Yalamanchili, R., VanArsdale, T., Ware, C., and Kieff, E. (1995). The Epstein-Barr virus transforming protein LMP1 engages signaling proteins for the tumor necrosis factor receptor family. *Cell* **80,** 389–399.

Murzin, A. G., Brenner, S. E., Hubbard, T., and Chothia, C. (1995). SCOP: A structural classification of protein database for the investigation of sequences and structures. *J. Mol. Biol.* **247,** 536–540.

Muzio, M., Chinnaiyan, A. M., Kischkel, F. C., O'Rourke, K., Shevchenko, A., Ni, J., Scaffidi, C., Bretz, J. D., Zhang, M., Gentz, R. *et al.* (1996). FLICE, a novelFADD-homologous ICE/CED-3-like protease, is recruited to the CD95 (Fas/APO-1) death-inducing signaling complex. *Cell* **85,** 817–827.

Muzio, M., Ni, J., Feng, P., and Dixit, V. M. (1997). IRAK (Pelle) family member IRAK-2 and MyD88 as proximal mediators of IL-1 signaling. *Science* **278,** 1612–1615.

Nagata, S. (1997). Apoptosis by death factor. *Cell* **88,** 355–365.

Nakano, H., Oshima, H., Chung, W., Williams-Abbott, L., Ware, C. F., Yagita, H., and Okumura, K. (1996). TRAF5, an activator of NF-kappaB and putative signal transducer for the lymphotoxin-beta receptor. *J. Biol. Chem.* **271,** 14661–14664.

Newton, R. C., and Decicco, C. P. (1999). Therapeutic potential and strategies for inhibiting tumor necrosis factor-a. *J. Med. Chem.* **42,** 2295–2314.

Ni, C. Z., Welsh, K., Leo, E., Chiou, C. K., Wu, H., Reed, J. C., and Ely, K. R. (2000). Molecular basis for CD40 signaling mediated by TRAF3. *Proc. Natl. Acad. Sci. USA* **97,** 10395–10399.

O'Neill, L. A., Fitzgerald, K. A., and Bowie, A. G. (2003). The Toll-IL-1 receptor adaptor family grows to five members. *Trends Immunol.* **24,** 286–290.

Orengo, C. A., and Thornton, J. M. (1993). Alpha plus beta folds revisited: Some favoured motifs. *Structure* **1,** 105–120.

Park, A., and Baichwal, V. R. (1996). Systematic mutational analysis of the death domain of the tumor necrosis factor receptor-1-associated protein TRADD. *J. Biol. Chem.* **271,** 9858–9862.

Park, Y. C., Burkitt, V., Villa, A. R., Tong, L., and Wu, H. (1999). Structural basis for self-association and receptor recognition of human TRAF2. *Nature* **398,** 533–538.

Park, Y. C., Ye, H., Hsia, C., Segal, D., Rich, R. L., Liou, H. -L., Myszka, D. G., and Wu, H. (2000). A novel mechanism of TRAF signaling revealed by structural and functional analyses of the TRADD-TRAF2 interaction. *Cell* **101,** 777–787.

Pennica, D., Nedwin, G. E., Hayflick, J. S., Seeburg, P. H., Derynck, R., Palladino, M. A., Kohr, W. J., Aggarwal, B. B., and Goeddel, D. V. (1984). Human tumour necrosis factor: Precursor structure, expression and homology to lymphotoxin. *Nature* **312,** 724–729.

Polekhina, G., House, C. M., Traficante, N., Mackay, J. P., Relaix, F., Sassoon, D. A., Parker, M. W., and Bowtell, D. D. (2002). Siah ubiquitin ligase is structurally related to TRAF and modulates TNF-alpha signaling. *Nat. Struct. Biol.* **9,** 68–75.

Pullen, S. S., Dang, T. T., Crute, J. J., and Kehry, M. R. (1999a). CD40 signaling through tumor necrosis factor receptor-associated factors (TRAFs). Binding site specificity and activation of downstream pathways by distinct TRAFs. *J. Biol. Chem.* **274,** 14246–14254.

Pullen, S. S., Labadia, M. E., Ingraham, R. H., McWhirter, S. M., Everdeen, D. S., Alber, T., Crute, J. J., and Kehry, M. R. (1999b). High-affinity interactions of tumor necrosis factor receptor-associated factors (TRAFs) and CD40 require TRAF trimerization and CD40 multimerization. *Biochemistry* **38,** 10168–10177.

Pullen, S. S., Miller, H. G., Everdeen, D. S., Dang, T. T., Crute, J. J., and Kehry, M. R. (1998). CD40-tumor necrosis factor receptor-associated factor (TRAF) interactions: Regulation of CD40 signaling through multiple TRAF binding sites and TRAF hetero-oligomerization. *Biochemistry* **37,** 11836–11845.

Qin, H., Srinivasula, S. M., Wu, G., Fernandes-Alnemri, T., Alnemri, E. S., and Shi, Y. (1999). Structural basis of procaspase-9 recruitment by the apoptotic protease-activating factor 1. *Nature* **399,** 549–557.

Regnier, C. H., Tomasetto, C., Moog-Lutz, C., Chenard, M. P., Wendling, C., Basset, P., and Rio, M. C. (1995). Presence of a new conserved domain in CART1, a novel member of the tumor necrosis factor receptor-associated protein family, which is expressed in breast carcinoma. *J. Biol. Chem.* **270,** 25715–25721.

Remy, I., Wilson, I. A., and Michnick, S. W. (1999). Erythropoietin receptor activation by a ligand-induced conformation change. *Science* **283,** 990–993.

Rost, B., and Sander, C. (1994). Combining evolutionary information and neural networks to predict protein secondary structure. *Proteins* **19,** 55–72.

Rothe, M., Pan, M. G., Henzel, W. J., Ayres, T. M., and Goeddel, D. V. (1995). The TNFR2-TRAF signaling complex contains two novel proteins related to baculoviral inhibitor of apoptosis proteins. *Cell* **83,** 1243–1252.

Rothe, M., Wong, S. C., Henzel, W. J., and Goeddel, D. V. (1994). A novel family of putative signal transducers associated with the cytoplasmic domain of the 75 kDa tumor necrosis factor receptor. *Cell* **78,** 681–692.

Rothe, M., Xiong, J., Shu, H. B., Williamson, K., Goddard, A., and Goeddel, D. V. (1996). I-TRAF is a novel TRAF-interacting protein that regulates TRAF-mediated signal transduction. *Proc. Natl. Acad. Sci. USA* **93,** 8241–8246.

Sandberg, M., Hammerschmidt, W., and Sugden, B. (1997). Characterization of LMP-1's association with TRAF1, TRAF2, and TRAF3. *J. Virol.* **71,** 4649–4656.

Sarin, A., Conan-Cibotti, M., and Henkart, P. A. (1995). Cytotoxic effect in TNF and lymphotoxin on T lymphoblasts. *J. Immunol.* **155,** 3716–3718.

Sato, T., Irie, S., and Reed, J. C. (1995). A novel member of the TRAF family of putative signal transducing proteins binds to the cytosolic domain of CD40. *FEBS Lett.* **358,** 113–118.

Seubert, N., Royer, Y., Staerk, J., Kubatzky, K. F., Moucadel, V., Krishnakumar, S., Smith, S. O., and Constantinescu, S. N. (2003). Active and inactive orientations of the transmembrane and cytosolic domains of the erythropoietin receptor dimer. *Mol. Cell* **12,** 1239–1250.

Shaulian, E., and Karin, M. (2002). AP-1 as a regulator of cell life and death. *Nat. Cell Biol.* **4,** E131–E136.

Shirai, T., Yamaguchi, H., Ito, H., Todd, C. W., and Wallace, R. B. (1985). Cloning and expression in *Escherichia coli* of the gene for human tumour necrosis factor. *Nature* **313,** 803–806.

Shu, H. B., Takeuchi, M., and Goeddel, D. V. (1996). The tumor necrosis factor receptor 2 signal transducers TRAF2 and c-IAP1 are components of the tumor necrosis factor receptor 1 signaling complex. *Proc. Natl. Acad. Sci. USA* **93,** 13973–13978.

Siegel, R. M., Frederiksen, J. K., Zacharias, D. A., Chan, F. K., Johnson, M., Lynch, D., Tsien, R. Y., and Lenardo, M. J. (2000). Fas preassociation required for apoptosis signaling and dominant inhibition by pathogenic mutations. *Science* **288,** 2354–2357.

Smith, C. A., Farrah, T., and Goodwin, R. G. (1994). The TNF receptor superfamily of cellular and viral proteins: Activation, costimulation and death. *Cell* **76,** 959–962.

Stanger, B. Z., Leder, P., Lee, T., Kim, E., and Seed, B. (1995). RIP: A novel protein containing a death domain that interacts with Fas/APO-1 (CD95) in yeast and causes cell death. *Cell* **81,** 513–523.

Stern, L. J., Brown, J. H., Jardetzky, T. S., Gorga, J. C., Urban, R. G., Strominger, J. L., and Wiley, D. C. (1994). Crystal structure of the human class II MHC protein HLA-DR1 complexed with an influenza virus peptide. *Nature* **368,** 215–221.

Sukits, S. F., Lin, L. L., Hsu, S., Malakian, K., Powers, R., and Xu, G. Y. (2001). Solution structure of the tumor necrosis factor receptor-1 death domain. *J. Mol. Biol.* **310,** 895–906.

Sutton, R. B., Davletov, B. A., Berghuis, A. M., Sudhof, T. C., and Sprang, S. R. (1995). Structure of the first C2 domain of synaptotagmin I: A novel Ca^{2+}/Phospholipid-binding fold. *Cell* **80,** 929–938.

Suzuki, N., Suzuki, S., Duncan, G. S., Millar, D. G., Wada, T., Mirtsos, C., Takada, H., Wakeham, A., Itie, A., Li, S. *et al.* (2002). Severe impairment of interleukin-1 and Toll-like receptor signalling in mice lacking IRAK-4. *Nature* **416,** 750–756.

Tainer, J. A., Getzoff, E. D., Beem, K. M., Richardson, J. S., and Richardson, D. C. (1982). Determination and analysis of the 2 Å-structure of copper, zinc superoxide dismutase. *J. Mol. Biol.* **160,** 181–217.

Telliez, J. B., Xu, G. Y., Woronicz, J. D., Hsu, S., Wu, J. L., Lin, L., Sukits, S. F., Powers, R., and Lin, L. L. (2000). Mutational analysis and NMR studies of the death domain of the tumor necrosis factor receptor-1. *J. Mol. Biol.* **300,** 1323–1333.

Tong, L., Qian, C., Massariol, M. J., Deziel, R., Yoakim, C., and Lagace, L. (1998). Conserved mode of peptidomimetic inhibition and substrate recognition of human cytomegalovirus protease. *Nat. Struct. Biol.* **5,** 819–826.

Trompouki, E., Hatzivassiliou, E., Tsichritzis, T., Farmer, H., Ashworth, A., and Mosialos, G. (2003). CYLD is a deubiquitinating enzyme that negatively regulates NF-kappaB activation by TNFR family members. *Nature* **424,** 793–796.

Tsao, D. H., McDonagh, T., Telliez, J. B., Hsu, S., Malakian, K., Xu, G. Y., and Lin, L. L. (2000). Solution structure of N-TRADD and characterization of the interaction of N-TRADD and C-TRAF2, a key step in the TNFR1 signaling pathway. *Mol. Cell* **5,** 1051–1057.

Uren, A. G., and Vaux, D. L. (1996). TRAF proteins and meprins share a conserved domain. *Trends Biochem. Sci.* **21,** 244–245.

VanArsdale, T. L., VanArsdale, S. L., Force, W. R., Walter, B. N., Mosialos, G., Kieff, E., Reed, J. C., and Ware, C. F. (1997). Lymphotoxin-β receptor signaling complex: Role of tumor necrosis factor receptor-associated factor 3 recruitment in cell death and activation of nuclear factor κB. *PNAS* **94,** 2460–2465.

Vandenabeele, P., Declercq, W., Vanhaesebroeck, B., Grooten, J., and Fiers, W. (1995). Both TNF receptors are required for TNF-mediated induction of apoptosis in PC60 cells. *J. Immunol.* **154,** 2904–2913.

Wang, A. M., Creasey, A. A., Ladner, M. B., Lin, L. S., Strickler, J., Van Arsdell, J. N., Yamamoto, R., and Mark, D. F. (1985). Molecular cloning of the complementary DNA for human tumor necrosis factor. *Science* **228,** 149–154.

Wang, C., Deng, L., Hong, M., Akkaraju, G. R., Inoue, J., and Chen, Z. J. (2001a). TAK1 is a ubiquitin-dependent kinase of MKK and IKK. *Nature* **412,** 346–351.
Wang, C. Y., Mayo, M. W., Korneluk, R. G., Goeddel, D. V., and Baldwin, A. S., Jr. (1998). NF-kappaB antiapoptosis: Induction of TRAF1 and TRAF2 and c-IAP1 and c-IAP2 to suppress caspase-8 activation. *Science* **281,** 1680–1683.
Wang, J., Chun, H. J., Wong, W., Spencer, D. M., and Lenardo, M. J. (2001b). Caspase-10 is an initiator caspase in death receptor signaling. *Proc. Natl. Acad. Sci. USA* **98,** 13884–13888.
Weber, C. H., and Vincenz, C. (2001). A docking model of key components of the DISC complex: Death domain superfamily interactions redefined. *FEBS Lett.* **492,** 171–176.
Weiss, T., Grell, M., Hessabi, B., Bourteele, S., Muller, G., Scheurich, P., and Wajant, H. (1997). Enhancement of TNF receptor p60-mediated cytotoxicity by TNF receptor p80: Requirement of the TNF receptor-associated factor-2 binding site. *J. Immunol.* **158,** 2398–2404.
Weiss, T., Grell, M., Siemienski, K., Muhlenbeck, F., Durkop, H., Pfizenmaier, K., Scheurich, P., and Wajant, H. (1998). TNFR80-dependent enhancement of TNFR60-induced cell death is mediated by TNFR-associated factor 2 and is specific for TNFR60. *J. Immunol.* **161,** 3136–3142.
Wesche, H., Gao, X., Li, X., Kirschning, C. J., Stark, G. R., and Cao, Z. (1999). IRAK-M is a novel member of the Pelle/interleukin-1 receptor-associated kinase (IRAK) family. *J. Biol. Chem.* **274,** 19403–19410.
Wolf, E., Kim, P. S., and Berger, B. (1997). MultiCoil: A program for predicting two- and three-stranded coiled coils. *Protein Science* **6,** 1179–1189.
Wu, H., and Arron, J. R. (2003). TRAF6, a molecular bridge spanning adaptive immunity, innate immunity and osteoimmunology. *Bioessays* **25,** 1096–1105.
Xiao, T., Towb, P., Wasserman, S. A., and Sprang, S. R. (1999). Three-dimensional structure of a complex between the death domains of Pelle and Tube. *Cell* **99,** 545–555.
Ye, H., Arron, J. R., Lamothe, B., Cirilli, M., Kobayashi, T., Shevde, N. K., Segal, D., Dzivenu, O. K., Vologodskaia, M., Yim, M. *et al.* (2002). Distinct molecular mechanism for initiating TRAF6 signalling. *Nature* **418,** 443–447.
Ye, H., Park, Y. C., Kreishman, M., Kieff, E., and Wu, H. (1999). The structural basis for the recognition of diverse receptor sequences by TRAF2. *Mol. Cell* **4,** 321–330.
Ye, H., and Wu, H. (2000). Thermodynamic characterization of the interaction between TRAF2 and receptor peptides by isothermal titration calorimetry. *PNAS* **97,** 8961–8966.
Yeh, W. C., Shahinian, A., Speiser, D., Kraunus, J., Billia, F., Wakeham, A., de la Pompa, J. L., Ferrick, D., Hum, B., Iscove, N. *et al.* (1997). Early lethality, functional NF-kappaB activation, and increased sensitivity to TNF-induced cell death in TRAF2-deficient mice. *Immunity* **7,** 715–725.
Yu, X., Sharma, K. D., Takahashi, T., Iwamoto, R., and Mekada, E. (2002). Ligand-independent dimer formation of epidermal growth factor receptor (EGFR) is a step separable from ligand-induced EGFR signaling. *Mol. Biol. Cell* **13,** 2547–2557.
Zapata, J. M., Pawlowski, K., Haas, E., Ware, C. F., Godzik, A., and Reed, J. C. (2001). TEFs: A diverse family of proteins containing TRAF domains. *J. Biol. Chem.* **276,** 24242–24252.
Zheng, L., Fisher, G., Miller, R. E., Peschon, J., Lynch, D. H., and Lenardo, M. J. (1995). Induction of apoptosis in mature T cells by tumour necrosis factor. *Nature* **377,** 348–351.

NKG2D AND RELATED IMMUNORECEPTORS

By ROLAND K. STRONG* AND BENJAMIN J. McFARLAND†

*Division of Basic Sciences, Fred Hutchinson Cancer Research Center, Seattle, Washington 98109; †Department of Chemistry and Biochemistry, Seattle Pacific University, Seattle, Washington 98119

Abstract

NK cells are crucial components of the innate immune system, capable of directly eliminating infected or tumorigenic cells and regulating downstream adaptive immune responses. Unlike T cells, where the key recognition event driving activation is mediated by the unique T cell receptor (TCR) expressed on a given cell, NK cells express multiple activating and inhibitory cell-surface receptors (NKRs), often with overlapping ligand specificities. NKRs display two ectodomain structural homologies, either immunoglobulin- or C-type lectin-like (CTLD). The CTLD immunoreceptor NKG2D is found on NK cells but is also widely expressed on T cells and other immune system cells, providing stimulatory or co-stimulatory signals. NKG2D drives target cell killing following engagement of diverse, conditionally expressed MHC class I-like protein ligands whose expression can signal cellular distress due to infection or transformation. The symmetric, homodimeric receptor interacts with its asymmetric, monomeric ligands in similar 2:1 complexes, with an equivalent surface on each NKG2D monomer binding extensively and intimately to distinct, structurally divergent surfaces on the ligands. Thus, NKG2D ligand-binding site recognition is highly degenerate, further demonstrated by NKG2D's ability to

ADVANCES IN PROTEIN CHEMISTRY, Vol. 68

0065-3233/04 $35.00

simultaneously accommodate multiple non-conservative allelic or isoform substitutions in the ligands. In TCRs, "induced-fit" recognition explains cross-reactivity, but structural, computational, thermodynamic and kinetic analyses of multiple NKG2D–ligand pairs show that rather than classical "induced-fit" binding, NKG2D degeneracy is achieved using distinct interaction mechanisms at each rigid interface: recognition degeneracy by "rigid adaptation." While likely forming similar complexes with their ligand (HLA-E), other NKG2x NKR family members do not require such recognition degeneracy.

I. T Cell Receptors and MHC Class I Proteins: Paradigms of Immunological Recognition

Cytotoxic responses by the cellular arm of the adaptive immune system are ultimately mediated by recognition events between $\alpha\beta$ T cell receptors (TCRs) on the surfaces of T cells and processed peptide fragments of endogenously expressed proteins, presented to TCRs as complexes on the cell surface with major histocompatibility complex class I proteins (pMHC) (Germain and Margulies, 1993). Thereby MHC presentation allows the immune system to monitor the proteome of a given cell for inappropriate protein expression associated with disease (e.g., tumorigenesis or infection). MHC class I proteins are integral-membrane, heterodimeric proteins with ectodomains consisting of a polymorphic heavy chain, comprising three extracellular domains ($\alpha1$, $\alpha2$, and $\alpha3$), associated with a non-polymorphic light chain, β_2-microglobulin (β_2m) (Bjorkman and Parham, 1999). Association with both peptide and β_2m is required for normal folding and cell-surface expression. The $\alpha1$ and $\alpha2$ domains together comprise the peptide- and TCR-binding "platform" domain; the $\alpha3$ and β_2m domains display C-type immunoglobulin folds. The canonical MHC class I platform fold comprises two long, roughly parallel, α-helices, interrupted by bends, arranged on an eight-stranded, anti-parallel β-sheet. These α-helices define the "walls" of the peptide binding groove. Crystal structures of TCR–pMHC complexes show that the TCR variable domains generally sit diagonally on the platform domain, making contacts to the peptide and the $\alpha1$ and $\alpha2$ domains (Fig. 1) (Garcia *et al.*, 1999). T cell activation requires an interaction between TCRs and appropriate target pMHC complexes in the context of appropriate co-receptor interactions (e.g., CD4 or CD8) and costimulatory signals from, for example, engagement of the CD28 receptor on T cells with CD80 or CD86 ligands on target cells (Lenschow *et al.*, 1996), all stabilized and organized by cell–cell adhesion interactions (Grakoui *et al.*,

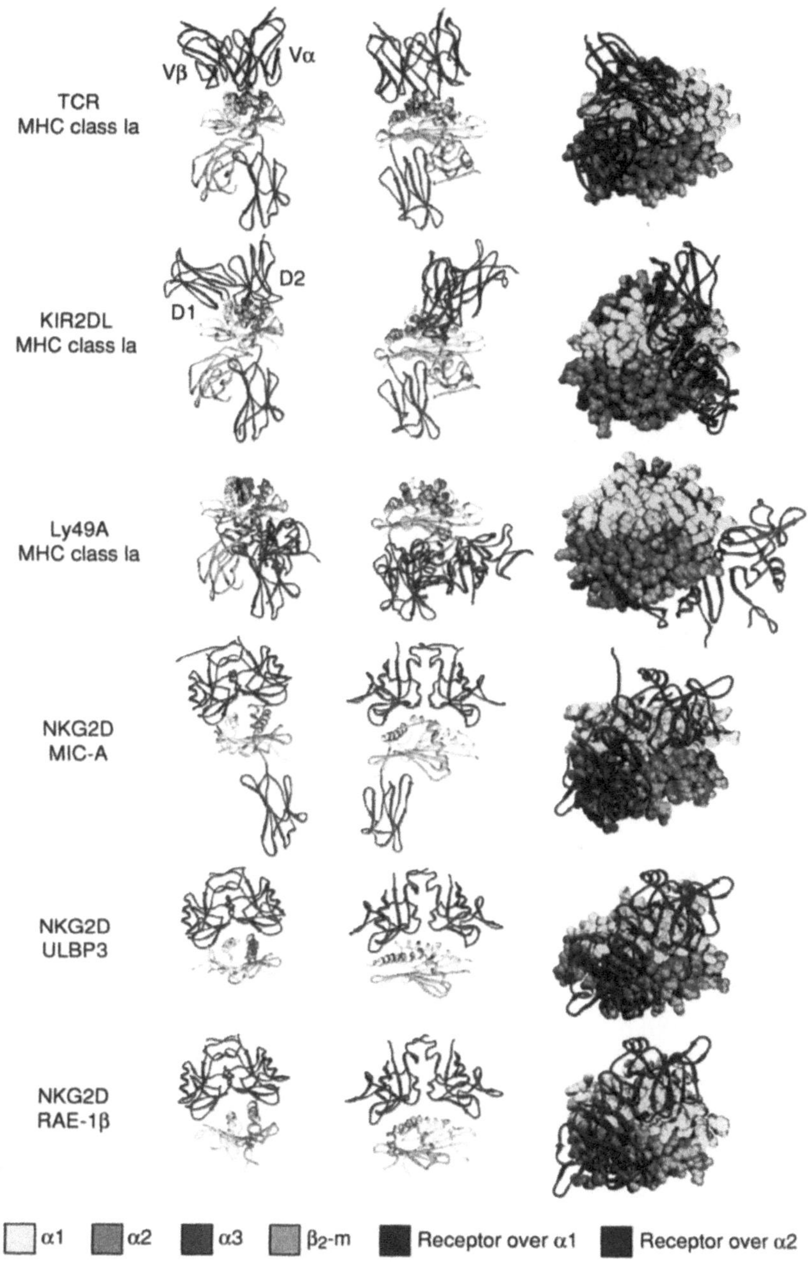

FIG. 1. (*continued*)

1999). Diverse cell-surface molecules that modulate T cell activation also include receptors first identified on natural killer (NK) cells that have since been found expressed on a range of cell types, including T cells (such as NKG2D, see later). The experimentally-observed cross-reactivity of TCR-pMHC interactions is best explained by conformational plasticity, or 'induced-fit' recognition (Koshland, 1958), where a flexible binding site can be molded to accommodate structural diversity across multiple ligands, typified by many antibody–antigen interactions (James *et al.*, 2003; Sundberg and Mariuzza, 2000; Wu *et al.*, 2002b).

II. NK Cells and Receptors

NK cells constitute an important component of the innate immune system, providing surveillance against cells undergoing tumorigenesis or infection (by viruses or internal pathogens), without requiring prior host sensitization. NK cells act to regulate innate and acquired immune responses through the release of various immune modulators, such as interferon-γ (IFN-γ), or by direct elimination of compromised cells (Trinchieri, 1989; Yokoyama and Plougastel, 2003). NK effector functions are regulated through a diverse array of cell-surface inhibitory and activating receptors (Table I) (Diefenbach and Raulet, 2003; Held *et al.*, 2003; Yokoyama and Plougastel, 2003). Different NKRs, with different MHC class I specificities, are expressed on overlapping, but distinct, subsets of NK cells. Many NK cell surface receptors (NKRs) are specific for classical

---

FIG. 1. Representative immunoreceptor–ligand complex structures. Views of a series of immunoreceptor–ligand complexes are shown, in a mix of ribbon and space-filling representations, highlighting the range of variation in these recognition events: $\alpha\beta$ TCR–classical pMHC class I (Garboczi *et al.*, 1996); KIR2DL–classical pMHC class I (Boyington *et al.*, 2000); Ly49A–classical pMHC class I (Tormo *et al.*, 1999); huNKG2D–MIC-A (Li *et al.*, 2001a); huNKG2D–ULBP3 (Radaev *et al.*, 2001); and muNKG2D–RAE-β (Li *et al.*, 2002). The views in the left-hand column are aligned looking down along the peptide-binding groove (when present). The views in the middle column are perpendicular to the dyad axis of the NKG2D receptor (a rotation of approximately 50° from the view in the left-hand column). The views in the right-hand column are oriented looking down onto the top (peptide-binding) surface of the platform domains of the MHC class I protein or homolog (a rotation of 90° from the view in the middle column). All the molecules in a column are aligned on the platform domain of the MHC class I protein or homolog. Domains are colored as indicated, with the receptors colored in blue and purple and the ligand proteins colored in yellow, orange, red and green; peptides, when present, are colored by atom-type (carbon: gray; oxygen: red; nitrogen: blue; and sulfur: yellow). (See Color Insert.)

TABLE I
Properties of Immunoreceptors

Receptor	Structure	Ligand	K_D (μM)	Signal
Human				
$\alpha\beta$ TCRs...	IgSF	pMHC class Ia	1–90	activation
CD8	IgSF	pMHC class Ia	65–200	(activation)
$\gamma\delta$ TCRs...	IgSF	prenyl pyrophosphates; alkamines	~0.001–10^3	activation
KIRs...	IgSF	pMHC class Ia	~10	activation or inhibition
LIR-1/ILT-2	IgSF	pMHC class Ia	15–100	inhibition
Nkp30, 44 & 46	IgSF	tumor antigens?; viral proteins?	?	activation
NKG2A–CD94	CTLD	HLA-E	0.36–≥34	inhibition
NKG2C–CD94	CTLD	HLA-E	2.3–≥56	activation
NKG2D	CTLD	(*Table II*)	(*Table II*)	activation
NKG2E–CD94	CTLD	HLA-E	?	activation
Mouse				
Ly49s...	CTLD	pMHC class Ia; viral proteins	~10	activation or inhibition
NKG2A–CD94	CTLD	Qa-1	?	inhibition
NKG2C–CD94	CTLD	Qa-1	?	activation
NKG2D	CTLD	(*Table II*)	(*Table II*)	activation
NKG2E–CD94	CTLD	Qa-1	?	activation

(HLA-A, -B, and -C) and non-classical (HLA-E) MHC class I proteins and occur in paired activating and inhibitory isoforms (Bakker *et al.*, 2000). Thus NK cell effector functions are regulated by integrating signals across the array of activating and inhibitory NKRs engaged upon interaction with target cell-surface NKR ligands (Diefenbach and Raulet, 2001; Held *et al.*, 2003; Raulet *et al.*, 2001), resulting in the elimination of cells with reduced MHC class I expression, a common consequence of infection or transformation (Lanier, 2000).

NKRs can be divided into two families based on structural homologies (Table I). The first family, including the human "killer cell immunoglobulin" receptors (KIRs) (Colonna, 1997) and consists of type I transmembrane glycoproteins containing one to three tandem immunoglobulin-like domains in the ectodomain. The second NKR family comprises homo- and heterodimeric type II transmembrane glycoproteins containing C-type lectin-*like* domains (CTLDs) in their ectodomains and includes the murine Ly49 family (Karlhofer *et al.*, 1992) and the human and murine-expressed NKG2x NKRs (x = A, B, C, D, E, F, and H) (Drickamer, 1999; Weis *et al.*,

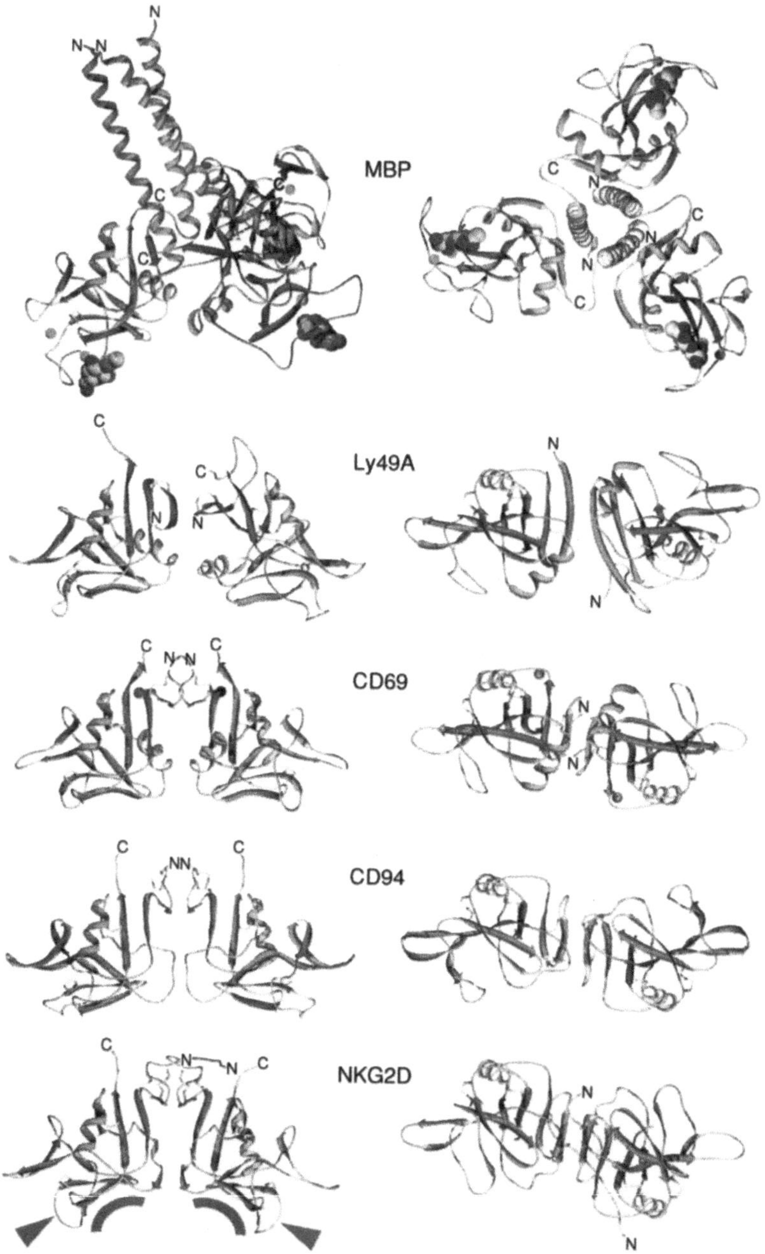
MBP
Ly49A
CD69
CD94
NKG2D

1998). C-type lectin-like receptors are structurally distinct from true, carbohydrate-binding C-type lectins and lack elements associated with specific carbohydrate recognition (Fig. 2) (Drickamer, 1993; Weis *et al.*, 1998). The canonical C-type lectin fold comprises two α-helices packed against two β-sheets, linked through two invariant disulfide bonds. A third conserved disulfide bond is found in an N-terminal extension defining the long-form animal C-type lectins; multiple disulfide bonds are found in the analogous extension common to CTLD NKRs (Fig. 2).

The number of KIR or Ly49 genes in a particular person or mouse strain varies widely, with around ten expressed in any given individual (Makrigiannis *et al.*, 2002; Uhrberg *et al.*, 2002). Single NK cells within a particular individual typically express from one to five different NKRs (Kubota *et al.*, 1999; Uhrberg *et al.*, 1997), yielding distinct KIR/Ly49-NKG2x combinations (Raulet *et al.*, 2001). How the particular repertoire of NKRs expressed on a particular NK cell is acquired is still unclear (Held *et al.*, 2003), though it has been demonstrated that NK cells do functionally adapt to the MHC class I environment of the host (Ohlen *et al.*, 1989).

Inhibitory NKRs contain "immunoreceptor tyrosine-based inhibition motifs" (ITIMs, typically V/IxYxxL sequences) in their endodomains; activating receptors associate with adaptor proteins (typically through complementary charge–charge interactions within transmembrane-spanning domains) containing "immunoreceptor tyrosine-based activation motifs" (ITAMs, typically YxxL/Ix6-8YxxL/I) or related sequence motifs (YxxM in

FIG. 2. Comparison of CTLD structures. Ribbon representations, colored by secondary structure (α-helices: yellow coils; β-strands: green arrows), are shown for a series of illustrative CTLD-containing immunoreceptors: Ly49A (Tormo *et al.*, 1999); CD69 (Llera *et al.*, 2001); CD94 (Boyington *et al.*, 1999); and huNKG2D (McFarland *et al.*, 2003). The structure of the archetype C-type lectin fold-containing protein, trimeric rat mannose binding protein (MBP), is shown at the top for comparison (Feinberg *et al.*, 2000). The structures are shown in two views, one perpendicular to the dyad axis of symmetry in the dimeric molecules (left) and one view from below, looking up onto the ligand binding sites (right; 90° from the view on the left for the dimeric molecules). The left-hand views are oriented so that the left CTLD or C-type lectin domains are superimposed, highlighting the differences in the fold of the short-form C-type lectins (MBP) and the CTLD-containing immunoreceptors, which are more homologous to long-form C-type lectins. N- and C-termini are labeled. A disaccharide ligand is shown (in a space-filling representation, colored by atom-type as in Fig. 1) for MBP; the NKG2D ligand-binding site is indicated with blue arcs. Bound ions are shown as spheres: calcium atoms (green) in the MBP structure; zinc atoms (gray) in CD69. The gray pointers mark the NKG2D stirrup loops, a structurally divergent feature of NKG2D that results in the distinctly saddle-shaped ligand-binding surface of this NKR. (See Color Insert.)

DAP10, for instance) (Diefenbach and Raulet, 2003; Held *et al.*, 2003). ITIMs with phosphorylated tyrosines signal inhibition through the recruitment and activation of the SHP-1 phosphatase; ITAMs with phosphorylated tyrosines signal activation through the recruitment of Syk or ZAP70 tyrosine kinases.

III. NKG2x NK Cell Receptors

The NKG2x NKR family can be further subdivided, structurally, functionally and by sequence relationship (Figs. 3 and 4), into two arms: the closely-related receptors NKG2A, B, C, E, F, and H and the more distantly related receptor NKG2D. NKG2D is a homodimeric, activating, CTLD-type immunoreceptor whose expression was first recognized on NK cells but was subsequently found on CD8-positive $\alpha\beta$ T cells, $\gamma\delta$ T cells and macrophages, making it one of the most widely distributed NKRs currently described (Bauer *et al.*, 1999; Wu *et al.*, 1999). The other members of the NKG2x family (A, B, C, E, and H) form obligate heterodimers with CD94, are highly homologous to each other (Figs. 3 and 4), are limited in expression to NK cells and are specific for the non-classical MHC class I protein HLA-E in humans or Qa-1 in mice (Table I) (Borrego *et al.*, 2002; Raulet *et al.*, 2001). The exception to this generalization may be NKG2F, which has not been demonstrated to be expressed on cell surfaces and is missing large, otherwise-conserved sections of the CTLD, but that does contain a cytoplasmic ITIM-like sequence (Plougastel and Trowsdale, 1997). HLA-E binds peptides, like the classical MHC class I (or class Ia) proteins, though with a much more restricted specificity (O'Callaghan *et al.*, 1998; Strong *et al.*, 2002), limited essentially to fragments of the leader sequences of MHC class I proteins (Lee *et al.*, 1998). Therefore, normal HLA-E cell-surface expression is an indirect check for the normal expression of MHC class I proteins. NKG2A/B and NKG2E/H are splice variants (Bellon *et al.*, 1999; Plougastel *et al.*, 1996), where the only resultant differences are that NKG2H has a longer (+16 residues), and different, C-terminal extension than NKG2E compared to the rest of the family; and that NKG2B has a truncated extracellular, N-terminal "arm" relative to NKG2A (which is also shorter than in any other family member). This arm spans the distance between CTLD and the transmembrane domain in NKG2x NKRs. The inhibitory NKG2x NKRs (A/B) have two ITIMs in their endodomains; the activating NKG2x NKRs (C and E/H, but not D), as well as many of the activating immunoglobulin-type NKRs, interact with the DAP12 adaptor (also known as KARAP) (Lanier *et al.*, 1998a,b).

NKG2D displays only limited sequence similarity to other NKG2x family members and CD94 (Figs. 3 and 4), has not been demonstrated to directly

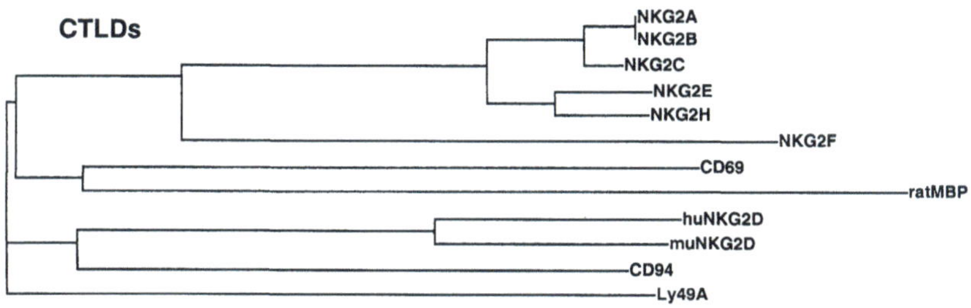

	NKG2A	NKG2B	NKG2C	NKG2D (human)	NKG2D (murine)	NKG2E	NKG2F	NKG2H	CD94	CD69	Ly49A	MBP
NKG2A	100	100	94	25	23	81	39	81	25	26	29	13
NKG2B	100*	100										
NKG2C	68	66	100	25	23	83	39	83	25	26	30	12
NKG2D (human)	8	33	18	100	72	20	17	20	32	21	25	7
NKG2D (murine)	8	33	9	42	100	22	15	27	35	21	27	7
NKG2E	73	50	94	10	10	100	39	88	24	24	28	11
NKG2F	54	66	68	8	8	68	100	39	23	17	15	3
NKG2H	73	50	94	10	10	100*	68	100	28	22	26	11
CD94	16	16	22	11	11	26	16	26	100	22	24	15
CD69	4	16	4	4	9	5	9	5	9	100	22	16
Ly49A											100	10

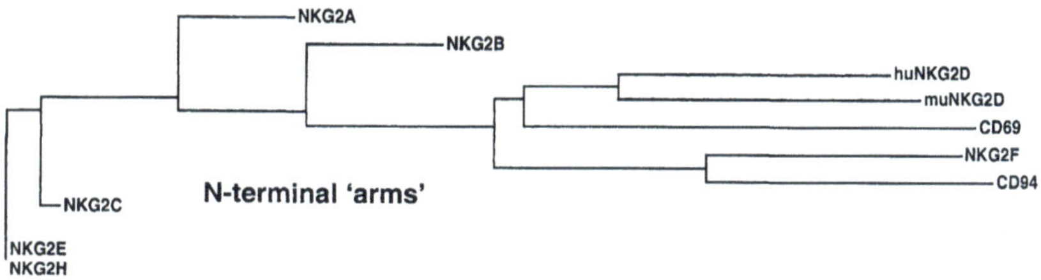

FIG. 3. CTLD/N-terminal arm sequence identities and phylogenies. Sequence phylogenies (top and bottom) and identities (middle) between the CTLDs (top and top-right half of middle panel) or the N-terminal arms (bottom and bottom-left half of middle panel) of the NKG2x NKRs, CD69, CD94, and Ly49A. In the middle panel, fewer values are shown for the CTLD of NKG2B, as this splice variant is identical to NKG2A in the CTLD, or for the N-terminal arm of Ly49A, which shows no meaningful similarity to the arms of the other receptors. Identities and phylogenies were calculated with CLUSTALW (Thompson *et al.*, 1994).

interact with MHC class I proteins and only forms obligate homodimers (Li *et al.*, 2001a; Steinle *et al.*, 2001). Human NKG2D (huNKG2D) engagement is signaled by recruitment of phosphatidylinositol 3-kinase and Grb2 through the adapter molecule DAP10 (Wu *et al.*, 1999, 2000), whereas

```
                120            130             140             150        160
NKG2A    HCGHCPEEWITYSNSCYYIGKERRTWEESLLACTSKNSS-LLSIDNEEEMK
NKG2B    HCGHCPEEWITYSNSCYYIGKERRTWEESLLACTSKNSS-LLSIDNEEEMK
NKG2C    HCGHCPEEWITYSNSCYYIGKERRTWEESLLACTSKNSS-LLSIDNEEEMK
NKG2E    PCGHCPEEWITYSNSCYYIGKERRTWEESLQACASKNSSSLLSIDNEEEMK
NKG2H    HCGHCPEEWITYSNSCYYIGKERRTWEESLQACASKNSSSLLSIDNEEEMK
NKG2F    PCGHCPEEWITYSNSCYYIGKERRTWEER---------------------

          *† §        † §         80          #90         100
CD94     DCCSCQEKWVGYRCNCYFISSEQKTWNESRHLCASKSS-LLQLQNTDELDF
            =    = ===                                   ==
                 ~~~    ~~~~~~       /////////

          †   §100  †    §         120      #  130          140
NKG2D    YCGPCPKNWICYKNNCYQFFDESKNWYESQASCMSQNAS-LLKVYSKEDQD
muNKG2D  YCGPCPNNWICHRNNCYQFFNEEKTWNQSQASCLSQNSS-LLKIYSKEEQD
         ======== = ==     ==
                 ~~~    ~~~~~~      //////////              \
-----------------------------------------------------------------

                170             180          190             200
NKG2A    FLSIISP--SSWIGVFRNSSHHPWVTMNG--LAFKHEIKDSDNAE---LNC
NKG2B    FLSIISP--SSWIGVFRNSSHHPWVTMNG--LAFKHEIKDSDNAE---LNC
NKG2C    FLASILP--SSWIGVFRNSSHHPWVTING--LAFKHKIKDSDNAE---LNC
NKG2E    FLASILP--SSWIGVFRNSSHHPWVTING--LAFKHEIKDSDHAE---RNC
NKG2H    FLASILP--SSWIGVFRNSSHHPWVTING--LAFKHEIKDSDHAE---RNC
NKG2F    ---------VCWP-VLRRT----------------------------LIC

             110            120          130          140        Δ
CD94     MQSSSQQ--FYWIGLSYSEEHTAWLWENG--SALSQYLFPSFETFN-TKNC
         =====
                  ~~~~~~~~~  ~~~~~~                              ~

                150            160          170          180      Δ
NKG2D    LLKLVKS--YHWMGLVHIPTNGSWQWEDG--SILSPNLLTIIEMQ--KGDC
muNKG2D  FLKLVKS--YHWMGLVQIPANGSWQWEDG--SSLSYNQLTLVEIP--KGSC
         = ==◇◇   ◇=                             ◇◇◇◇◇◇   ◇
         \\\      ~~~~ ~~~~      ~~~~            ~~~             ~
-----------------------------------------------------------------

          210         220         230
NKG2A    AVLQVNRLKSAQCGSS-IIYHCKHKL------------------------
NKG2B    AVLQVNRLKSAQCGSS-IIYHCKHKL------------------------
NKG2C    AVLQVNRLKSAQCGSS-MIYHCKHKL------------------------
NKG2E    AMLHVRGLISDQCGSSRIIZRRGF-IMLTRLVLNS---------------
NKG2H    AMLHVRGLISDQCGSSRIIVSISFRIKALELAVHQIKFYICSNRNDIMIA
NKG2F    FL------------------------------------------------

              160   Δ    170 #
CD94     IAYNPNGNALDESCEDKNRYICKQQLI-----------------------
         ~~~~   ~~~~~       ~~~~~

         190        200Δ      #
NKG2D    ALYASSFKGYIENCSTPNTYICMQRTV-----------------------
muNKG2D  AVYGSSFKAYTEDCANLNTYICMKRAV-----------------------
          ◇  ◇  ◇  ◇ ◇◇◇    ◇  ◇
         ~~~~   ~~~~~      ~~~~~~~
```

different splice variants of murine NKG2D (muNKG2D) have been reported to utilize both DAP10 and DAP12, perhaps in functionally-distinct contexts (Diefenbach *et al.*, 2002; Gilfillan *et al.*, 2002).

IV. HuNKG2D Ligands: MIC-A/B

HuNKG2D ligands (Table II) include the closely related proteins MIC-A and MIC-B (MHC class I chain-related) (Bahram and Spies, 1996; Bahram *et al.*, 1994; Groh *et al.*, 1996) and the ULBPs (human cytomegalovirus [CMV] UL16-binding proteins) (Cosman *et al.*, 2001). All are distant MHC class I homologs that do not function in conventional peptide antigen presentation. HuNKG2D–MIC recognition events stimulate effector responses from NK cells (calcium fluxing, production of IFN-γ, GM-CSF, TNF-α and -β, and MIP-1β) as well as $\gamma\delta$ T cells and may positively modulate CD8-positive $\alpha\beta$ T cell responses, thus serving a co-stimulatory function (Bauer *et al.*, 1999; Groh *et al.*, 1998). On macrophages, stimulation through huNKG2D triggers TNF-α production and release of nitric oxide (Diefenbach *et al.*, 2000). On NK cells, stimulation through NKG2D alone is sufficient to trigger effector functions (Jamieson *et al.*, 2002). The NKG2D activation signal can override inhibitory signals that would otherwise prevent activation (Bauer *et al.*, 1999; Cerwenka *et al.*, 2000; Diefenbach *et al.*, 2000, 2001) but apparently not in all contexts (Pende *et al.*, 2001).

Unlike the widely and constitutively expressed classical and non-classical MHC class I proteins, MIC-A and MIC-B are induced only in response to cellular stress on intestinal epithelium, epithelially-derived tumors and vascular endothelium (Groh *et al.*, 1996, 1999). While MIC-A and MIC-B are quite similar to each other (~84% identical; Fig. 5) (Bahram and Spies, 1996; Bahram *et al.*, 1996), they have diverged significantly from the MHC class I family as a whole, with identities of approximately 28% to 35% domain-by-domain when aligned with the human MHC class I proteins. MIC-A and -B are highly polymorphic, with over fifty MIC-A and thirteen

---

Fig. 4. NKG2x NKR sequence alignments. Sequences of the NKG2x NKRs and CD94 have been aligned using CLUSTALW (Thompson *et al.*, 1994). Sequences have been numbered from the initiator methionine in the leader peptide, but only the residues in the mature ectodomain CTLDs are shown. Cysteines have been highlighted and disulfide bond partners have been indicated with matching symbols (*, †, §, #, Δ), but only when based on crystallographic data. Below each sequence, receptor dimer contacts (=), ligand contacts (◇) and secondary structure element (α-helix: /; β-strand: ~; 3_{10}-helix: \) are indicated with symbols.

TABLE II
NKG2D Ligands and Affinities

Ligand	Domain structure	K_D (μM)	Expression	Inducers
Human				
MIC-A	(α1α2)α3-TM	0.3–0.94	intestinal epithelium; tumors	cellular stress; tumorigenesis; infection
MIC-B	(α1α2)α3-TM	0.79	intestinal epithelium; tumors	cellular stress; tumorigenesis; infection
ULBP1	(α1α2)-GPI	1.1	kidney? thyroid?	?
ULBP2	(α1α2)-GPI	?	?	?
ULBP3	(α1α2)-GPI	4.0	kidney?	?
ULBP4	(α1α2)-GPI	?	?	?
Mouse				
RAE-1α	(α1α2)-GPI	0.42–0.59	onco-fetal	retinoic acid; carcinogens
RAE-1β	(α1α2)-GPI	0.57–1.9	onco-fetal	retinoic acid; carcinogens
RAE-1γ	(α1α2)-GPI	0.35–0.38	onco-fetal	retinoic acid; carcinogens
RAE-1δ	(α1α2)-GPI	0.73–1.0	onco-fetal	retinoic acid; carcinogens
RAE-1ϵ	(α1α2)-GPI	?	?	retinoic acid?
RAE-1B6	(α1α2)-GPI	0.028–0.034	?	retinoic acid?
H60	(α1α2)-TM	0.014–0.027	?	carcinogens
MULT1	(α1α2)-TM	0.0015–0.0056	?	?

MIC-B alleles recognized, numbers that continue to increase (Stephens, 2001). The polymorphisms are spread over the extracellular domains of the proteins, and are predominantly the result of single amino acid substitutions that generate dimorphic positions (Fig. 5). Many of these changes are non-conservative and the pattern of sequence variation is wholly distinct from that for the classical MHC class I proteins—a pattern not readily rationalizable in terms of known interactions with any of its receptors (Holmes *et al.*, 2002). MIC-A/B proteins are conserved in most mammals except rodents. MIC proteins do not require either peptide or β_2m for stability or cell-surface expression and apparently do not bind any other ligand in the shallow pocket that represents the only remnant of the peptide-binding groove of true MHC class I proteins (Groh *et al.*, 1996; Holmes *et al.*, 2002). Human tumors are capable of evading NKG2D-mediated immunosurveillance by shedding soluble forms of MIC proteins that down-regulate the expression of NKG2D on effector cells (Groh *et al.*, 2002; Salih *et al.*, 2002).

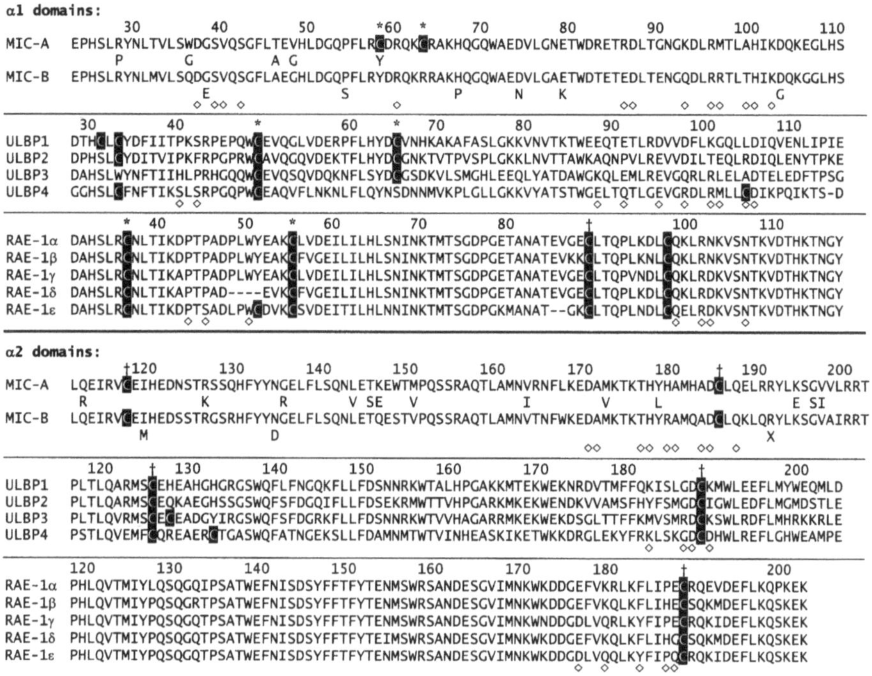

FIG. 5. Platform sequence alignments of the MHC class I-like ligands of NKG2D. Sequences of MIC-A and -B, the ULBPs and the RAE-1s have been aligned, divided by family and domain, using CLUSTALW (Thompson *et al.*, 1994). Note that the alignments across families are only very approximate at these levels of sequence identity. Sequences have been numbered from the initiator methionine in the leader peptide, but only the residues in the mature proteins have been shown. Cysteines have been highlighted, and disulfide bond partners have been indicated with matching symbols (*, †). For the MIC sequences, allelic substitutions have been indicated by the additional residues shown below the sequences (deletions are indicated with an 'X'). Diamonds below the sequences indicate NKG2D contact positions, based on the known complex structures (MIC-A*001, ULBP3, and RAE-1β).

V. HuNKG2D Ligands: ULBPs

ULBPs are homologous to the α1α2 peptide-binding platform domains of MHC class I proteins but lack α3 domains and are anchored in the membrane by GPI-linkages. ULBP1, 2, 3, and 4 (Table II, Fig. 6) are 20% to 27% identical in sequence to MIC-A, MIC-B or classical MHC class I proteins (Cosman *et al.*, 2001). The functional significance of the huNKG2D–ULBP interaction remains to be determined, though human CMV UL16–ULBP and UL16–MIC binding may block huNKG2D–ULBP and huNKG2D–MIC interactions, thus potentially representing a viral

	ULBP1	ULBP2	ULBP3	ULBP4
ULBP1	100	59	54	35
ULBP2		100	54	30
ULBP3			100	33
ULBP4				100

	RAE-1α	RAE-1β	RAE-1γ	RAE-1δ	RAE-1ε
RAE-1α	100	94	94	93	89
RAE-1β		100	92	95	89
RAE-1γ			100	92	92
RAE-1δ				100	89
RAE-1ε					100

FIG. 6. ULBP and RAE-1 sequence relationships. Sequence identities between members of the ULBP (left) or RAE-1 (right) families are tabulated. Identities were calculated with CLUSTALW (Thompson *et al.*, 1994).

strategy to mask these antigens, preventing activation through NKG2D and limiting anti-viral innate immune responses (Cosman *et al.*, 2001). Alternately, UL16 binding may act by retaining NKG2D, MIC-B, and ULBP 1 and 2 (but apparently not other NKG2D ligands) in the endoplasmic reticulum, preventing their cell-surface expression and function (Welte *et al.*, 2003).

VI. MuNKG2D Ligands: RAE-1s, H60, and MULT1

Rodents lack any recognizable homologs of MIC-A/B, but muNKG2D ligands do include the RAE-1 (*r*etinoic *a*cid *e*arly inducible) family of proteins, H60 and the recently described MULT1 (Table II) (Carayannopoulos *et al.*, 2002a; Cerwenka *et al.*, 2000; Diefenbach *et al.*, 2000; Zou *et al.*, 1996). Like the ULBPs, RAE-1 and H60 are homologous to the platform domains of MHC class I proteins (RAE-1 is 19% to 20% identical to a bovine MHC class I protein) (Zou *et al.*, 1996), lack $\alpha 3$ domains, and are also anchored in the membrane by GPI-linkages. RAE-1 and H60 show only weak homology to each other (approximately 24%) or to MIC-A and MIC-B (approximately 20%) (Cerwenka *et al.*, 2000). The RAE-1 family comprises five highly-homologous isoforms ($\geq$89% identical; Fig. 6), RAE-1α, β, γ, δ, and ϵ, which are highly expressed during embryonic development and upregulated on multiple tumor types, but are rare in normal adult tissues (Carayannopoulos *et al.*, 2002b; Diefenbach *et al.*, 2000, 2001; Girardi *et al.*, 2001; Nomura *et al.*, 1996; Zou *et al.*, 1996). It has been shown that tumors expressing RAE-1 molecules can be recognized by NK cells and rejected (Diefenbach *et al.*, 2001). Like huNKG2D–MIC stimulation of NK cells, RAE-1 mediated rejection can override inhibitory signals from the expression of self MHC class I proteins on tumor cells. H60 was originally identified as an immunodominant minor histocompatibility antigen

(Malarkannan *et al.*, 1998, 2000). Though differentially expressed in inbred mouse strains, H60 transcripts were found present at low levels in embryonic tissue and on activated thymoblasts, but at higher levels on macrophages and dendritic cells in the spleen and some tumor cells (Diefenbach and Raulet, 2001; Diefenbach *et al.*, 2000; Malarkannan *et al.*, 1998, 2000). Little is known about the function of MULT1, but it is apparently widely and constitutively transcribed (Carayannopoulos *et al.*, 2002a).

Therefore NK cells mediate potent anti-tumor and anti-viral responses, either through *i*) recognition of the loss of expression of the normal complement of classical and non-classical MHC class I proteins on cell surfaces or by *ii*) recognition of the induced expression of cell-surface markers of cellular "distress" (responses to tumorigenesis or infection). These mechanisms can also contribute to significant NK-mediated graft-versus-leukemia responses during non-myeloablative allogeneic stem cell transplantation (Farag *et al.*, 2002).

VII. NKG2D–Ligand Complexation

The symmetric NKG2D homodimers bind their asymmetric, monomeric ligands (MIC-A, ULBP3 and RAE-1β) in a 2:1 molar stoichiometry (Fig. 1). Equivalent binding sites on each NKG2D monomer contribute nearly equally to an extensive interface (buried solvent accessible surface areas from 1681 to 2282Å^2) where each receptor monomer binds a distinct ligand surface (Fig. 1). The interfaces encompass a mix of bonding interactions (Figs. 7 and 8), where neither electrostatic nor hydrophobic terms dominate. All three NKG2D complexes are quite similar overall, despite the dissimilarity in detail between the structures of the ligand proteins (ligand structural differences are large enough to almost preclude meaningful rmsd calculations [Li *et al.*, 2002]). The saddle-shaped NKG2D homodimer sits astride the platform domain of the MHC class I-like ligands, with each NKG2D monomer primarily contacting either the $\alpha 1$ or $\alpha 2$ sub-domain of each ligand. Shape complementarities (Lawrence and Colman, 1993) are quite high (0.63 to 0.72) and sufficient to exclude water molecules from the interfaces. The NKG2D footprint on its MHC class I-like ligands (Fig. 1) overlaps the footprints of $\alpha\beta$ TCRs and KIR NKRs on MHC class Ia proteins, but is distinct from that of murine Ly49 NKRs or LIR-1 (Willcox *et al.*, 2003) on their MHC class Ia ligands.

In these complexes, each NKG2D monomer–ligand subdomain ($\alpha 1$ or $\alpha 2$) pair is referred to as a "half-site." The footprints of the ligands on each of the six NKG2D half-sites essentially overlap, showing that NKG2D truly utilizes a single binding site consisting of residues from the body of the

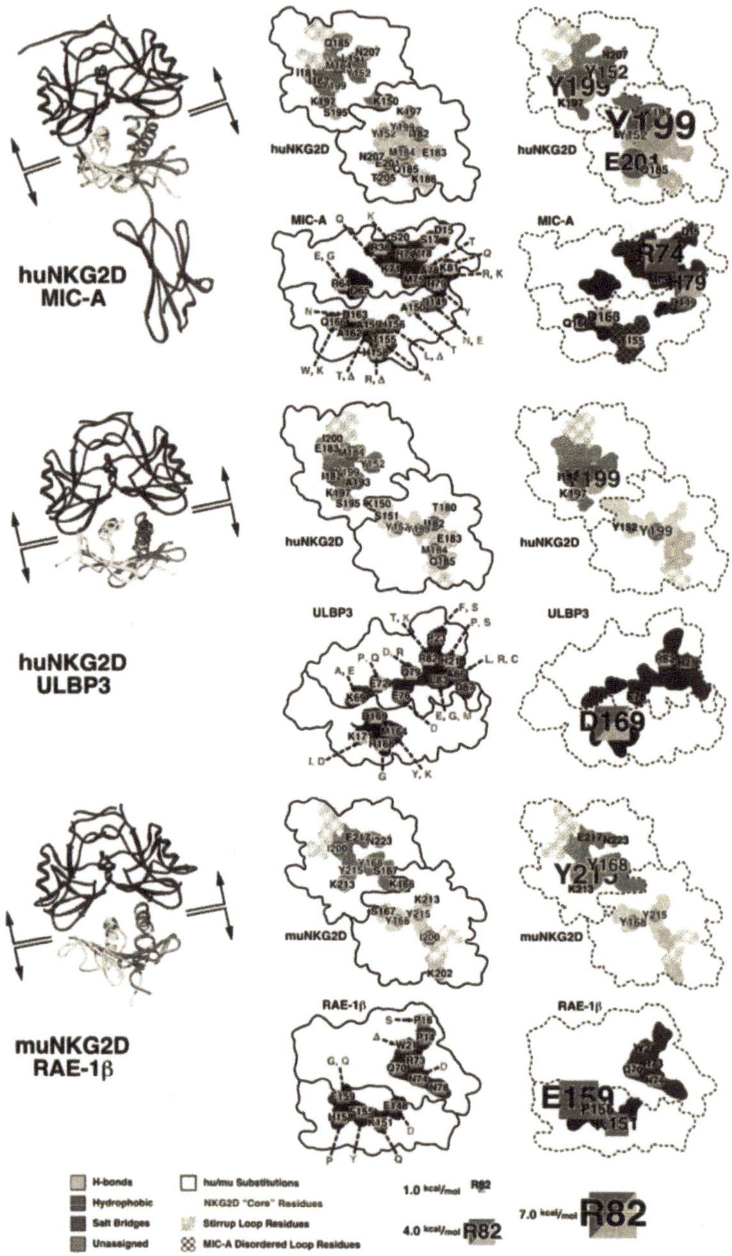
huNKG2D
MIC-A
huNKG2D
ULBP3
muNKG2D
RAE-1β
huNKG2D
MIC-A
ULBP3
muNKG2D
RAE-1β
H-bonds
Hydrophobic
Salt Bridges
Unassigned
hu/mu Substitutions
NKG2D "Core" Residues
Stirrup Loop Residues
MIC-A Disordered Loop Residues
1.0 kcal/mol R82
4.0 kcal/mol R82
7.0 kcal/mol R82

NKG2D CTLD and one loop (Fig. 7). This loop, referred to as the "stirrup" loop (Li *et al.*, 2001b) (Fig. 1), is the most extended element of NKG2D that contacts ligand. The conformation of this loop, distinct from other CTLD structures (Fig. 2), gives the ligand-binding surface on NKG2D its distinctive concave curvature. Stirrup loop sequence differences between muNKG2D and huNKG2D result in different overall curvatures, contributing to altered ligand preferences between these two orthologs at a very coarse structural level: huNKG2D will bind the narrow platforms of MIC and ULBP proteins, but not the broader RAE-1s (Fig. 7); the more splayed muNKG2D homodimer will bind human ligands as well, consistent with the conservation of the key ligand-binding residues between the human and murine receptors (McFarland *et al.*, 2003).

The single NKG2D binding site has therefore evolved to recognize at least six different surfaces, predominantly on the $\alpha1$ or $\alpha2$ domains of MIC-A, ULBP3 and RAE-1β, with dramatically different shapes. The degree of this extreme recognition degeneracy is further magnified as many of the very non-conservative sequence differences and deletions between MIC-A and -B alleles, and ULBP and RAE-1 isoforms, map to NKG2D-contacting residues on the ligand proteins (Fig. 7) (Holmes *et al.*, 2002; Li *et al.*, 2002; Radaev *et al.*, 2001). NKG2D's interaction with the highly divergent, and, therefore, likely structurally-distinct, ligand H60 is also yet to be characterized, but would be predicted to display yet additional examples of NKG2D recognition degeneracy. The diversity of interfacial

FIG. 7. Schematization of NKG2D–ligand interfaces. Residues making contacts in the NKG2D–ligand complexes are represented schematically on outlines of the separated proteins involved in the interaction: huNKG2D–MIC-A (top) (Li *et al.*, 2001a), huNKG2D–ULBP3 (middle) (Radaev *et al.*, 2001) and muNKG2D–RAE-1β (bottom) (Li *et al.*, 2002). On the left, the plane of the separation is indicated on ribbon representations of the complex structures, colored by domain as in Fig. 1. The interfaces are schematized by mapping the position of the contact residues (indicated by gnomons colored to reflect the nature of the bonds as indicated) onto outlines of the receptor (top) and the ligand (bottom). The overall footprint of one binding partner on the other is shown, colored by the domain making the footprint. Allelic and isoform substitutions affecting contact residues are indicated (conservative substitutions in green, non-conservative substitutions in red), as are portions of the receptors from stirrup loop residues (cross-hatched areas) and the portion of MIC-A from disordered loop residues (checkerboard area). The right-hand pair of interface schematics (dotted outlines) shows the same contact residue mapping, but with residue gnomons now scaled (as indicated) by calculated $\Delta\Delta G$ value (McFarland *et al.*, 2003); residues with $\Delta\Delta G$ values below the cut-off for the definition of binding "hotspots" (1 kcal/mole) are not shown. Gnomons with red outlines highlight NKG2D ligand contacting residues where sequence differences occur between murine and huNKG2D; NKG2D sequence positions that are ligand contacts in all six half-sites (NKG2D "core" residues) are indicated with blue labeling. (See Color Insert.)

Human NKG2D	MIC-A Contact	Human NKG2D	ULBP3 Contact	Murine NKG2D	RAE-1β Contact
A-chain		A-chain		A-chain	
K150	x	K150	E76°‡	K166	x
S151	x	S151	Q79*	S167	Q70
Y152	K71°, R74*, M75°	Y152	Q79°, R82*, L83°	Y168	W21, Q70, R73
T180	x	T180	K69*	T196	x
I181	x	I181	x	L197	x
I182	H79°	I182	A86°, D87°	V198	x
E183	K81*	E183	H21*	E199	x
M184	V18*°, R74°, A78°	M184	H21*°, R82°, A86°	I200	P14
Q185	V18*	Q185	P23°	P201	x
K186	D15*, S17*	K186	x	K202	P16
L191	x	L191	x	V207	x
A193	x	A193	x	G209	x
S195	x	S195	x	S211	x
K197	D149‡	K197	x	K213	N78
Y199	M75°, H79*	Y199	L83°	Y215	R73, N74
I200	x	I200	x	T216	x
E201	R74‡	E201	x	E217	x
T205	S20*	T205	x	N221	x
N207	R38*	N207	x	N223	x

Human NKG2D	MIC-A Contact	Human NKG2D	ULBP3 Contact	Murine NKG2D	RAE-1β Contact
B-chain		B-chain		B-chain	
K150	A150*	K150	x	K166	E148‡
S151	x	S151	x	S167	E148
Y152	H156°, A159°	Y152	M164°	Y168	K151, F155
T180	x	T180	x	T196	x
I181	Q166*	I181	x	L197	x
I182	A162°, Q166°	I182	R168°	V198	x
E183		E183	K171*	E199	x
M184	H158°, A162°	M184	R168°, K171*	I200	H158
Q185	H158*	Q185	x	P201	x
K186	x	K186	x	K202	x
L191	T155°	L191	x	V207	x
A193	x	A193	L83°	G209	x
S195	R64*	S195	E72*	S211	x
K197	D65‡	K197	D169‡	K213	E159‡
Y199	A159°, D163*	Y199	M164°, R168°, D169*	Y215	F155, E159
I200	x	I200	R168*	T216	x
E201	x	E201	x	E217	K151‡
T205	x	T205	x	N221	x
N207	T155*	N207	x	N223	K151

Hydrophobic/van der Waals interaction: ° ; hydrogen bond: * ; salt bridge: ‡

≤ 1 kcal/mol ≤ 3 kcal/mol ≤ 5 kcal/mol **> 5 kcal/mol**

FIG. 8. Tabulations of NKG2D–ligand contacts. Residues involved in receptor–ligand contacts are shown, separated by NKG2D monomer. Absence of a contact in one complex that is present in another is indicated ("x"). The calculated $\Delta\Delta G$ value for each residue is reflected in the font size as indicated. The nature of the bonding interaction is shown with symbols (°, *, ‡) as indicated (note that the resolution of the muNKG2D–RAE-1β complex was limited enough to restrict bond assignments).

contacts ensures that NKG2D recognition degeneracy is not the product of a dominantly hydrophobic or electrostatic binding site, relatively geometry-insensitive bonds that otherwise enable degenerate recognition in other systems.

VIII. NKG2D–Ligand Recognition Degeneracy: "Rigid Adaptation" Rather Than "Induced-Fit"

In contrast to TCR–pMHC and other NK receptor–ligand interactions, and consistent with comparisons of bound and unbound NKG2D structures (McFarland *et al.*, 2003), thermodynamic and kinetic analyses of four NKG2D–ligand pairs (MIC-A*001, MIC-B*005, ULBP1 and RAE-1β) also show that the enthalpic and entropic terms of binding, heat capacities, association rates and activation energy barriers are comparable to typical, rigid protein–protein interactions and distinct from the values associated with classical definitions of induced-fit binding (Figs. 9 and 10) (McFarland and Strong, 2003). NKG2D degeneracy is alternatively achieved by employing distinct interaction mechanisms at each rigid interface. At the center of the NKG2D binding site lie two conserved tyrosine residues (152 and 199 in huNKG2D and 168 and 215 in muNKG2D) that constitute the dominant binding-energy 'hotspots' (Clackson and Wells, 1995) in each complex half-site (Fig. 7) (McFarland *et al.*, 2003). These two tyrosines are held fairly rigidly in the NKG2D structure, where the only significant conformational change observed is utilization of a single alternate rotamer by Tyr152 in two of the total of eight independent crystallographic views of the NKG2D monomer (two NKG2D monomers in each complex structure plus one monomer each in the free muNKG2D and huNKG2D structures) (McFarland *et al.*, 2003). The conformational plasticity associated with induced-fit binding often encompasses backbone movements of six ångstroms or more, well beyond the scale of the side-chain "wiggle" observed among NKG2D structures. As has been seen in many antibody combining sites (Nikula *et al.*, 1995; Padlan *et al.*, 1990), these tyrosines make multifarious interactions among the distinct ligand surfaces: conserved and non-conserved hydrogen bonds, differential hydrophobic interactions with a range of residues, ring/ring-stacking interactions and even cation-π bonds.

NKG2D's extreme recognition degeneracy is therefore achieved by investing a significant proportion of the binding energy in core residues that, while rigidly constrained, are capable of making specific, yet disparate, interactions with the divergent ligand binding surfaces. These core interactions are placed within the context of extensive, water-excluding, highly shape-complementary interfaces, where additional electrostatic, hydrogen

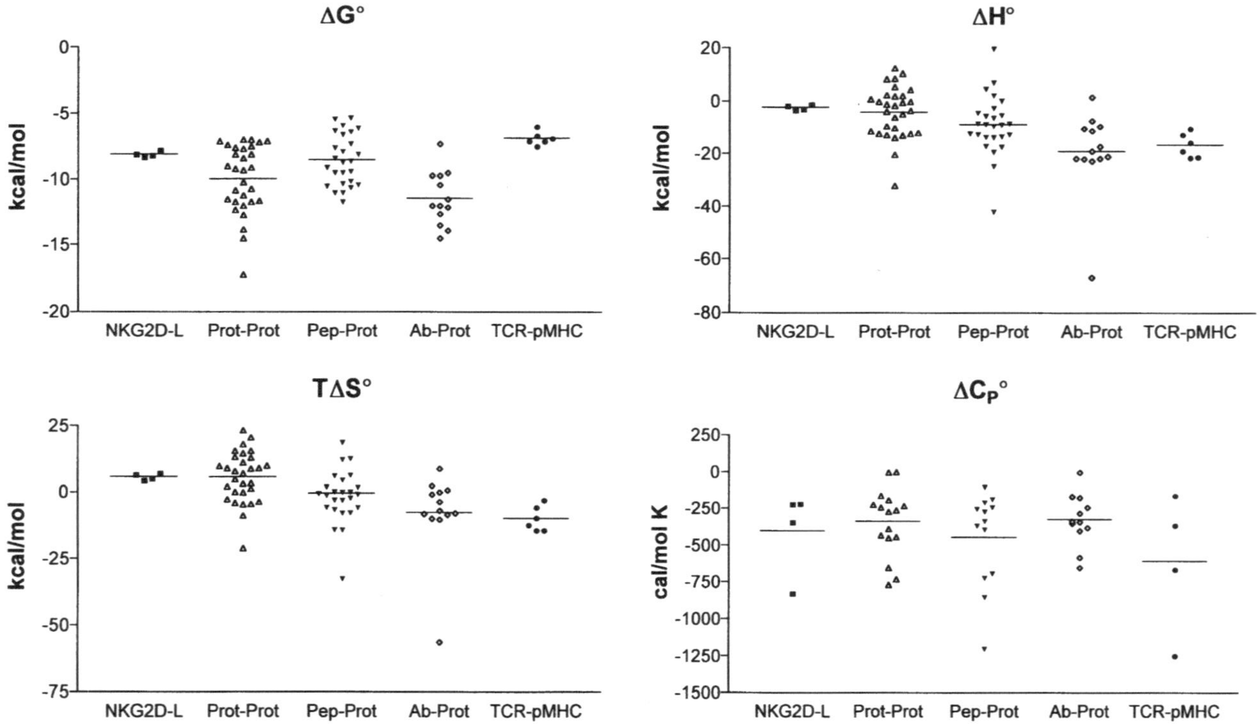

FIG. 9. Thermodynamics of receptor–ligand interactions. Experimentally derived values for ΔG°, ΔH°, TΔS°, and ΔC_P° are shown for, from left to right: NKG2D–ligand interactions, rigid protein–rigid protein interactions (excluding antibodies), peptide–protein interactions, antibody–protein interactions (where the antibody is known to utilize induced-fit recognition) and $\alpha\beta$ TCR–pMHC interactions (Anikeeva *et al.*, 2003; Boniface *et al.*, 1999; Garcia *et al.*, 2001; McFarland and Strong, 2003; Stites, 1997; Willcox *et al.*, 1999). Horizontal lines represent average values for each class. Values were measured in the range of 287.3 to 303 K.

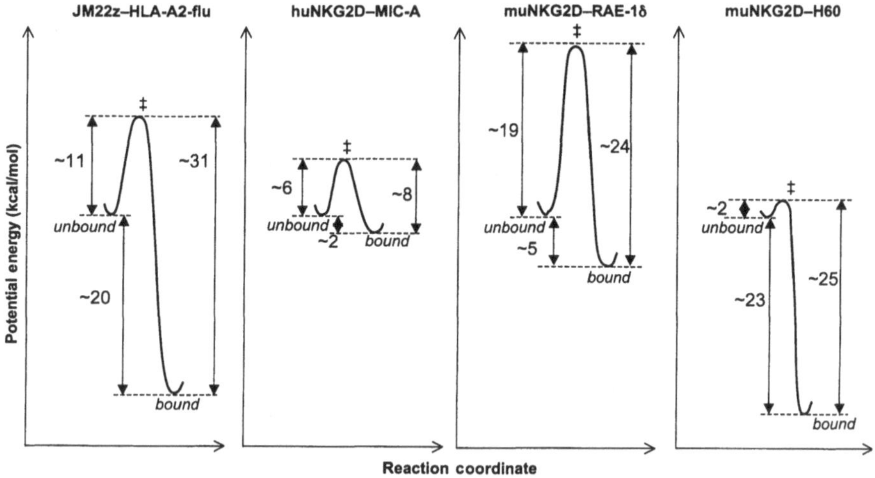

FIG. 10. Immunoreceptor–ligand binding energetics. Reaction energy profiles, to scale, of four representative immunoreceptor-ligand interactions: JM22z $\alpha\beta$ TCR–HLA-A2–flu pMHC, (Willcox *et al.*, 1999) huNKG2D–MIC-A, (McFarland and Strong, 2003) muNKG2D–RAE-1β and –H60 (O'Callaghan *et al.*, 2001). All kinetic values were originally derived from SPR-determined association and dissociation rate constants over a range of temperatures. E_a^{ass}, energy of association, is the transition from unbound state to the high-energy intermediate; E_a^{diss}, energy of dissociation, is the transition from the high-energy intermediate to the bound state; ‡, high-energy intermediate.

and van der Waals bonds contribute to the overall affinity while minimizing the dominance of any single peripheral contact (Figs. 8 and 11). The extent of the interfaces contributes to affinity and specificity by enabling multiple peripheral bonds to add to affinity, but also by requiring that potential target ligands stringently exclude deleterious steric clashes, both on the scale of individual side-chains and in the overall shape of the NKG2D binding saddle (McFarland *et al.*, 2003).

Thus NKG2D has evolved to utilize a recognition mechanism that is capable of specifically binding to diverse ligands while tolerating considerable variation in ligand interfaces. The latter phenomenon may allow the immune system to fine-tune the NKG2D activation threshold through subtle alteration of the kinetics and affinities of particular interactions in specific contexts, allowing modulation of NKG2D signals through peripheral ligand sequence variation. This extreme degenerate recognition is achieved within an essentially rigid receptor binding site structure by a "rigid adaptation" mechanism that complements the function of NKG2D (McFarland and Strong, 2003). NKG2D is a dominantly activating

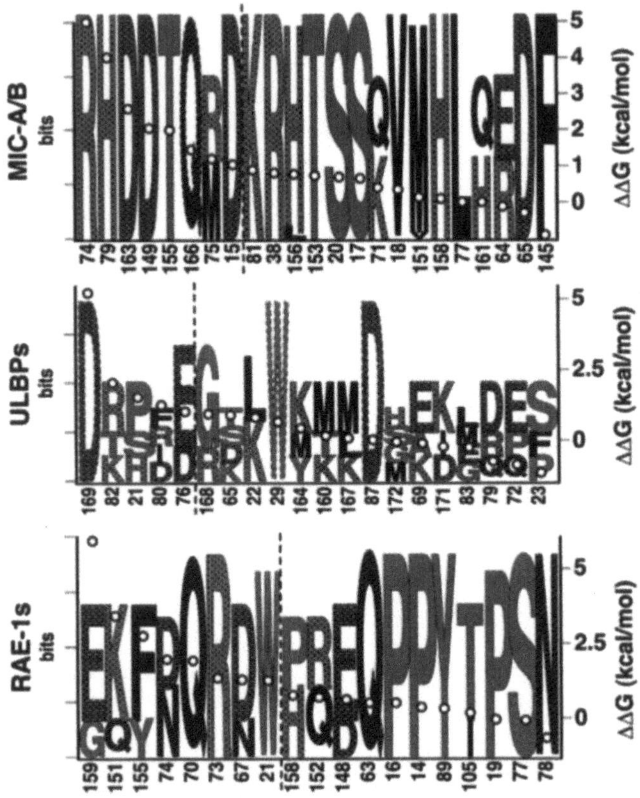

FIG. 11. Sequence variation at NKG2D–ligand contact residues. The positions, in the sequence, of MIC (top), ULBP (middle) or RAE-1 (bottom) residues contacting NKG2D are plotted from left to right, in descending order of their associated contribution to NKG2D binding (ΔΔG), which is plotted as open circles (right-hand ordinate). Sequence variation at NKG2D contacts due to allelic (MIC-A/B) and isoform (ULBP and RAE-1) substitutions is shown as sequence "logos." (Schneider and Stephens, 1990) Amino acids are shown in single-letter coding, patterned by type. The vertical dashed line indicates the 1 kcal/mole ΔΔG cut-off, the accepted definition of the threshold for a binding "hotspot." The predicted pattern is readily apparent for the MIC proteins, where increasing sequence variation correlates with decreasing ΔΔG value. This pattern is superficially reversed for the RAE-1s, though not with closer inspection: the sequence substitutions associated with higher ΔΔG values are either quite conservative or do not affect the contacts to backbone atoms. There is little pattern to the ULBPs, reflecting the proposition that the low sequence conservation (59 to 30%) among family members results in structural changes significant enough to invalidate the transfer of NKG2D contact maps from one family member to the next.

immunoreceptor, where NKG2D engagement delivers strongly activating signals to effector cells that can override many, if not all, inhibitory signals. Utilizing rigid adaptation recognition in this context has the distinct advantage that the cross-reactivity inherent in induced-fit binding mechanisms (Boniface *et al.*, 1999; James *et al.*, 2003; Wu *et al.*, 2002b) is prevented, while at the same time enabling recognition degeneracy to expand the repertoire of potential NKG2D ligands, extending NKG2D functionality to a variety of contexts. Inappropriate ligand engagement by NKG2D, through cross-reactivity, would result in the elimination of inappropriate target cells, with potentially serious physiological consequences. Thus, the immune system has developed an elegant system delivering broad utility while minimizing potentially deleterious responses, mirroring the functionally distinct recognition mechanisms utilized by TCRs or antibodies.

IX. NKG2D: Open Questions

While the degree of sequence variation tolerated by NKG2D at ligand interfaces is remarkable, the picture for the ULBP ligands becomes more complicated. The sequence conservation among MIC proteins ($\geq$84%) and RAE-1s ($\geq$89%) is sufficiently high that the structures among family members are likely conserved to the degree that these ligands interact with NKG2D in very similar ways. Therefore the assumption underlying Fig. 11, that receptor contact maps are valid across family members, is likely correct. However, the sequence conservation among the ULBPs ($\geq$30%) is low enough that the underlying structures are almost certainly significantly different, invalidating the assumptions that the NKG2D interactions, and, therefore, contact maps, would be conserved across ULBP family members. This prediction is borne out by the analysis shown in Fig. 11, where the correlation between $\Delta\Delta G$ value and degree of conservation (demonstrated by the MIC ligands) breaks down for the ULBPs. The conclusion is that the ULBP family may show considerable variation in structure and, therefore, interactions with NKG2D.

The NKG2D structures also present another conundrum: How is ligand engagement signaled through the ectodomain? The most flexible part of the receptor is the N-terminal stalk of the ectodomain between the CTLD and the membrane-spanning domain (NKG2x receptors are type II transmembrane proteins). These arms are among the more variable elements in the NKG2x NKR family (Fig. 12). Although the various crystallization constructs used in the crystallographic analyses encompass most, if not all, of this region, at most only about a quarter, and typically only a few residues, of the stalk is ordered in any of the five different crystal structures of NKG2D (human or mouse, free or complexed). However, while

```
   NKG2A  ----PSTLIQRHNNSSLNTRTQKARHCG 24
   NKG2B  ----PS------------------RHCG 6
   NKG2C  ----IPFLEQ--NNSSPNTRTQKARHCG 22
   NKG2E  -------LEQ--NNSSPNTRTQKARPCG 19
   NKG2F  --PCIGVLEQ--NSFSLNRRMQKARHCG 24
   NKG2H  -------LEQ--NNSSPNTRTQKARPCG 19

   NKG2D  -AIWSAVFLNSLFNQEVQIPLTES-YCG 26
 muNKG2D  -AIFKETFQPVLCNKEVPVSSREG-YCG 26

    CD94  KNSFTKLSIEPAFTPGPNIELQKDSDCC 28

    CD69  ------SVGQYNCPGQYTFSMPSDSHVS 22
```

FIG. 12. Comparison of NKG2x N-terminal ectodomain arm sequences. Alignments were calculated with CLUSTALW (Thompson *et al.*, 1994).

extremely flexible, the stalks do not contribute to "induced-fit" recognition because they are distal to the ligand binding sites. Electron density is observed for more of the flexible N-terminal stalk in the free human NKG2D structure (McFarland *et al.*, 2003) than in any other structure. In this most fully resolved view, the stalk displays no defined secondary structure, and the only contacts between stalks of the same homodimer are van der Waals bonds near the interface between monomers in the homodimer. The extreme flexibility of the N-terminal stalk, and the lack of any obvious, consistent associations between stalks or stalk and CTLD, leaves us without an obvious structural mechanism for signaling ligand engagement to the cell's interior. The crystallographic analyses suggest that the receptor does not multimerize in any way relevant to signaling. No proteins have been identified that associate with any part, such as the unstructured arm, of the NKG2D ectodomain on the surface of effector cells that may contribute to a signal transduction mechanism.

X. NKG2D: IMPLICATIONS FOR NKG2x–CD94 RECOGNITION

A complex structure between NKG2x–CD94 receptors and HLA-E ligands has been modeled, in some detail, based on the NKG2D–MIC-A complex structure (Li *et al.*, 2001a; Strong *et al.*, 2002), extending a lower-resolution model based on the crystal structure of CD94 (Boyington *et al.*, 1999). In this model, CD94 overlies the $\alpha 1$ domain of HLA-E, with a small hydrophobic patch on CD94 matching a similar patch on HLA-E (residues in the $\alpha 1$ domain and bound peptide at P8). There is no comparable hydrophobic patch on the $\alpha 2$ domain of HLA-E underlying the presumed position of the NKG2 moiety, but compensatory changes occur in the sequences of both NKG2A and NKG2C, thus accommodating this difference. The flatter NKG2x–CD94 binding surface complements the wider surface of HLA-E relative to the NKG2D ligands, where the platform interhelical distance is broadened by the bound, MHC class Ia-derived peptide. The side-chain of a conserved arginine at the P5 position in the peptide would also be able to exchange hydrogen bonds with residues from either NKG2x or CD94 at the homodimer interface. Fewer peptide side-chains are accessible in HLA-E complexes (mostly the P4, P5, and P8 residues, with P1 and P6 to a lesser extent) than MHC class Ia proteins due to the deeper, more encompassing, peptide groove. Residues in a loop of CD94 at the heterodimer interface are in position to reach into the peptide binding groove, hydrogen bonding to the peptide backbone at either P4 or P5. The comparable loop in NKG2D and Ly49A does not extend as far downward, toward the binding saddle, as in CD94. The restructuring of the $\alpha 2$ helix in CD94 into an extended loop (at the base of the homodimer interface in Fig. 2) accommodates the presence of a peptide in complex with the MHC protein. Four acidic residues in CD94 dominate a patch at the predicted interface matching a cluster of positively-charged residues on HLA-E and P5-arginine in the peptide. Using the CD94 homodimer and the structure of NKG2D to model an NKG2A–CD94 heterodimer results in a binding site on NKG2A that is dominated by charged and polar residues which would overlie a similarly charged surface on HLA-E, suggesting that the NKG2A–HLA-E interaction may be more similar in character to the KIR2DL2–pMHC and Ly49A–pMHC interactions, which are dominated by complementary charge–charge interactions, than to NKG2D–MIC-A binding. This is consistent with the different recognition mechanisms employed by NKG2D and the other receptors of the NKG2x family: where NKG2D displays extreme recognition degeneracy, the NKG2x–CD94 receptors are expected to display much more typical, highly-specific, protein–protein recognition since their ligand repertoire is so much more limited. However, these models are not good enough to completely delineate the role peptide plays in recognition. It is also clear that the peptide can have indirect effects on receptor interactions; substitutions at the P2 position can markedly affect the thermal stability of HLA-E, mostly through the introduction of cavities, that subsequently affects both cell-surface expression levels and receptor interactions (Strong *et al.*, 2002).

XI. MIC and $\gamma\delta$ TCRs

Aside from its role as a ligand for NKG2D, MIC proteins are also directly recognized by $\gamma\delta$ TCRs of the Vδ1 subset (Wu *et al.*, 2002a). Unlike $\alpha\beta$ TCRs, which only interact with peptide fragments of protein antigens presented as complexes with MHC class I molecules, $\gamma\delta$ TCRs are proposed to interact directly with intact antigens, apparently without the requirement of extensive processing (Allison and Garboczi, 2002; Davis *et al.*, 1998; Morita *et al.*, 1995; Porcelli *et al.*, 1991; Schild *et al.*, 1994)–although considerably less is known about the functional details of this class of TCR. In humans, $\gamma\delta$ T cells can be functionally divided on the basis of the Vδ gene utilization of the expressed $\gamma\delta$ TCR. Vδ2Vγ9 T cells predominate in the peripheral blood and are thought to provide anti-bacterial defenses by directly recognizing soluble, mycobacterially-derived prenyl pyrophosphate and alkamine compounds (Allison and Garboczi, 2002; Bukowski *et al.*, 1995, 1999; Constant *et al.*, 1994; Tanaka *et al.*, 1995). Both on the basis of the distribution of CDR sequence variation and the crystal structure of a Vδ2Vγ9 TCR, $\gamma\delta$ TCRs are proposed to interact with and recognize ligands more like antibodies than $\alpha\beta$ TCRs (Allison and Garboczi, 2002; Allison *et al.*, 2001; Rock *et al.*, 1994). The scarcer Vδ1-bearing T cells are enriched in the epithelial compartment (Deusch *et al.*, 1991; Spencer *et al.*, 1989), paralleling the restricted tissue distribution of MIC proteins. Previously, it had been shown that Vδ1 $\gamma\delta$ T cell lines recognize and kill MIC-bearing targets, and that this interaction could be blocked by anti-$\gamma\delta$ TCR antibodies (Bahram *et al.*, 1994; Groh *et al.*, 1996, 1998, 1999). But, since these cells also express MIC-specific NKG2D receptors, it was not clear that a direct interaction between MIC and Vδ1 $\gamma\delta$ TCR occurred and drove activation prior to the aforementioned report (Wu *et al.*, 2002a). The possibility also exists that Vδ1 $\gamma\delta$ TCR–NKG2D–MIC form receptor–co-receptor–ligand complexes analogous to $\alpha\beta$ TCR–CD8–pMHC complexes, possibly accommodated by the length and flexibility of the NKG2D N-terminal arm. This could require that the $\gamma\delta$ TCR interact with MIC at a site not overlapping with the NKG2D-interaction site.

In conclusion, though our understanding of the roles that NKG2D plays in mediating responses of the innate and adaptive immune systems is continuing to expand, it is already clear that this immunoreceptor uses unique, almost unprecedented recognition machinery to accomplish these tasks. Therefore, the principles of NKG2D immunorecognition represent a wholly distinct paradigm from that of TCR–ligand recognition, which together define two poles of protein–protein interactions employed to accomplish the disparate functions of the immune system.

References

Allison, T. J., and Garboczi, D. N. (2002). Structure of gd T cell receptors and their recognition of non-peptide antigens. *Mol. Immunol.* **38,** 1051–1061.

Allison, T. J., Winter, C. C., Fournié, J.-J., Bonneville, M., and Garboczi, D. N. (2001). Structure of a human $\gamma\delta$ T-cell antigen receptor. *Nat. Immunol.* **411,** 820–824.

Anikeeva, N., Lebedeva, T., Krogsgaard, M., Tetin, S. Y., Martinez-Hackert, E., and Kalams, S. A. *et al.* (2003). Distinct molecular mechanisms account for the specificity of two different T-cell receptors. *Biochem.* **42**(16), 4709–4716.

Bahram, S., and Spies, T. A. (1996). Nucleotide sequence of a human *MHC* class I *MICB* cDNA. *Immunogene.* **43,** 230–233.

Bahram, S., Bresnahan, M., Geraghty, D. E., and Spies, T. A. (1994). A second lineage of mammalian major histocompatibility complex class I genes. *Proc. Natl. Acad. Sci. USA* **91,** 6259–6263.

Bahram, S., Mizuki, N., Inoko, H., and Spies, T. A. (1996). Nucleotide sequence of the human MHC class I *MICA* gene. *Immunogene.* **44,** 80–81.

Bakker, A. B., Wu, J., Phillips, J. H., and Lanier, L. L. (2000). NK cell activation: Distinct stimulatory pathways counterbalancing inhibitory signals. *Human Immunol.* **61**(1), 18–27.

Bauer, S., Groh, V., Wu, J., Steinle, A., Phillips, J. H., and Lanier, L. L. *et al.* (1999). Activation of NK cells and T cells by NKG2D, a receptor for stress-inducible MICA. *Science* **285**(5428), 727–729.

Bellon, T., Heredia, A. B., Llano, M., Minguela, A., Rodriguez, A., and Lopez-Botet, M. *et al.* (1999). Triggering of effector functions on a CD8+ T cell clone upon the aggregation of an activatory CD94/kp39 heterodimer. *J. Immunol.* **162**(7), 3996–4002.

Bjorkman, P. J., and Parham, P. (1999). Structure, function and diversity of class I major histocompatibility complex molecules. *Ann. Rev. Biochem.* **90,** 253–288.

Boniface, J. J., Reich, Z., Lyons, D. S., and Davis, M. M. (1999). Thermodynamics of T cell receptor binding to peptide-MHC: Evidence for a general mechanism of molecular scanning. *Proc. Natl. Acad. Sci. USA* **96**(20), 11446–11451.

Borrego, F., Kabat, J., Kim, D.-K., Lieto, L., Maasho, K., and Peña, J. *et al.* (2002). Structure and function of major histocompatibility complex (MHC) class I specific receptors expressed on human natural killer (NK) cells. *Mol. Immunol.* **1063,** 1–24.

Boyington, J. C., Riaz, A. N., Patamawenu, A., Coligan, J. E., Brooks, A. G., and Sun, P. D. (1999). Structure of CD94 reveals a novel C-type lectin fold: Implications for the NK cell-associated CD94/NKG2 receptors. *Immunity* **10**(1), 75–82.

Boyington, J. C., Motyka, S. A., Schuck, P., Brooks, A. G., and Sun, P. D. (2000). Crystal structure of an NK cell immunoglobulin-like receptor in complex with its class I MHC ligand. *Nature* **405**(6786), 537–543.

Bukowski, J. F., Morita, C. T., and Brenner, M. B. (1999). Human gamma delta T cells recognize alkylamines derived from microbes, edible plants, and tea: Implications for innate immunity. *Immunity* **11**(1), 57–65.

Bukowski, J. F., Morita, C. T., Tanaka, Y., Bloom, B. R., Brenner, M. B., and Band, H. (1995). V gamma 2V delta 2 TCR-dependent recognition of non-peptide antigens and Daudi cells analyzed by TCR gene transfer. *J. Immunol.* **154**(3), 998–1006.

Carayannopoulos, L. N., Naidenko, O. V., Fremont, D. H., and Yokoyama, W. M. (2002a). Cutting edge: Murine UL16-binding protein-like transcript 1: A newly described transcript encoding a high-affinity ligand for murine NKG2D. *J. Immunol.* **169**(8), 4079–4083.

Carayannopoulos, L. N., Naidenko, O. V., Kinder, J., Ho, E. L., Fremont, D. H., and Yokoyama, W. M. (2002b). Ligands for murine NKG2D display heterogeneous binding behavior. *Eur. J. Immunol.* **32**(3), 597–605.

Cerwenka, A., Bakker, A. B., McClanahan, T., Wagner, J., Wu, J., and Phillips, J. H. *et al.* (2000). Retinoic acid early inducible genes define a ligand family for the activating NKG2D receptor in mice. *Immunity* **12**(6), 721–727.

Clackson, T., and Wells, J. A. (1995). A hot spot of binding energy in a hormone-receptor interface. *Science* **267**(5196), 383–386.

Colonna, M. (1997). Immunoglobulin superfamily inhibitory receptors: From natural killer cells to antigen-presenting cells. *Res. Immunol.* **148**(3), 169–171.

Constant, P., Davodeau, F., Peyrat, M. A., Poquet, Y., Puzo, G., and Bonneville, M. *et al.* (1994). Stimulation of human gamma delta T cells by nonpeptidic mycobacterial ligands. *Science* **264**(5156), 267–270.

Cosman, D., Mullberg, J., Sutherland, C. L., Chin, W., Armitage, R., and Fanslow, W. *et al.* (2001). ULBPs, novel MHC class I-related molecules, bind to CMV glycoprotein UL16 and stimulate NK cytotoxicity through the NKG2D receptor. *Immunity* **14**(2), 123–133.

Davis, M. M., Boniface, J. J., Reich, Z., Lyons, D., Hampl, J., and Arden, B. *et al.* (1998). Ligand recognition by alpha beta T cell receptors. *Annu. Rev. Immunol.* **16,** 523–544.

Deusch, K., Luling, F., Reich, K., Classen, M., Wagner, H., and Pfeffer, K. (1991). A major fraction of human intraepithelial lymphocytes simultaneously expresses the gamma/delta T cell receptor, the CD8 accessory molecule and preferentially uses the V delta 1 gene segment. *Eur. J. Immunol.* **21**(4), 1053–1059.

Diefenbach, A., and Raulet, D. H. (2001). Strategies for target cell recognition by natural killer cells. *Immunol. Rev.* **181,** 170–184.

Diefenbach, A., and Raulet, D. H. (2003). Innate immune recognition by stimulatory immunoreceptors. *Curr. Opin. Immunol.* **15**(1), 37–44.

Diefenbach, A., Jamieson, A. M., Liu, S. D., Shastri, N., and Raulet, D. H. (2000). Ligands for the murine NKG2D receptor: Expression by tumor cells and activation of NK cells and macrophages. *Nat. Immunol.* **1**(2), 119–126.

Diefenbach, A., Jensen, E. R., Jamieson, A. M., and Raulet, D. H. (2001). Rae1 and H60 ligands of the NKG2D receptor stimulate tumour immunity. *Nature* **413,** 165–171.

Diefenbach, A., Tomasello, E., Lucas, M., Jamieson, A. M., Hsia, J. K., and Vivier, E. *et al.* (2002). Selective associations with signaling proteins determine stimulatory versus costimulatory activity of NKG2D. *Nat. Immunol.* **3**(12), 1142–1149.

Drickamer, K. (1993). Ca^{2+}-dependent carbochydrate-recognition domains in animal proteins. *Curr. Opin. Struct. Biol.* **3,** 393–400.

Drickamer, K. (1999). C-type lectin-like domains. *Curr. Opin. Struct. Biol.* **9**(5), 585–590.

Farag, S. S., Fehniger, T. A., Ruggeri, L., Velardi, A., and Caligiuri, M. A. (2002). Natural killer cell receptors: New biology and insights into the graft-versus-leukemia effect. *Blood* **100**(6), 1935–1947.

Feinberg, H., Park-Snyder, S., Kolatkar, A. R., Heise, C. T., Taylor, M. E., and Weis, W. I. (2000). Structure of a C-type carbohydrate recognition domain from the macrophage mannose receptor. *J. Biol. Chem.* **275**(28), 21539–21548.

Garboczi, D. N., Ghosh, P., Utz, U., Fan, Q. R., Biddison, W. E., and Wiley, D. C. (1996). Structure of the complex between human T-cell receptor, viral peptide and HLA-A2. *Nature* **384**(6605), 134–141.

Garcia, K. C., Degano, M., Speir, J. A., and Wilson, I. A. (1999). Emerging principles for T cell receptor recognition of antigen in cellular immunity. *Rev. Immunogene.* **21,** 75–90.

Garcia, K. C., Radu, C. G., Ho, J., Ober, R. J., and Ward, E. S. (2001). Kinetics and thermodynamics of T cell receptor-autoantigen interactions in murine experimental autoimmune encephalomyelitis. *Proc. Natl. Acad. Sci. USA* **98**(12), 6818–6823.

Germain, R. N., and Margulies, D. H. (1993). The biochemistry and cell biology of antigen processing and presentation. *Ann. Rev. Immunol.* **11,** 403–450.

Gilfillan, S., Ho, E. L., Cella, M., Yokoyama, W. M., and Colonna, M. (2002). NKG2D recruits two distinct adapters to trigger NK cell activation and costimulation. *Nat. Immunol.* **3**(12), 1150–1155.

Girardi, M., Oppenheim, D. E., Steele, C. R., Lewis, J. M., Glusac, E., and Filler, R. *et al.* (2001). Regulation of cutaneous malignancy by gammadelta T cells. *Science* **294**(5542), 605–609.

Grakoui, A., Bromley, S. K., Sumen, C., Davis, M. M., Shaw, A. S., and Allen, P. M. *et al.* (1999). The immunological synapse: A molecular machine controlling T cell activation. *Science* **285**(5425), 221–227.

Groh, V., Bahram, S., Bauer, S., Herman, A., Beauchamp, M., and Spies, T. (1996). Cell stress-regulated human major histocompatibility complex class I gene expressed in gastrointestinal epithelium. *Proc. Natl. Acad. Sci. USA* **93,** 12445–12450.

Groh, V., Rhinehart, R., Secrist, H., Bauer, S., Grabstein, K. H., and Spies, T. (1999). Broad tumor-associated expression and recognition by tumor-derived gamma delta T cells of MICA and MICB. *Proc. Natl. Acad. Sci. USA* **96**(12), 6879–6884.

Groh, V., Steinle, A., Bauer, S., and Spies, T. (1998). Recognition of stess-induced MHC molecules by intestinal epithelial $\gamma\delta$ T cells. *Science* **279,** 1737–1740.

Groh, V., Wu, J., Yee, C., and Spies, T. (2002). Tumour-derived soluble MIC ligands impair expression of NKG2D and T-cell activation. *Nature* **419**(6908), 734–738.

Holmes, M. A., Li, P., Petersdorf, E. W., and Strong, R. K. (2002). Structural studies of allelic diversity of the MHC class I homolog MIC-B, a stress-inducible ligand for the activating immunoreceptor NKG2D. *J. Immunol.* **169**(3), 1395–1400.

Held, W., Coudert, J. D., and Zimmer, J. (2003). The NK cell receptor repertoire: Formation, adaptation and exploitation. *Curr. Opin. Immunol.* **15**(2), 233–237.

James, L. C., Roversi, P., and Tawfik, D. S. (2003). Antibody multispecificity mediated by conformational diversity. *Science* **299**(5611), 1362–1367.

Jamieson, A. M., Diefenbach, A., McMahon, C. W., Xiong, N., Carlyle, J. R., and Raulet, D. H. (2002). The role of the NKG2D immunoreceptor in immune cell activation and natural killing. *Immunity* **17**(1), 19–29.

Karlhofer, F. M., Ribaudo, R. K., and Yokoyama, W. M. (1992). MHC class I alloantigen specificity of Ly-49+ IL-2-activated natural killer cells. *Nature* **358**(6381), 66–70.

Koshland D. E., Jr. (1958). Mechanism of transfer enzymes. *In* "The Enzymes," Revised ed., (Boyer, P. *et al.* Eds.), pp. 305–315. Academic Press, New York.

Kubota, A., Kubota, S., Lohwasser, S., Mager, D. L., and Takei, F. (1999). Diversity of NK cell receptor repertoire in adult and neonatal mice. *J. Immunol.* **163**(1), 212–216.

Lanier, L. L. (2000). Turning on natural killer cells. *J. Exp. Med.* **191**(8), 1259–1262.

Lanier, L. L., Corliss, B., Wu, J., and Phillips, J. H. (1998a). Association of DAP12 with activating CD94/NKG2C NK cell receptors. *Immunity* **8**(6), 693–701.

Lanier, L. L., Corliss, B. C., Wu, J., Leong, C., and Phillips, J. H. (1998b). Immunoreceptor DAP12 bearing a tyrosine-based activation motif is involved in activating NK cells. *Nature* **391**(6668), 703–707.

Lawrence, M., and Colman, P. M. (1993). Shape complementarity at protein/protein interfaces. *J. Mol. Biol.* **234,** 946–950.
Lee, N., Goodlett, D. R., Ishitani, A., Marquardt, H., and Geraghty, D. E. (1998). HLA-E surface expression depends on binding of TAP-dependent peptides derived from certain HLA class I signal sequences. *J. Immunol.* **160**(10), 4951–4960.
Lenschow, D. J., Walunas, T. L., and Bluestone, J. A. (1996). CD28/B7 system of T cell costimulation. *Ann. Rev. Immunol.* **14,** 233–258.
Li, P., McDermott, G., and Strong, R. K. (2002). Crystal structures of RAE-1beta and its complex with the activating immunoreceptor NKG2D. *Immunity* **16**(1), 77–86.
Li, P., Morris, D. L., Willcox, B. E., Steinle, A., Spies, T., and Strong, R. K. (2001a). Complex structure of the activating immunoreceptor NKG2D and its MHC class I-like ligand MICA. *Nat. Immunol.* **2,** 443–451.
Li, P., Morris, D. L., Willcox, B. E., Steinle, A., Spies, T., and Strong, R. K. (2001b). Complex structure of the activating immunoreceptor NKG2D and its MHC class I-like ligand MICA. *Nat. Immunol.* **2**(5), 443–451.
Llera, A. S., Viedma, F., Sanchez-Madrid, F., and Tormo, J. (2001). Crystal structure of the C-type lectin-like domain from the human hematopoietic cell receptor CD69. *J. Biol. Chem.* **276**(10), 7312–7319.
Makrigiannis, A. P., Pau, A. T., Schwartzberg, P. L., McVicar, D. W., Beck, T. W., and Anderson, S. K. (2002). A BAC contig map of the Ly49 gene cluster in 129 mice reveals extensive differences in gene content relative to C57BL/6 mice. *Genomics* **79**(3), 437–444.
Malarkannan, S., Horng, T., Eden, P., Gonzalez, F., Shih, P., and Brouwenstijn, N. *et al.* (2000). Differences that matter: Major cytotoxic T cell-stimulating minor histocompatibility antigens. *Immunity* **13**(3), 333–344.
Malarkannan, S., Shih, P. P., Eden, P. A., Horng, T., Zuberi, A. R., and Christianson, G. *et al.* (1998). The molecular and functional characterization of a dominant minor H antigen, H60. *J. Immunol.* **161**(7), 3501–3509.
McFarland, B. J., and Strong, R. K. (2003). Thermodynamic analysis of degenerate recognition by the NKG2D immunoreceptor: Not 'induced fit' but 'rigid adaptation'. *Immunity* **19**(6), 803–812.
McFarland, B. J., Kortemme, T., Yu, S. F., Baker, D., and Strong, R. K. (2003). Symmetry Recognizing Asymmetry. Analysis of the Interactions between the C-Type Lectin-like Immunoreceptor NKG2D and MHC Class I-like Ligands. *Structure (Camb)* **11**(4), 411–422.
Morita, C. T., Beckman, E. M., Bukowski, J. F., Tanaka, Y., Band, H., and Bloom, B. R. *et al.* (1995). Direct presentation of nonpeptide prenyl pyrophosphate antigens to human gamma delta T cells. *Immunity* **3**(4), 495–507.
Nikula, T. K., Bocchia, M., Curcio, M. J., Sgouros, G., Ma, Y., and Finn, R. D. *et al.* (1995). Impact of the high tyrosine fraction in complementarity determining regions: Measured and predicted effects of radioiodination on IgG immunoreactivity. *Mol. Immunol.* **32**(12), 865–872.
Nomura, M., Zou, Z., Joh, T., Takihara, Y., Matsuda, Y., and Shimada, K. (1996). Genomic structures and characterization of Rae1 family members encoding GPI-anchored cell surface proteins and expressed predominantly in embryonic mouse brain. *J. Biochem.* **120**(5), 987–995.
O'Callaghan, C. A., Cerwenka, A., Willcox, B. E., Lanier, L. L., and Bjorkman, P. J. (2001). Molecular competition for NKG2D: H60 and RAE1 compete unequally for NKG2D with dominance of H60. *Immunity* **15,** 201–211.

O'Callaghan, C. A., Tormo, J., Willcox, B. E., Braud, V. M., Jakobsen, B. K., and Stuart, D. I. *et al.* (1998). Structural features impose tight peptide binding specificity in the nonclassical MHC molecule HLA-E. *Mol. Cell* **1**(4), 531–541.

Ohlen, C., Kling, G., Hoglund, P., Hansson, M., Scangos, G., and Bieberich, C. *et al.* (1989). Prevention of allogeneic bone marrow graft rejection by H-2 transgene in donor mice. *Science* **246**(4930), 666–668.

Padlan, E. A. (1990). On the nature of antibody combining sites: Unusual structural features that may confer on these sites an enhanced capacity for binding ligands. *Proteins* **7**(2), 112–124.

Pende, D., Cantoni, C., Rivera, P., Vitale, M., Castriconi, R., and Marcenaro, S. *et al.* (2001). Role of NKG2D in tumor cell lysis mediated by human NK cells: Cooperation with natural cytotoxicity receptors and capability of recognizing tumors of nonepithelial origin. *Eur. J. Immunol.* **31**(4), 1076–1086.

Plougastel, B., Jones, T., and Trowsdale, J. (1996). Genomic structure, chromosome location, and alternative splicing of the human NKG2A gene. *Immunogenetics* **44**(4), 286–291.

Plougastel, B., and Trowsdale, J. (1997). Cloning of NKG2-F, a new member of the NKG2 family of human natural killer cell receptor genes. *Eur. J. Immunol.* **27**(11), 2835–2839.

Porcelli, S., Brenner, M. B., and Band, H. (1991). Biology of the human gamma delta T-cell receptor. *Immunol. Rev.* **120,** 137–183.

Radaev, S., Rostro, B., Brooks, A. G., Colonna, M., and Sun, P. D. (2001). Conformational plasticity revealed by the cocrystal structure of NKG2D and its class I MHC-like ligand ULBP3. *Immunity* **15**(6), 1039–1049.

Raulet, D. H., Vance, R. E., and McMahon, C. W. (2001). Regulation of the natural killer cell receptor repertoire. *Ann. Rev. Immunol.* **19,** 291–330.

Rock, E. P., Sibbald, P. R., Davis, M. M., and Chien, Y. H. (1994). CDR3 length in antigen-specific immune receptors. *J. Exp. Med.* **179**(1), 323–328.

Salih, H. R., Rammensee, H. G., and Steinle, A. (2002). Cutting edge: Down-regulation of MICA on human tumors by proteolytic shedding. *J. Immunol.* **169**(8), 4098–4102.

Schild, H., Mavaddat, N., Litzenberger, C., Ehrich, E. W., Davis, M. M., and Bluestone, J. A. *et al.* (1994). The nature of major histocompatibility complex recognition by gamma delta T cells. *Cell* **76**(1), 29–37.

Schneider, T. D., and Stephens, R. M. (1990). Sequence logos: A new way to display consensus sequences. *Nucleic Acids Res.* **18**(20), 6097–6100.

Spencer, J., Isaacson, P. G., Diss, T. C., and MacDonald, T. T. (1989). Expression of disulfide-linked and non-disulfide-linked forms of the T cell receptor gamma/delta heterodimer in human intestinal intraepithelial lymphocytes. *Eur. J. Immunol.* **19**(7), 1335–1338.

Steinle, A., Li, P., Morris, D. L., Groh, V., Lanier, L. L., and Strong, R. K. *et al.* (2001). Interactions of human NKG2D with its ligands MICA, MICB, and homologs of the mouse RAE-1 protein family. *Immunogenetics* **53,** 279–287.

Stephens, H. A. (2001). MICA and MICB genes: Can the enigma of their polymorphism be resolved? *Trends Immunol.* **22**(7), 378–385.

Stites, W. E. (1997). Protein-protein interactions: Interface structure, binding thermodynamics, and mutational analysis. *Chem. Rev.* **97**(5), 1233–1250.

Strong, R. K., Holmes, M. A., Li, P., Braun-Jones, L., Lee, N., and Geraghty, D. E. (2002). HLA-E allelic variants: Correlating differential expression, peptide affinities, crystal structures and thermal stabilities. *J. Biol. Chem.* **278**(7), 5082–5090.

Sundberg, E. J., and Mariuzza, R. A. (2000). Luxury accommodations: The expanding role of structural plasticity in protein-protein interactions. *Struct. Fold Des.* **8**(7), R137–142.

Tanaka, Y., Morita, C. T., Nieves, E., Brenner, M. B., and Bloom, B. R. (1995). Natural and synthetic non-peptide antigens recognized by human gamma delta T cells. *Nature* **375**(6527), 155–158.

Thompson, J. D., Higgins, D. G., and Gibson, T. J. (1994). CLUSTAL W: Improving the sensitivity of progressive multiple sequence alignment through sequence weighting, position-specific gap penalties and weight matrix choice. *Nucleic Acids Res.* **22**(22), 4673–4680.

Tormo, J., Natarajan, K., Margulies, D. H., and Mariuzza, R. A. (1999). Crystal structure of a lectin-like natural killer cell receptor bound to its MHC class I ligand. *Nature* **402,** 623–631.

Trinchieri, G. (1989). Biology of natural killer cells. *Adv. Immunol.* **47,** 187–376.

Uhrberg, M., Parham, P., and Wernet, P. (2002). Definition of gene content for nine common group B haplotypes of the Caucasoid population: KIR haplotypes contain between seven and eleven KIR genes. *Immunogenetics* **54**(4), 221–229.

Uhrberg, M., Valiante, N. M., Shum, B. P., Shilling, H. G., Lienert-Weidenbach, K., and Corliss, B. *et al.* (1997). Human diversity in killer cell inhibitory receptor genes. *Immunity* **7**(6), 753–763.

Weis, W. I., Taylor, M. E., and Drickamer, K. (1998). The C-type lectin superfamily in the immune system. *Immunol. Rev.* **163,** 19–34.

Welte, S. A., Sinzger, C., Lutz, S. Z., Singh-Jasuja, H., Sampaio, K. L., and Eknigk, U. *et al.* (2003). Selective intracellular retention of virally induced NKG2D ligands by the human cytomegalovirus UL16 glycoprotein. *Eur. J. Immunol.* **33**(1), 194–203.

Willcox, B. E., Gao, G. F., Wyer, J. R., Ladbury, J. E., Bell, J. I., and Jakobsen, B. K. *et al.* (1999). TCR binding to peptide-MHC stabilizes a flexible recognition interface. *Immunity* **10**(3), 357–365.

Willcox, B. E., Thomas, L. M., and Bjorkman, P. J. (2003). Crystal structure of HLA-A2 bound to LIR-1, a host and viral major histocompatibility complex receptor. *Nat. Immunol.* **4**(9), 913–919.

Wu, J., Cherwinski, H., Spies, T., Phillips, J. H., and Lanier, L. L. (2000). DAP10 and DAP12 form distinct, but functionally cooperative, receptor complexes in Natural Killer cells. *J. Exp. Med.* **192,** 1059–1067.

Wu, J., Groh, V., and Spies, T. (2002a). T cell antigen receptor engagement and specificity in the recognition of stress-inducible MHC class I-related chains by human epithelial $\gamma\delta$ T cells. *J. Immunol.* **169,** 1236–1240.

Wu, J., Song, Y., Bakker, A. B., Bauer, S., Spies, T., and Lanier, L. L. *et al.* (1999). An activating immunoreceptor complex formed by NKG2D and DAP10. *Science* **285**(5428), 730–732.

Wu, L. C., Tuot, D. S., Lyons, D. S., Garcia, K. C., and Davis, M. M. (2002b). Two-step binding mechanism for T-cell receptor recognition of peptide MHC. *Nature* **418**(6897), 552–556.

Yokoyama, W. M., and Plougastel, B. F. (2003). Immune functions encoded by the natural killer gene complex. *Nat. Rev. Immunol.* **3**(4), 304–316.

Zou, Z., Nomura, M., Takihara, Y., Yasunaga, T., and Shimada, K. (1996). Isolation and characterization of retinoic acid-inducible cDNA clones in F9 cells: A novel cDNA family encodes cell surface proteins sharing partial homology with MHC class I molecules. *J. Biochem.* **119**(2), 319–328.

IONOTROPIC GLUTAMATE RECEPTOR RECOGNITION AND ACTIVATION

By ROBERT E. OSWALD

Department of Molecular Medicine, Cornell University, Ithaca, New York 14853

Abstract

Ionotropic glutamate receptors are the major excitatory neurotransmitters in mammalian brain but are found throughout the animal kingdom as well as in plants and bacteria. A great deal of progress in understanding the structure of these essential neurotransmitter receptors has been made since the first examples were cloned and sequenced in 1989. The atomic structure of the ligand-binding domain of several ionotropic glutamate receptors has been determined, and a great deal of progress has been made in relating the structural properties of the binding site to the function of the intact receptor. In addition, the identification of glutamate receptors from a wide variety of organisms ranging from several types of bacteria to *Arabidopsis* to a range of animal species has made glutamate receptors a molecular laboratory for studying the evolution of proteins. The fact that glutamate receptors are a particularly ancient intercellular

ADVANCES IN PROTEIN CHEMISTRY, Vol. 68

0065-3233/04 $35.00

signaling molecule suggests a potential role in the transition from single celled to multicellular organisms. This review focuses on the structure and dynamics of ionotropic glutamate receptors and their relation to the function and evolution of these proteins.

I. Introduction

Since the late 1950s, glutamate and aspartate have been known to excite neurons (Curtis *et al.*, 1959). Because most central neurons are excited by glutamate, this was originally interpreted as a nonspecific effect. Although soon afterwards, Curtis and Watkins (1963) identified N-methyl-D-aspartic acid (NMDA) as a potent activator of a depolarizing current, the existence of glutamate receptors as important neurotransmitter receptors was not fully accepted until the 1980s. The realization that glutamate activates a number of pharmacologically distinct subtypes (Ascher and Nowak, 1988; Collingridge and Lester, 1989; Monaghan *et al.*, 1989) and plays an important role in long-term potentiation (Bliss and Collingridge, 1993) led to the recognition that most excitatory neurotransmission in the central nervous system is mediated by glutamate receptors. Glutamate receptors are now recognized as important mediators of a wide range of neuronal functions (Dingledine *et al.*, 1999) and are known to play a role in a number of neurodegenerative diseases and epilepsy (Gillessen *et al.*, 2002). In addition, glutamate receptors seem to have functional roles outside of the nervous system (insulin secretion, bone resorption, cardiac pacemaking, tactile sensation [Ault and Hildebrand, 1993; Carlton *et al.*, 1995; Chenu *et al.*, 1998; Gill *et al.*, 1998; Inagaki *et al.*, 1995; Jorgensen *et al.*, 1995; Patton *et al.*, 1998; Weaver *et al.*, 1996]) and are found in plants (Davenport, 2002; Kang and Turano, 2003; Kim *et al.*, 2001; Lam *et al.*, 1998) and bacteria (Chen *et al.*, 1999; Kuner *et al.*, 2003; Mayer *et al.*, 2001). Both glutamate receptors that have an integral ion channel and those that activate G-proteins are known. The glutamate receptors with integral ion channels (ionotropic glutamate receptors, iGluRs) consist of four subunits surrounding a central ion channel (Dingledine *et al.*, 1999) and are responsible for a large portion of the fast excitatory neurotransmission in mammalian brain. The glutamate receptors that activate G-proteins (metabotropic glutamate receptors, mGluRs) have transmembrane segments that are similar to other G-protein coupled receptors, but extracellular ligand binding domains that have some structural homology to domains in iGluRs (O'Hara *et al.*, 1993).

More than 20 different mammalian iGluR subunits have been identified. These form receptors that fall into three major categories based on the characteristics of activation by different agonists: (1) kainate receptors

(GluR5-7; KA1 and KA2), (2) AMPA (α-amino-3-hydroxy-5-methyl-4-isoxazole propionate) receptors (GluR1-4, also referred to as GluRA-D), and (3) NMDA receptors (NR1, NR2A-D, NR3A-B). Kainate receptors are activated by the seaweed toxin, kainate, and by glutamate. Both produce a full activation and rapid desensitization. AMPA receptors are fully activated and desensitized by AMPA, quisqualate, and glutamate; and kainate produces a partial, less desensitizing activation of these receptors. NMDA receptors are comprised of at least two different subunit types (NR1 and NR2) and require the binding of both glycine and glutamate for full activation, with glycine binding to the NR1 subunit and glutamate binding to the NR2 subunit. A vast number of compounds with varying efficacies and specificities have been developed that can activate iGluRs (Brauner-Osborne *et al.*, 2000), although highly specific drugs targeted to specific subunits have not yet been reported. Despite differences in pharmacology and some activation properties, all iGluRs have similar structures. This review will focus on recent developments in the understanding of ligand binding and channel activation in ionotropic glutamate receptors and the relationship to the structure and dynamics of these essential neurotransmitter receptors.

II. Beginnings of Structure: Cloning, Mutagenesis, Transmembrane Topology, and Homology Models

The first iGluRs were cloned in 1989 (Gregor *et al.*, 1989; Hampson *et al.*, 1989; Hollmann *et al.*, 1989) and subsequently a large number of other subtypes were cloned, sequenced, and expressed (Bettler *et al.*, 1990; Boulter *et al.*, 1990; Keinänen *et al.*, 1990; Moriyoshi *et al.*, 1991). Studies of heterologously expressed iGluRs confirmed pharmacological studies that several different receptor subtypes exist, but the variety of different gene products was not anticipated by functional studies. In addition to the subtype diversity illustrated by the number of different genes, alternative splicing and RNA editing introduce additional diversity. In several cases, these posttranscriptional events have important impacts on function. For example, an alternatively spliced region known as flip/flop (Sommer *et al.*, 1990) can affect desensitization in that the "flip" version desensitizes at least 4-times slower than the "flop" version of the receptor (Mosbacher *et al.*, 1994). Likewise, RNA editing, which is mediated by a complementary intronic sequence (Melcher *et al.*, 1996) and which changes one base to produce a change in the amino acid sequence, can affect recovery from desensitization of AMPA receptors at the R/G site (Lomeli *et al.*, 1994) and can affect ion channel permeability in kainate (Sommer *et al.*, 1991; Swanson *et al.*, 1996) and AMPA (Seeburg, 1996; Swanson *et al.*, 1996,

1997b) receptors. In addition to posttranscriptional modifications, posttranslational modifications such as phosphorylation can affect various aspects of receptor function (Leonard and Hell, 1997; Moss *et al.*, 1993; Nakazawa *et al.*, 1995; Raymond *et al.*, 1993, 1994; Tan *et al.*, 1994; Wang *et al.*, 1993). Thus, although glutamate can indeed excite most central neurons, the rich diversity of function from a relatively large family of genes, posttranscription modifications, and posttranslational modifications provide a wide range of specific functions.

Following the cloning of the first glutamate receptors, the proposal was made, based on sequence and hydropathy plots, that these receptors are structurally homologous to nicotinic acetylcholine receptors, with four transmembrane domains and with C- and N-termini on the extracellular side of the membrane (Gasic and Hollmann, 1992). Heterologous expression in *Xenopus* oocytes and HEK293 cells along with site directed mutagenesis pinpointed a number of important properties of glutamate receptor subtypes and indicated important residues responsible for function (Hollmann and Heinemann, 1994). However, the initial breakthrough for the understanding of glutamate receptor structure came with the realization that portions of the sequence were related to bacterial amino acid binding proteins (Nakanishi *et al.*, 1990; O'Hara *et al.*, 1993). The N-terminal portion of the sequence was found to be related to one class of amino acid binding proteins (typified by the leucine-isoleucine-valine binding protein [LIVBP]) and two portions of the sequence were related to the lysine-arginine-ornithine binding protein (LAOBP). Interestingly, these two portions of the sequence were found on regions of the protein thought at the time to be on opposite sides of the membrane. However, N-glycosylation sites were found on what was previously thought to be intracellular segments (Roche *et al.*, 1994; Taverna *et al.*, 1994), and subsequently, the current model of three transmembrane segments and a reentrant pore loop (Fig. 1) was proposed first for a kainate receptor (Wo and Oswald, 1994) and subsequently for two AMPA receptors (Bennett and Dingledine, 1995; Hollmann *et al.*, 1994). As discussed later, the channel domain shows some homology to potassium channels (Wo and Oswald, 1995b; Wood *et al.*, 1995) and the C-terminal domain is variable with a wealth of different functions. Overall, however, the modular structure of glutamate receptors was clear as shown in Fig. 1 (Wo and Oswald, 1995b). This modular structure is further illustrated by iGluRs in lower vertebrates and bacteria. The kainate binding protein of lower vertebrates (chick, frog, fish) lacks the LIVBP domain (Gregor *et al.*, 1988; Wada *et al.*, 1989; Wo and Oswald, 1994) and bacterial iGluRs lack both the LIVBP domain and the C-terminal domain (Chen *et al.*, 1999).

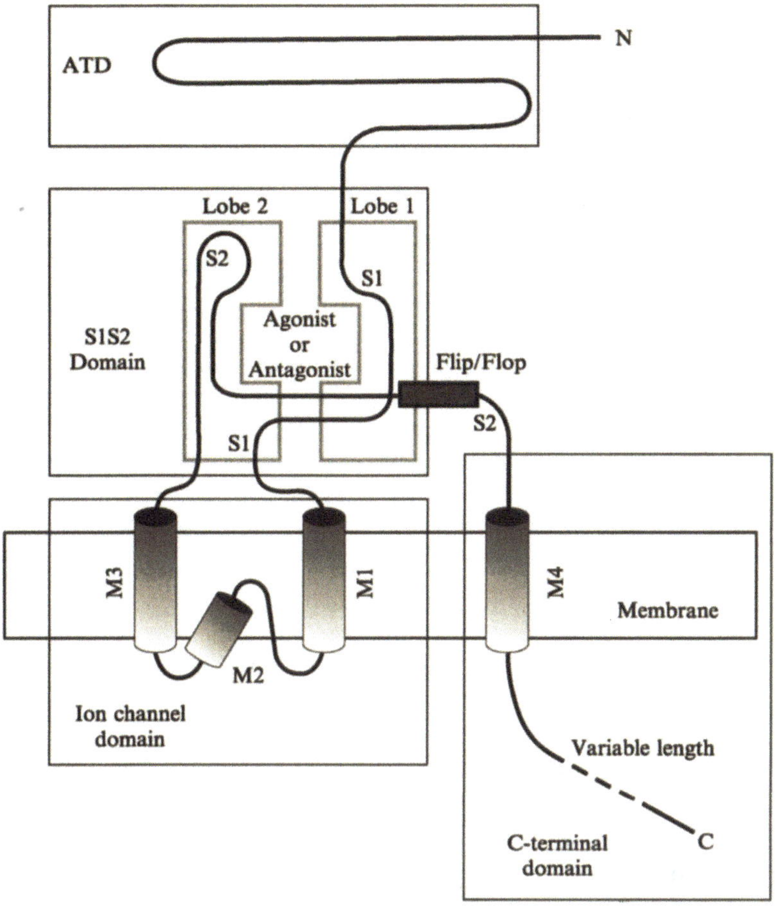

FIG. 1. Domain structure of ionotropic glutamate receptors (iGluRs) showing the four major components of mammalian iGluRs and the transmembrane topology. Bacterial iGluRs do not have the ATD and the C-terminal domain, and kainate binding proteins from lower vertebrates do not have the ATD. Note that S1 and S2 refer to segments of the primary sequence; whereas, Lobes 1 and 2 refer to protein folding units. Both Lobes 1 and 2 are made up of portions of both S1 and S2.

The interplay between homology models (Laube *et al.*, 1997; Paas *et al.*, 1996; Sutcliffe *et al.*, 1996) and site-directed mutagenesis (*e.g.*, Kuryatov *et al.*, 1994; Laube *et al.*, 1997; Paas *et al.*, 1996; Stern-Bach *et al.*, 1994; Swanson *et al.*, 1997a; Wo and Oswald, 1995a, 1996; Wood *et al.*, 1997) provided an increasingly clear view of the structure of the agonist-binding site. However, the second breakthrough was the preparation of soluble versions of the agonist-binding site of GluR4 by Keinänen and collaborators (Kuusinen *et al.*, 1995). This provided a largely monomeric, soluble protein suitable for further structural studies. The most exciting aspect of this preparation was that the binding properties were essentially indistinguishable from the intact membrane-bound protein, so that the isolated binding domain could be used as a structural model of the intact receptor. The original expression systems produced relatively low levels of protein in a native state. A higher level of expression was achieved by Gouaux and collaborators using GluR2 by expressing the protein in a denatured form in bacteria and then refolding to the native state (Chen and Gouaux, 1997). Again the protein had the same binding properties as the intact receptor. With the availability of a high yield expression system, the GluR2 binding domain (S1S2) was crystallized and its structure was determined (Armstrong *et al.*, 1998). At the same time, numerous spectroscopic studies were pursued to provide additional structural and dynamic information. The remainder of this review concentrates on the structure of ionotropic glutamate receptors and attempts to draw, where possible, correlations between the structure and function of the protein. The discussion of the structure is organized by the modular domains shown in Fig. 1.

III. The Amino Terminal Domain (ATD)

No structural information is currently available on the amino terminal domain (ATD), although sequence similarities suggest that it may be related to the LIVBP binding protein and domains of mGluRs (O'Hara *et al.*, 1993). If so, this would suggest that this domain is a bilobed structure, although a ligand that binds to this structure has not been established. LIVBP (Sack *et al.*, 1989a,b) and the mGluR domain (Kunishima *et al.*, 2000) have some overall structural similarity to the agonist binding domain (S1S2) that will be described in the following section. However, the LIVBP structure consists of two lobes each with α-helices and a central β-core, with the β-sheets consisting of six parallel strands. In contrast, the β-core of the S1S2 domain consists of 5 mixed parallel/antiparallel β-strands.

A number of functions have been assigned to the ATD. For example, in NMDA receptors, it has been suggested to be involved in modulation by ions and small molecules (Fayyazuddin *et al.*, 2000; Masuko *et al.*, 1999; Zheng *et al.*, 2001) and has been implicated in desensitization (Krupp *et al.*, 1998; Zheng *et al.*, 2001). Likewise, evidence exists for its role in oligomeric assembly of subunits (Ayalon and Stern-Bach, 2001; Kuusinen *et al.*, 1999; Leuschner and Hoch, 1999) and an anterograde trafficking signal has been identified near the N-terminus (Xia *et al.*,

2002). Given these well-documented effects, the finding by Keinänen and collaborators (Pasternack *et al.*, 2002) that removal of the entire domain of GluR4 (GluRD) had no effect on cell surface expression, ligand binding, channel activation, or desensitization was somewhat surprising. Partial deletions of the domain did not affect total GluR4 expression but led to a complete loss of surface expression. This would suggest that the ATD has no role in many of the major functions of iGluRs. However, the functional importance of the ATD cannot be completely ruled out. More subtle regulation by drugs and ions in NMDA receptors may still be important, and removal of the N-terminus of GluR4 may remove a trafficking signal that would direct the protein to the cell surface. Perhaps full deletion of the domain exposes a cryptic trafficking signal that allows the protein to be directed to the cell surface. In fact, both glutamate receptors from bacteria (Chen *et al.*, 1999) and kainate binding proteins from lower vertebrates (Wo *et al.*, 1999) lack an ATD but are expressed on the cell surface. Thus, iGluRs seem to function without an ATD, but the ATD may exert more subtle regulation, particularly *in vivo.*

IV. Structure and Dynamics of the Glutamate-Binding Domain

Crystal structures have provided dramatic, high-resolution pictures of the structure of the glutamate-binding (S1S2) domain and have led to suggestions as to the control of channel gating and desensitization. However, understanding the function of the protein will require additional approaches more suited to examining protein dynamics. In addition to crystallography, the S1S2 domain of AMPA receptors has also been studied by UV and IR spectroscopy (Cheng *et al.*, 2002; Deming *et al.*, 2003; Jayaraman *et al.*, 2000a,b; Madden *et al.*, 2001), NMR spectroscopy (McFeeters and Oswald, 2002; McFeeters *et al.*, 2002; Zeng *et al.*, 2002), isothermal titration calorimetry (Madden *et al.*, 2000), small-angle X-ray scattering (Abele *et al.*, 1999) and molecular dynamics simulations (Arinaminpathy *et al.*, 2002; Mendieta *et al.*, 2001). A combination of these approaches provides interesting insights into function, but many questions remain.

A. *Crystallography*

Crystal structures are currently available for the S1S2 domain of GluR2 (Armstrong and Gouaux, 2000; Armstrong *et al.*, 1998, 2003; Hogner *et al.*, 2002, 2003; Jin and Gouaux, 2003; Jin *et al.*, 2002, 2003; Kasper *et al.*, 2002; Lunn *et al.*, 2003; Sun *et al.*, 2002), GluR0 (Mayer *et al.*, 2001), and NMDA NR1 (Furukawa and Gouaux, 2003) in a variety of agonist and antagonist

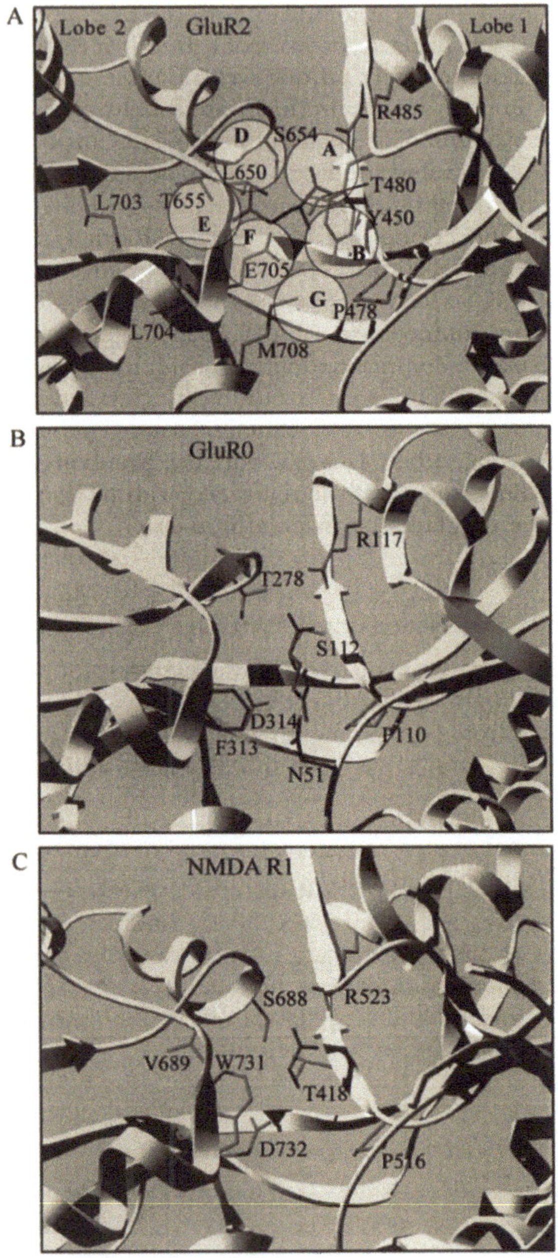

FIG. 2. The ligand binding sites of the S1S2 domain from (A) GluR2 (Armstrong and Gouaux, 2000), (B) GluR0 (Mayer *et al.*, 2001), and (C) NMDA R1 (Jin and

bound forms and with several mutations. This body of work provided the first detailed view of the binding site of a neurotransmitter receptor and has revealed clues to how the structure can be related to function.

1. Ligand Binding

The first structure of an S1S2 domain was GluR2 bound to kainate (Armstrong *et al.*, 1998), followed by a number of other ligands bound to the domain (Armstrong and Gouaux, 2000; Armstrong *et al.*, 2003; Hogner *et al.*, 2002, 2003; Jin and Gouaux, 2003; Jin *et al.*, 2002, 2003; Kasper *et al.*, 2002; Lunn *et al.*, 2003; Sun *et al.*, 2002). In addition, structures of GluR0 (Mayer *et al.*, 2001) and NMDA R1 S1S2 (Furukawa and Gouaux, 2003) domains have been solved. In all cases, the ligand binds in a cleft between the two lobes of the protein, making contacts with both lobes. As described by Gouaux and collaborators (Armstrong and Gouaux, 2000), perhaps the best way to understand the binding site is as a series of subsites, which may or may not be occupied by functional groups of the ligand. These are illustrated in Fig. 2A for the GluR2 S1S2 domain. Subsite A interacts with the α-carboxyl group and its primary substituent is R485, which has been shown in a number of mutational studies to be essential for ligand binding (Keinanen *et al.*, 1997; Laube *et al.*, 1997; Paas *et al.*, 1996). The backbone NH of both T480 and S654 contribute hydrogen bonds. Similar roles are played by R523, T518 and S688 in NR1, and R117, S112, and T278 in GluR0. Thus, binding to the α-carboxyl seems to be highly conserved. Subsite B interacts with the α-amino group of the ligand and is likewise highly conserved. In GluR2, this subsite is made up of the sidechain of T480, the backbone carbonyl oxygen of P478, and the carboxylate oxygen of E705. The site consists of T518, P516, and D732 in NR1, and S112, P110, and D314 in GluR0. Thus the α-substituents of glutamateric ligands are similar with essentially identical interactions with subsites A and B. These ligands, however, differ markedly in their "sidechains" and interaction with the other subsites (D, E, F, and G) in the binding cleft.

Subsites D and E in GluR2 are made up of the backbone NHs of S654 and T655, the hydroxyl of T655, two water molecules, and the carbonyl oxygen of L703. The γ-carboxyls of both glutamate and kainate interact with these hydrogen-bonding donors. It is, however, this portion of the structure that helps distinguish the binding selectivity of NR1 from GluR2.

Gouaux, 2003). The ligands (glutamate for GluR2 and GluR0 and glycine for NMDA R1) are shown in red. The relevant sidechains that make up the binding site are shown in green and are labeled. The subsites referred to in the text are shown as semitransparent circles. (See Color Insert.)

The residue equivalent to T655 (GluR2) is V689 in NR1. As predicted by homology modeling (Chohan *et al.*, 2000) and demonstrated by the NR1 S1S2 structure (Furukawa and Gouaux, 2003), the loss of the hydrogen-bonding donor prevents the binding of glutamate to this subunit but allows the binding of glycine to this site. Also, the bulky indole ring of W731 in NR1, which is in the same position as L704 in GluR2, further favors the binding of glycine relative to glutamate. However, in the case of GluR0, F313 is in a position similar to W731 in NR1, yet glutamate binds to and activates GluR0. F313 (GluR0) makes van der Waals contact with the γ-carboxyl of glutamate but prevents glutamate from binding in the same mode as GluR2; however, glutamate can take on a more extended conformation and interact with Asn 51 in Lobe 1. Thus, in GluR0, although contacts are made with Lobe 2, the major interaction of the γ-substituent is with Lobe 1 rather than Lobe 2. This may be particularly interesting from the point of view of the evolution of glutamate receptors as this is also the orientation of amino acids in the binding site of GlnBP and LAOBP, both bacterial amino acid binding proteins. When AMPA is bound to GluR2, the α-substituents are identical to that of the glutamate-bound structure but the sidechain portion of the ligand is rotated toward Lobe 1 into subsites E, G, and F, with M708 adopting an alternative rotomeric state to open up subsite G. A water molecule then fills subsite D. Thus the orientation of the α-substituents is similar between agonists, between subtypes, and (to the extent known) between eukaryotes and prokaryotes. The γ-substituents, however, bind in different manners, presumably giving rise to different functional properties.

2. Lobe Closure

Although both homology modeling (Sutcliffe *et al.*, 1996) and the first crystal structure (Armstrong *et al.*, 1998) provided suggestions that the two lobes of the binding domain could exist in different orientations depending upon the bound ligand, this was spectacularly confirmed (Armstrong *et al.*, 2000) by the structures of the GluR2 S1S2 domain in the apo form and bound to DNQX (antagonist), kainate (partial agonist), AMPA (full agonist) and glutamate (full agonist). Clearly, the two lobes of the S1S2 domain could close upon the agonist, but this appeared to be different for the partial agonist, kainate, than for the full agonists, AMPA and glutamate. That is, the apo form and the DNQX form both show no functional activity and have indistinguishable lobe orientations. When bound to kainate, the two lobes close by perhaps 12° (this should probably be thought of as an approximate value because only one crystal state for kainate is available and the various crystal forms of the apo state show some variability, perhaps due to inherent flexibility). Full agonists close the lobe further by approximately 8°. Thus the degree of lobe closure can be, to a first approximation, related to the efficacy of the ligand. Using integrated current measurements with homomeric GluR2, this correlation has held reasonably well over a variety of different partial agonists (Armstrong *et al.*, 2003; Hogner *et al.*, 2002). In the case of kainate, the lobes are held partially open by a steric clash between the kainate isopropenyl group and Y450 and L650. In fact, mutation of L650 to a threonine increases the efficacy of kainate and results in a full lobe closure (Armstrong *et al.*, 2003).

An important question is whether the degree of lobe closure is a static property of an individual ligand-bound form or whether the orientation of the two lobes is more dynamic. That is, does the agonist-bound form of this domain exist in both the open and closed lobe forms (perhaps favoring the closed lobe) and do the antagonist-bound and apo forms exist partially in the closed lobe forms (favoring the open lobe). The crystal structures may then represent a selection of one of the possible conformers. Several lines of evidence suggest that at least in some cases the lobes may show dynamic behavior. The bacterial GluR0 S1S2 domain exhibits a lobe closure in the presence of glutamate and serine that is slightly more open than the closed form of GluR2 bound to a full agonist. On the other hand, the apo form of GluR0 S1S2 is also in the closed form (Mayer *et al.*, 2001). Likewise, the GluR2 L650T mutation when bound to AMPA exhibits both a partially closed and a fully closed state in different crystal forms. In the case of NMDA NR1 S1S2, the lobe closure observed with the partial agonist D-cycloserine is the same as the full agonist, glycine. D-cycloserine may induce both a partially closed and fully closed state and may have just crystallized in the fully closed state. Alternatively, the NMDA S1S2 domain may exhibit fully closed and open forms in dynamic equilibrium, and the stabilization of the closed state may be greater for full agonists than for partial agonists.

Considering other GluRs and periplasmic binding proteins, domain flexibility has been suggested by a variety of crystal structures. The structure of the mGluR1 ligand binding domain, which is probably structurally related to the ATD of GluR2, may be instructive (Kunishima *et al.*, 2000). Both the glutamate-bound and apo forms exhibit both closed-lobe and open-lobe forms, and it is presumed that glutamate stabilizes the closed-lobe form. This would shift the equilibrium toward the active, closed-lobe form but would not preclude the existence of the open-lobe form in the presence of glutamate. In the case of another two-lobed binding protein, the maltodextin-binding protein, both open and closed forms can be observed in various crystal forms bound to maltooligosaccharides, suggesting a dynamic equilibrium between open and closed forms (Duan *et al.*, 2001). When bound to β-cyclodextrin, the crystal structure is exclusively in an open form (Sharff *et al.*, 1993). However, solution NMR studies using residual dipolar coupling (Evenas *et al.*, 2001; Skrynnikov *et al.*, 2000) and diffusion anisotropy (Hwang *et al.*, 2001), indicate that the lobe closure is at least 10° greater in solution than in the crystal state. This may be due to an intermediate closed state in the presence of β-cyclodextrin or may represent an average structure between the open and closed forms. In any event, most of the available evidence suggests that bilobed binding proteins are likely to be in dynamic equilibrium between open and closed forms (and perhaps intermediate forms), with the binding of agonist shifting the equilibrium toward the closed or intermediate forms. This notion will be considered in more detail later in the section relating structure to function.

B. *UV Spectroscopy*

Although UV spectroscopy is generally of low resolution for studying protein structure and dynamics, it has been used to investigate an important question concerning the use of the S1S2 domain to interpret the behavior of the intact membrane-bound glutamate receptor. The AMPA antagonist, CNQX, has a characteristic UV spectrum that changes upon binding the GluR4 S1S2 domain. Jayaraman and collaborators (Deming *et al.*, 2003) have shown that the changes in the CNQX UV spectrum induced by binding to the membrane-bound homomeric GluR4 receptor expressed in HEK293 cells are identical to those seen when CNQX is bound to the soluble GluR4 S1S2 domain. This suggests that the electronic environment of the binding site is preserved in the soluble binding protein. This, in combination with the similar binding affinity for intact receptors as for the soluble binding domains (Chen and Gouaux, 1997; Kuusinen *et al.*, 1995), provides a strong justification for the use of the isolated S1S2 domain as a model system to study the glutamate-binding domain of the intact receptor.

C. *Fluorescence Spectroscopy*

The tryptophan intrinsic fluorescence of the GluR4 S1S2 domain has been used to determine the rate of conformational transitions upon agonist binding (Abele *et al.*, 2000). Although this domain has four tryptophans such that the structural basis of the changes in fluorescence cannot be ascertained, the change in fluorescence as a function of time can be accurately measured. The results were interpreted as a fast binding interaction ("docking") with Lobe 1 followed by a slower transition presumably related to lobe closure ("locking"). The "docking" rate approached diffusion control (10^7 to 10^8 $M^{-1}s^{-1}$), and the "locking" step was slower (at least for kainate as the agonist) with a rate on the order of 10^3 s^{-1}. A detailed structural analysis of these transitions would require a better understanding of the contribution of each tryptophan in the structure (or fewer and better defined fluorescent reporters), but the kinetic information is of value in demonstrating that a binding event can be followed by a slower transition possibly resulting in the channel open state or the desensitized state.

D. *IR Spectroscopy*

Fourier transform infrared (IR) spectroscopy has provided sensitive measures of changes upon agonist binding. The crystal structures of the GluR2 S1S2 domain do not show any significant changes in secondary structure upon agonist binding; the structural transition (except for subtle rearrangements in the binding pocket) is largely a change in lobe orientation. Based on this, when the IR spectra of unbound S1S2 and glutamate-bound S1S2 in D_2O are compared, one might expect no change in the amide I modes associated with turns, α-helices, and β-sheets. However, glutamate binding results in an increase in all three secondary structural elements. Kainate binding, on the other hand, results in only an increase in β-sheet content, and CNQX and DNQX result in a modest increase in α-helical content (Jayaraman *et al.*, 2000a). One interpretation of these data would be that the agonists or antagonists actually induce new secondary structure, but if so, one might expect this to be evident in the crystal structure. A more likely interpretation is that the protein exhibits internal dynamics even in the β-sheet core (see section on NMR spectroscopy). The binding of glutamate, which seems to result in a closed lobe conformation, may also decrease the internal dynamics (either amplitude or change in kinetics) of the secondary structure such that an increase in the IR signature for that element is observed. Kainate would be less stabilizing and the antagonists even less so. These changes in dynamics are not evident from the B-factors of the crystal structures, but depending upon the time scale, the motions may or may not be reflected in the B-factors. Likewise, the hydrogen bond formed between the sidechain of C426 and the carbonyl of either A478 or I477 is unaltered upon binding kainate (assessed by monitoring the frequency of the S-H stretching band at 2563 cm^{-1}) but strengthened in the presence of glutamate. As discussed in Section VII, these changes in S1S2 flexibility may have important functional consequences.

E. NMR Spectroscopy

Although NMR spectroscopy can be used to determine *de novo* the three-dimensional structure of a protein, perhaps the most important use of this technique for understanding the S1S2 domain of glutamate receptors is in studying the dynamics of the protein and the degree of lobe closure and as a moderate-throughput screen for glutamate receptor specific drugs (McFeeters and Oswald, 2002; McFeeters *et al.*, 2002; Zeng *et al.*, 2002). The GluR2 S1S2 domain is ideally suited for these measurements because of the quality of the spectra and the fact that the sequence specific assignments of the backbone resonances have been made (McFeeters *et al.*, 2002).

Despite its relatively large size (263 amino acids), the S1S2 domain of GluR2 can be readily studied by NMR spectroscopy due to the fact that it is a monomer in solution and is highly stable. The dynamic properties of this domain have been studied using relaxation measurements (McFeeters and Oswald, 2002). In addition to the global tumbling of the protein, the internal motion of individual bond vectors can be measured on timescales ranging from ps to ms (Palmer *et al.*, 2001). The most straightforward measurements are the N-H bond vectors, which report on the motion of the backbone of the protein. Although often ignored because of the inherent beauty of crystal structures, protein flexibility has been known for many years to play important roles in protein function (Frauenfelder *et al.*, 2001). Examples include access to enzyme active sites (Nicholson *et al.*, 1995), interfaces for protein-protein interactions (Loh *et al.*, 1999, 2001), and pathways transmitting allosteric signals from one portion of a protein to another (Luque *et al.*, 2002). Interestingly, the S1S2 domain of GluR2 exhibits internal motion on both a rapid time scale (ps-ns) and a time scale similar to functional transitions in the intact protein (μs-ms). The slower motions are detected as a consequence of the fact that the spins are experiencing different environments on the chemical shift time scale, giving rise to chemical exchange. Several potentially important observations have resulted from studying exchange behavior of the backbone N-H bonds. As noted in the section on crystal structures, the binding site for agonists and antagonists, while containing a number of subsites, can actually be thought of as being comprised of two distinct halves. As described previously, the half associated largely with Lobe 1 binds the α-substituents of glutamate and is mostly invariant between structures with different ligands bound. The other half (associated with Lobe 2) binds the γ-substituents of glutamate and differs considerably between structures. Dynamics measurements by NMR (Fig. 3) also show a clear distinction between these two halves of the binding site. The residues

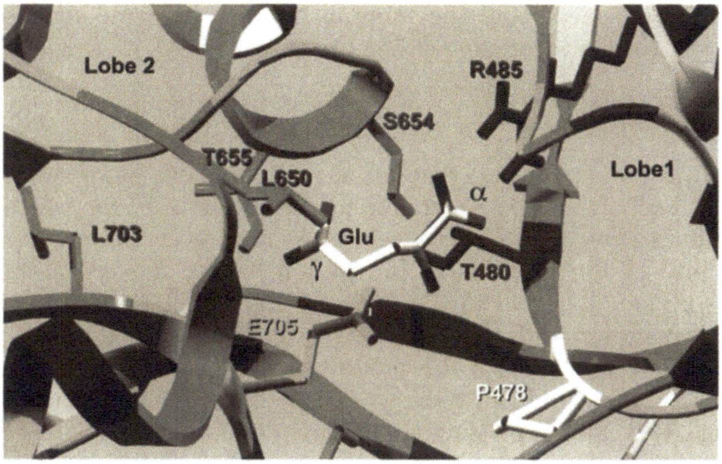

FIG. 3. The binding site of S1S2 GluR2 showing the amino acids important for interaction with glutamate with color coding according to the dynamics properties measured by NMR spectroscopy (McFeeters and Oswald, 2002). Blue and cyan represent residues that show only fast motions (ps-ns); whereas, residues colored yellow and green exhibit slower motions on the timescale of μs-ms. (See Color Insert.)

interacting with the α-substituents show very little internal motion on the physiological time scale, while those associated with the γ-substituents all show chemical exchange behavior. A simple interpretation of these results is that the portion of the binding site on Lobe 1 functions as a lock-and-key, with glutamateric effectors binding to a rigid structure. Once bound, the more flexible, Lobe 2 portion of the structure adapts to ligand and maintains a generally more flexible interaction. This could perhaps be viewed as an induced fit mechanism, which in turn impacts the functional response (see Section VII).

In addition to the binding pocket, several other portions of the protein exhibit potentially interesting motion. Both lobes have a core consisting of a β-sheet. The core of Lobe 1 exhibits very little internal motion on the μs-ms timescale. In contrast, the Lobe 2 β-sheet exhibits considerable chemical exchange. Although Lobe 2 tends to show greater B-factors in the crystal structure than Lobe 1, the β-core in Lobe 2 exhibits relatively low B-factors that are unchanged upon ligand binding. Given the fact that the timescale of the motions that give rise to B-factors is not well defined, this may only indicate that excess flexibility in the β-core is only observable by using techniques sensitive to motions in the order of μs or ms. However, motion detected by chemical exchange may have functional consequences in that the γ-portion of the binding site is linked to the channel domain through this β-sheet, and the motions are on the same time scale as transitions from the open-to-closed states measured in single channel recording.

When the two lobes close around a ligand in the binding site, very little space is available for the exit of the ligand in the absence of lobe opening. Indeed, most of the residues that line any potential exit pathway show little chemical exchange. The exception to this is Helix F, for which all of the residues show flexibility on the μs-ms time scale. Interestingly, this is also one of the portions of the protein where the crystal structure exhibits greater B-factors in the presence than in the absence of glutamate (Armstrong and Gouaux, 2000). Preliminary estimates of the exchange rate suggest that the motion of this helix may be concerted, suggesting that it could provide an exit pathway in the absence of lobe opening. This may represent an independent mechanism of receptor desensitization.

F. Small-Angle X-Ray Scattering

The radius of gyration was determined in the presence and absence of glutamate for a somewhat extended S1S2 domain of GluR4 (Abele *et al.*, 1999). No difference was observed, suggesting the lack of a major conformational change (e.g., lobe closure) upon agonist binding. More recent studies (D Madden, personal communication) with the S1S2 domain of GluR2, however, have shown that the radius of gyration does change with agonist binding in a manner consistent with the degree of lobe closure measured using X-ray crystallography. This provides strong support for the notion that the lobe closure observed in the crystal structures is not an artifact of the crystallization.

G. Molecular Dynamics

Although the timescale of molecular dynamics is generally short relative to the function of glutamate receptors, simulations have provided useful insights into the structure and dynamics of the binding domain. Mendieta *et al.* (2001) used activated molecular dynamics to study the transition between the open and closed lobe form of S1S2. Only a small force constant was used to bias the simulation from the closed form to the open form, suggesting that the energy barrier between the two forms is not large. Secondly, they found that the β-sheet in the Lobe 2 became more disordered in the open form than in the closed form, a finding consistent with the increased dynamics in Lobe 2 relative to Lobe 1 observed in NMR experiments. Arinaminpathy *et al.* (2002) employed 2–5 ns simulations without a bias driving the lobes open or closed. During the time of the simulations, no transitions in either direction were observed, but fluctuations of one lobe relative to the other were observed. Using the crystal structures for the apo (Armstrong and Gouaux, 2000), kainate-bound (Armstrong *et al.*, 1998) and glutamate-bound (Armstrong and Gouaux, 2000) structures as a starting point and superimposing just Lobe 1 every 200 ps, the difference in the movement of Lobe 2 relative to Lobe 1 could be observed. Clearly, the apo form showed the greatest movement between lobes, followed by kainate and then glutamate. In other words, the full agonist constrained the movement of the two lobes to a much greater extent than the partial agonist or the absence of agonist. As will be discussed later, these differences may be functionally important.

Recently Kurnakova and collaborators (Speranskiy and Kurnikova, 2003) have used a hybrid molecular dynamics/high level quantum mechanical description to model vibrational frequencies in the S1S2 binding site and to calculate binding energies. The agreement with vibrational frequencies measured by IR spectroscopy has been remarkable, suggesting that this approach will lead to additional insights concerning aspects of the ligand binding site that are difficult to access experimentally.

V. Structure of the Ion Channel

Most clues as to the structure of the ion channel domains have come from comparisons with K^+ channels (Fig. 4). Sequence homology with the P-loop of K^+ channels was recognized soon after the delineation of the transmembrane topology (Wo and Oswald, 1995b; Wood *et al.*, 1995), and a consensus motif (aromatic/aromatic/8X/G/3–5X/P) for the P-loop of cyclic nucleotide-gated channels, K^+ channels and glutamate receptors was identified (Wo and Oswald, 1995b). However, the P-loop dips into the structure from the cytoplasmic side of the membrane in the case of glutamate receptors whereas it extends into the structure from the extracellular side in the case of K^+ and cyclic nucleotide-gated channels. Thus to a first approximation, glutamate receptors might be thought of as an inverted K^+ channel-like structure. Scanning mutagenesis of the K^+ channel and various glutamate receptor subtypes have suggested a similar structure. The pattern of residues exposed to the ion channel pore is similar across subunits as judged from cysteine scanning mutagenesis in the NMDA (Kuner *et al.*, 1996) and GluR4 (Kuner *et al.*, 2001) subunits and from alanine and tryptophan mutagenesis in the GluR6 subunit (Panchenko *et al.*, 2001). Likewise, it is similar to the pattern found from cysteine scanning mutagenesis in several K^+ channels (e.g., Kurz

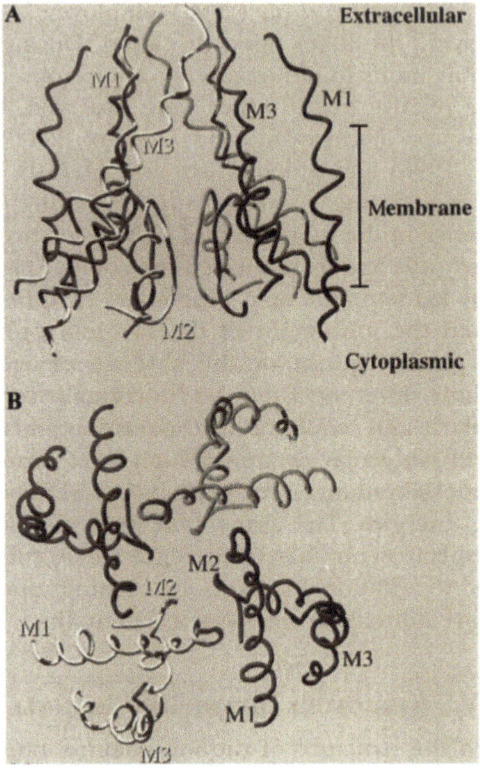

FIG. 4. A homology model of the channel domain of an NMDA receptor (Chohan *et al.*, 2000) based on the structure of the KcsA channel (Doyle *et al.*, 1998). (A) The side view of the channel is shown with only the M1, M2, and M3 regions and each of the four subunits in a different color. Note that M2 is a reentrant pore loop that enters the pore from the cytoplasmic face of the membrane. (B) A view of the channel from the extracellular side of the membrane is shown. Both selectivity and gating is controlled at least partially by the M2 region. (See Color Insert.)

et al., 1995; Pascual *et al.*, 1995). With the availability of the structure of several K^+ channels (Doyle *et al.*, 1998; Jiang *et al.*, 2002a, 2003), homology models of glutamate receptors have been developed that are both consistent with the scanning mutagenesis work and most functional measurements (Chohan *et al.*, 2000; Kuner *et al.*, 2003; Tikhonov *et al.*, 2002). The current view is that the tip of the pore loop is a polar residue two positions N-terminal to the G in the consensus motif. A helix extends from the cytoplasm to the tip of the helix and returns to the cytoplasm in an extended conformation (Fig. 4). Interestingly, a solution and solid-state NMR study of a peptide derived from the portion of NR1 M2 region extending to just past the glycine in the consensus motif was obtained in DPC micelles (Opella *et al.*, 1999). The structure was entirely helical and seemed to traverse the membrane, although the angle relative to the membrane could not be determined. Although the authors interpreted the results as evidence against a reentrant pore loop, in fact the results are largely consistent with the view that the first portion of the P-loop is helical. In the absence of the remainder of the structure, the helix extended one turn and the electrostatics of the helix dipole and the charge of the N- and C-termini likely forced the transmembrane orientation. Thus the homology modeling and experimental evidence are consistent with an ion channel domain that is similar to an inverted K^+ channel.

In addition to M2, M3 seems to play an important role in gating. This was first illustrated in the Lurcher mutation, where the A654T mutation in M3 of the δ_2 subunit converts receptors containing this subunit from inactive to constitutively active (Zuo *et al.*, 1997). Similar mutations in other glutamate receptor subunits also produce constitutively active channels (Kohda *et al.*, 2000). This would suggest that M3 is involved in some way with the gating process. Interestingly, M3 is the most highly conserved membrane-associated domain among glutamate receptors (Chiu *et al.*, 1999), and homology exists between M3 and the helix downstream of the P-loop in K^+ channels (Kuner *et al.*, 2003). More detailed mutagenesis studies of M3 suggest that it plays a major role in the gating process (Jones *et al.*, 2002; Sobolevsky *et al.*, 2003). In the K^+ channel, the inner helix (i.e., the helix downstream from the P-loop) seems to control the gate at a hinge created by a glycine (Jiang *et al.*, 2002b). The helix bends at this hinge, producing a 12 Å pore. Although the current evidence would suggest that M3 is involved in gating the iGluR ion channel, mutations of the two glycines in M3 have no effect on function (Sobolevsky *et al.*, 2003). This finding and the fact that the gate itself seems to be near the M2 loop in glutamate receptors but near the cytoplasmic side of the inner helix in K^+ channels would suggest that the details of gating differ between glutamate receptors and K^+ channels but that the overall structure is probably well conserved.

VI. The C-Terminal Domain

The C-terminal portion of glutamate receptor subunits is the most diverse region of the protein, ranging from short sequences of less than 50 amino acids to the large signaling complexes of NMDA R2 subunits. In contrast to the S1S2 domain, the structures of the C-terminal domains are less well understood. The relatively short C-terminal domains of AMPA and kainate receptors do not adopt a stable structure when expressed as soluble proteins (McFeeters and Oswald, unpublished observations). As of yet, the structures of the more complex C-terminal domains of NMDA R2 subunits have not been solved; however, at least a portion of the NMDA R1 subunit assumes a stable structure in complex with calmodulin (Kweon and Oswald, unpublished observations). Nevertheless, this domain is clearly important in terms of regulating the function of the glutamate receptors, for localizing the receptor to synaptic complexes, and for adding an additional dimension to the electrical signaling; that is, coupling the ion channel flux to the signal transduction machinery of the cell. A detailed discussion of the functional role of this important part of iGluRs is beyond the scope of this review but has been reviewed in detail elsewhere (Dingledine *et al.*, 1999; Sheng and Kim, 2002).

VII. Relating Structure, Function, and Dynamics: Channel Gating and Desensitization

A. *Binding and Channel Opening*

Although the structure of an intact, membrane-bound glutamate receptor is not yet available, the wealth of information on individual domains can provide important clues as to mechanisms of channel activation and desensitization (Jin *et al.*, 2003; Sun *et al.*, 2002). If one makes the assumption that glutamate receptors are evolutionarily related to K^+ channels, the presumed mechanisms of channel gating in the K^+ channel may shed light on the process for glutamate receptors. As described previously, glutamate receptors can be seen to a first approximation as a K^+ channel inverted 180° relative to the membrane. This would place the reentrant loop on the cytoplasmic side of the channel and the control mechanism on the extracellular side. As discussed previously, most evidence suggests that the ion channel pore of K^+ channels has at least some similarity to the glutamate receptor ion channel.

Potassium channels use a variety of mechanisms to control the gating of the channel. That which is probably most relevant to glutamate receptors is the RCK domain. The SCOP database (Murzin *et al.*, 1995) classifies the RCK domains of *E. coli* and *M. thermautotrophicus* K^+ channels, the LIV binding protein (which is a structural homolog of the ATD domain of glutamate receptors), and the S1S2 domain of GluR2 and GluR0 in the class of alpha/beta proteins. The RCK domains are NAD(P)-binding Rossmann-fold domains (parallel 6 stranded β-sheet) with a peripheral subdomain (C-terminal alpha/beta domain). Both LIVBP and S1S2 are considered periplasmic binding protein-like and have two alpha/beta/alpha domains.

Like the RCK domains, LIVBP has parallel 6-stranded β-sheets; whereas the S1S2 domains have 5-stranded mixed β-sheets. Thus, RCK domains are not close structural homologs to glutamate receptor extracellular domains, but some features are shared. In the case of the MthK channel, eight RCK domains are associated with one tetrameric channel (Fig. 5A). Four of the RCK domains have an N-terminal portion that forms the ion channel (RCK-A) and four RCK domains lack a channel domain (RCK-B). The eight RCK domains have been referred to as the "gating ring." RCK-A and RCK-B domains interact through both a flexible and fixed interface, with the flexible interface associated with a calcium-binding site. The flexible interface is formed across both the Rossman-fold lobe and the peripheral subdomain of both RCK-A and RCK-B. The fixed interface is between Rossman-fold lobes but, in the case of a given RCK-A domain, it is formed with a different RCK-B domain from that which forms the flexible interface. Binding of calcium in the vicinity of the flexible interface results in a remodeling of that interface. This in turn tilts the unit formed by the fixed interface, expanding the gating ring and pulling the helices of the channel to the open position.

Glutamate receptors may undergo an analogous process that is strongly suggested by a series of crystal structures. In this case, four S1S2 domains are associated with four channel domains. The interaction between subunits seems to occur through the J helix in Lobe 1, and this can be thought of as analogous to the "fixed" interface of the K^+ channel (Fig. 5B). The "flexible" interface in this case would be intramolecular, the cleft between Lobe 1 and Lobe 2. Thus, Lobe 1 would loosely be analogous to RCK-B, and Lobe 2 would play the role of RCK-A. Binding of agonist closes the gap between the two lobes, doing work on the channel and, by a currently unknown mechanism, leads to channel opening. The details of channel gating may well differ between the K^+ channel and glutamate receptors, but the overall process appears to be analogous.

B. *Details of Channel Activation and Partial Agonism*

With the development of homology models, lobe closure was suggested to be a mechanism for channel gating (Sutcliffe *et al.*, 1996); however, the crystal structures added an important detail. That is, the degree of lobe closure differed not only between agonists and antagonists but also between agonists and partial agonists (Armstrong and Gouaux, 2000). At least in the case of GluR2 S1S2, the degree to which the lobe is closed is correlated to the activation of the channel as measured by integrated current measurements in which desensitization has been inhibited by cyclothiazide (Armstrong and Gouaux, 2000; Armstrong *et al.*, 2003;

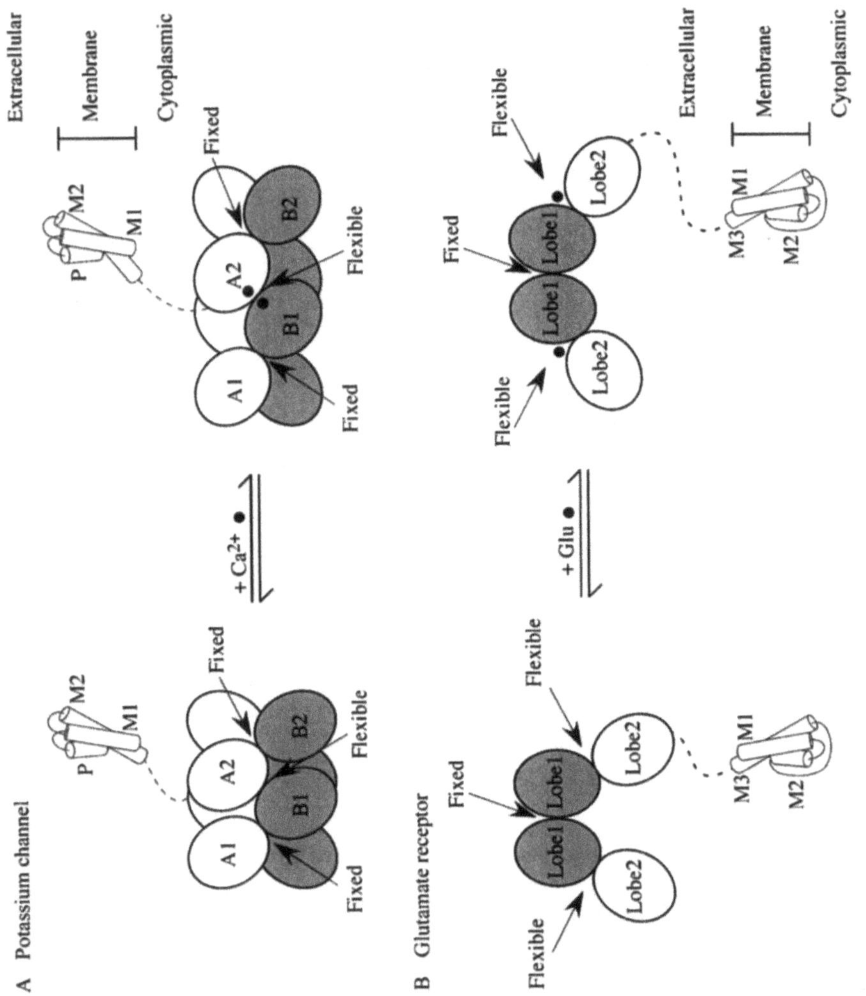
A Potassium channel
B Glutamate receptor
Extracellular
Membrane
Cytoplasmic
P
M1
M2
M3
A1
A2
B1
B2
Fixed
Flexible
+ Ca2+
+ Glu
Lobe1
Lobe2

Hogner *et al.*, 2002). This is a compelling result that lends considerable credibility to comparisons between crystal structures of an isolated domain and the function of the intact multimeric, membrane-bound receptor. However, a number of factors including the number of channels activated, the single channel amplitude, the probability of channel opening, and desensitization determine the amplitude of integrated current measurements. Ultimately, the degree of lobe closure should be correlated with a single observable parameter in order to make mechanistic conclusions. Perhaps the most logical parameter would be the single channel conductance in that the size of the conductance path (and hence the rate of ion flow through the channel) could be controlled by the relative movement of the two lobes of the S1S2 domain. The major problem is that most of the structural work has been done with the GluR2, for which the single channel conductance is low and the openings are brief. Despite these difficulties, Jin *et al.* (2003) measured single channel currents from GluR2 (flip/Q, L483Y) expressed in HEK cells in the presence of glutamate (full agonist) and a series of substituted willardiine compounds (partial agonists). Although the data were heavily filtered, the major conclusion was that both agonists and partial agonists shared a series of conductance levels but that the full agonists preferentially activated larger conductances and the partial agonists preferentially activated lower conductances. The question then is how to relate these results on a presumed tetrameric receptor to lobe closure in an isolated S1S2 domain. The authors developed a model in which each subunit contributes a gate and the number of open gates determines the conductance level (Fig. 6A). A similar

---

FIG. 5. (A) Mechanism for potassium ion channel gating as proposed by MacKinnon and collaborators (Jiang *et al.*, 2002a,b). The RCK domains are shown as ovals with the RCK-A domains shown as open ovals and the RCK-B domains as filled ovals. Only one channel domain (helices labeled P, M1, and M2) is shown for clarity, but four make up one ion channel. The channel domain and the RCK-A domain are on the same polypeptide. The orientation of the intermolecular flexible interface is rearranged in response to calcium binding which does work on the tether to M2 of the channel domain. The M2 helix is bent at a glycine hinge and the channel is opened. (B) A hypothesis for the linkage between channel gating and glutamate binding in iGluRs as proposed by Gouaux, Mayer, and collaborators (Armstrong *et al.*, 2003; Jin *et al.*, 2003; Mayer *et al.*, 2001). Only two of four S1S2 domains are shown (One S1S2 domain is made up of the ovals labeled Lobe 1 and Lobe2) and one of four channel domains is shown. In this case, the flexible interface is intramolecular and the movement of Lobe 1 relative to Lobe 2 does work on the M3 helix leading to channel opening. A glycine hinge as described for the potassium channel is probably not the mechanism of channel opening. Note that in iGluR notation, M2 is the same as the P loop in potassium channels and M2 corresponds to M3.

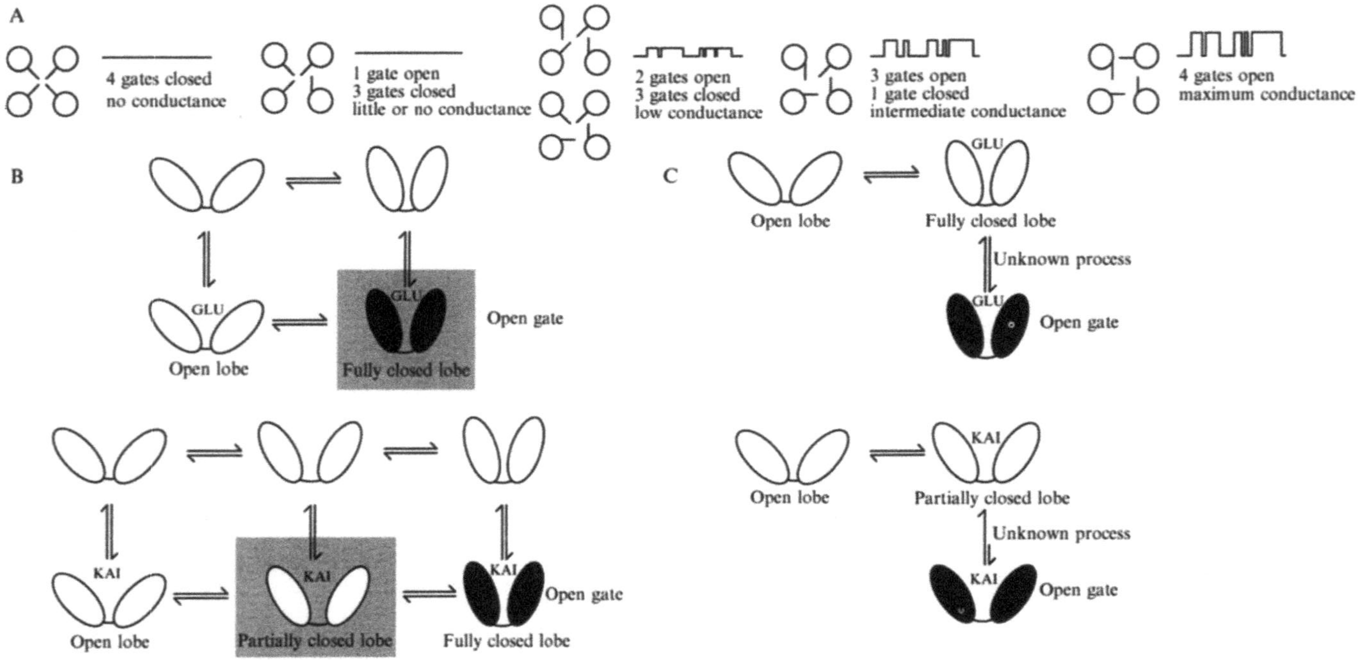

FIG. 6. (A) Hypothesis for the control of channel conductance based on the number of individual subunit gates that open. This model was first proposed by Rosenmund *et al.* (1998). Two possible mechanisms of channel activation influenced by individual S1S2 domains: (B) The S1S2 domain can exist in a number of closed, open and partially open states. The favored state is shown in a shaded box, and the open gate form with black-filled lobes. A full agonist, such as glutamate (GLU) can bind and the lobes can be either open or fully closed. The fully closed form is favored and in this state, and the channel gate for that subunit opens. The opening of the gate for each of the four subunits results in an incremental increase in conductance as shown in (A). A partial agonist, such as kainate (KAI) can

mechanism was proposed earlier by Rosenmund *et al.* (1998). A simple binomial model with an efficacy factor accounted well for all of the data. Interestingly, the efficacy factor correlated with the degree of lobe closure, and the interpretation was that the greater the degree of lobe closure, the more likely that the gate for that subunit would open. This has the effect of converting a continuous variable to a binary variable that can be used to analyze the single channel results. In essence, this suggests that the lobe closure places the system in an enabled state, and that other processes actually control whether or not the gate actually opens, essentially lowering the energy barrier to gate opening by different degrees depending upon the angle of lobe closure.

A number of questions remain: (1) Are the single channel data sufficiently resolved to support the conclusions? Given the difficulty of the experiment, a relatively small number of heavily filtered channels were observed. However, recordings of better resolved AMPA receptors support the general conclusion that full agonists share similar conductance states with partial agonists, with the full agonists favoring larger conductances (Margot, 1998; Swanson *et al.*, 1997b). (2) What is the structural change directly linked to channel opening? This question is difficult to approach without information on the structure and dynamics of the intact receptor, but assuming that the controlling factor is the S1S2 domain, two scenarios, both of which invoke dynamic processes, seem plausible (Fig. 6B and C).

The first (Fig. 6B) assumes that a channel gate is opened only by a full closure of the S1S2 lobe. Each subunit contributes a gate, and the number of gates that are opened determines the conductance level as in Fig. 6A. The S1S2 lobe could close to varying degrees, and the fully closed state would be favored with full agonists but partial agonists would assume the fully closed state less frequently. That is, a partially closed state (closed channel gate) would be favored by partial agonists. This model would retain a direct link between the closing of S1S2 lobes and the opening of the ion channel, and is consistent with both the open and closed states found for ligand-bound forms of other bilobed proteins.

---

bind and the lobes can be open, partially closed or fully closed. The favored state is the partially closed form, but only the fully closed form results in an open gate. Thus, more gates on average are closed, resulting in a lower conductance. (C) Both partially open and fully open lobes can lead to opening of the gate. The processes controlling gate opening are not defined, but the barrier leading to gate opening is lower for the fully closed lobes than for the partially closed lobes. The process by which a gate opens is undefined but may be related to dynamic processes in Lobe 2 (McFeeters and Oswald, 2002, 2003).

The second scenario (Fig. 6C) assumes that lobe closure would increase the probability that the gate would open. The actual opening of the gate would then be controlled by other processes within the S1S2 domain or in other parts of the subunit. The connection between the agonist binding site and the transmembrane segments is formed by Lobe 2 and presumably flexible linkers between Lobe 2 and the transmembrane segments (M1 and M3). The Lurcher mutation near the cytoplasmic end of M3 in δ2-containing receptors results in channels that open spontaneously (Zuo *et al.*, 1997), indicating the importance of this region in gating. Also, the NMR dynamics studies described previously suggest a flexible connection between the γ-portion of the binding site and the linkers to the channel domain. Flexibility on the μs-ms time scale was detected, which is relevant to channel gating. In this model, the opening of the gate would require the additional motion within Lobe 2. Full closure of the lobes may require less internal motion to open the gate, and partial closure would require more internal motion. One might then expect that openings induced by full agonists would be less likely to be interrupted by brief closures (or brief transitions to lower conductance states), and openings by partial agonists would be more affected by transient closures or transitions to other conductance levels. Such details could not be readily observed in the heavily filtered homomeric GluR2 single channel records, but in homomeric GluR3 (Margot, 1998) and various GluR4-containing receptors (Swanson *et al.*, 1997b), rapid transitions between conductance levels (or brief closures) are more common for partial than for full agonists. Thus, a combination of lobe closure and dynamic processes within the S1S2 domain may be responsible for the full range of channel opening dynamics.

C. *Desensitization*

A defining characteristic of glutamate receptors is the rate of desensitization. Particularly in the case of GluR2, the channel opens very rapidly (less than 1 ms) and can desensitize in a few ms, presumably with glutamate still bound (Li *et al.*, 2003). Again, the crystal structures of GluR2 S1S2 domain shed important light on this process. Gouaux and collaborators (Sun *et al.*, 2002) have shown that mutations or allosteric modulators that inhibit desensitization promote the stability of a dimer interface as measured by analytical ultracentrifugation and crystallography. One interpretation of this is that the "fixed" interface is crucial for maintaining an open channel and that desensitization is associated with the dissociation of this interface. Although confirmation of this compelling hypothesis will require understanding the structure of the intact protein, the overall process is supported by all of the data and is consistent with the notions of gating developed from the K^+ channel structure.

Like the K^+ channel, however, desensitization is probably mediated by multiple mechanisms, some of which are associated with the S1S2 domain and others that are not. For example, in the case of the NMDA receptor, calmodulin can bind to the NR1 subunit and produce a Ca^{2+}-dependent inactivation (Ehlers *et al.*, 1996; Zhang *et al.*, 1998). Although little structural information is available concerning this process, it is unlikely to be mediated by the S1S2 domain, and calmodulin could simply be occluding the channel from the cytoplasmic face. Also, in GluR6, the rate of desensitization is modified by mutations in the cytoplasmic loops of the channel domain which are presumably not related to the S1S2 interface (Kornreich and Oswald, unpublished observations).

VIII. Evolution of Glutamate Receptors

The metazoan radiation, which gave rise to multicellular organisms, required communication between cells in the form of hormones, neurotransmitters, and their receptors. Clearly, glutamate receptors are an ancient form of communication in that representatives of this class of receptors are found from bacteria to plants to mammals. Indeed, a genealogical analysis of ionotropic glutamate receptor sequences from plants and animals suggests that the coassembly of the S1S2 domain and the ion channel predates the divergence of plants and animals (Chiu *et al.*, 1999). Thus it is logical to look for traces of the evolution of receptors in bacteria. Although the idea that glutamate receptors arose from a splicing of bacterial genes was proposed almost 10 years ago (Wo and Oswald, 1995b), the sequencing of a number of bacterial genomes has provided a new level of understanding. Glutamate receptors are comprised of four domains, three of which have clear homologs in bacteria. Interestingly, bacterial glutamate receptors consist only of two of these domains (the ion channel domain and the agonist binding domain). However, both of these domains suggest an evolutionary relationship between glutamate receptors and bacterial proteins.

The sequence similarity between eukaryotic K^+ channels and glutamate receptor ion channels is relatively low (Wo and Oswald, 1995b); however, as described previously, scanning mutagenesis suggests that the structures are similar. In fact, the low similarity between eukaryotic K^+ channels and the ion channels of glutamate receptors prompted Chiu *et al.* (1999) to suggest that the two classes of structures arose from convergent evolution. However, bacterial K^+ channels and bacterial glutamate receptors show a much higher sequence similarity (Kuner *et al.*, 2003). In particular, a striking homology exists between the P-loops of the first bacterial glutamate receptor to be identified (GluR0, Chen *et al.*, 1999) and KcsA. The original suggestion was that glutamate receptors evolved from a K^+ channel-like protein (Wo and Oswald, 1995b); however, Kuner *et al.* (2003) have raised the possibility that the similarities to K^+ channels may be a result of lateral gene transfer from mammals to bacteria. However, scanning a large number of bacterial genomes has provided evidence for a range of different GluR0 sequences, some of which are quite similar to K^+ channels, others that are closer to NMDA receptors, and some that are similar to *Arabidopsis* glutamate receptors (Kuner *et al.*, 2003). Thus, although substantial differences exist between glutamate receptor ion channels and K^+ channels (e.g., orientation with respect to the membrane, probable location of the gate), the sequence and presumed structural similarities between K^+ channels and glutamate receptors most likely arises from an evolutionary link.

The other portion of glutamate receptors that seems to arise from a bacterial protein is the agonist-binding domain (S1S2). Mayer *et al.* (2001) have compared the structures of the bacterial GluR0 S1S2 domain to mammalian GluR2 S1S2 and the bacterial glutamine binding protein (GlnBP). Although the sequences are not well conserved, the structures are very similar (Mayer *et al.*, 2001); however, GluR0 S1S2 resembles GlnBP more closely than it does GluR2 S1S2. This is illustrated in several ways. The first is that two insertions are present in GluR2 relative to GluR0 and GlnBP. This consists of a flexible loop (Loop 2) and an additional helix (Helix G) that are not present in GluR0 and GlnBP. Perhaps even more interesting is the orientation of the ligand in the binding site. As described previously, the α-substituents have highly conserved binding interactions across all known structures. However, both GluR0 (F313) and GlnBP (H156) have an aromatic residue that prevents binding of the γ-carboxyl to subsites D and E (as in GluR2 S1S2), resulting in interaction with Lobe 1 and a more extended ligand conformation. Thus, like the ion channel, the agonist-binding domain of GluR0 is closely related to a homologous bacterial protein.

Given the striking structural similarities between plant and animal GluRs, GluR0 and the homologous bacterial K^+ channels and periplasmic binding proteins, our original suggestion in 1995 (Wo and Oswald, 1995b) that glutamate receptors arose from splicing of bacterial genes to form a protein suitable for intercellular communication remains plausible. Thus, this ancient form of intercellular communication may have been one of the factors that made possible the transition from single celled to multicellular organisms.

IX. Summary

Our understanding of ionotropic glutamate receptors has grown dramatically in the last 10–15 years, from the initial indications that these receptors are of pivotal importance to brain function to the understanding of the structure of the glutamate-binding domain at atomic resolution. Important challenges lie ahead. Currently, our knowledge of the structure of the intact membrane-bound protein is based solely on the atomic structure of the binding domain and homology models supplemented with site-directed mutagenesis. An experimentally determined structure of the intact receptor would provide a much more powerful context for studying the structural basis of glutamate receptor function. Likewise, although a wealth of important studies have addressed the function of the intracellular domains of glutamate receptors (particularly the NMDA NR2 subtypes), understanding the structural basis of the role of this portion of the protein in glutamate receptor function and the integration with signal transduction processes is critical. Because of the role of glutamate receptors in a wide variety of neurological diseases, the structure-based design of glutamateric drugs will be crucial for new generations of therapeutic agents directed toward individual subtypes.

Acknowledgments

The author would like to thank numerous collaborators for many productive discussions (Rob McFeeters, Michael Sutcliffe, Galen Wo, Bruce Kornreich, Jeehye Kweon, Ahmed Ahmed, Adrienne Loh, Gregory Weiland, Linda Nowak). Support was generously provided by the National Science Foundation (IBN-0323874).

References

Abele, R., Keinanen, K., and Madden, D. R. (2000). Agonist-induced isomerization in a glutamate receptor ligand-binding domain. A kinetic and mutagenetic analysis. *J. Biol. Chem.* **275,** 21355–21363.

Abele, R., Svergun, D., Keinanen, K., Koch, M. H., and Madden, D. R. (1999). A molecular envelope of the ligand-binding domain of a glutamate receptor in the presence and absence of agonist. *Biochemistry* **38,** 10949–10957.

Arinaminpathy, Y., Sansom, M. S., and Biggin, P. C. (2002). Molecular Dynamics Simulations of the Ligand-Binding Domain of the Ionotropic Glutamate Receptor GluR2. *Biophys. J.* **82,** 676–683.

Armstrong, N., and Gouaux, E. (2000). Mechanisms for activation and antagonism of an AMPA-sensitive glutamate receptor: Crystal structures of the GluR2 ligand binding core. *Neuron* **28,** 165–181.

Armstrong, N., Mayer, M., and Gouaux, E. (2003). Tuning activation of the AMPA-sensitive GluR2 ion channel by genetic adjustment of agonist-induced conformational changes. *Proc. Natl. Acad. Sci. USA* **100,** 5736–5741.

Armstrong, N., Sun, Y., Chen, G. Q., and Gouaux, E. (1998). Structure of a glutamate-receptor ligand-binding core in complex with kainate. *Nature* **395,** 913–917.

Ascher, P., and Nowak, L. (1988). Quisqualate- and kainate-activated channels in mouse central neurons in culture. *J. Physiol.* **399,** 227–245.

Ault, B., and Hildebrand, L. M. (1993). Activation of nociceptive reflexes by peripheral kainate receptors. *J. Pharmacol. Exp. Ther.* **265,** 927–932.

Ayalon, G., and Stern-Bach, Y. (2001). Functional assembly of AMPA and kainate receptors is mediated by several discrete protein-protein interactions. *Neuron* **31,** 103–113.

Bennett, J. A., and Dingledine, R. (1995). Topology profile for a glutamate receptor: Three transmembrane domains and a channel-lining re-entrant membrane loop. *Neuron* **14,** 373–384.

Bettler, B., Boulter, J., Hermans-Borgmeyer, I., O'Shea-Greenfield, A., Deneris, E. S., Moll, C., Borgmeyer, U., Hollman, M., and Heinemann, S. (1990). Cloning of a novel glutamate receptor subunit, GluR5: Expression in the nervous system during development. *Neuron* **5,** 583–595.

Bliss, T. V. P., and Collingridge, G. L. (1993). A synaptic model of memory: Long-term potentiation in the hippocampus. *Nature* **361,** 31–39.

Boulter, J., Hollmann, M., O'Shea-Greenfield, A., Hartley, M., Deneris, E., Maron, C., and Heinemann, S. (1990). Molecular cloning and functional expression of glutamate receptor subunit genes. *Science* **249,** 1033–1037.

Brauner-Osborne, H., Egebjerg, J., Nielsen, E. O., Madsen, U., and Krogsgaard-Larsen, P. (2000). Ligands for glutamate receptors: Design and therapeutic prospects. *J. Med. Chem.* **43,** 2609–2645.

Carlton, S. M., Hargett, G. L., and Coggeshall, R. E. (1995). Localization and activation of glutamate receptors in unmyelinated axons of rat glabrous skin. *Neurosci. Lett.* **197,** 25–28.

Chen, G. Q., Cui, C., Mayer, M. L., and Gouaux, E. (1999). Functional characterization of a potassium-selective prokaryotic glutamate receptor. *Nature* **402,** 817–821.

Chen, G. Q., and Gouaux, E. (1997). Overexpression of a glutamate receptor (GluR2) ligand binding domain in Escherichia coli: Application of a novel protein folding screen. *Proc. Natl. Acad. Sci. USA* **94,** 13431–13436.

Cheng, Q., Thiran, S., Yernool, D., Gouaux, E., and Jayaraman, V. (2002). A vibrational spectroscopic investigation of interactions of agonists with GluR0, a prokaryotic glutamate receptor. *Biochemistry* **41,** 1602–1608.

Chenu, C., Serre, C. M., Raynal, C., Burt-Pichat, B., and Delmas, P. D. (1998). Glutamate receptors are expressed by bone cells and are involved in bone resorption. *Bone* **22,** 295–299.

Chiu, J., DeSalle, R., Lam, H. M., Meisel, L., and Coruzzi, G. (1999). Molecular evolution of glutamate receptors: A primitive signaling mechanism that existed before plants and animals diverged. *Mol. Biol. Evol.* **16,** 826–838.

Chohan, K. K., Wo, Z. G., Oswald, R. E., and Sutcliffe, M. J. (2000). Structural insights into NMDA ionotropic glutamate receptors via molecular modeling. *J. Mol. Model.* **6,** 16.

Collingridge, G. L., and Lester, R. A. (1989). Excitatory amino acid receptors in the vertebrate central nervous system. *Pharmacol. Rev.* **41,** 143–210.

Curtis, D. R., Phillis, J. W., and Watkins, J. C. (1959). Chemical excitation of spinal neurones. *Nature* **183,** 611–612.

Curtis, D. R., and Watkins, J. C. (1963). Acidic amino acids with strong excitatory actions on mammalian neurones. *J. Physiol.* **166,** 1–14.

Davenport, R. (2002). Glutamate receptors in plants. *Ann. Bot. (Lond.)* **90,** 549–557.

Deming, D., Cheng, Q., and Jayaraman, V. (2003). Is the isolated ligand binding domain a good model of the domain in the native receptor? *J. Biol. Chem.* **278,** 17589–17592.

Dingledine, R., Borges, K., Bowie, D., and Traynelis, S. (1999). The glutamate receptor ion channels. *Pharmacol. Rev.* **51,** 7–61.

Doyle, D. A., Cabral, J. M., Pfuetzner, R. A., Kuo, A. L., Gulbis, J. M., Cohen, S. L., Chait, B. T., and MacKinnon, R. (1998). The structure of a potassium channel: Molecular basis of K+ conduction and selectivity. *Science* **280,** 69–77.

Duan, X., Hall, J. A., Nikaido, H., and Quiocho, F. A. (2001). Crystal structures of the maltodextrin/maltose-binding protein complexed with reduced oligosaccharides: Flexibility of tertiary structure and ligand binding. *J. Mol. Biol.* **306,** 1115–1126.

Ehlers, M. D., Zhang, S., Bernhadt, J. P., and Huganir, R. L. (1996). Inactivation of NMDA receptors by direct interaction of calmodulin with the NR1 subunit. *Cell* **84,** 745–755.

Evenas, J., Tugarinov, V., Skrynnikov, N. R., Goto, N. K., Muhandiram, R., and Kay, L. E. (2001). Ligand-induced structural changes to maltodextrin-binding protein as studied by solution NMR spectroscopy. *J. Mol. Biol.* **309,** 961–974.

Fayyazuddin, A., Villarroel, A., Le Goff, A., Lerma, J., and Neyton, J. (2000). Four residues of the extracellular N-terminal domain of the NR2A subunit control high-affinity Zn2+ binding to NMDA receptors. *Neuron* **25,** 683–694.

Frauenfelder, H., McMahon, B. H., Austin, R. H., Chu, K., and Groves, J. T. (2001). The role of structure, energy landscape, dynamics, and allostery in the enzymatic function of myoglobin. *Proc. Natl. Acad. Sci. USA* **98,** 2370–2374.

Furukawa, H., and Gouaux, E. (2003). Mechanisms of activation, inhibition and specificity: Crystal structures of the NMDA receptor NR1 ligand-binding core. *EMBO J.* **22,** 2873–2885.

Gasic, G. P., and Hollmann, M. (1992). Molecular neurobiology of glutamate receptors. *Annu. Rev. Physiol.* **54,** 507–536.

Gill, S. S., Pulido, O. M., Mueller, R. W., and McGuire, P. F. (1998). Molecular and immunochemical characterization of the ionotropic glutamate receptors in the rat heart. *Brain Res. Bull.* **46,** 429–434.

Gillessen, T., Budd, S. L., and Lipton, S. A. (2002). Excitatory amino acid neurotoxicity. *Adv. Exp. Med. Biol.* **513,** 3–40.

Gregor, P., Eshhar, N., Ortega, A., and Teichberg, V. (1988). Isolation, immunochemical characterization and localization of the kainate subclass of glutamate receptor from chick cerebellum. *EMBO J.* **7,** 2673–2679.

Gregor, P., Mano, I., Maoz, I., McKeown, M., and Teichberg, V. (1989). Molecular structure of the chick cerebellar kainate-binding subunit of a putative glutamate receptor. *Nature* **342,** 689–692.

Hampson, D. R., Wheaton, K. D., Dechesne, C. J., and Wenthold, R. J. (1989). Identification and characterization of the ligand binding subunit of a kainic acid receptor using monoclonal antibodies and peptide mapping. *J. Biol. Chem.* **264,** 13329–13335.

Hogner, A., Greenwood, J. R., Liljefors, T., Lunn, M. L., Egebjerg, J., Larsen, I. K., Gouaux, E., and Kastrup, J. S. (2003). Competitive antagonism of AMPA receptors by ligands of different classes: Crystal structure of ATPO bound to the GluR2 ligand-binding core, in comparison with DNQX. *J. Med. Chem.* **46,** 214–221.

Hogner, A., Kastrup, J., Jin, R., Liljefors, T., Mayer, M., Egebjerg, J., Larsen, I., and Gouaux, E. (2002). Structural Basis for AMPA Receptor Activation and Ligand Selectivity: Crystal Structures of Five Agonist Complexes with the GluR2 Ligand-binding Core. *J. Mol. Biol.* **322,** 93–109.

Hollmann, M., and Heinemann, S. (1994). Cloned glutamate receptors. *Annu. Rev. Neurosci.* **17,** 31–108.

Hollmann, M., Maron, C., and Heinemann, S. (1994). N-Glycosylation site tagging suggests a three transmembrane domain topology for the glutamate receptor GluR1. *Neuron* **13,** 1331–1343.

Hollmann, M., O'Shea-Greenfield, A., Rogers, S. W., and Heinemann, S. (1989). Cloning by functional expression of a member of the glutamate receptor family. *Nature* **342,** 643–648.

Hwang, P. M., Skrynnikov, N. R., and Kay, L. E. (2001). Domain orientation in beta-cyclodextrin-loaded maltose binding protein: Diffusion anisotropy measurements confirm the results of a dipolar coupling study. *J. Biomol. NMR* **20,** 83–88.

Inagaki, N., Kuromi, H., Gonoi, T., Okamoto, Y., Ishida, H., Seino, Y., Kaneko, T., Iwanaga, T., and Seino, S. (1995). Expression and role of ionotropic glutamate receptors in pancreatic islet cells. *FASEB J.* **9,** 686–691.

Jayaraman, V., Keesey, R., and Madden, D. R. (2000a). Ligand–protein interactions in the glutamate receptor. *Biochemistry* **39,** 8693–8697.

Jayaraman, V., Thiran, S., and Madden, D. R. (2000b). Fourier transform infrared spectroscopic characterization of a photolabile precursor of glutamate. *FEBS Lett.* **475,** 278–282.

Jiang, Y., Lee, A., Chen, J., Cadene, M., Chait, B. T., and MacKinnon, R. (2002a). Crystal structure and mechanism of a calcium-gated potassium channel. *Nature* **417,** 515–522.

Jiang, Y., Lee, A., Chen, J., Cadene, M., Chait, B. T., and MacKinnon, R. (2002b). The open pore conformation of potassium channels. *Nature* **417,** 523–526.

Jiang, Y., Lee, A., Chen, J., Ruta, V., Cadene, M., Chait, B. T., and MacKinnon, R. (2003). X-ray structure of a voltage-dependent K(+) channel. *Nature* **423,** 33–41.

Jin, R., Banke, T. G., Mayer, M. L., Traynelis, S. F., and Gouaux, E. (2003). Structural basis for partial agonist action at ionotropic glutamate receptors. *Nat. Neurosci.* **6,** 803–810.

Jin, R., and Gouaux, E. (2003). Probing the Function, Conformational Plasticity, and Dimer-Dimer Contacts of the GluR2 Ligand-Binding Core: Studies of 5-Substituted Willardiines and GluR2 S1S2 in the Crystal. *Biochemistry* **42,** 5201–5213.

Jin, R., Horning, M., Mayer, M. L., and Gouaux, E. (2002). Mechanism of activation and selectivity in a ligand-gated ion channel: Structural and functional studies of GluR2 and quisqualate. *Biochemistry* **41,** 15635–15643.

Jones, K. S., VanDongen, H. M., and VanDongen, A. M. (2002). The NMDA receptor M3 segment is a conserved transduction element coupling ligand binding to channel opening. *J. Neurosci.* **22,** 2044–2053.

Jorgensen, M., Tygesen, C. K., and Andersen, P. H. (1995). Ionotropic glutamate receptors–focus on non-NMDA receptors. *Pharmacol. Toxicol.* **76,** 312–319.

Kang, J., and Turano, F. J. (2003). The putative glutamate receptor 1.1 (AtGLR1.1) functions as a regulator of carbon and nitrogen metabolism in Arabidopsis thaliana. *Proc. Natl. Acad. Sci. USA* **100,** 6872–6877.

Kasper, C., Lunn, M. L., Liljefors, T., Gouaux, E., Egebjerg, J., and Kastrup, J. S. (2002). GluR2 ligand-binding core complexes: Importance of the isoxazolol moiety and 5-substituent for the binding mode of AMPA-type agonists. *FEBS Lett.* **531,** 173–178.

Keinanen, K., Arvola, M., Kuusinen, A., and Johnson, M. (1997). Ligand recognition in glutamate receptors: Insights from mutagenesis of the soluble alpha-amino-3-hydroxy-5-methyl-4-isoxazole propionic acid (AMPA)-binding domain of glutamate receptor type D (GluR-D). *Biochem. Soc. Trans.* **25,** 835–838.

Keinänen, K., Wisden, W., Sommer, B., Werner, P., Herb, A., Verdoorn, T. A., Sakmann, B., and Seeburg, P. H. (1990). A family of AMPA-selective glutamate receptors. *Science* **249,** 556–560.

Kim, S. A., Kwak, J. M., Jae, S. K., Wang, M. H., and Nam, H. G. (2001). Overexpression of the AtGluR2 gene encoding an Arabidopsis homolog of mammalian glutamate receptors impairs calcium utilization and sensitivity to ionic stress in transgenic plants. *Plant Cell Physiol.* **42,** 74–84.

Kohda, K., Wang, Y., and Yuzaki, M. (2000). Mutation of a glutamate receptor motif reveals its role in gating and delta2 receptor channel properties. *Nat. Neurosci.* **3,** 315–322.

Krupp, J. J., Vissel, B., Heinemann, S. F., and Westbrook, G. L. (1998). N-terminal domains in the NR2 subunit control desensitization of NMDA receptors. *Neuron* **20,** 317–327.

Kuner, T., Beck, C., Sakmann, B., and Seeburg, P. H. (2001). Channel-lining residues of the AMPA receptor M2 segment: Structural environment of the Q/R site and identification of the selectivity filter. *J. Neurosci.* **21,** 4162–4172.

Kuner, T., Seeburg, P. H., and Guy, H. R. (2003). A common architecture for K+ channels and ionotropic glutamate receptors? *Trends Neurosci.* **26,** 27–32.

Kuner, T., Wollmuth, L. P., Karlin, A., Seeburg, P. H., and Sakmann, B. (1996). Structure of the NMDA receptor channel M2 segment inferred from the accessibility of substituted cysteines. *Neuron* **17,** 343–352.

Kunishima, N., Shimada, Y., Tsuji, Y., Sato, T., Yamamoto, M., Kumasaka, T., Nakanishi, S., Jingami, H., and Morikawa, K. (2000). Structural basis of glutamate recognition by a dimeric metabotropic glutamate receptor. *Nature* **407,** 971–977.

Kuryatov, A., Laube, B., Betz, H., and Kuhse, J. (1994). Mutational analysis of the glycine-binding site of the NMDA receptor: Structural similarity with bacterial amino acid-binding proteins. *Neuron* **12,** 1291–1300.

Kurz, L. L., Zuhlke, R. D., Zhang, H. J., and Joho, R. H. (1995). Side-chain accessibilities in the pore of a K+ channel probed by sulfhydryl-specific reagents after cysteine-scanning mutagenesis. *Biophys. J.* **68,** 900–905.

Kuusinen, A., Abele, R., Madden, D. R., and Keinanen, K. (1999). Oligomerization and ligand-binding properties of the ectodomain of the alpha-amino-3-hydroxy-5-methyl-4-isoxazole propionic acid receptor subunit GluRD. *J. Biol. Chem.* **274,** 28937–28943.

Kuusinen, A., Arvola, M., and Keinänen, K. (1995). Molecular dissection of the agonist binding site of an AMPA receptor. *EMBO J.* **14,** 6327–6332.

Lam, H. M., Chiu, J., Hsieh, M. H., Meisel, L., Oliveira, I. C., Shin, M., and Coruzzi, G. (1998). Glutamate-receptor genes in plants. *Nature* **396,** 125–126.

Laube, B., Hirai, H., Sturgess, M., Betz, H., and Kuhse, J. (1997). Molecular determinants of agonist discrimination by NMDA receptor subunits: Analysis of the glutamate binding site on the NR2B subunit. *Neuron* **18,** 493–503.

Leonard, A. S., and Hell, J. W. (1997). Cyclic AMP-dependent protein kinase and protein kinase C phosphorylate N-methyl-D-aspartate receptors at different sites. *J. Biol. Chem.* **272,** 12107–12115.

Leuschner, W. D., and Hoch, W. (1999). Subtype-specific assembly of alpha-amino-3-hydroxy-5-methyl-4-isoxazole propionic acid receptor subunits is mediated by their n-terminal domains. *J. Biol. Chem.* **274,** 16907–16916.

Li, G., Pei, W., and Niu, L. (2003). Channel opening kinetics of GluR2Qflip AMPA receptor: A laser-pulse photolysis study. *Biochem.* **42,** 12358–12366.

Loh, A. P., Guo, W., Nicholson, L. K., and Oswald, R. E. (1999). Backbone dynamics of inactive, active, and effector-bound Cdc42Hs from measurements of ^{15}N relaxation parameters at multiple field strengths. *Biochem.* **38,** 12547–12557.

Loh, A. P., Pawley, N., Nicholson, L. K., and Oswald, R. E. (2001). An increase in side chain entropy facilitates effector binding: NMR characterization of the side chain methyl group dynamics in Cdc42Hs. *Biochem.* **40,** 4590–4600.

Lomeli, H., Mosbacher, J., Melcher, T., Hoger, T., Geiger, J. R., Kuner, T., Monyer, H., Higuchi, M., Bach, A., and Seeburg, P. H. (1994). Control of kinetic properties of AMPA receptor channels by nuclear RNA editing. *Science* **266,** 1709–1713.

Lunn, M. L., Hogner, A., Stensbol, T. B., Gouaux, E., Egebjerg, J., and Kastrup, J. S. (2003). Three-dimensional structure of the ligand-binding core of GluR2 in complex with the agonist (S)-ATPA: Implications for receptor subunit selectivity. *J. Med. Chem.* **46,** 872–875.

Luque, I., Leavitt, S. A., and Freire, E. (2002). The linkage between protein folding and functional cooperativity: Two sides of the same coin? *Annu. Rev. Biophys. Biomol. Struct.* **31,** 235–256.

Madden, D. R., Abele, R., Andersson, A., and Keinanen, K. (2000). Large-scale expression and thermodynamic characterization of a glutamate receptor agonist-binding domain. *Eur. J. Biochem.* **267,** 4281–4289.

Madden, D. R., Thiran, S., Zimmermann, H., Romm, J., and Jayaraman, V. (2001). Stereochemistry of quinoxaline antagonist binding to a glutamate receptor investigated by Fourier transform infrared spectroscopy. *J. Biol. Chem.* **276,** 37821–37826.

Margot, K. C. (1998). "Single channel characterization of a human GluR3-type glutamate receptor-channel." MS dissertation, Cornell University Publications.

Masuko, T., Kashiwagi, K., Kuno, T., Nguyen, N. D., Pahk, A. J., Fukuchi, J., Igarashi, K., and Williams, K. (1999). A regulatory domain (R1-R2) in the amino terminus of the N-methyl-D-aspartate receptor: Effects of spermine, protons, and ifenprodil, and structural similarity to bacterial leucine/isoleucine/valine binding protein. *Mol. Pharmacol.* **55,** 957–969.

Mayer, M. L., Olson, R., and Gouaux, E. (2001). Mechanisms for ligand binding to GluR0 ion channels: Crystal structures of the glutamate and serine complexes and a closed apo state. *J. Mol. Biol.* **311,** 815–836.

McFeeters, R. L., and Oswald, R. E. (2002). Structural mobility of the extracellular ligand-binding core of an ionotropic glutamate receptor. Analysis of NMR relaxation dynamics. *Biochem.* **41,** 10472–10481.

McFeeters, R. L., and Oswald, R. E. (2004). Emerging Structural Explanations of Ionotropic Glutamate Receptor Function. *FASEB J.* **18,** 428–438.

McFeeters, R. L., Swapna, G. V. T., Montelione, G. T., and Oswald, R. E. (2002). Semi-automated backbone resonance assignments of the extracellular ligand-binding domain of an ionotropic glutamate receptor. *J. Biomol. NMR* **22,** 297–298.

Melcher, T., Maas, S., Herb, A., Sprengel, R., Seeburg, P. H., and Higuchi, M. (1996). A mammalian RNA editing enzyme. *Nature* **379,** 460–464.

Mendieta, J., Ramirez, G., and Gago, F. (2001). Molecular dynamics simulations of the conformational changes of the glutamate receptor ligand-binding core in the presence of glutamate and kainate. *Proteins* **44,** 460–469.

Monaghan, D. T., Bridges, R. J., and Cotman, C. W. (1989). The excitatory amino acid receptors: Their classes, pharmacology, and distinct properties in the function of the central nervous system. *Annu. Rev. Pharmacol. Toxicol.* **29,** 365–402.

Moriyoshi, K., Masu, M., Ishii, T., Shigemoto, R., Mizuno, N., and Nakanishi, S. (1991). Molecular cloning and characterization of the rat NMDA receptor. *Nature* **354,** 31.

Mosbacher, J., Schoepfer, R., Monyer, H., Burnashev, N., Seeburg, P. H., and Ruppersberg, J. P. (1994). A molecular determinant for submillisecond desensitization in glutamate receptors. *Science* **266,** 1059–1062.

Moss, S. J., Blackstone, C. D., and Huganir, R. L. (1993). Phosphorylation of recombinant non-NMDA glutamate receptors on serine and tyrosine residues. *Neurochem. Res.* **18,** 105–110.

Murzin, A. G., Brenner, S. E., Hubbard, T., and Chothia, C. (1995). SCOP: A structural classification of proteins database for the investigation of sequences and structures. *J. Mol. Biol.* **247,** 536–540.

Nakanishi, N., Schneider, N. A., and Axel, R. (1990). A family of glutamate receptor genes: Evidence for the formation of heteromultimeric receptors with distinct channel properties. *Neuron* **5,** 569–581.

Nakazawa, K., Mikawa, S., Hashikawa, T., and Ito, M. (1995). Transient and persistent phosphorylation of AMPA-type glutamate receptor subunits in cerebellar Purkinje cells. *Neuron* **15,** 697–709.

Nicholson, L. K., Yamazaki, T., Torchia, D. A., Grzesiek, S., Bax, A., Stahl, S. J., Kaufman, J. D., Wingfield, P. T., Lam, P. Y., and Jadhav, P. K. *et al.* (1995). Flexibility and function in HIV-1 protease. *Nat. Struct. Biol.* **2,** 274–280.

O'Hara, P. J., Sheppard, P. O., Thøgersen, H., Venezia, D., Haldeman, B. A., McGrane, V., Houamed, K. M., Thomsen, C., Gilbert, T. L., and Mulvihill, E. R. (1993). The ligand-binding domain in metabotropic glutamate receptors is related to bacterial periplasmic binding proteins. *Neuron* **11,** 41–52.

Opella, S. J., Marassi, F. M., Gesell, J. J., Valente, A. P., Kim, Y., Oblatt-Montal, M., and Montal, M. (1999). Structures of the M2 channel-lining segments from nicotinic acetylcholine and NMDA receptors by NMR spectroscopy. *Nat. Struct. Biol.* **6,** 374–379.

Paas, Y., Eisenstein, M., Medevielle, F., Teichberg, V. I., and Devillers-Thiery, A. (1996). Identification of the amino acid subsets accounting for the ligand binding specificity of a glutamate receptor. *Neuron* **17,** 979–990.

Palmer, A. G., 3rd, Kroenke, C. D., and Loria, J. P. (2001). Nuclear magnetic resonance methods for quantifying microsecond-to-millisecond motions in biological macromolecules. *Methods Enzymol.* **339,** 204–238.

Panchenko, V. A., Glasser, C. R., and Mayer, M. L. (2001). Structural similarities between glutamate receptor channels and K^+ channels examined by scanning mutagenesis. *J. Gen. Physiol.* **117,** 345–360.

Pascual, J. M., Shieh, C. C., Kirsch, G. E., and Brown, A. M. (1995). K^+ pore structure revealed by reported cysteines and inner and outer surfaces. *Neuron* **14,** 1055–1063.

Pasternack, A., Coleman, S. K., Jouppila, A., Mottershead, D. G., Lindfors, M., Pasternack, M., and Keinanen, K. (2002). Alpha-amino-3-hydroxy-5-methyl-4-isoxazolepropionic acid (AMPA) receptor channels lacking the N-terminal domain. *J. Biol. Chem.* **277,** 49662–49667.

Patton, A. J., Genever, P. G., Birch, M. A., Suva, L. J., and Skerry, T. M. (1998). Expression of an N-methyl-D-aspartate-type receptor by human and rat osteoblasts and osteoclasts suggests a novel glutamate signaling pathway in bone. *Bone* **22,** 645–649.

Raymond, L. A., Blackstone, C. D., and Huganir, R. L. (1993). Phosphorylation and modulation of recombinant GluR6 glutamate receptors by cAMP-dependent protein kinase. *Nature* **361,** 637–641.

Raymond, L. A., Tingley, W. G., Blackstone, C. D., Roche, K. W., and Huganir, R. L. (1994). Glutamate receptor modulation by protein phosphorylation. *J. Physiol. Paris* **88,** 181–192.

Roche, K. W., Raymond, L. A., Blackstone, C., and Huganir, R. L. (1994). Transmembrane topology of the glutamate receptor subunit GluR6. *J. Biol. Chem.* **269,** 11679–11682.

Rosenmund, C., Stern-Bach, Y., and Stevens, C. F. (1998). The tetrameric structure of a glutamate receptor channel. *Science* **280,** 1596–1599.

Sack, J. S., Saper, M. A., and Quiocho, F. A. (1989a). Periplasmic binding protein structure and function. Refined X-ray structures of the leucine/isoleucine/valine-binding protein and its complex with leucine. *J. Mol. Biol.* **206,** 171–191.

Sack, J. S., Trakhanov, S. D., Tsigannik, I. H., and Quiocho, F. A. (1989b). Structure of the L-leucine-binding protein refined at 2.4 A resolution and comparison with the Leu/Ile/Val-binding protein structure. *J. Mol. Biol.* **206,** 193–207.

Seeburg, P. H. (1996). The role of RNA editing in controlling glutamate receptor channel properties. *J. Neurchem.* **66,** 1–5.

Sharff, A. J., Rodseth, L. E., and Quiocho, F. A. (1993). Refined 1.8-Å structure reveals the mode of binding of beta-cyclodextrin to the maltodextrin binding protein. *Biochem.* **32,** 10553–10559.

Sheng, M., and Kim, M. J. (2002). Postsynaptic signaling and plasticity mechanisms. *Science* **298,** 776–780.

Skrynnikov, N. R., Goto, N. K., Yang, D., Choy, W. Y., Tolman, J. R., Mueller, G. A., and Kay, L. E. (2000). Orienting domains in proteins using dipolar couplings measured by liquid-state NMR: Differences in solution and crystal forms of maltodextrin binding protein loaded with beta-cyclodextrin. *J. Mol. Biol.* **295,** 1265–1273.

Sobolevsky, A. I., Yelshansky, M. V., and Wollmuth, L. P. (2003). Different gating mechanisms in glutamate receptor and K^+ channels. *J. Neurosci.* **23,** 7559–7568.

Sommer, B., Keinänen, K., Verdoorn, T. A., Wisden, W., Burnashev, N., Herb, A., Köhler, M., Takagi, T., Sakmann, G., and Seeburg, P. H. (1990). Flip and flop: A cell-specific functional switch in glutamate-operated channels of the CNS. *Science* **249,** 1580–1584.

Sommer, B., Köhler, M., Sprengel, R., and Seeburg, P. H. (1991). RNA editing in brain controls a determinant of ion flow in glutamate-gated channels. *Cell* **67,** 11–19.

Speranskiy, K., and Kurnikova, M. (2003). Composite MD/Continuum modeling of the glutamate receptor GluR2. Analysis of ligand binding energy and vibrational spectra. *Biophys. J.* Abstracts, 1661P.

Stern-Bach, Y., Bettler, B., Hartley, M., Sheppard, P. O., O'Hara, P. J., and Heinemann, S. F. (1994). Agonist selectivity of glutamate receptors is specified by two domains structurally related to bacterial amino acid-binding proteins. *Neuron* **13,** 1345–1357.

Sun, Y., Olson, R., Horning, M., Armstrong, N., Mayer, M., and Gouaux, E. (2002). Mechanism of glutamate receptor desensitization. *Nature* **417,** 245–253.

Sutcliffe, M. J., Wo, Z. G., and Oswald, R. E. (1996). Three-dimensional models of non-NMDA glutamate receptors. *Biophys. J.* **70,** 1575–1589.

Swanson, G. T., Feldmeyer, D., Kaneda, M., and Cull-Candy, S. G. (1996). Effect of RNA editing and subunit co-assembly on single-channel properties of recombinant kainate receptors. *J. Physiol. (Lond.)* **492,** 129–142.

Swanson, G. T., Gereau, R. W. T., Green, T., and Heinemann, S. F. (1997a). Identification of amino acid residues that control functional behavior in GluR5 and GluR6 kainate receptors. *Neuron* **19,** 913–926.

Swanson, G. T., Kamboj, S. K., and Cull-Candy, S. G. (1997b). Single-channel properties of recombinant AMPA receptors depend on RNA editing, splice variation, and subunit composition. *J. Neurosci.* **17,** 58–69.

Tan, S. E., Wenthold, R. J., and Soderling, T. R. (1994). Phosphorylation of AMPA-type glutamate receptors by Ca^{++}/calmodulin-dependent protein kinase II and PKC in cultured hippocampal neurons. *J. Neurosci.* **14,** 1123–1129.

Taverna, F. A., Wang, L. Y., MacDonald, J. F., and Hampson, D. R. (1994). A transmembrane model for an ionotropic glutamate receptor predicted on the basis of the location of asparagine-linked oligosaccharides. *J. Biol. Chem.* **269,** 14159–14164.

Tikhonov, D. B., Mellor, J. R., Usherwood, P. N., and Magazanik, L. G. (2002). Modeling of the pore domain of the GLUR1 channel: Homology with K^+ channel and binding of channel blockers. *Biophys. J.* **82,** 1884–1893.

Wada, K., Dechesne, C. J., Shimasaki, S., King, R. G., Kusano, K., Buonanno, A., Hampson, D. R., Banner, C., Wenthold, R. J., and Nakatani, Y. (1989). Sequence and expression of a frog brain complementary DNA encoding a kainate-binding protein. *Nature* **342,** 684–689.

Wang, L. Y., Taverna, F. A., Huang, X. P., MacDonald, J. F., and Hampson, D. R. (1993). Phosphorylation and modulation of a kainate receptor (GluR6) by cAMP-dependent protein kinase. *Science* **259,** 1173–1175.

Weaver, C. D., Yao, T. L., Powers, A. C., and Verdoorn, T. A. (1996). Differential expression of glutamate receptor subtypes in rat pancreatic islets. *J. Biol. Chem.* **271,** 12977–12984.

Wo, Z. G., Chohan, K. K., Chen, H., Sutcliffe, M. J., and Oswald, R. E. (1999). Cysteine mutagenesis and homology modeling of the ligand-binding site of a kainate-binding protein. *J. Biol. Chem.* **274,** 37210–37218.

Wo, Z. G., and Oswald, R. E. (1994). Transmembrane topology of two kainate receptor subunits revealed by N-glycosylation. *Proc. Natl. Acad. Sci. USA* **91,** 7154–7158.

Wo, Z. G., and Oswald, R. E. (1995a). A topological analysis of goldfish kainate receptors predicts three transmembrane segments. *J. Biol. Chem.* **270,** 2000–2009.

Wo, Z. G., and Oswald, R. E. (1995b). Unraveling the modular design of glutamate-gated ion channels. *Trends Neurosci.* **18,** 161–168.

Wo, Z. G., and Oswald, R. E. (1996). Ligand binding characteristics and related structural features of goldfish kainate receptors: Identification of a conserved disulfide bond and three residues important for ligand binding. *Mol. Pharmacol.* **50,** 770–780.

Wood, M. W., VanDongen, H. M., and VanDongen, A. M. (1997). An alanine residue in the M3-M4 linker lines the glycine binding pocket of the N-methyl-D-aspartate receptor. *J. Biol. Chem.* **272,** 3532–3537.

Wood, M. W., VanDongen, H. M. A., and VanDongen, A. M. J. (1995). Structural conservation of ion conduction pathways in K channels and glutamate receptors. *Proc. Natl. Acad. Sci. USA* **92,** 4882–4886.

Xia, H., von Zastrow, M., and Malenka, R. C. (2002). A novel anterograde trafficking signal present in the N-terminal extracellular domain of ionotropic glutamate receptors. *J. Biol. Chem.* **277,** 47765–47769.

Zeng, L., Lu, L., Muller, M., Gouaux, E., and Zhou, M. M. (2002). Structure-based functional design of chemical ligands for AMPA-subtype glutamate receptors. *J. Mol. Neurosci.* **19,** 113–116.

Zhang, S., Ehlers, M. D., Bernhardt, J. P., Su, C. T., and Huganir, R. L. (1998). Calmodulin mediates calcium-dependent inactivation of N-methyl-D-aspartate receptors. *Neuron* **21,** 443–453.

Zheng, F., Erreger, K., Low, C. M., Banke, T., Lee, C. J., Conn, P. J., and Traynelis, S. F. (2001). Allosteric interaction between the amino terminal domain and the ligand binding domain of NR2A. *Nat. Neurosci.* **4,** 894–901.

Zuo, J., De Jager, P. L., Takahashi, K. A., Jiang, W., Linden, D. J., and Heintz, N. (1997). Neurodegeneration in Lurcher mice caused by mutation in delta2 glutamate receptor gene. *Nature* **388,** 769–773.

CHEMOKINE-RECEPTOR INTERACTIONS: GPCRS, GLYCOSAMINOGLYCANS AND VIRAL CHEMOKINE BINDING PROTEINS

By ELAINE K. LAU, SAMANTHA ALLEN, ANDRO R. HSU, AND TRACY M. HANDEL

Department of Molecular and Cell Biology
University of California
Berkeley, California 94720

Abstract

A key feature of the immune system is the migration of leukocytes throughout the organism in an effort to patrol for infectious pathogens, tissue damage, and other physiological insults. This remarkable surveillance system is controlled by a family of proteins called *chemokines* (chemoattractant cytokines), and their respective receptors. Originally discovered because of their role in cell recruitment during inflammation, it is now well recognized that chemokines are also involved in other diverse processes including lymphocyte development and homing, organogenesis, and neuronal communication. While chemokines have evolved largely for host protection, their ability to induce cell damage and inappropriate cell recruitment, can lead to disease. Thus, there is considerable interest in

0065-3233/04 $35.00

developing antagonists. In this review we emphasize what is known about the structural biology of chemokines, chemokine receptors, and interactions with cell surface glycosaminoglycans. We also briefly describe their role in certain diseases and strategies for interfering with chemokine function that have emerged from mechanistic and structural understanding of their function. Finally we discuss viral mechanisms for sabotaging or manipulating the chemokine system, in part to illustrate the level of molecular mimicry that viruses have achieved and the evolutionary pressure imposed on the immune system by these pathogens.

I. Introduction

A key feature of the immune system is the migration of leukocytes throughout the organism in an effort to patrol for infectious pathogens, tissue damage, and other physiological insults. This remarkable surveillance system is controlled by a family of proteins called chemokines (chemoattractant cytokines), and their respective receptors (Baggiolini, 1998; Baggiolini *et al.*, 1997; Murphy, 1994; Olson *et al.*, 2002; Ono *et al.*, 2003). Chemokines were originally discovered because of their role in cell recruitment during inflammation, but it is now well recognized that chemokines are also involved in other diverse processes including lymphocyte development and homing (Butcher *et al.*, 1996; Kunkel *et al.*, 2002), organogenesis (Muller *et al.*, 2003), and neuronal communication (Bajetto *et al.*, 2001). They recruit a diversity of cell types, primarily of hematopoietic origin; however, other cells such as endothelial, epithelial, and smooth muscle cells, fibroblasts, and many cells in the CNS, express functional receptors. The functional diversity of chemokines derives from the combinatorial complexity of the chemokine superfamily, their spatial, temporal and cellular expression, receptor-binding patterns, signal transduction, regulatory mechanisms, and their cooperation with other macromolecules such as adhesion proteins.

The role of chemokines in tissue-specific homing of lymphocytes and emigration of leukocytes to inflamed tissue are perhaps the best-understood processes. The current paradigm features many sequential interactions involving adhesion molecules, and glycosaminoglycans in addition to the chemoattractants (Weber, 2003). Figure 1 illustrates the migration of a leukocyte from the blood, across the endothelium and basement membrane, and into the affected tissue in the case of an inflammatory response. First, selectins on the endothelium interact with their mucin counterreceptors on the leukocyte through labile interactions, slowing and inducing rolling of the leukocyte along the endothelial surface (Dwir *et al.*, 2004; Springer,

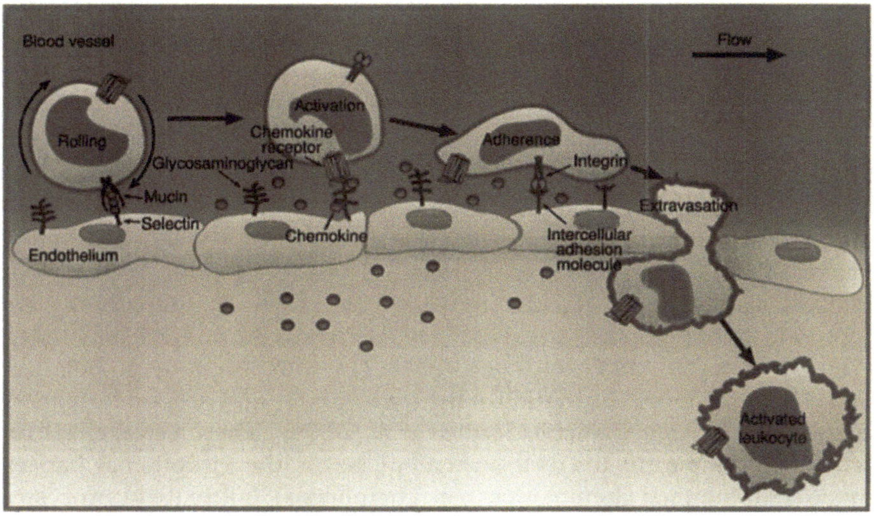

FIG. 1. Illustration of the migration of a leukocyte from blood to extravascular tissue in response to chemokine production. Interaction of selectins on the endothelial surface with mucins on the leukocyte causes reversible tethering and a rolling behavior of the leukocyte. Chemokines are shown bound to the surface of the endothelial cell via interactions with glycosaminoglycans (GAGs); however, whether they simultaneously bind to both GAGs and chemokine receptors, as depicted in the figure, is unknown. In addition to GAGs, a promiscuous non-signaling chemokine binding protein (DARC) is thought to be involved in presentation of chemokines on endothelial cell surfaces. Tissue specific expression of these presentation molecules may play a role in the migratory patterns of receptor bearing cells. Subsequent binding of chemokines to chemokine receptors on leukocytes stimulates a wide variety of signaling pathways, the consequences of which include increased adhesiveness of the cells via activation of integrins, and cytoskeletal changes involved in cell movement. Migration toward subluminal chemokines occurs via one of two routes, either between the intercellular junction (shown here) or transcellularly through a pore in the endothelial cells. (See Color Insert.)

1994, 1995). Chemokines then come into play. In response to a variety of signals such as proinflammatory cytokines or exogenous mitogenic factors, they are secreted either directly from the endothelial cells, or from extravascular cells and trancytosed in caveolae from the abluminal to luminal surface of the endothelia cell (Middleton *et al.*, 2002). Retention and presentation of chemokines on the surface of endothelial cells is then mediated by glycosaminoglycans (Ali *et al.*, 2000; Middleton *et al.*, 1997) and possibly the G-protein coupled receptor, DARC (Dawson *et al.*, 2000; Middleton *et al.*, 2002). These interactions are thought to provide a mechanism for keeping a high concentration of chemokine localized to the site of production, and directional cues for the leukocyte.

Upon accumulation on the endothelial cell surface, chemokines interact with G-protein-coupled receptors (GPCRs) on the leukocyte, and activate complex signal transduction cascades that are only beginning to be understood in an integrated way (Ono *et al.*, 2003; Thelen, 2001). These pathways involve heterotrimeric G-proteins, adenylyl cyclase, phospholipases, protein tyrosine, serine/threonine and lipid kinases, rho family GTPases, and second messengers. G-protein independent pathways like JAK/STAT may also be activated (Mellado *et al.*, 2001, 2002) (see Section II.E. later). Profound physiological and morphological changes follow. Activation of integrin adhesiveness on the leukocyte for their Ig superfamily receptors leads to firm attachment of the leukocytes to the endothelial surface (Fig. 1) (Alon *et al.*, 2003; Springer, 1995). The cells also undergo polarization with lamellipodia formation at the leading edge and uropod formation at the trailing edge (Sanchez-Madrid *et al.*, 1999). These cytoskeletal rearrangements permit net cell movement across the endothelial barrier towards sublumenal chemokines (e.g., chemotaxis). Subsequent processes induced by some chemokines include degranulation, protease release, the respiratory burst, and induction of apoptosis to resolve the abnormal cellular pathology (Baggiolini *et al.*, 1989, 1990, 1993). Desensitization and internalization of the receptor terminate the response.

While chemokines have evolved largely for host protection, their ability to induce cell damage and inappropriate cell recruitment, can lead to disease (Baggiolini, 2001; D'Ambrosio *et al.*, 2003; Gerard *et al.*, 2001; Proudfoot, 2002). Thus, there is considerable interest in developing antagonists. In this review we emphasize what is known about the structural biology of chemokines, chemokine receptors, and interactions with cell surface glycosaminoglycans. We also briefly describe their role in certain diseases and strategies for interfering with chemokine function that have emerged from mechanistic and structural understanding of their function. Finally we discuss viral mechanisms for sabotaging or manipulating the chemokine system, in part to illustrate the level of molecular mimicry that viruses have achieved, and the evolutionary pressure imposed on the immune system by these pathogens.

II. The Chemokine Superfamily

A. *Chemokine Family Organization and Nomenclature*

There are currently 18 known chemokine receptors and 45 ligands although these numbers do not include alternative splice variants, which exist for both the ligands (Baird *et al.*, 1999) and receptors (Gupta *et al.*,

1999). Traditionally, chemokines and their receptors have been divided into four families based on the pattern of cysteine residues in the ligands (CXC, CC, C, and CX3C). The new nomenclature for the receptors and ligands is based on these subdivisions (receptors are designated CXCRN, CCRN, XCRN, and CX3CRN, while the ligands are CXCLN, CCLN, XCLN, and CX3CLN, where N = a number) (Bacon *et al.*, 2002; Murphy *et al.*, 2000). In this review, the ligands will be referenced by the old name/ followed by the new name. Another useful classification is based on broad functional criteria. the inflammatory chemokines are expressed from circulating leukocytes and related cells only upon activation, while the homeostatic or housekeeping chemokines are constitutively expressed in specific locations like bone marrow, thymus, spleen, lymph node, and skin.

Most of the ligands are 70–125 amino acid soluble proteins that tend to be highly basic and contain between one and three disulfide bonds. Two exceptions are fractalkine/CX3CL1 and CXCL16; their chemokine domains are fused to long C-terminal mucin-like stalks that are tethered to the cell membrane via a single transmembrane helix. They are also released in soluble form after proteolytic cleavage within the stalk. Surface immobilization via the stalk enables these proteins to act directly as firm adhesion molecules in the absence of signal transduction and integrin activation (Goda *et al.*, 2000; Haskell *et al.*, 1999; Shimaoka *et al.*, 2004). The stalk may also provide an alternative immobilization/presentation mechanism in lieu of cell surface glycosaminoglycans (see Section III).

The receptors belong to the Class A family of GPCRs which are characterized by their high homology to rhodopsin. They also contain disulfides in their extracellular domains, providing some constraints on their topological organization. One disulfide connects the N-terminus to the third extracellular loop, while the second links the first and second extracellular loops. Many chemokines, especially those of the inflammatory class, are ligands of the same "shared" receptor. As an example, the CCR1 ligands include CCL3, CCL5, CCL7, CCL8, CCL13, CCL14, CCL15, CCL16, CCL23, and possibly others. Additionally, many ligands bind multiple receptors, and depending on receptor context, act as agonists or antagonists. For instance, all four MCPs and eotaxin/CCL11 are ligands of CCR2, and at least eotaxin/CCL11, MCP-2/CCL8, MCP-3/CCL7, and MCP-4/CCL13 interact with other receptors (eotaxin/CCL11 is an antagonist of CCR2 and an agonist of CCR3 and CCR5 [Ogilvie *et al.*, 2001, 2003]). Similarly, I-TAC/CXCL11, MIG/CXCL9, and IP10/CXCL10 are agonists of CXCR3 and antagonists of CCR3 (Loetscher *et al.*, 2001). Because of their role in inflammation, such "promiscuous" interactions may have evolved under evolutionary pressures from different pathogen: host encounters (see Section IV). By contrast, other receptors such as

CXCR4, have a single ligand (e.g., SDF-1/CXCL12), possibly because these proteins are involved in crucial processes to the organism that are subject to significantly less rapid evolutionary change, like development (Homey *et al.*, 2002). These observations raise some important questions. Does the presence of multiple ligands for a single receptor represent functional redundancy (i.e., a backup plan in case of a defective ligand), or fine-tuning of the immune response (Devalaraja *et al.*, 1999; Mantovani, 1999)? Other structurally-related questions include what determines ligand-receptor specificity and what causes a given ligand to be an agonist of one receptor and an antagonist of another?

B. Chemokine Structures

Structures of several chemokines have been solved by NMR and X-ray crystallography (Blaszczyk *et al.*, 2000; Booth *et al.*, 2002; Clore *et al.*, 1990; Crump *et al.*, 1998; Handel *et al.*, 1996; Hoover *et al.*, 2000, 2002; Kim *et al.*, 1996; Kuloglu *et al.*, 2001; Liwang *et al.*, 1999; Lodi *et al.*, 1994; Mayer *et al.*, 2000; Mizoue *et al.*, 1999; Perez-Canadillas *et al.*, 2001; Rajarathnam *et al.*, 2001; Skelton *et al.*, 1995; Swaminathan *et al.*, 2003; Ye *et al.*, 2001). These studies revealed that despite low sequence homology amongst many chemokines, they adopt a remarkably conserved tertiary structure consisting of a (usually) disordered N-terminus, a long irregular "N-loop" that ends in a single turn of 3_{10} helix, a three stranded β-sheet and a C-terminal helix (Fig. 2A). Some chemokines have an extra ~20 amino acids following the C-terminal helix, and in the case of

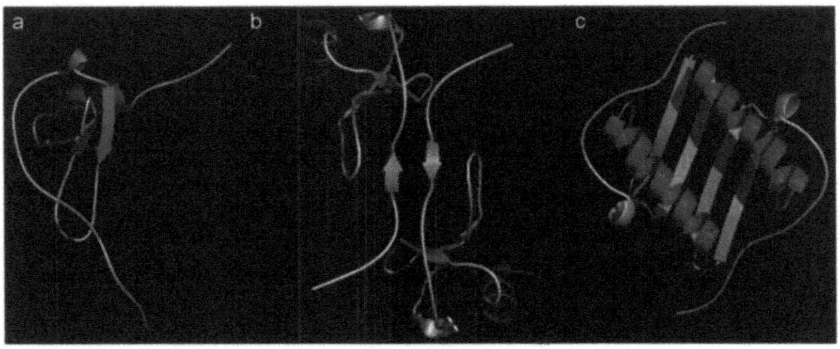

FIG. 2. Ribbon diagram of (A) the MCP-1/CCL2 monomer, (B) the MCP-1/CCL2 dimer (from PDB code 1DOM), and (C) the IL-8/CXCL8 dimer (PDB code 1IL8). Figures were generated with Pymol (DeLano, 2002). The β-sheets are colored blue, the C-terminal helices are red, the 3_{10} helices are yellow, and loops, turns and the unstructured N-termini are white. (See Color Insert.)

XCL1/Lymphotactin, this domain is also unstructured (Kuloglu *et al.*, 2001). Oligomerization of chemokines is also frequently observed. CC chemokines have been shown to dimerize through residues near their disordered N-termini surrounding the first two cysteines (Fig. 2B). By contrast, CXC chemokines dimerize predominantly through residues in the first strand of β-sheet (Fig. 2C). Thus the paradigm in the field has been to associate the dimerization motif in Fig. 2B with CC chemokines and that of Fig. 2C with CXC chemokines, but as discussed later (Section III), this may not be strictly correct. Furthermore, several of the chemokines reported to be monomers may also oligomerize when examined under more physiological conditions than the low pH conditions frequently used for the structure determination(s).

C. *Chemokine Oligomerization*

Despite the abundance of chemokines that oligomerize, several studies indicate that chemokines bind their receptors as monomers, at least in the context of chemotaxis. Ian Clark-Lewis and coworkers made a synthetic variant of IL-8/CCL8 in which Leu25 was N-methylated ([Nme-^{25}L]-IL-8), with the expectation that the modification would inhibit dimerization by disrupting hydrogen bonding between the two central strands shown in Fig. 2C. Nevertheless, [Nme-^{25}L]-IL-8 stimulated elastase release and chemotaxis of neutrophils as efficiently as the wild type protein (Rajarathnam *et al.*, 1994). Similarly, a Pro8 to Ala substitution in CCL2/MCP-1 ([P8A]-MCP-1) and subsequently in CCL3/MIP-1β ([P8A]-MIP-1β) also disrupted dimerization (Kim *et al.*, 2001; Paavola *et al.*, 1998); nevertheless these proteins induce cell migration with equivalent efficacy and potency as the WT counterparts. However, additional studies indicate that oligomerization may be necessary for functions other than chemotaxis. RANTES/CCL5 is a chemokine that has a tendency to extensively aggregate into high-order oligomeric structures. However the mutant [E66S]-RANTES only forms dimers in solution, and while perfectly functional in chemotaxis and calcium flux assays, it was unable to induce T-cell activation and proinflammatory responses through low affinity tyrosine kinase pathway(s) (Appay *et al.*, 1999). As discussed later, oligomerization is also involved in GAG binding (Section III.A).

D. *Receptor-Binding Epitopes on Chemokines*

In addition to determining structures, a large number of studies have defined receptor-binding epitopes on chemokines. The seminal structure–function studies were done on IL-8/CCL8. By generating synthetic

N-terminal truncation mutants, these studies were the first to demonstrate the crucial role of the N-terminus in both binding and signaling; for example, IL-8/CCL8 (residues 7–72) had a greater than 10,000-fold loss of binding affinity on neutrophils, and was completely inactive (Moser *et al.*, 1993). Within the N-terminus, three residues (Glu4, Leu5, Arg6) were discovered to be particularly important as truncation or mutation to alanine significantly impaired receptor binding and neutrophil stimulation. This tripeptide motif was subsequently shown to have functional importance as it confers CXCR1 and CXCR2 ligands with angiogenic activity on vascular endothelium (Loetscher and Clark-Lewis, 2001). Later studies also showed that specific amino acids in the N-loop (Y13, Phe17, Phe21) were important for receptor binding and specificity, but that the C-terminal helix was not (Clark-Lewis *et al.*, 1994).

In an effort to further understand the basis for the diversity and specificity in chemokine-receptor interactions, we subsequently undertook a large-scale mutagenesis study of MCP-1/CCL2 (Hemmerich *et al.*, 1999; Jarnagin *et al.*, 1999). Comparison of our results to the IL-8/CCL8 data illustrates some of the common and unique features of these two chemokine-receptor systems. The MCP-1/CCL2 study involved a panel of alanine or other (e.g., charge reversal) mutants which were characterized in terms of their ability to bind CCR2 expressing cells, and to induce receptor activation as measured by calcium flux, inhibition of adenylyl cyclase, and chemotaxis. Figure 3A summarizes the results by highlighting receptor-binding epitopes on the surface of the protein. Of all the epitopes, we found that Y13 contributes most to receptor binding; when mutated to Ala, the affinity for CCR2 was reduced by a factor of $\sim$100, compared to 57 pM for the WT protein. As for IL-8/CCL8, the N-terminus of MCP-1/CCL2 contributes to binding affinity. However, in contrast to the large affinity contributed by the IL-8/CXCL8 N-terminus, deletion of seven N-terminal amino acids, a variant referred to as "7ND" (Zhang *et al.*, 1995), showed only a 7-fold reduction in binding. All other receptor-binding epitopes involved basic Arg or Lys amino acids (R18, K19, R24, K35, K38, K49), and these were not confined to the N-loop but were more broadly distributed over the whole face of the chemokine. Of the alanine mutants, R24A showed the largest effect on binding affinity ($\sim$35-fold reduction); however, mutation to Glu (R24E) caused a 1600-fold reduction, suggesting that R24 interacts with an acidic region of the receptor. Perhaps the most interesting finding was that for all the mutants involving basic residues, the binding affinity correlated with the EC50 for chemotaxis (EC50 is the agonist concentration which produces 50% of the maximal response). However for 7ND and Y13A there was no correlation as these mutants did not induce migration, despite their ability

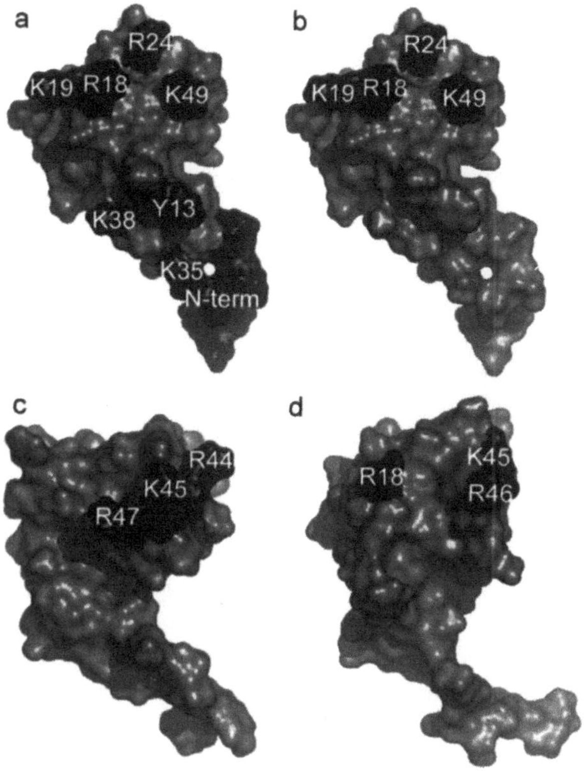

FIG. 3. (A) Surface topology model of the MCP-1/CCL2 monomer, from the NMR structure (PDB code 1DOM) with receptor binding residues highlighted in dark grey and labeled. Mutagenesis approaches can only reliably reveal highly exposed receptor binding epitopes; thus it is possible that additional interactions are made, for example with the hydrophobic pocket that is flanked by the two basic patches, as reflected in the structure of IL-8 complexed to a peptide analogue from CXCR4 (see later). (B) Same as "a" except GAG binding epitopes are highlighted. The epitopes are all basic and include R18, K19, R24, K49, and to a lesser extent, K58 and H66. H66 and K58 are not visible in this view. (C) The RANTES/CCL5 monomer with the GAG-binding epitopes highlighted (PDB code 1RTO). (D) The MIP-1β/CCL4 monomer with the GAG-binding epitopes highlighted (PDB code 1HUM). The figure illustrates the variability in the distribution of GAG binding residues amongst chemokines, suggesting that chemokines may discriminate between different carbohydrate sequences, as has been observed from *in vitro* binding studies.

to bind the receptor with very high affinity. Thus these epitopes are referred to as active triggering residues (Jarnagin *et al.*, 1999; Loetscher *et al.*, 2001).

The N-terminus is also important for receptor binding and function of all other chemokines characterized to date (Loetscher and Clark-Lewis, 2001). In fact, there is currently no example of a chemokine where the N-terminus is not essential for receptor activation. Collectively, the data have lead to a two-site model of receptor activation (Pakianathan *et al.*, 1997; Wells *et al.*, 1996), an arguably oversimplified but nevertheless useful concept, in which residues beyond the N-terminus of the chemokine (the first site, also referred to as the "core or docking domain") provide a large portion of the binding energy and orient the N-terminus (the second site) to induce the appropriate conformational changes in the receptor (Fig. 4).

Given that deletion of even a few amino acids can affect chemokine function, it was not surprising when the discovery was made that certain proteases (e.g., matrix metalloproteases, and the serine protease CD26) cleave the N-termini of chemokines and modulate their activity (Struyf *et al.*, 2003). Proteases may therefore play important roles in the inactivation of chemokines, or conversion of agonists into antagonists during an inflammatory response (Struyf *et al.*, 2003). Truncation can also lead to increased agonism, or changes in cell specificity. For example, deletion of the first amino acid of MCP-1/CCL2 converts it from a basophil chemoattractant to a potent stimulator of eosinophil migration (Weber *et al.*, 1996). Thus proteolysis *in vivo* adds an additional dimension of complexity, and a mechanism for regulating the function of these proteins.

Extension of the N-terminus also affects some chemokines. This was discovered as a consequence of expressing RANTES/CCL5 and MCP-1/CCL2 in bacteria, wherein the initiating methionine was retained. Both [Met]-MCP-1 (Hemmerich *et al.*, 1999) and especially [Met]-RANTES (Proudfoot *et al.*, 1996), bind their receptors with high affinity; however, they show little or no ability to induce cell migration. RANTES/CCL5 with a synthetic aminooxypentane group added to the N-terminus ([AOP]-RANTES) is another interesting example (Simmons *et al.*, 1997). This protein is a potent antagonist of HIV entry into cells, in part because it not only has a high affinity for CCR5, but also because it causes desensitization and internalization of the receptor, and prolonged down-modulation of the cell surface expression of CCR5 (Signoret *et al.*, 2000).

In the case of two chemokines with C-terminal tails (e.g., lymphotactin/XCL1 and MIG/CXCL9), truncation studies have shown that these domains are also critical for function (Clark-Lewis *et al.*, 2003; Hedrick *et al.*, 1997). Interestingly, like the N-terminal triggering domain, in XCL1, the C-terminal tail is unstructured.

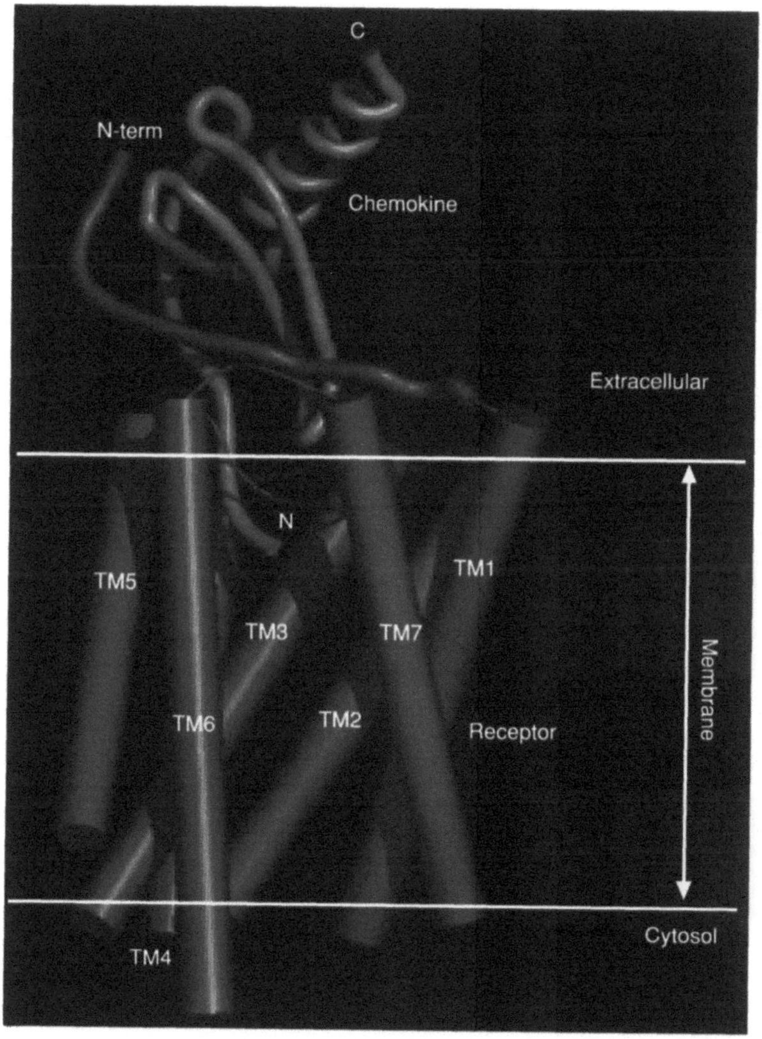

FIG. 4. Schematic illustration of chemokine-receptor interactions. The transmembrane helices of the receptor, shown as blue tubes, were derived from the structure of rhodopsin (PDB code 1L9H). The chemokine (pink) and the N-terminal extracellular domain of the receptor were derived from the IL8-CXCR1 peptide complex shown in Fig. 5 (PDB code 1ILP). Except for the N-terminus, the loops of the receptor are not displayed. The figure illustrates the relative size of the receptor and ligand, and the interaction of the receptor N-terminus along one face of the chemokine. The interaction orients the N-terminal signal domain towards the receptor; as displayed here it is oriented towards the helical bundle, which may or may not be correct for some chemokines. (See Color Insert.)

E. Chemokine-Binding Epitopes on Receptors

In contrast to the ligands, much less is known about the receptors because of the inherent difficulties associated with determining structures of membrane proteins, particularly GPCRs. Several studies involving receptor chimeras, and truncation and point mutants, have implicated a role for the receptor N-termini in ligand binding (Blanpain *et al.*, 2003; Hemmerich *et al.*, 1999; Ho *et al.*, 1999; Katancik *et al.*, 1997; Monteclaro *et al.*, 1997; Quan *et al.*, 1996; Suzuki *et al.*, 1994; Wu *et al.*, 1996; Xanthou *et al.*, 2003). Additionally ligand-binding and NMR structural studies have been done using peptides from receptor N-termini (Booth *et al.*, 2002; Handel *et al.*, 2004; Mayer, Stone *et al.*, 2000; Mizoue *et al.*, 1999; Skelton *et al.*, 1999). In general, it has been observed that affinities between the receptor peptides and chemokines are weak (millimolar to high micromolar) and that the receptor domain in isolation is unstructured. However, receptor binding does occur on the face of the chemokine shown in Fig. 3A, complimenting the mutagenesis studies. In one very elegant study, a peptide analog of the N-terminus of CXCR1 was fused to a hexanoic acid group giving it an IC50 of 13 μM for displacement of IL-8/CXCL8 from the intact CXCR1 receptor (Skelton *et al.*, 1999). The interaction was sufficiently strong to allow solution of the structure of a complex between CXCL8 and the CXCR1 N-terminal analog. The structure shows the receptor lying in an extended fashion along a hydrophobic cleft (Fig. 5). While most of the contacts are hydrophobic, three negatively charged residues of the peptide interact with basic epitopes on the IL-8/CXCL8 surface. As the authors point out, because the peptide lacks the rest of the receptor, and has been modified, it raises the question of the relevance to the interaction with the intact receptor. However, the structure is consistent with all the biochemical data from mutagenesis studies.

It is also worth noting that many of the receptors have predicted tyrosine sulfation sites in their N-termini (Farzan *et al.*, 1999, 2002; Fong *et al.*, 2002; Preobrazhensky *et al.*, 2000). The canonical sequence contains patterns of tyrosines flanked by aspartic acid residues, such as $D_{25}Y_{26}D_{27}Y_{28}$ in CCR2b. Tyrosine sulfation has been positively demonstrated for CCR5, CCR2b, CX3CR1, and CXCR4, and it enhances chemokine binding and HIV entry. These motifs create highly acidic patches on the receptors which complement the basic epitopes on the ligands such as those identified for MCP-1/CCL2 (Fig. 3A). For example, given the effect of the R24E mutation on MCP-1:CCR2 binding (Section II.D), we predicted that the tyrosine sulfation motif of CCR2 interacts with the basic patch involving R24. Similarly, there is a potential tyrosine sulfation site in the CXCR1 N-terminus, and if real, the tyrosine sulfate would be in an electrostatically favorable position for interacting with two basic residues (Lys11 and Lys15) as shown in the CCL8:CXCR1 peptide complex (Fig. 5).

Figure 4 shows a very crude model of a chemokine interacting with its receptor based on the structure of rhodopsin fused to the structure of the CCL8:CXCR1 peptide complex; it represents the current but unfortunately relatively minimal level of understanding of the receptors. The key features to note are the interaction of the receptor N-terminus with the chemokine, and the positioning of the chemokine N-terminal signaling domain towards the receptor, although the contacts between the signaling domain and the receptor are largely unknown. However, in a recent study, the N-terminus of MIP-1α was labeled with benzophenone, a photoaffinity cross-linking agent, and used to covalently label CCR1 (Zoffmann *et al.*, 2002). After enzymatic and chemical cleavage of the

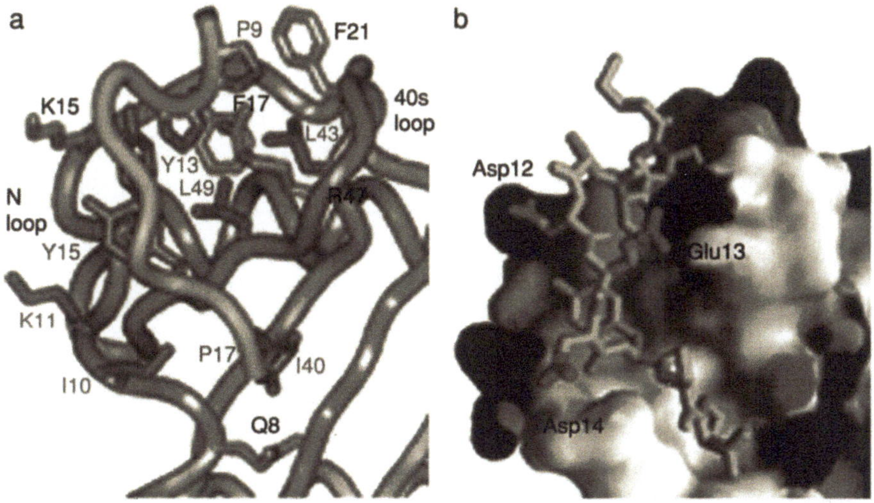

FIG. 5. View of the structure of a modified peptide fragment from CXCR1 in complex with IL-8/CCL8. (A) The monomeric subunits of IL-8 are colored pink and grey, the peptide is cyan. The IL-8 residues surrounding the binding cleft and the peptide residues buried in the interface are shown as stick representations. (B) IL-8/CXCL8 is colored according to its electrostatic potential. The peptide is cyan with the terminal atoms of three acidic residues colored red. Note the interactions of the peptide within the hydrophobic pocket of IL-8 and those between the peptide acidic groups and the basic residues of IL-8/CXCL8. Whether or not Y15 is tyrosine sulfated is unclear, but Y15 is imbedded in a canonical tyrosine sulfation motif. The orientation is the same as in "a". Reprinted from *Structure* with copyright permission from Elsevier: "Skelton, N. J., Quan, C., Reilly, D., and Lowman, H. (1999). Structure of a CXC chemokine receptor fragment in complex with interleukin-8. *Structure* **7,** 157–168." (See Color Insert.)

complex, it was possible to show that the MIP-1α/CCL3 N-terminus interacts with the second extracellular loop between transmembrane helix 2 and 3, within a region spanning amino acids 178–194. This is in contrast to the commonly held notion that the N-terminal signaling domain inserts into the helices, as is thought for small molecule antagonists of chemokine receptors (Dragic *et al.*, 2000; Mirzadegan *et al.*, 2000; Onuffer *et al.*, 2002, 2003). On the other hand, in a study of MIP-1α/CCL3 and RANTES/CCL5 binding to CCR5, it was concluded that the N-termini of the chemokines interact with the transmembrane helical bundle (Blanpain *et al.*, 2003). If both studies are accurate, it suggests variability in the orientation of the signaling trigger amongst different chemokine-receptor complexes, which could contribute to differences in signaling output.

Given the relatively low (micromolar-millimolar) affinity of N-terminal receptor peptides for the chemokines, and the size of the ligands relative to the receptor (Fig. 4), it is likely that other extracellular loops are involved. Again, this has been borne out by studies using receptor chimeras, and by mutagenesis studies of several receptors. However, to truly understand how these proteins function at the molecular level, high resolution structural information will be required. Two interesting approaches have been reported for exploring the role of other extracellular domains of the receptor: (a) linking the extracellular domains together, and (b) finding other scaffolds suitable for their presentation. In the latter study, the N-terminal domain and the third extracellular loop of CCR3 were fused to the N-terminus and a loop region of the soluble B1 domain of Streptococcal protein G, to orient them in approximately the same way as expected for the intact receptor (Datta *et al.*, 2003). The design of the so-called "CROSS" is illustrated in Fig. 6. After optimization, micromolar (2.9 μM) binding affinity was observed between eotaxin/CCL11 and the CROSS. This affinity was significantly improved over the 135 μM and 179 μM affinities determined for constructs containing only the third extracellular loop or the N-terminus, respectively. Importantly, affinities of CCL11 mutants for the soluble GPCR mimic correlated with those found in the intact receptor.

Along the same lines, another group grafted the extracellular domains of CCR5 onto bacteriorhodopsin (BR) (Abdulaev *et al.*, 2002). Amazingly, addition of just the N-terminal domain conferred significant HIV coreceptor activity, comparable to that of the intact receptor as monitored by cell fusion assays. Whether any of the CCR5-BR chimeras also bound the CCR5 ligands, was not addressed. However, since BR is easily expressed, purified, and has been crystallized at high resolution, in principle this approach could lead to chimeras conductive to crystallization, and further insight into recognition of chemokines by their receptors.

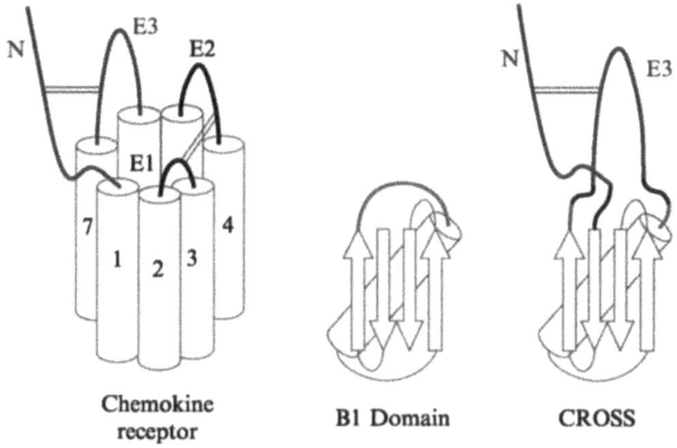

FIG. 6. Schematic representation of the CROSS protein design. Left. Predicted topology of chemokine receptors. Membrane spanning helices are shown as cylinders, the extracellular loops are displayed as lines, and the disulfide bonds between the extracellular domains are double lines. Center. The B1 domain of protein G (PDB code 2GB1). β-strands are shown as arrows and the helix as a cylinder. The loop to be replaced by extracellular loop three (E3) of CCR3 is highlighted in green. Right. CROSS protein showing the chemokine receptor elements (red) displayed on the B1 scaffold. The N-terminus is attached via a $(Gly)_3$ linker (blue). The E3 loop from the chemokine receptor (red) is inserted between 2nd β-strand and the α-helix with a $(Gly)_2$ linker at each junction. Reprinted from *Protein Science* with copyright permission from Cold Spring Harbor Laboratory Press: "Datta, A., and Stone, M. (2003) Soluble mimics of a chemokine receptor: Chemokine binding by receptor elements juxtaposed on a soluble scaffold. *Protein Sci.* **12,** 2482–2491." (See Color Insert.)

F. Receptor Dimerization

Another notable feature of chemokine-receptor structural biology is that several receptors (CCR2, CXCR4, and CCR5) have been reported to undergo dimerization (Mellado *et al.*, 2001, 2002). Dimerization of GPCRs is widely accepted, however, the functional role remains somewhat controversial, especially in the chemokine field (Thelen *et al.*, 2001). Nevertheless, dimerization is common to many cytokine activated receptors (e.g., tyrosine kinase) and other GPCR classes (Bai, 2004; Breitwieser, 2004; Terrillon *et al.*, 2004). In principle, it could provide a tremendous mechanism for creating diversity in chemokine signaling and functional output.

The classical view of GPCR signal transduction involves activation of trimeric G-proteins, and in the case of chemokine receptors, Gαi is key, as indicated by their sensitivity to pertussis toxin. However, evolving theories of dimerization-induced signal transduction add an intermediate step involving JAK/STAT pathways (Rodriguez-Frade *et al.*, 2001). The currently proposed mechanism is that agonist binding causes receptor dimerization and phosphorylation of a tyrosine in a DRY motif, conserved in the second intracellular loops of many chemokine receptors. Janus kinases (JAKs) are then recruited to the receptor followed by STAT transcription factors, and as a result of subsequent conformational changes, activation of Gαi ensues. Notably, a variant of CCR2b, containing a Y139F mutation in the DRY box, retained the ability to dimerize, but it was incapable of promoting Gαi mediated effects such as calcium mobilization and chemotaxis. This result is thought to be due to the inability of the mutant to be tyrosine phosphorylated, activate the JAK/STAT pathway, and in turn the G-protein. The implication is that G-protein activation is dependent on dimerization and the JAK/STAT pathway, at least in the case of the receptors studied thus far.

Sequence analysis and molecular modeling has also been used to identify potential amino acids involved in the dimerization of CCR5 (Hernanz-Falcon *et al.*, 2004). In this study, a large region of TM1 and TM4 was predicted to comprise the dimer interface with I52 and V150 making particularly important contacts (Fig. 7). To test this hypothesis, a double

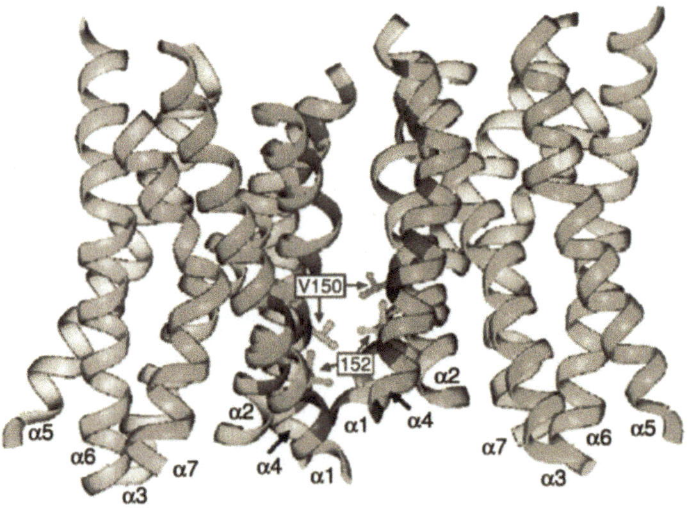

FIG. 7. Structural model of CCR5 dimer association presented as a ribbon representation. TM1, TM2, and TM4 (labeled $\alpha1$, $\alpha2$ and $\alpha4$) participate directly in the interaction surface. The positions of the mutated residues, V150 and 152, are indicated on each of the subunits. Reprinted from *Nature Immunology*, with copyright permission from the Nature Publishing Group: "Hernanz, P., *et al.* (2004). Identification of amino acid residues crucial for chemokine receptor dimerization. *Nature Immunology* **5**, 216–223." (See Color Insert.)

mutant of CCR5 was generated. Although it expressed as well as the WT receptor, the mutant could not dimerize or trigger signaling. Importantly, synthetic peptides encompassing I52 and V150 functioned as dominant negative inhibitors and blocked WT CCR5 activation.

In addition to homodimerization, heterodimerization has also been observed between CCR5 and CCR2, leading to increased sensitivity to the ligands and activation of unique signaling pathways (Mellado, Rodriguez-Frade *et al.*, 2001). Heterodimerization is not surprising given the homology between receptor transmembrane domains. While further studies are required to unequivocally establish relevance (Thelen and Baggiolini, 2001), in principle, different modes of oligomerization, which may not be restricted to dimers or members of the chemokine receptor family, could add significantly to the complexity of chemokine pharmacology, ranging from ligand recognition to signaling specificity and receptor trafficking. Receptor oligomerization may also need to be considered in drug discovery programs since it could affect binding of small molecule antagonists.

III. Interaction of Chemokines and Glycosaminoglycans

In addition to binding GPCRs, chemokines have been hypothesized to bind to cell surface GAGs (Middleton *et al.*, 1997; Rot, 1993). In theory, in the absence of such an interaction, chemokines would diffuse away from the site of production, become diluted to concentrations below the threshold for binding, thereby masking directional information for cell recruitment. Indeed, chemokines have been shown to bind to purified glycosaminoglycans and to endothelial cell surface GAGs *in vitro* (Hoogewerf *et al.*, 1997; Kuschert *et al.*, 1999) and *in vivo* (Middleton *et al.*, 1997; Rot, 1993). From a structural perspective, GAG-binding might be anticipated since chemokines tend to be highly basic proteins, and GAGs are long linear polysulfated carbohydrates. However, until recently, there was no formal proof that GAG-binding was functionally relevant.

A. *The Relevance of GAG Binding and Oligomerization*

To investigate this issue, GAG-deficient mutants of three chemokines (RANTES/CCL5, MCP-1/CCL2 and MIP-1β/CCL4) were designed that have severely impaired GAG binding, but retain the ability to induce robust chemotaxis *in vitro*. These mutants were then used to test the relevance of GAG binding *in vivo* using an intra-peritoneal recruitment assay in mice (Proudfoot *et al.*, 2003). In all three cases, whereas the wild type proteins caused significant cell migration into the peritoneum, the mutants did not recruit, suggesting that the GAG binding is indeed important for function.

Another hypothesis that remained untested for years, is that oligomerization of chemokines might be related to GAG binding. Accordingly, in the *in vivo* study, monomeric mutants of the same three chemokines were also tested in the peritoneal recruitment model (Proudfoot *et al.*, 2003). These proteins, [P8A]-MCP1, [P8A]-MIP-1β (discussed in Section II.C), and an N-methylated variant of RANTES, Nme-^{7}T-RANTES, were all capable of *in vitro* chemotaxis that was as efficacious and potent as the WT proteins. Nevertheless, they could not induce migration *in vivo*, suggesting that in addition to GAG-binding, oligomerization is required for function *in vivo*.

Functional coupling between GAG binding and oligomerization has been demonstrated in several ways. First, it was shown that several chemokines (RANTES/CCL5, MCP-1/CCL2, IL-8/CXCL8, and MIP-1α/CCL3) oligomerized on heparin sepharose beads by the accumulation of ^{125}I-labeled chemokine with increasing concentration of cold chemokine (Hoogewerf *et al.*, 1997; Proudfoot *et al.*, 2003). Secondly, addition of size-defined heparin as small as an octasaccharide, induced formation of MCP-1/CCL2 tetramers as demonstrated by analytical ultracentrifugation and crosslinking studies (Lau *et al.*, 2004). The role of tetramers as a minimal aggregation state for *in vivo* function was also suggested by results of the intra-peritoneal recruitment assay with RANTES/CCL5 mutants that could only form monomers, dimers or tetramers (Nme-^{7}T-RANTES, E66S-RANTES, and E26S-RANTES, respectively) (Proudfoot *et al.*, 2003). All three of these proteins had WT chemotaxis promoting abilities *in vitro*, but when tested *in vivo*, only the tetramer caused cellular recruitment. These data raised the question, what do the tetramers and/or other oligomeric forms look like?

B. Proposed Structure of the GAG Bound State of MCP-1/CCL2

Previously, the structure of MCP-1/CCL2 was solved by NMR and subsequently by X-ray crystallography using the NMR structure as a starting model for molecular replacement (Lubkowski *et al.*, 1997). As described in Section II.B, the solution structure revealed a dimer similar to dimerization motifs observed for other CC chemokines. In the case of the X-ray structure, two different crystal forms were isolated from the same droplet. One form contained a dimer similar to the solution structure (Fig. 2B), while the other form contained a tetramer as shown in Fig. 8. In the tetramer, there is a primary CC dimer interface that consists of the

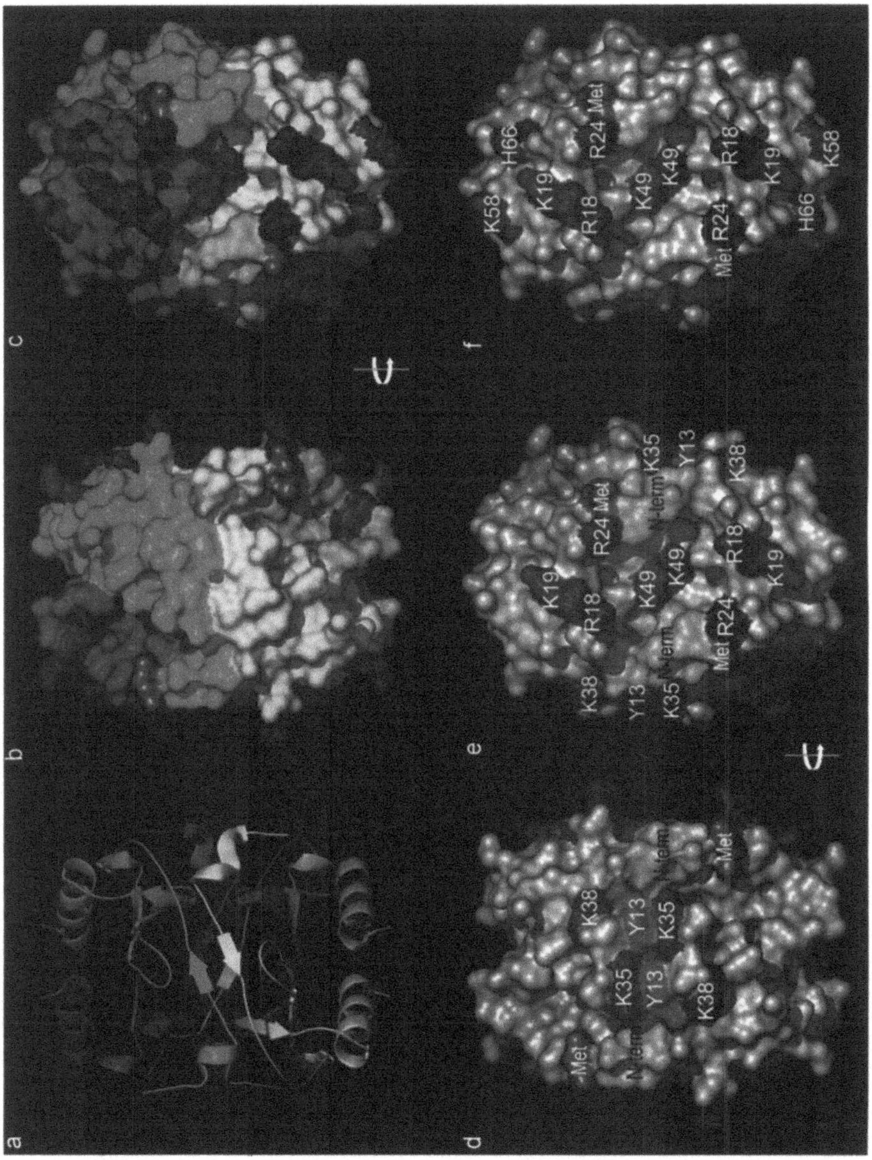

FIG. 8. *(continued)*

subunits colored green/yellow, and red/blue, respectively. A second CXC-like dimer interface is observed between the red/green, and yellow/blue subunits, respectively (Fig. 8A and B). Thus this structure contains elements of both CC and CXC dimers. Finally, a third interface is made by interactions between all subunits (Fig. 8C).

This structure most likely represents the tetramer of MCP-1 induced by the interaction with heparin. When mapped onto the tetramer, the GAG-binding hotspots form a continuous pattern of basic residues that effectively encircle the tetramer, providing an interaction surface well-suited for binding long linear polysulfated glycosaminoglycans like heparin (Fig. 8B and C) (Lau *et al.*, 2004). The binding site is formed only when all four subunits come together (Fig. 8C), giving a plausible explanation for how GAG binding induces tetramer formation. The dimensions of the binding site are also consistent with the fact that an octasaccharide is the shortest GAG capable of inducing tetramer formation (for calibration, a dodecasaccharide is approximately the length of the binding site visible in Fig. 8C).

In support of this structural model, Platelet Factor 4 (PF4/CXCR4) forms a similar tetrameric structure, stable on its own in solution (Zhang *et al.*, 1994). Based on electrostatic surface calculations, it is even more basic than MCP-1, and in the tetrameric state has a similar putative binding surface. Furthermore, PF4 and other chemokines like IL-8 have been reported to have the highest affinity for long, >dp20, saccharide chains, which could effectively wrap around the tetramer (dp = degree of polymerization) (Kuschert *et al.*, 1999; Stringer *et al.*, 1997).

FIG. 8. Structural views of the MCP-1/CCL2 tetramer (PDB code 1DOL) with each subunit color-coded red, green yellow, or blue, and functional epitopes highlighted as described: (A) Ribbon diagram showing the CC dimer subunits in front and back (green/yellow and red/blue). Green/red and yellow/blue subunits interact through a CXC-like motif at the top and bottom of the structure. (B) Surface topology model of the MCP-1/CCL2 tetramer in the same orientation as "a". GAG binding epitopes are colored in purple. (C) Same as "b", but rotated 90° to show the unique interface involving all four subunits. The GAG binding epitopes form a continuous basic strip when the tetramer is formed. (D) Same orientation as "a" and "b" but with the receptor (CCR2) binding epitopes highlighted. The N-terminal signal residues are more protected, and Y13, also required for chemotaxis, is largely buried, compared to the monomer structure of Figure 3a. (E) Same orientation as "c", but with the receptor binding epitopes highlighted along the interface involving all four subunits for comparison to the GAG binding site in "f". (F) Same orientation as "e", but with GAG binding residues highlighted. Note there is considerable overlap between the GAG and receptor binding sites. Thus in the context of a presentation function, oligomerization could enable simultaneous interaction with GAGs and the receptor. The burial of the signaling triggers could reflect an inhibitory function of GAG binding as has been observed *in vitro*, or protection from proteolysis. (See Color Insert.)

Interferon–γ-inducible protein 10 kDa (IP-10/CXCL10) is a chemokine that in solution, exists in equilibrium between monomer and dimer. A solution structure of a monomeric variant has been solved by NMR (Booth *et al.*, 2002). However, like MCP-1, it has also been crystallized in different space groups, revealing tetrameric oligomerization states (Swaminathan *et al.*, 2003). One of the forms is similar to the PF4 and MCP-1 tetramers. The other two tetramers form a novel twelve-stranded β-sheet containing structure (Fig. 9), which the authors postulate could represent structures induced by binding of different GAGs.

Given that GAG binding is a critical component of chemokine function, the message emerging from these studies is that yet another level of complexity and diversity in chemokine function may be derived from interactions with GAGs. Indeed, chemokines show substantial differences in their affinities for the various types of GAGs (Kuschert *et al.*, 1999; Patel *et al.*, 2001). Figure 3B–D shows the GAG binding epitopes of MCP-1, MIP-1β and RANTES, mapped onto the structure of the monomer. In the case of IL-8/CXCL8, the GAG binding site is in the C-terminal helix (not shown), and would not be visible in the orientation presented in the figure. Thus there is considerable epitope variation at the level of tertiary structure (Proudfoot *et al.*, 2003). In addition, different oligomerization states (dimer, tetramer, other) may add to the specificity of the GAG interaction, as postulated for IP-10/CXCL10 and MCP-1/CCL2. RANTES/CCL5 may also adopt structurally distinct oligomeric form(s) that favor interaction with specific GAGs because if one does a least squares fit of the RANTES/CCL5 monomer onto the MCP-1/CCL2 or PF4/CXCL4 tetramer, the distribution of the GAG binding epitopes is entirely different (Handel, unpublished). MIP-1α also oligomerizes extensively in solution, but it has been reported to preferentially bind heparan sulfate as a dimer (Stringer *et al.*, 2002). This finding makes structural sense because heparan sulfate is a less sulfated GAG than heparin, and contains regions of low sulfation separated by patches of high sulfation that complement the distribution of basic GAG binding epitopes in the context of the MIP-1α dimer. For similar reasons, although heparin induces tetramer formation of MCP-1, it would not be surprising if the dimeric form was involved in binding to heparan sulfate (Lau *et al.*, 2004).

C. *Functional Implications*

While the structural biology of chemokine:GAG interactions is only beginning to emerge, a quote from a review on heparan sulfate glycosaminoglycans (HSGAGs) adds perspective on how diverse these interactions may be, compared to other biopolymers (Shriver *et al.*, 2002):

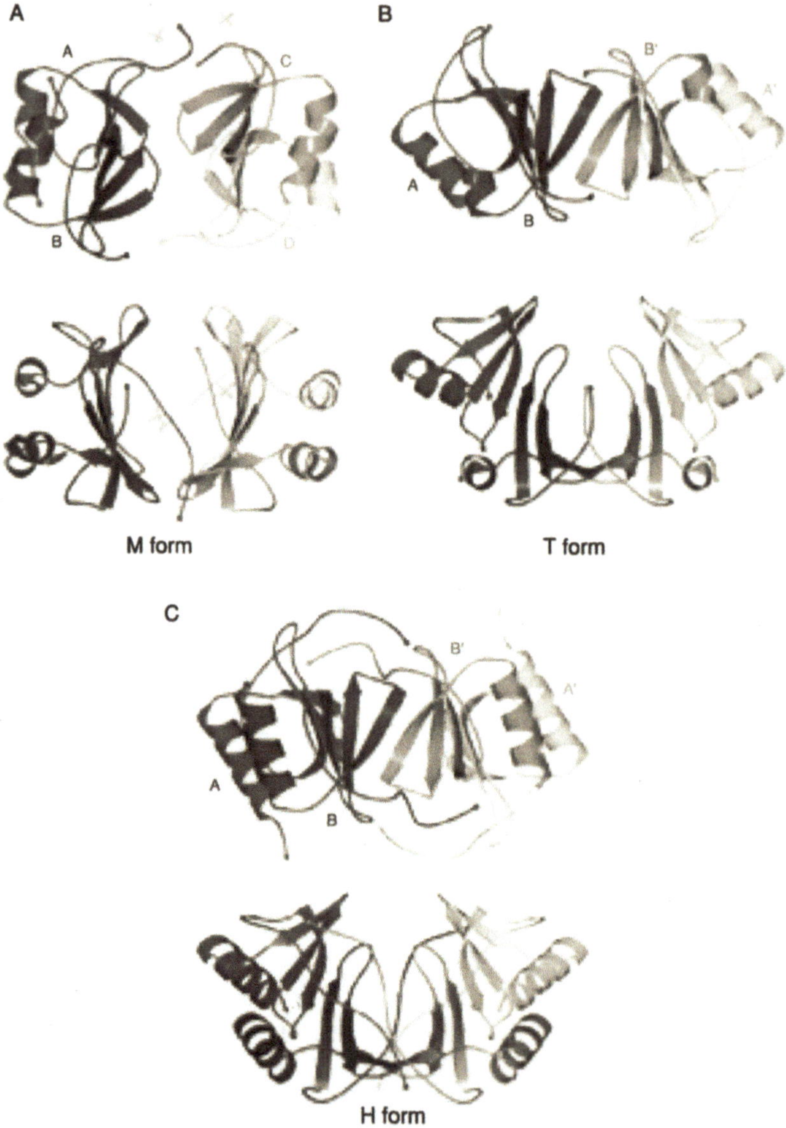

FIG. 9. Ribbon structures of the three IP-10/CXCL10 tetramers. Two orthogonal views are shown for each. The M-form is similar to the PF4/CXCL8 and MCP-1/CCL2 structures that contain elements of CXC and CC dimers. The T and H forms contain a novel 12-stranded β-sheet. Reprinted from *Structure* with copyright permission from Elsevier. "Swaminathan, G. *et al.* (2003). Crystal structures of oligomeric forms of the IP-10/CXCL10 chemokine. *Structure* **11,** 521–532." (See Color Insert.)

> "For DNA, there are 4^6 or 4096 possible sequences. In contrast, for a hexapeptide there are many more possibilities, 20^6 or 64 million possible permutations. However for HSGAGs, a polymer of six disaccharide units could have a total of over 12 billion possible sequences, a staggering degree of variation...!"

Given this level of complexity, chemokine:GAG interactions may play an enormous role in cellular localization, both in normal physiological processes as well as pathophysiological processes such as inflammation and cancer. For example, it is now recognized that cells, such as tumor cells, dynamically regulate their carbohydrate surfaces to modulate extracellular signals; these include growth promoting and growth inhibiting carbohydrate sequences that can be immobilized on the cell surface through protein cores, or shed (Liu *et al.*, 2002; Sasisekharan *et al.*, 2002). Along the same line, a comparison of the receptor binding-epitopes of MCP-1/CCL2 (Fig. 3A [monomer] and 8D, E [tetramer]) to the GAG-binding epitopes (Fig. 3B [monomer] and 8F [tetramer]) suggests other potential roles for GAG binding beyond localization and presentation. In the context of the monomer, the signaling triggers (Y13 and the N-terminus) are exposed and available for interaction with the receptor (Fig. 3A). By contrast, in the tetramer (and in fact the dimer) they are much more buried (Fig. 8D). Thus GAG binding could serve as a negative regulator of chemokine function. It could also provide a mechanism for protecting the chemokine N-terminus from proteolytic degradation as discussed in Section II.D. Nevertheless, at this point, the extent of the functional implications and the structural diversity of chemokine:GAG interactions are largely unknown. Thus chemokine:GAG interactions comprise a whole new paradigm of chemokine structural biology that remains to be thoroughly explored. That these interactions are fundamentally important for chemokine function is underscored by the fact that viruses and other pathogens target these interactions as a mechanism to subvert the host immune system (Section IV). For example, certain bacterial pathogens such as Yersinia and Staphylococcus recruit heparin and other sulfated polysaccharides as universal binding sites for interfering with a wide variety of immunologically important proteins, including chemokines (Duensing *et al.*, 1999).

IV. Viral Chemokines and Receptors, Chemokine Mimics, and Chemokine Binding Proteins

Viruses, especially large DNA viruses like poxviruses and herpesviruses, have evolved several mechanisms to evade, subvert, and manipulate the host immune system, ensuring their survival and spread (Murphy, 2001; Nash *et al.*, 1999; Seet *et al.*, 2003; Wells *et al.*, 1997). Not surprisingly, within their arsenal of survival tactics are mechanisms to both utilize and interfere with the chemokine system (Murphy, 2001). HIV uses host chemokine receptors, especially CXCR4 and CCR5, as portals into cells, eventually depleting the $CD4^+$ T-cell population in late stage AIDs disease. Other viruses have captured host chemokines and chemokine receptors, and structurally optimized them for their own purposes. These include receptors that act broadly to scavenge host chemokines, and chemokines that act as host receptor antagonists. In addition to immune evasion mechanisms, certain viral chemokine-related proteins are involved in more dynamic ways to enhance cell to cell transfer (e.g., cytomegalovirus US28), and to coordinate cellular reprogramming by activating or silencing certain signaling pathways (Rosenkilde *et al.*, 2001). For example, ORF74 from herpesvirus 8 has high constitutive signaling properties, which results in secretion of VEGF (vascular endothelial growth factor). It is also involved in angiogenesis through interaction with angiomodulatory chemokines. These properties contribute to its transforming capabilities, which include the induction of the highly vascularized Karposis's sarcoma-like lesions in transgenic mice (Holst *et al.*, 2001).

A. *Viral Chemokine Binding Proteins*

In addition to chemokine/receptor related genes, many viruses produce chemokine binding proteins (CBPs) that have no homology to chemokine receptors, yet have an extraordinary capacity to scavenge a broad range of chemokines with very high affinity, and block their interaction with host receptors and GAGs. M3 from murine-γ-herpesvirus 68 is currently the only published structure of a viral CBP in complex with a human chemokine (MCP-1/CCL2), but it recapitulates in a remarkable way, many of the interactions of chemokines that have been described in the previous sections of this review (Alexander *et al.*, 2002). Because this complex will be described elsewhere in this volume in great detail, the structural details are discussed here only briefly to point out how thoroughly the virus blocks chemokine function.

M3 has two domains: an N-terminal domain and a C-terminal domain, each composed of elaborated β-sandwiches with closest structural homologies to the C-terminal receptor binding domain of diptheria toxin, and the V-type Ig-fold, respectively. It exits in solution as a head to tail dimer, which leads to the formation of a cleft between the N-terminal domain of one subunit and the C-terminal domain of the other. Chemokines bind within each cleft, thus forming a 2:2 complex. Figure 10A shows a surface

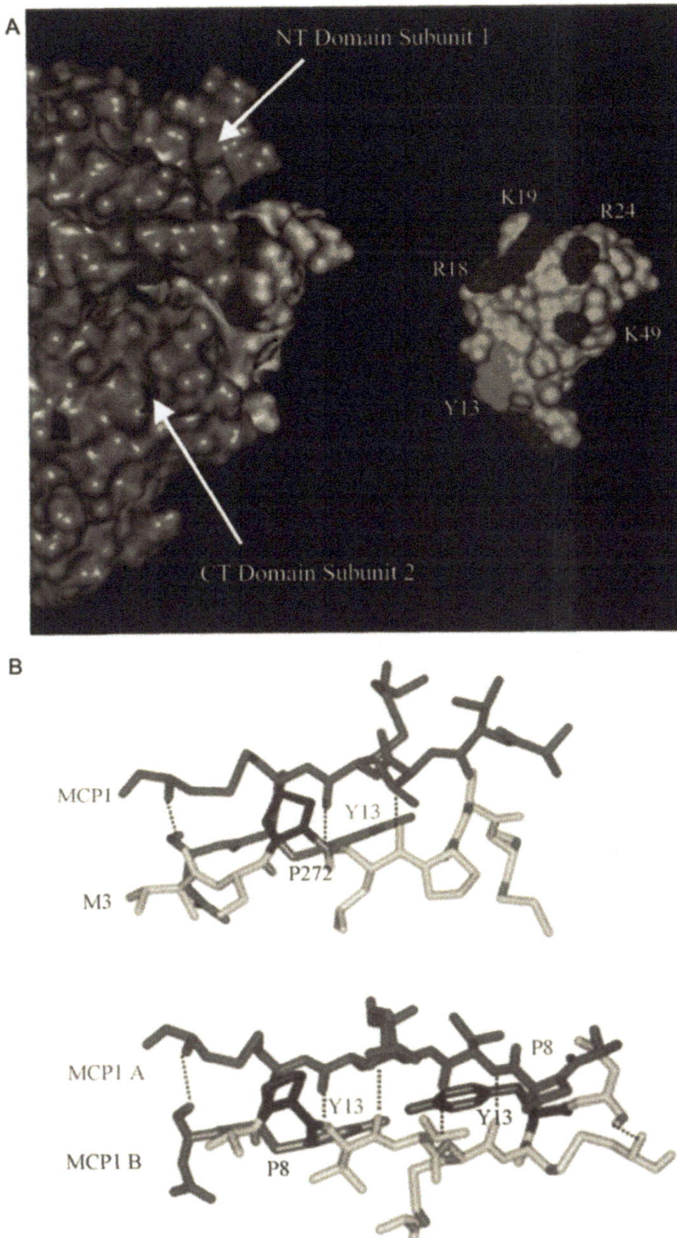

FIG. 10. (*continued*)

topology model of the M3:MCP-1 complex, focused in on one of the clefts. MCP-1 has been duplicated and pulled apart to reveal the binding surface. It is exactly the same surface as displayed in Fig. 3A, and M3 uses all of the same epitopes for engagement as identified from the mutagenesis study of MCP-1 (Section II.D). The N-terminal domain of M3 is especially acidic and binds to the most basic patch of MCP-1 involving R18, K19, R24 and K49. Additionally, it makes extensive hydrophobic contacts with the adjacent hydrophobic pocket, similar to where the N-terminal CXCR1 peptide binds IL-8 (Fig. 5). The C-terminal domain is particularly intriguing. It forms a β-strand interaction with MCP-1 that is structurally analogous to the strand formed in the MCP-1 homodimer (Fig. 10B). This interaction buries the signaling trigger, Y13, and abuts the other CCR2 binding epitopes, K35 and K38. Thus, without any structural homology, it effectively blocks the interaction with the host receptor by completely masking the core domain, it also simultaneously blocks interactions with GAGs, and it prevents chemokine oligomerization, all critical features of chemokine function.

Another remarkable aspect of this protein is its ability to bind many different chemokines across all four families with nanomolar affinities (van Berkel *et al.*, 2000). While no common chemokine sequence motifs have been identified thus far, the promiscuity of M3 has been partly attributed to structural plasticity as a consequence of its oligomeric framework, which can undergo structural rearrangements, and the use of flexible loops as primary contact regions. Future studies of additional M3:chemokine complexes will undoubtedly provide further insight into the brilliant design of this highly evolved viral stealth weapon.

FIG. 10. (A) Surface topology model of γ-herpesvirus M3 in complex with human MCP-1/CCL2. Only $\sim$1/2 of the structure is displayed, focusing in on one of the chemokine binding clefts. M3 is colored purple, MCP-1/CCL2 is cyan with CCR2 receptor binding epitopes, defined by mutagenesis, highlighted in green (Y13) and blue (basic Lys of Arg residues). The N-terminal domain of M3 interacts with the basic patch at the top of MCP-1, as well as the adjacent hydrophobic region immediately below (see also Fig. 3a). The C-terminal domain of M3 pairs with the dimer interface of one subunit of MCP-1/CCL2, forming a β-strand similar to that observed in the MCP-1 homodimer. Note Pro272 of M3; it occupies a structurally homologous orientation as the Pro8 in the MCP-1 dimer that is required for homodimerization. (B) (bottom) Stick model of the dimer interface in MCP-1. One subunit is pink with P8 and Y13 sidechains highlighted in red. The other subunit is cyan with P8 and Y13 sidechains highlighted in dark blue. (top) M3:MCP-1 interactions in the C-terminal domain. MCP-1 is pink with Y13 highlighted in red. M3 is cyan with P272 highlighted in dark blue. (See Color Insert.)

B. Viral Chemokine Mimics: HIV-1 TAT

A second example of molecular mimicry that impinges on the chemokine field involves the HIV protein, TAT (Albini *et al.*, 1998; Brigati *et al.*, 2003; Noonan *et al.*, 2000; Vene *et al.*, 2000). TAT was originally discovered as a transcriptional transactivator of the HIV LTR, but it also transactivates many host cellular genes including cytokines, class I MHC, extracellular matrix proteins, and proto-oncogenes (Vene *et al.*, 2000). However, in addition to its role in the nucleus, it can be released from cells, and is known to interact with cell surface receptors involved in cell movement: integrin adhesion proteins, the VEGF receptor, and apropos to this review, chemokine receptors (CCR2, CCR3, and CXCR4). TAT also binds to CD26, a protease involved in modulating chemokine activity as described in Section II.D. All of these molecular interactions are thought to greatly contribute to immune suppression of the host and spread of HIV. In particular, it has been proposed that these interactions could induce or augment angiogenesis, tumorigenesis, and metastasis (Benelli *et al.*, 2000; Brigati *et al.*, 2003; Vene *et al.*, 2000). With respect to chemokine receptor interactions, TAT induces migration of many cells types including endothelial cells, monocytes and T-cells, by acting on chemokine receptors (Vene *et al.*, 2000), and stimulating the production IL-8/CXCL8 (Benelli *et al.*, 2000). Among other things, its ability to stimulate cell migration has been proposed to mobilize uninfected cells to the TAT-secreting infected cells, in order to facilitate further infection. It has also been proposed that TAT may be involved in controlling the transition from R5 (CCR5 using) to X4 (CXCR4 using) strains of HIV, where X4 infections are thought to accelerate $CD4^+$ T-cell depletion and the onset of disease (Ghezzi *et al.*, 2000; Murphy, 2001).

Despite the diversity of its interactions, TAT is only a small 86–101 amino acid protein (Brigati *et al.*, 2003). Its structure has been qualitatively divided into four regions: a cysteine rich domain, a basic domain, the core, and a domain containing a RGD sequence. The RGD sequence is involved in binding to integrins. The chemokine-binding domain resides in the cysteine-rich core domain (residues 24–51), which is also quite basic, and has been described as the "chemokine-like" region. Interestingly, within this small 28-residue peptide, there is a CCF motif that is crucial for its chemotactic properties; this motif is analogous to CCF motifs in several CC chemokines and corresponds to the Y13 signal trigger in MCP-1/CCL2 (the CCL2 sequence is $C_{11}C_{12}\mathbf{Y}_{13}$, but $C_{11}C_{12}\mathbf{F}_{13}$ also functions like WT in MCP-1/CCL2 [Hemmerich *et al.*, 1999]). Apart from these overall sequence similarities, by standard sequence comparisons there is very little homology. It has been reported as a largely unstructured protein in solution and thus structural plasticity probably plays a role in it ability to encode so many functions within such a small sequence. Considering the simplicity of HIV (15 proteins and an RNA), it seems that HIV has done an amazing job to evolve TAT to function incredibly efficiently, including mimicry and induction of several chemokines.

V. Chemokines and Disease

In addition to the role of chemokines in viral life cycles and viral disease, inappropriate expression of either receptor or chemokine has now been linked to a large number of classically defined inflammatory disease such as asthma/allergic inflammation, rheumatoid arthritis, and transplant rejection (for excellent reviews on chemokines implicated in various diseases, see Baggiolini, 2001; D'Ambrosio *et al.*, 2003; Gerard and Rollins, 2001; Horuk *et al.*, 2000; and Houshmand *et al.*, 2003). This is not surprising given the potential tissue damage that can result from chemokine-induced effector functions. However, other diseases also involve chemokines including cardiovascular disease, neuroinflammation (multiple sclerosis, Alzheimer's, HIV-induced neurotoxicity), and cancer. Two examples are given to illustrate mechanisms by which chemokines contribute to disease progression. Subsequently we review possible strategies for interfering with chemokine function that are based on our present understanding of chemokine structural biology.

A. *Atherosclerosis and Cancer*

Atherosclerosis is a particularly instructive example, with compelling evidence of the pathogenic role of chemokines from animal models and knockout mice (Gerard, Rollins *et al.*, 2001). MCP-1/CCL2 and its receptor CCR2 are the chemoattractants most implicated in atherosclerosis (Aiello *et al.*, 1999; Boring *et al.*, 1998; Gosling *et al.*, 1999; Gu *et al.*, 1998; Rollins, 2001). Using a rodent model of atherosclerosis, it was shown that LDL receptor/MCP-1-deficient mice had 83% less lipid deposition throughout their aortas, and a correspondingly lower aortic macrophage content, than controls fed a high cholesterol diet (Gu *et al.*, 1998). Likewise, atherosclerotic lesions were markedly decreased in CCR2−/− mice (Boring *et al.*, 1998).

Current theories are that arterial damage caused by hypertension and/or hypercholesterolemia induces release of MCP-1/CCL2 from vascular endothelium or smooth muscle cells (Gerard and Rollins, 2001). The presence of chemokine then causes migration of CCR2-bearing monocytes into the subendothelium where they differentiate into macrophages and take up permanent residence. Ingestion of lipid and cholesterol converts them into foam cells, thereby contributing to the fatty deposition. Continued production of cytokines such as growth factors and/or chemoattractants by the endothelial and foam cells may also contribute to further cellular infiltration and vessel wall thickening. Thus atherosclerosis is in part a progressive inflammatory disorder.

The role of chemokines in cancer has only recently begun to be appreciated (Murphy, 2001; Vicari *et al.*, 2002; Zlotnik, 2004). Often beginning as genetic abnormalities of growth control and apoptosis, chemoattractants may play key roles in survival and spread by shaping the tumor microenvironment, and by promoting metastasis and possibly "organogenesis" at secondary sites (Ben-Baruch, 2003). The original investigations of a role of chemokines in cancer were brought about by the observation that leukocytes, particularly tumor associated macrophages (TAMs) and neutrophils, concentrate within the tumor environment (Sica *et al.*, 2002). The present hypothesis is that tumors secrete chemokines to recruit these cells, and rather than providing anti-tumor responses, the leukocytes provide rich sources of cytokines, growth factors and proteases. Among these proteins are angiogenic chemokines that promote vascularization (e.g., the ELR chemokines like IL-8/CXCL8), and matrix metalloproteases that may aid metastasis by remodeling the extracellular matrix.

Recently it was demonstrated that certain chemokine receptors (e.g., CXCL4/CCR7) are highly expressed in breast cancer cells, and that their respective ligands (CXCL12/CCL21) are found at very high levels in tissues known to be the primary sites of metastasis (Muller *et al.*, 2001). The tissue specific expression of these chemokines provides one possible explanation for the clearly non-random nature of metastasis. Importantly, neutralizing antibodies against CXCL12/CCR4 significantly impaired metastasis to lymph node and lung in mice (Muller *et al.*, 2001). If this correlation holds true in humans and also for other cancers, it suggests that targeting chemokines may provide revolutionary treatments by interfering with the metastatic process, the main cause of death in cancer patients (Dowsland *et al.*, 2003).

B. Strategies for Disrupting Chemokine Function for the Treatment of Disease

From what is currently known about various aspects of chemokine structural biology, several types of strategies can be envisioned for interfering with chemokine function. The most widespread approach involves screening for small molecule antagonists that block chemokine binding and activation of their receptors (Horuk, 2003; Onuffer and Horuk, 2002). These compounds are typically competitive inhibitors that bind within the transmembrane helices. Particularly useful strategies for defining the ligand-binding domains and modeling the agonist-receptor interaction have exploited comparisons of antagonist potencies of the same receptor from different species (Onuffer *et al.*, 2003).

Non-competitive small molecule inhibitors that prevent conformational changes required to activate downstream signal transduction might also be therapeutically valuable. For example, if it does prove to be a critical phenomenon for receptor activation, blocking receptor dimerization with small molecules could be an effective approach for receptor antagonism.

While small molecule antagonist approaches are historically favored, there are clear challenges. For one, cross-reactivity with other GPCRs is a problem; this is not unanticipated given the highly hydrophobic nature of GPCR transmembrane helices. Conversely, there is a significant lack of cross-species reactivity, which precludes pre-clinical testing in common (rodent) animal models (Horuk, 2003). Alternative strategies include replacing the rodent receptors with human forms, but this assumes that all the pharmacology would be identical, or the use of primate models, which is expensive (Horuk, 2003). Fortunately, protein and peptide based therapeutics are also options, especially those based on the chemokines themselves, because they are guaranteed to have a high degree of specificity. Indeed, knowledge of the localization of the activation trigger to the N-terminus of chemokines has guided the production of many mutant chemokine antagonists of receptors that function well *in vivo.* For example, 7ND (the N-terminal MCP-1 deletion mutant) (Zhang, Rollins *et al.*, 1995) has proven effective in models of cardiovascular and renal disease (Egashira, 2003; Kitamoto *et al.*, 2002). AOP-RANTES, an N-terminal extension mutant, is also a particularly effective inhibitor of HIV infection (Simmons *et al.*, 2000). Other mutants have been discovered by phage display, by mutation, and by selection of N-terminal modified RANTES mutants (Hartley *et al.*, 2003).

There are also lessons to learn from viruses as some of these proteins, like the chemokine binding protein M3, are exceptionally potent in suppressing inflammation *in vivo* (Jensen *et al.*, 2003). Thus, they hold promise as therapeutic agents targeted against the chemokine ligands (Seet *et al.*, 2003), and their broad specificity for chemokines may also be advantageous. An interesting antagonist was designed to mimic HIV TAT. It is a nine-residue arginine peptide stabilized by d-amino acids (Doranz *et al.*, 1997; Murphy, 2001). This peptide inhibits HIV entry into CXCR4 expressing T-cells with an IC50 of 3 nM, and thus may prove usefully in delaying the onset of full-blown AIDs. Of course, one concern with any foreign or modified protein/peptide is immunogenicity, but viral proteins by their very nature may be less susceptible to recognition by the host immune response.

Perhaps one of the most novel arenas for drug discovery involves interfering with chemokine:GAG interactions. GAG mutants, like those described in Section III, can apparently act as dominant negative inhibitors as demonstrated for a RANTES variant (Johnson *et al.*, 2004). In principle, oligomerization mutants could also act in a similar manner since oligomerization is required for GAG-binding, at least for some chemokines. Even better, specific low molecular weight carbohydrate sequences that compete for binding of chemokines to cell surface glycosaminoglycans should also have potent anti-inflammatory activities, that compare with those of heparin, but without the undesirable effects such as bleeding and thrombocytopenia (Capila *et al.*, 2002; Sundaram *et al.*, 2003) or immunogenicity. However, to design and develop carbohydrate-based therapeutics will require a much deeper understanding of protein:GAG interactions and new sophisticated methods for sequencing (Keiser *et al.*, 2001) and synthesizing GAGs because of their non-template nature, unlike DNA and proteins.

References

Abdulaev, N. G., Strassmaier, T. T., Ngo, T., Chen, R., Luecke, H., Oprian, D. D., and Ridge, K. D. (2002). Grafting segments from the extracellular surface of CCR5 onto a bacteriorhodopsin transmembrane scaffold confers HIV-1 coreceptor activity. *Structure (Camb)* **10,** 515–525.

Aiello, R. J., Bourassa, P. A., Lindsey, S., Weng, W., Natoli, E., Rollins, B. J., and Milos, P. M. (1999). Monocyte chemoattractant protein-1 accelerates atherosclerosis in apolipoprotein E-deficient mice. *Arterioscler Thromb. Vasc. Biol.* **19,** 1518–1525.

Albini, A., Ferrini, S., Benelli, R., Sforzini, S., Giunciuglio, D., Aluigi, M. G., Proudfoot, A. E., Alouani, S., Wells, T. N., Mariani, G., Rabin, R. L., Farber, J. M., and Noonan, D. M. (1998). HIV-1 Tat protein mimicry of chemokines. *Proc. Natl. Acad. Sci. USA* **95,** 13153–13158.

Alexander, J. M., Nelson, C. A., van Berkel, V., Lau, E. K., Studts, J. M., Brett, T. J., Speck, S. H., Handell, T. M., Virgin, H. W., and Fremont, D. H. (2002). Structural basis of chemokine sequestration by a herpesvirus decoy receptor. *Cell* **111,** 343–356.

Ali, S., Palmer, A. C., Banerjee, B., Fritchley, S. J., and Kirby, J. A. (2000). Examination of the function of RANTES, MIP-1alpha, and MIP-1beta following interaction with heparin-like glycosaminoglycans. *J. Biol. Chem.* **275,** 11721–11727.

Alon, R., Grabovsky, V., and Feigelson, S. (2003). Chemokine induction of integrin adhesiveness on rolling and arrested leukocytes local signaling events or global stepwise activation? *Microcirculation* **10,** 297–311.

Appay, V., Brown, A., Cribbes, S., Randle, E., and Czaplewski, L. G. (1999). Aggregation of RANTES is responsible for its inflammatory properties. Characterization of nonaggregating, noninflammatory RANTES mutants. *J. Biol. Chem.* **274,** 27505–27512.

Bacon, K., Baggiolini, M., Broxmeyer, H., Horuk, R., Lindley, L., Mantovani, A., Maysushima, K., Murphy, P., Nomiyama, H., Oppenheim, J., Rot, A., Schall, T., Tsang, M., Thorpe, R., Van Damme, J., Wadhwa, M., Yoshie, O., Zlotnik, A., and Zoon, K. (2002). Chemokine/chemokine receptor nomenclature. *J. Interferon Cytokine Res.* **22,** 1067–1068.

Baggiolini, M. (1998). Chemokines and leukocyte traffic. *Nature* **392,** 565–568.

Baggiolini, M. (2001). Chemokines in pathology and medicine. *J. Intern. Med.* **250,** 91–104.

Baggiolini, M., Boulay, F., Badwey, J. A., and Curnutte, J. T. (1993). Activation of neutrophil leukocytes: Chemoattractant receptors and respiratory burst. *FASEB J.* **7,** 1004–1010.

Baggiolini, M., Deranleau, D. A., Dewald, B., Thelen, M., von Tscharner, V., and Wymann, M. P. (1989). [Properties and activation mechanism of neutrophilic leukocytes]. *Arzneimittelforschung* **39,** 177–180.

Baggiolini, M., Dewald, B., and Moser, B. (1997). Human chemokines: An update. *Annu. Rev. Immunol.* **15,** 675–705.

Baggiolini, M., and Wymann, M. P. (1990). Turning on the respiratory burst. *Trends Biochem. Sci.* **15,** 69–72.

Bai, M. (2004). Dimerization of G-protein-coupled receptors: Roles in signal transduction. *Cell Signal.* **16,** 175–186.

Baird, J. W., Nibbs, R. J., Komai-Koma, M., Connolly, J. A., Ottersbach, K., Clark-Lewis, I., Liew, F. Y., and Graham, G. J. (1999). ESkine, a novel beta-chemokine, is differentially spliced to produce secretable and nuclear targeted isoforms. *J. Biol. Chem.* **274,** 33496–33503.

Bajetto, A., Bonavia, R., Barbero, S., Florio, T., and Schettini, G. (2001). Chemokines and their receptors in the central nervous system. *Front Neuroendocrinol.* **22,** 147–184.

Ben-Baruch, A. (2003). Host microenvironment in breast cancer development: Inflammatory cells, cytokines and chemokines in breast cancer progression: Reciprocal tumor-microenvironment interactions. *Breast Cancer Res.* **5,** 31–36.

Benelli, R., Barbero, A., Ferrini, S., Scapini, P., Cassatella, M., Bussolino, F., Tacchetti, C., Noonan, D. M., and Albini, A. (2000). Human immunodeficiency virus transactivator protein (Tat) stimulates chemotaxis, calcium mobilization, and activation of human polymorphonuclear leukocytes: Implications for Tat-mediated pathogenesis. *J. Infect. Dis.* **182,** 1643–1651.

Blanpain, C., Doranz, B. J., Bondue, A., Govaerts, C., De Leener, A., Vassart, G., Doms, R. W., Proudfoot, A., and Parmentier, M. (2003). The core domain of chemokines binds CCR5 extracellular domains while their amino terminus interacts with the transmembrane helix bundle. *J. Biol. Chem.* **278,** 5179–5187.

Blaszczyk, J., Coillie, E. V., Proost, P., Damme, J. V., Opdenakker, G., Bujacz, G. D., Wang, J. M., and Ji, X. (2000). Complete crystal structure of monocyte chemotactic protein-2, a CC chemokine that interacts with multiple receptors. *Biochem.* **39,** 14075–14081.

Booth, V., Keizer, D. W., Kamphuis, M. B., Clark-Lewis, I., and Sykes, B. D. (2002). The CXCR3 binding chemokine IP-10/CXCL10: Structure and receptor interactions. *Biochem.* **41,** 10418–10425.

Boring, L., Gosling, J., Cleary, M., and Charo, I. F. (1998). Decreased lesion formation in CCR2-/- mice reveals a role for chemokines in the initiation of atherosclerosis. *Nature* **394,** 894–897.

Breitwieser, G. E. (2004). G protein-coupled receptor oligomerization: Implications for G protein activation and cell signaling. *Circ. Res.* **94,** 17–27.

Brigati, C., Giacca, M., Noonan, D. M., and Albini, A. (2003). HIV Tat, its TARgets and the control of viral gene expression. *FEMS Microbiol. Lett.* **220,** 57–65.

Butcher, E. C., and Picker, L. J. (1996). Lymphocyte homing and homeostasis. *Science* **272,** 60–66.

Capila, I., and Linhardt, R. J. (2002). Heparin-protein interactions. *Angew. Chem. Int. Ed. Engl.* **41,** 391–412.

Clark-Lewis, I., Dewald, B., Loetscher, M., Moser, B., and Baggiolini, M. (1994). Structural requirements for interleukin-8 function identified by design of analogs and CXC chemokine hybrids. *J. Biol. Chem.* **269,** 16075–16081.

Clark-Lewis, I., Mattioli, I., Gong, J. H., and Loetscher, P. (2003). Structure-function relationship between the human chemokine receptor CXCR3 and its ligands. *J. Biol. Chem.* **278,** 289–295.

Clore, G., Appella, M. E., Yamada, M., Matsushima, K., and Gronenborn, A. M. (1990). Three-dimensional structure of interleukin 8 in solution. *Biochem.* **29,** 1689–1696.

Crump, M. P., Rajarathnam, K., Kim, K. S., Clark-Lewis, I., and Sykes, B. D. (1998). Solution structure of eotaxin, a chemokine that selectively recruits eosinophils in allergic inflammation. *J. Biol. Chem.* **273,** 22471–22479.

D'Ambrosio, D., Panina-Bordignon, P., and Sinigaglia, F. (2003). Chemokine receptors in inflammation: An overview. *J. Immunol. Methods* **273,** 3–13.

Datta, A., and Stone, M. J. (2003). Soluble mimics of a chemokine receptor: Chemokine binding by receptor elements juxtaposed on a soluble scaffold. *Protein Sci.* **12,** 2482–2491.

Dawson, T. C., Lentsch, A. B., Wang, Z., Cowhig, J. E., Rot, A., Maeda, N., and Peiper, S. C. (2000). Exaggerated response to endotoxin in mice lacking the Duffy antigen/receptor for chemokines (DARC). *Blood* **96,** 1681–1684.

DeLano, W. L. (2002). The PyMOL Molecular Graphics System. DeLano Scientific, San Carlos, CA.

Devalaraja, M. N., and Richmond, A. (1999). Multiple chemotactic factors: Fine control or redundancy? *Trends Pharmacol. Sci.* **20,** 151–156.

Doranz, B. J., Grovit-Ferbas, K., Sharron, M. P., Mao, S. H., Goetz, M. B., Daar, E. S., Doms, R. W., and O'Brien, W. A. (1997). A small-molecule inhibitor directed against the chemokine receptor CXCR4 prevents its use as an HIV-1 coreceptor. *J. Exp. Med.* **186,** 1395–1400.

Dowsland, M. H., Harvey, J. R., Lennard, T. W., Kirby, J. A., and Ali, S. (2003). Chemokines and breast cancer: A gateway to revolutionary targeted cancer treatments? *Curr. Med. Chem.* **10,** 579–592.

Dragic, T., Trkola, A., Thompson, D. A., Cormier, E. G., Kajumo, F. A., Maxwell, E., Lin, S. W., Ying, W., Smith, S. D., Sakmar, T. P., and Moore, J. P. (2000). A binding pocket for a small molecule inhibitor of HIV-1 entry within the transmembrane helices of CCR5. *Proc. Natl. Acad. Sci. USA* **97,** 5639–5644.

Duensing, T. D., Wing, J. S., and van Putten, J. P. (1999). Sulfated polysaccharide-directed recruitment of mammalian host proteins: A novel strategy in microbial pathogenesis. *Infect. Immun.* **67,** 4463–4468.

Dwir, O., Grabovsky, V., Alon, R., and Feigelson, S. (2004). Selectin acidity modulation by chemokines at subsecond endothelial contacts: A novel regulatory level of leukocyte trafficking. *Ernst. Schering. Res. Found Workshop* **10,** 109–135.

Egashira, K. (2003). Molecular mechanisms mediating inflammation in vascular disease: Special reference to monocyte chemoattractant protein-1. *Hypertension* **41,** 834–841.

Farzan, M., Babcock, G. J., Vasilieva, N., Wright, P. L., Kiprilov, E., Mirzabekov, T., and Choe, H. (2002). The role of post-translational modifications of the CXCR4 amino terminus in stromal-derived factor 1 alpha association and HIV-1 entry. *J. Biol. Chem.* **277,** 29484–29489.

Farzan, M., Mirzabekov, T., Kolchinsky, P., Wyatt, R., Cayabyab, M., Gerard, N. P., Gerard, C., Sodroski, J., and Choe, H. (1999). Tyrosine sulfation of the amino terminus of CCR5 facilitates HIV-1 entry. *Cell* **96,** 667–676.

Fong, A. M., Alam, S. M., Imai, T., Haribabu, B., and Patel, D. D. (2002). CX3CR1 tyrosine sulfation enhances fractalkine-induced cell adhesion. *J. Biol. Chem.* **277,** 19418–19423.

Gerard, C., and Rollins, B. J. (2001). Chemokines and disease. *Nat. Immunol.* **2,** 108–115.

Ghezzi, S., Noonan, D. M., Aluigi, M. G., Vallanti, G., Cota, M., Benelli, R., Morini, M., Reeves, J. D., Vicenzi, E., Poli, G., and Albini, A. (2000). Inhibition of CXCR4-dependent HIV-1 infection by extracellular HIV-1 Tat. *Biochem. Biophys. Res. Commun.* **270,** 992–996.

Goda, S., Imai, T., Yoshie, O., Yoneda, O., Inoue, H., Nagano, Y., Okazaki, T., Imai, H., Bloom, E. T., Domae, N., and Umehara, H. (2000). CX3C-chemokine, fractalkine-enhanced adhesion of THP-1 cells to endothelial cells through integrin-dependent and -independent mechanisms. *J. Immunol.* **164,** 4313–4320.

Gosling, J., Slaymaker, S., Gu, L., Tseng, S., Zlot, C. H., Young, S. G., Rollins, B. J., and Chard, I. F. (1999). MCP-1 deficiency reduces susceptibility to atherosclerosis in mice that overexpress human apolipoprotein B. *J. Clin. Invest.* **103,** 773–778.

Gu, L., Okada, Y., Clinton, S. K., Gerard, C., Sukhova, G. K., Libby, P., and Rollins, B. J. (1998). Absence of monocyte chemoattractant protein-1 reduces atherosclerosis in low density lipoprotein receptor-deficient mice. *Mol. Cell* **2,** 275–281.

Gupta, S. K., and Pillarisetti, K. (1999). Cutting edge: CXCR4-Lo: Molecular cloning and functional expression of a novel human CXCR4 splice variant. *J. Immunol.* **163,** 2368–2372.

Handel, T. M., and Domaille, P. J. (1996). Heteronuclear (1H, 13C, 15N) NMR assignments and solution structure of the monocyte chemoattractant protein-1 (MCP-1) dimer. *Biochem.* **35,** 6569–6584.

Handel, T. M., and Lau, E. K. (2004). Chemokine structure and receptor interactions. *Ernst. Schering. Res. Found Workshop* **45,** 101–124.

Hartley, O., Dorgham, K., Perez-Bercoff, D., Cerini, F., Heimann, A., Gaertner, H., Offord, R. E., Pancino, G., Debre, P., and Gorochov, G. (2003). Human immunodeficiency virus type 1 entry inhibitors selected on living cells from a library of phage chemokines. *J. Virol.* **77,** 6637–6644.

Haskell, C. A., Cleary, M. D., and Charo, I. F. (1999). Molecular uncoupling of fractalkine-mediated cell adhesion and signal transduction. Rapid flow arrest of CX3CR1-expressing cells is independent of G-protein activation. *J. Biol. Chem.* **274,** 10053–10058.

Hedrick, J. A., Saylor, V., Figueroa, D., Mizoue, L., Xu, Y., Menon, S., Abrams, J., Handel, T., and Zlotnik, A. (1997). Lymphotactin is produced by NK cells and attracts both NK cells and T cells *in vivo. J. Immunol.* **158,** 1533–1540.

Hemmerich, S., Paavola, C., Bloom, A., Bhakta, S., Freedman, R., Grunberger, D., Krstenansky, J., Lee, S., McCarley, D., Mulkins, M., Wong, B., Pease, J., Mizoue, L., Mirzadegan, T., Polsky, I., Thompson, K., Handel, T. M., and Jarnagin, K. (1999). Identification of residues in the monocyte chemotactic protein-1 that contact the MCP-1 receptor, CCR2. *Biochem.* **38,** 13013–13025.

Hernanz-Falcon, P., Rodriguez-Frade, J. M., Serrano, A., Juan, D., del Sol, A., Soriano, S. F., Roncal, F., Gomez, L., Valencia, A., Martinez, A. C., and Mellado, M. (2004).

Identification of amino acid residues crucial for chemokine receptor dimerization. *Nat. Immunol.* **5,** 216–223.

Ho, H. H., Du, D., and Gershengorn, M. C. (1999). The N terminus of Kaposi's sarcoma-associated herpesvirus G protein-coupled receptor is necessary for high affinity chemokine binding but not for constitutive activity. *J. Biol. Chem.* **274,** 31327–31332.

Holst, P. J., Rosenkilde, M. M., Manfra, D., Chen, S. C., Wiekowski, M. T., Holst, B., Cifire, F., Lipp, M., Schwartz, T. W., and Lira, S. A. (2001). Tumorigenesis induced by the HHV8-encoded chemokine receptor requires ligand modulation of high constitutive activity. *J. Clin. Invest.* **108,** 1789–1796.

Homey, B., Muller, A., and Zlotnik, A. (2002). Chemokines: Agents for the immunotherapy of cancer? *Nat. Rev. Immunol.* **2,** 175–184.

Hoogewerf, A. J., Kuschert, G. S., Proudfoot, A. E., Borlat, F., Clark-Lewis, I., Power, C. A., and Wells, T. N. (1997). Glycosaminoglycans mediate cell surface oligomerization of chemokines. *Biochem.* **36,** 13570–13578.

Hoover, D. M., Boulegue, C., Yang, D., Oppenheim, J. J., Tucker, K., Lu, W., and Lubkowski, J. (2002). The structure of human macrophage inflammatory protein-3alpha/CCL20. Linking antimicrobial and CC chemokine receptor-6-binding activities with human beta-defensins. *J. Biol. Chem.* **277,** 37647–37654.

Hoover, D. M., Mizoue, L. S., Handel, T. M., and Lubkowski, J. (2000). The crystal structure of the chemokine domain of fractalkine shows a novel quaternary arrangement. *J. Biol. Chem.* **275,** 23187–23193.

Horuk, R. (2003). Development and evaluation of pharmacological agents targeting chemokine receptors. *Methods* **29,** 369–375.

Horuk, R., and Ng, H. P. (2000). Chemokine receptor antagonists. *Med. Res. Rev.* **20,** 155–168.

Houshmand, P., and Zlotnik, A. (2003). Therapeutic applications in the chemokine superfamily. *Curr. Opin. Chem. Biol.* **7,** 457–460.

Jarnagin, K., Grunberger, D., Mulkins, M., Wong, B., Hemmerich, S., Paavola, C., Bloom, A., Bhakta, S., Diehl, F., Freedman, R., McCarley, D., Polsky, I., Ping-Tsou, A., Kosaka, A., and Handel, T. M. (1999). Identification of surface residues of the monocyte chemotactic protein 1 that affect signaling through the receptor CCR2. *Biochem.* **38,** 16167–16177.

Jensen, K. K., Chen, S. C., Hipkin, R. W., Wiekowski, M. T., Schwarz, M. A., Chou, C. C., Simas, J. P., Alcami, A., and Lira, S. A. (2003). Disruption of CCL21-induced chemotaxis *in vitro* and *in vivo* by M3, a chemokine-binding protein encoded by murine gammaherpesvirus 68. *J. Virol.* **77,** 624–630.

Johnson, Z., Kosco-Vibois, M., Herren, S., Cirillo, R., Muzio, V., Smailbegovis, A., Rose, M., Lever, R., Page, C., Wells, T., and Proudfoot, A. E. I. (2004). Interference with heparin binding and oligomerization creates a novel anti-inflammatory strategy targeting the chemokine system. *Proc. Natl. Acad. Sci. USA,* submitted.

Katancik, J. A., Sharma, A., Radel, S. J., and De Nardin, E. (1997). Mapping of the extracellular binding regions of the human interleukin-8 type B receptor. *Biochem. Biophys. Res. Commun.* **232,** 663–668.

Keiser, N., Venkataraman, G., Shriver, Z., and Sasisekharan, R. (2001). Direct isolation and sequencing of specific protein-binding glycosaminoglycans. *Nat. Med.* **7,** 123–128.

Kim, K. S., Rajarathnam, K., Clark-Lewis, I., and Sykes, B. D. (1996). Structural characterization of a monomeric chemokine: Monocyte chemoattractant protein-3. *FEBS Lett.* **395,** 277–282.

Kim, S., Jao, S., Laurence, J. S., and LiWang, P. J. (2001). Structural comparison of monomeric variants of the chemokine MIP-1 beta having differing ability to bind the receptor CCR5. *Biochem.* **40,** 10782–10791.

Kitamoto, S., and Egashira, K. (2002). Gene therapy targeting monocyte chemoattractant protein-1 for vascular disease. *J. Atheroscler. Thromb.* **9,** 261–265.

Kuloglu, E. S., McCaslin, D. R., Kitabwalla, M., Pauza, C. D., Markley, J. L., and Volkman, B. F. (2001). Monomeric solution structure of the prototypical 'C' chemokine lymphotactin. *Biochem.* **40,** 12486–12496.

Kunkel, E., and Butcher, E. (2002). Homeostatic chemokines and the targeting of regional immunity. *Adv. Exp. Med. Biol.* **512,** 65–72.

Kuschert, G. S. V., Coulin, F., Power, C. A., Proudfoot, A. E. I., Hubbard, R. E., Hoogewerf, A. J., and Wells, T. N. C. (1999). Glycosaminoglycans interact selectively with chemokines and modulate receptor binding and cellular responses. *Bio. Chem.* **38,** 12959–12968.

Lau, E. K., Kungl, A., Paavola, C., Johnson, Z., Borlat, F., Proudfoot, A. E. I., and Handel, T. M. (2004). Identification of the glycosaminoglycan binding site of the CC chemokine, MCP-1; Implications for structure and function *in vivo. J. Biol. Chem* **279,** 22294–305.

Liu, D., Shriver, Z., Qi, Y., Venkataraman, G., and Sasisekharan, R. (2002). Dynamic regulation of tumor growth and metastasis by heparan sulfate glycosaminoglycans. *Semin. Thromb. Hemost.* **28,** 67–78.

Liwang, A. C., Wang, Z. X., Sun, Y., Peiper, S. C., and Liwang, P. J. (1999). The solution structure of the anti-HIV chemokine vMIP-II. *Protein. Sci.* **8,** 2270–2280.

Lodi, P. J., Garrett, D. S., Kuszewski, J., Tsang, M. L., Weatherbee, J. A., Leonard, W. J., Gronenborn, A. M., and Clore, G. M. (1994). High-resolution solution structure of the beta chemokine hMIP-1 beta by multidimensional NMR. *Science* **263,** 1762–1767.

Loetscher, P., and Clark-Lewis, I. (2001). Agonistic and antagonistic activities of chemokines. *J. Leukoc. Biol.* **69,** 881–884.

Loetscher, P., Pellegrino, A., Gong, J. H., Mattioli, I., Loetscher, M., Bardi, G., Baggiolini, M., and Clark-Lewis, I. (2001). The ligands of CXC chemokine receptor 3, I-TAC, Mig, and IP10, are natural antagonists for CCR3. *J. Biol. Chem.* **276,** 2986–2991.

Lubkowski, J., Bujacz, G., Boque, L., Domaille, P. J., Handel, T. M., and Wlodawer, A. (1997). The structure of MCP-1 in two crystal forms provides a rare example of variable quaternary interactions. *Nat. Struct. Biol.* **4,** 64–69.

Mantovani, A. (1999). The chemokine system: Redundancy for robust outputs. *Immunol. Today* **20,** 254–257.

Mayer, K. L., and Stone, M. J. (2000). NMR solution structure and receptor peptide binding of the CC chemokine eotaxin-2. *Biochem.* **39,** 8382–8395.

Mellado, M., Martinez, A. C., Rodriguez-Frade, J. M., Vila-Coro, A. J., Martinez, C., and Manes, S. (2002). Analysis of G-protein-coupled receptor dimerization following chemokine signaling. *Methods* **27,** 349–357.

Mellado, M., Rodriguez-Frade, J. M., Vila-Coro, A. J., Fernandez, S., Martin de Ana, A., Jones, D. R., Toran, J. L., and Martinez, A. C. (2001). Chemokine receptor homo- or heterodimerization activates distinct signaling pathways. *EMBO J.* **20,** 2497–2507.

Mellado, M., Vila-Coro, A. J., Martinez, C., Rodriguez-Frade, J. M., Manes, S., and Martinez, A. C. (2001). Receptor dimerization: A key step in chemokine signaling. *Cell Mol. Biol. (Noisy-le-grand)* **47,** 575–582.

Middleton, J., Neil, S., Wintle, J., Clark-Lewis, I., Moore, H., Lam, C., Auer, M., Hub, E., and Rot, A. (1997). Transcytosis and surface presentation of IL-8 by venular endothelial cells. *Cell* **91,** 385–395.

Middleton, J., Patterson, A. M., Gardner, L., Schmutz, C., and Ashton, B. A. (2002). Leukocyte extravasation: Chemokine transport and presentation by the endothelium. *Blood* **100,** 3853–3860.

Mirzadegan, T., Diehl, F., Ebi, B., Bhakta, S., Polsky, I., McCarley, D., Mulkins, M., Weatherhead, G. S., Lapierre, J. M., Dankwardt, J., Morgans, D., Jr., Wilhelm, R., and Jarnagin, K. (2000). Identification of the binding site for a novel class of CCR2b chemokine receptor antagonists: Binding to a common chemokine receptor motif within the helical bundle. *J. Biol. Chem.* **275,** 25562–25571.

Mizoue, L. S., Bazan, J. F., Johnson, E. C., and Handel, T. M. (1999). Solution structure and dynamics of the CX3C chemokine domain of fractalkine and its interaction with an N-terminal fragment of CX3CR1. *Biochemistry* **38,** 1402–1414.

Monteclaro, F. S., and Charo, I. F. (1997). The amino-terminal domain of CCR2 is both necessary and sufficient for high affinity binding of monocyte chemoattractant protein 1. Receptor activation by a pseudo-tethered ligand. *J. Biol. Chem.* **272,** 23186–23190.

Moser, B., Dewald, B., Barella, L., Schumacher, C., Baggiolini, M., and Clark-Lewis, I. (1993). Interleukin-8 antagonists generated by N-terminal modification. *J. Biol. Chem.* **268,** 7125–7128.

Muller, A., Homey, B., Soto, H., Ge, N., Catron, D., Buchanan, M. E., McClanahan, T., Murphy, E., Yuan, W., Wagner, S. N., Barrera, J. L., Mohar, A., Verastegui, E., and Zlotnik, A. (2001). Involvement of chemokine receptors in breast cancer metastasis. *Nature* **410,** 50–56.

Muller, G., Hopken, U. E., and Lipp, M. (2003). The impact of CCR7 and CXCR5 on lymphoid organ development and systemic immunity. Concerted action of the chemokine and lymphotoxin system in secondary lymphoid-organ development. *Immunol. Rev.* **195,** 117–135.

Murphy, P. M. (1994). The molecular biology of leukocyte chemoattractant receptors. *Annu. Rev. Immunol.* **12,** 593–633.

Murphy, P. M. (2001). Chemokines and the molecular basis of cancer metastasis. *N. Engl. J. Med.* **345,** 833–835.

Murphy, P. M. (2001). Viral exploitation and subversion of the immune system through chemokine mimicry. *Nat. Immunol.* **2,** 116–122.

Murphy, P. M., Baggiolini, M., Charo, I. F., Hebert, C. A., Horuk, R., Matsushima, K., Miller, L. H., Oppenheim, J. I., and Power, C. A. (2000). International union of pharmacology. XXII. Nomenclature for chemokine receptors. *Pharmacol. Rev.* **52,** 145–176.

Nash, P., Barrett, J., Cao, J. X., Hota-Mitchell, S., Lalani, A. S., Everett, H., Xu, X. M., Robichaud, J., Hnatiuk, S., Ainslie, C., Seet, B. T., and McFadden, G. (1999). Immunomodulation by viruses: The myxoma virus story. *Immunol. Rev.* **168,** 103–120.

Noonan, D., and Albini, A. (2000). From the outside in: Extracellular activities of HIV Tat. *Adv. Pharmacol.* **48,** 229–250.

Ogilvie, P., Bardi, G., Clark-Lewis, I., Baggiolini, M., and Uguccioni, M. (2001). Eotaxin is a natural antagonist for CCR2 and an agonist for CCR5. *Blood* **97,** 1920–1924.

Ogilvie, P., Paoletti, S., Clark-Lewis, I., and Uguccioni, M. (2003). Eotaxin-3 is a natural antagonist for CCR2 and exerts a repulsive effect on human monocytes. *Blood* **102,** 789–794.

Olson, T. S., and Ley, K. (2002). Chemokines and chemokine receptors in leukocyte trafficking. *Am. J. Physiol. Regul. Integr. Comp. Physiol.* **283,** R7–R28.

Ono, S. J., Nakamura, T., Miyazaki, D., Ohbayashi, M., Dawson, M., and Toda, M. (2003). Chemokines: Roles in leukocyte development, trafficking, and effector function. *J. Allergy Clin. Immunol.* **111,** 1185–1199; quiz 1200.

Onuffer, J., McCarrick, M. A., Dunning, L., Liang, M., Rosser, M., Wei, G. P., Ng, H., and Horuk, R. (2003). Structure function differences in nonpeptide CCR1 antagonists for human and mouse CCR1. *J. Immunol.* **170,** 1910–1916.

Onuffer, J. J., and Horuk, R. (2002). Chemokines, chemokine receptors and small-molecule antagonists: Recent developments. *Trends Pharmacol. Sci.* **23,** 459–467.

Paavola, C. D., Hemmerich, S., Grunberger, D., Polsky, I., Bloom, A., Freedman, R., Mulkins, M., Bhakta, S., McCarley, D., Wiesent, L., Wong, B., Jarnagin, K., and Handel, T. M. (1998). Monomeric monocyte chemoattractant protein-1 (MCP-1) binds and activates the MCP-1 receptor CCR2B. *J. Biol. Chem.* **273,** 33157–33165.

Pakianathan, D. R., Kuta, E. G., Artis, D. R., Skelton, N. J., and Hebert, C. A. (1997). Distinct but overlapping epitopes for the interaction of a CC-chemokine with CCR1, CCR3 and CCR5. *Biochemistry* **36,** 9642–9648.

Patel, D. D., Koopmann, W., Imai, T., Whichard, L. P., Yoshie, O., and Krangel, M. S. (2001). Chemokines have diverse abilities to form solid phase gradients. *Clin. Immunol.* **99,** 43–52.

Perez-Canadillas, J. M., Zaballos, A., Gutierrez, J., Varona, R., Roncal, F., Albar, J. P., Marquez, G., and Bruix, M. (2001). NMR solution structure of murine CCL20/MIP-3alpha, a chemokine that specifically chemoattracts immature dendritic cells and lymphocytes through its highly specific interaction with the beta-chemokine receptor CCR6. *J. Biol. Chem.* **276,** 28372–28379.

Preobrazhensky, A. A., Dragan, S., Kawano, T., Gavrilin, M. A., Gulina, I. V., Chakravarty, L., and Kolattukudy, P. E. (2000). Monocyte chemotactic protein-1 receptor CCR2B is a glycoprotein that has tyrosine sulfation in a conserved extracellular N-terminal region. *J. Immunol.* **165,** 5295–5303.

Proudfoot, A. E. (2002). Chemokine receptors: Multifaceted therapeutic targets. *Nat. Rev. Immunol.* **2,** 106–115.

Proudfoot, A. E., Power, C. A., Hoogewerf, A. J., Montjovent, M. O., Borlat, F., Offord, R. E., and Wells, T. N. (1996). Extension of recombinant human RANTES by the retention of the initiating methionine produces a potent antagonist. *J. Biol. Chem.* **271,** 2599–2603.

Proudfoot, A. E. I., Handel, T. M., Johnson, Z., Lau, E. K., LiWang, P., Clark-Lewis, I., Borlat, F., Wells, T. N. C., and Kosco-Vilbois, M. H. (2003). Glycosaminoglycan binding and oligomerization are essential for the *in vivo* activity of certain chemokines. *Proc. Natl. Acad. Sci. USA* **100,** 1885–1890.

Quan, J. M., Martin, T. R., Rosenberg, G. B., Foster, D. C., Whitmore, T., and Goodman, R. B. (1996). Antibodies against the N-terminus of IL-8 receptor A inhibit neutrophil chemotaxis. *Biochem. Biophys. Res. Commun.* **219,** 405–411.

Rajarathnam, K., Li, Y., Rohrer, T., and Gentz, R. (2001). Solution structure and dynamics of myeloid progenitor inhibitory factor-1 (MPIF-1), a novel monomeric CC chemokine. *J. Biol. Chem.* **276,** 4909–4916.

Rajarathnam, K., Sykes, B. D., Kay, C. M., Dewald, B., Geiser, T., Baggiolini, M., and Clark-Lewis, I. (1994). Neutrophil activation by monomeric interleukin-8. *Science* **264,** 90–92.

Rodriguez-Frade, J. M., Mellado, M., and Martinez, A. C. (2001). Chemokine receptor dimerization: Two are better than one. *Trends Immunol.* **22,** 612–617.

Rollins, B. J. (2001). Chemokines and atherosclerosis: What Adam Smith has to say about vascular disease. *J. Clin. Invest.* **108,** 1269–1271.

Rosenkilde, M. M., Waldhoer, M., Luttichau, H. R., and Schwartz, T. W. (2001). Virally encoded 7TM receptors. *Oncogene* **20,** 1582–1593.

Rot, A. (1993). Neutrophil attractant/activation protein-1 (interleukin-8) induces *in vitro* neutrophil migration by haptotactic mechanism. *Eur. J. Immunol.* **23,** 303–306.

Sanchez-Madrid, F., and del Pozo, M. A. (1999). Leukocyte polarization in cell migration and immune interactions. *EMBO J.* **18,** 501–511.

Sasisekharan, R., Shriver, Z., Venkataraman, G., and Narayanasami, U. (2002). Roles of heparan-sulphate glycosaminoglycans in cancer. *Nat. Rev. Cancer* **2,** 521–528.

Seet, B. T., Johnston, J. B., Brunetti, C. R., Barrett, J. W., Everett, H., Cameron, C., Sypula, J., Nazarian, S. H., Lucas, A., and McFadden, G. (2003). Poxviruses and immune evasion. *Annu. Rev. Immunol.* **21,** 377–423.

Shimaoka, T., Nakayama, T., Fukumoto, N., Kume, N., Takahashi, S., Yamaguchi, J., Minami, M., Hayashida, K., Kita, T., Ohsumi, J., Yoshie, O., and Yonehara, S. (2004). Cell surface-anchored SR-PSOX/CXC chemokine ligand 16 mediates firm adhesion of CXC chemokine receptor 6-expressing cells. *J. Leukoc. Biol.* **75,** 267–274.

Shriver, Z., Liu, D., and Sasisekharan, R. (2002). Emerging views of heparan sulfate glycosaminoglycan structure/activity relationships modulating dynamic biological functions. *Trends Cardiovasc. Med.* **12,** 71–77.

Sica, A., Saccani, A., and Mantovani, A. (2002). Tumor-associated macrophages: A molecular perspective. *Int. Immunopharmacol.* **2,** 1045–1054.

Signoret, N., Pelchen-Matthews, A., Mack, M., Proudfoot, A. E., and Marsh, M. (2000). Endocytosis and recycling of the HIV coreceptor CCR5. *J. Cell Biol.* **151,** 1281–1294.

Simmons, G., Clapham, P. R., Picard, L., Offord, R. E., Rosenkilde, M. M., Schwartz, T. W., Buser, R., Wells, T. N., and Proudfoot, A. E. (1997). Potent inhibition of HIV-1 infectivity in macrophages and lymphocytes by a novel CCR5 antagonist. *Science* **276,** 276–279.

Simmons, G., Reeves, J. D., Hibbitts, S., Stine, J. T., Gray, P. W., Proudfoot, A. E., and Clapham, P. R. (2000). Co-receptor use by HIV and inhibition of HIV infection by chemokine receptor ligands. *Immunol. Rev.* **177,** 112–126.

Skelton, N. J., Aspiras, F., Ogez, J., and Schall, T. J. (1995). Proton NMR assignments and solution conformation of RANTES, a chemokine of the C-C type. *Biochemistry* **34,** 5329–5342.

Skelton, N. J., Quan, C., Reilly, D., and Lowman, H. (1999). Structure of a CXC chemokine-receptor fragment in complex with interleukin-8. *Structure Fold Des.* **7,** 157–168.

Springer, T. A. (1994). Traffic signals for lymphocyte recirculation and leukocyte emigration: The multistep paradigm. *Cell* **76,** 301–314.

Springer, T. A. (1995). Traffic signals on endothelium for lymphocyte recirculation and leukocyte emigration. Traffic signals for lymphocyte recirculation and leukocyte emigration: The multistep paradigm. *Annu. Rev. Physiol.* **57,** 827–872.

Stringer, S. E., Forster, M. J., Mulloy, B., Bishop, C. R., Graham, G. J., and Gallagher, J. T. (2002). Characterization of the binding site on heparan sulfate for macrophage inflammatory protein 1alpha. *Blood* **100,** 1543–1550.

Stringer, S. E., and Gallagher, J. T. (1997). Specific binding of the chemokine platelet factor 4 to heparan sulfate. *J. Biol. Chem.* **272,** 20508–20514.

Struyf, S., Proost, P., and Van Damme, J. (2003). Regulation of the immune response by the interaction of chemokines and proteases. *Adv. Immunol.* **81,** 1–44.

Sundaram, M., Qi, Y. W., Shriver, Z., Liu, D. F., Zhao, G. L., Venkataraman, G., Langer, R., and Sasisekharan, R. (2003). Rational design of low-molecular weight heparins with improved *in vivo* activity. *Proc. Natl. Acad. Sci. USA* **100,** 651–656.

Suzuki, H., Prado, G. N., Wilkinson, N., and Navarro, J. (1994). The N terminus of interleukin-8 (IL-8) receptor confers high affinity binding to human IL-8. *J. Biol. Chem.* **269,** 18263–18266.

Swaminathan, G. J., Holloway, D. E., Colvin, R. A., Campanella, G. K., Papageorgiou, A. C., Luster, A. D., and Acharya, K. R. (2003). Crystal structures of oligomeric forms of the IP-10/CXCL10 chemokine. *Structure (Camb)* **11,** 521–532.

Terrillon, S., and Bouvier, M. (2004). Roles of G-protein-coupled receptor dimerization. *EMBO Rep.* **5,** 30–34.

Thelen, M. (2001). Dancing to the tune of chemokines. *Nat. Immunol.* **2,** 129–134.

Thelen, M., and Baggiolini, M. (2001). Is dimerization of chemokine receptors functionally relevant? *Sci. STKE* **2001,** PE34.

van Berkel, V., Barrett, J., Tiffany, H. L., Fremont, D. H., Murphy, P. M., McFadden, G., Speck, S. H., and Virgin, H. I. (2000). Identification of a gammaherpesvirus selective chemokine binding protein that inhibits chemokine action. *J. Virol.* **74,** 6741–6747.

Vene, R., Benelli, R., Noonan, D. M., and Albini, A. (2000). HIV-Tat dependent chemotaxis and invasion, key aspects of tat mediated pathogenesis. *Clin. Exp. Metastasis* **18,** 533–538.

Vicari, A. P., and Caux, C. (2002). Chemokines in cancer. *Cytokine Growth Factor Rev.* **13,** 143–154.

Weber, C. (2003). Novel mechanistic concepts for the control of leukocyte transmigration: Specialization of integrins, chemokines, and junctional molecules. *J. Mol. Med.* **81,** 4–19.

Weber, M., Uguccioni, M., Baggiolini, M., Clark-Lewis, I., and Dahinden, C. A. (1996). Deletion of the NH2-terminal residue converts monocyte chemotactic protein 1 from an activator of basophil mediator release to an eosinophil chemoattractant. *J. Exp. Med.* **183,** 681–685.

Wells, T. N., Power, C. A., Lusti-Narasimhan, M., Hoogewerf, A. J., Cooke, R. M., Chung, C. W., Peitsch, M. C., and Proudfoot, A. E. (1996). Selectivity and antagonism of chemokine receptors. *J. Leukoc. Biol.* **59,** 53–60.

Wells, T. N., and Schwartz, T. W. (1997). Plagiarism of the host immune system: Lessons about chemokine immunology from viruses. *Curr. Opin. Biotechnol.* **8,** 741–748.

Wu, L., Ruffing, N., Shi, X., Newman, W., Soler, D., Mackay, C. R., and Oin, S. (1996). Discrete steps in binding and signaling of interleukin-8 with its receptor. *J. Biol. Chem.* **271,** 31202–31209.

Xanthou, G., Williams, T. J., and Pease, J. E. (2003). Molecular characterization of the chemokine receptor CXCR3: Evidence for the involvement of distinct extracellular domains in a multi-step model of ligand binding and receptor activation. *Eur. J. Immunol.* **33,** 2927–2936.

Ye, J., Mayer, K. L., Mayer, M. R., and Stone, M. J. (2001). NMR solution structure and backbone dynamics of the CC chemokine eotaxin-3. *Biochem.* **40,** 7820–7831.

Zhang, X., Chen, L., Bancroft, D. P., Lai, C. K., and Maione, T. E. (1994). Crystal structure of recombinant human platelet factor 4. *Biochem.* **33,** 8361–8366.

Zhang, Y., and Rollins, B. J. (1995). A dominant negative inhibitor indicates that monocyte chemoattractant protein 1 functions as a dimer. *Mol. Cell. Biol.* **15,** 4851–4855.

Zlotnik, A. (2004). Chemokines and cancer. *Ernst. Schering. Res. Found Workshop* 53–58.

Zoffmann, S., Chollet, A., and Galzi, J. L. (2002). Identification of the extracellular loop 2 as the point of interaction between the N terminus of the chemokine MIP-1alpha and its CCR1 receptor. *Mol. Pharmacol.* **62,** 729–736.

CHEMOTAXIS RECEPTORS AND SIGNALING

By AARON F. MILLER AND JOSEPH J. FALKE

Department of Chemistry and Biochemistry,
University of Colorado, Boulder, Colorado 80309

Abstract

Chemotaxis is an important cellular response common in biology. In many chemotaxing cells the signal that regulates movement is initiated by G protein-coupled receptors on the cell surface that bind specific chemoattractants. These receptors share important structural similarities with other G protein-coupled receptors, including rhodopsin, which currently serves as the best starting point for modeling their structures. However, the chemotaxis receptors also share a number of relatively unique structural features that are less common in other GPCRs. The chemoattractant ligands of chemotaxis receptors exhibit a broad variety of sizes and chemical properties, ranging from small molecules and peptides to protein

ADVANCES IN PROTEIN CHEMISTRY, Vol. 68

0065-3233/04 $35.00

ligands. As a result, different chemotaxis receptors have evolved specialized mechanisms for the early steps of ligand binding and receptor activation. The mechanism of transmembrane signaling is currently under intensive study and several alternate mechanisms proposing different conformational rearrangements of the transmembrane helices have been proposed. Some chemotaxis receptors are proposed to form dimers, and in certain cases dimer formation is proposed to play a role in transmembrane signaling. In principle the structural and dynamical changes that occur during transmembrane signaling could be specialized for different receptors, or could be broadly conserved. Extensive mutagenesis studies have been carried out, and have begun to identify critical residues involved in ligand binding, receptor activation, and transmembrane signaling.

I. Introduction to Chemotaxis

The migration of leukocytes occupies a fundamental role in the response to infection, tissue damage and in the normal development of the immune system in mammals (Mackay, 2001). Invading pathogens and tissue damage are sources of exogenous and endogenous soluble chemoattractants that serve to recruit leukocytes and epithelial cell types to assist in the immune response and wound healing (Baggiolini, 1998). Such chemotaxis is initiated by attractant binding to specific G protein-coupled, tyrosine kinase, and serine/threonine kinase receptors that are expressed on the surfaces of a variety of cell types including leukocytes and epithelial cells (Gallin *et al.*, 1999; Locati *et al.*, 1999; Parent *et al.*, 1999; Thelen, 2001). The present review focuses on mammalian G protein-coupled receptors (GPCRs) that regulate chemotaxis, the signaling pathways they activate, and the mechanisms responsible for ligand-induced receptor activation.

Chemical attractants that activate GPCRs exhibit a wide range of sizes and molecular properties, from small formylated peptides like N-formyl-Met-Leu-Phe (fMLF) (Le *et al.*, 2002) and larger proteins like complement factor 5a (C5a) (Wetsel, 1995) to lipids like platelet activating factor (PAF) or leukotriene B_4 (LTB_4). Historically, prior to the identification of the chemokine protein interleukin-8 (CXCL8) (Yoshimura *et al.*, 1987) and the cloning of its receptors (Holmes *et al.*, 1991; Murphy *et al.*, 1991), most of the work in the field of leukocyte chemotaxis focused on the functions of the "classic" chemoattractants, fMLF (Marasco *et al.*, 1984), PAF (Rediske *et al.*, 1992) and C5a (Murphy, 1994). Currently, most research focuses on the chemoattractant receptors for the peptide and protein ligands fMLF, C5a and the chemokines (Fig. 1), in part because the

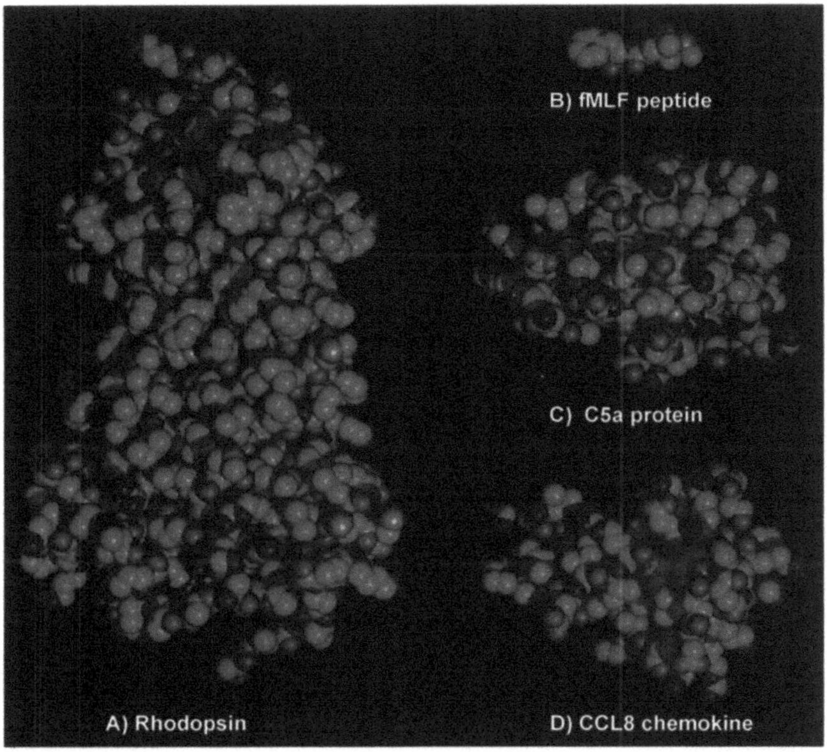

FIG. 1. Space-filling models illustrating the relative sizes of a G protein-coupled receptor and three chemoattractant ligands. Shown are (A) rhodopsin, with its G protein-coupling surface oriented downward (Palczewski *et al.*, 2000), (B) the chemoattractant peptide formyl-Met-Leu-Phe, (C) the chemoattractant protein C5a (Zhang *et al.*, 1997), and (D) the chemoattractant chemokine CCL8 (or interleukin 8) (Baldwin *et al.*, 1991). (See Color Insert.)

ligands for these receptors are soluble and straightforward to use in binding assays, while the insolubility of lipid attractants like PAF makes their chemoattractant receptors more difficult to study.

The receptors activated by classic chemoattractants and chemokines have long been known to be GPCRs that regulate cytoplasmic GTP binding activity (Bokoch *et al.*, 1984; Feltner *et al.*, 1986; Goldman *et al.*, 1985; Koo *et al.*, 1983; Smith *et al.*, 1986; Thelen, 2001; Verghese *et al.*, 1986; Wilde *et al.*, 1989). Since chemotaxis GPCRs share structural similarities, bind their ligands in the extracellular compartment, and regulate cytoplasmic heterotrimeric G proteins, there are certain basic mechanistic features shared by these receptors. At the same time, it is likely that other mechanistic features differ among chemotaxis GPCRs due to their specialization for different ligands and pathways. For example, different chemotaxis GPCRs can activate different G protein isoforms and signaling events, which requires specialization of their G protein recognition sites. In principle, the mechanical and conformational signals that regulate these different G protein recognition sites could be either universal or highly specialized.

II. Signal Transduction Events

Chemotaxis is a fundamental yet complex cellular response involving signaling pathways that link cell surface chemoreceptors to the regulation of dynamic cytoskeletal and adhesion elements that direct cellular shape change, polarization and movement (Maghazachi, 1999; Parent *et al.*, 1999; Thelen, 2001). Together these signaling events yield the systematic movement of a cell up a concentration gradient of attractant (for reviews see, Firtel *et al.*, 2000; Wilkinson *et al.*, 1997). This important cellular process is found in a wide range of organisms, from the relatively simple organisms such as the slime mold *Dictyostelium discoideum* to the leukocytes of the most complex mammalian immune systems.

In a typical mammalian GPCR-regulated chemotaxis pathway, signaling begins with the binding of chemoattractant to the extracellular surface of the GPCR. Subsequently the receptor generates a transmembrane signal that enables its cytoplasmic region to catalyze the exchange of GTP for GDP at the α subunit of a membrane-bound, heterotrimeric G protein that also possess β and γ subunits. The α subunit is reversibly bound to the tightly associated $\beta\gamma$ heterodimer (Sprang, 1997). Upon GTP occupancy, the α subunit dissociates from both receptor and $\beta\gamma$ thus liberating the α and $\beta\gamma$ components to interact with effector molecules elsewhere in the cell (Fig. 2). Current evidence suggests that the chemoattractant GPCRs predominantly regulate the $G\alpha_i$ family of α subunits (Bokoch, 1995; Neptune *et al.*, 1997), although there are examples of chemotaxis receptors CXCR1, CXCR2 and C5aR coupling to $G\alpha_q$ family members (Amatruda *et al.*, 1993; Wu *et al.*, 1993). For example, an abundant source of $G\alpha_i$ in humans is the circulating neutrophil that chemotaxes toward bacterial infections and other insults (Bokoch *et al.*, 1988). $G\alpha_i$ family members are characterized by their ability to inhibit cAMP synthesis, their GTPase activity, and their sensitivity to inactivation by *Bordetella pertussis* toxin (Fig. 2). Following receptor-catalyzed dissociation of $G\alpha_i$ from $G\beta\gamma$, the $G\alpha_i$ may play a direct role in chemotaxis by activating Src family tyrosine kinases (Fig. 3) (Ma *et al.*, 2000). Such tyrosine kinase activity is known to follow activation of chemoattractant GPCRs, thereby providing a plausible link between $G\alpha_i$

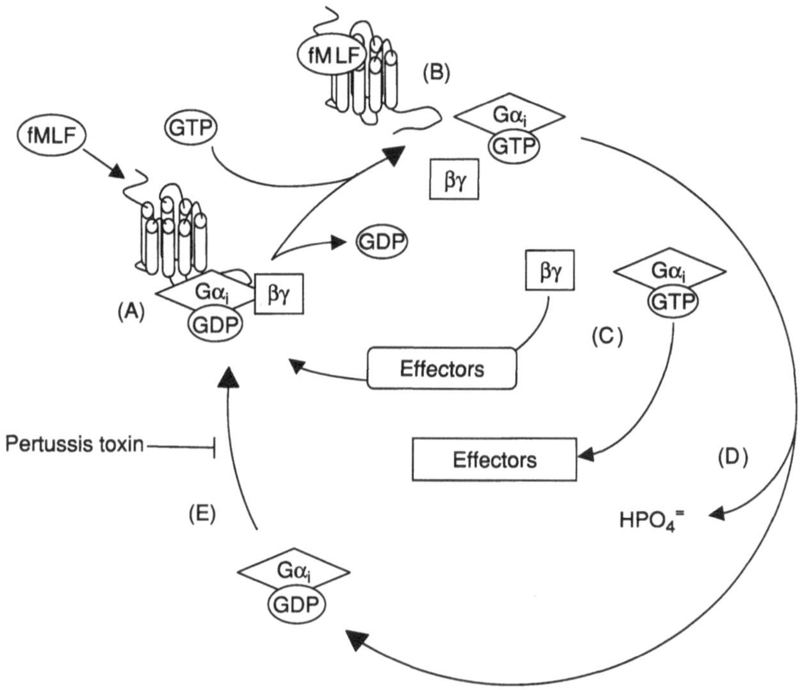

FIG. 2. Minimal components of the G protein-coupling cycle in a schematic chemotaxis pathway. (A) Chemotaxis G protein-coupled receptors reside in the plasma membrane associated with specific heterotrimeric G proteins. (B) Upon binding of a chemoattractant such as fMLF, the receptor catalyzes the exchange of GTP for GDP in the α subunit of the G protein, thereby dissociating the GTP-bound α subunit from the $\beta\gamma$ subunit complex. Chemotaxis GPCRs typically utilize the α_i isoform of the α subunit. (C) Subsequently the dissociated GTP-bound α_i subunit and the $\beta\gamma$ subunit complex each dock to effectors elsewhere in the cell. (D) The α_i subunit possesses intrinsic GTPase activity that hydrolyzes the bound GTP to GDP, thereby regenerating the GDP-bound α_i subunit that reassociates with the $\beta\gamma$ subunit complex and with the receptor. *Bordetella pertussis* toxin covalently and specifically modifies the α_i isoform of the α subunit and prevents its association with receptor (see text for additional discussion and references).

activation and the mitogen activated protein kinase cascades believed to play a role in chemotaxis (Knall *et al.*, 1997; Nick *et al.*, 1997).

Better studied are the chemotaxis signaling events triggered by the liberated $\beta\gamma$ subunits upon GPCR-catalyzed dissociation of the G protein (Fig. 3). In the current model, the $\beta\gamma$ complex directly docks to and activates phosphatidylinositol-3-kinase γ (P13Kγ), a membrane-associated enzyme that phosphorylates the inositol headgroup of phosphatidylinositol

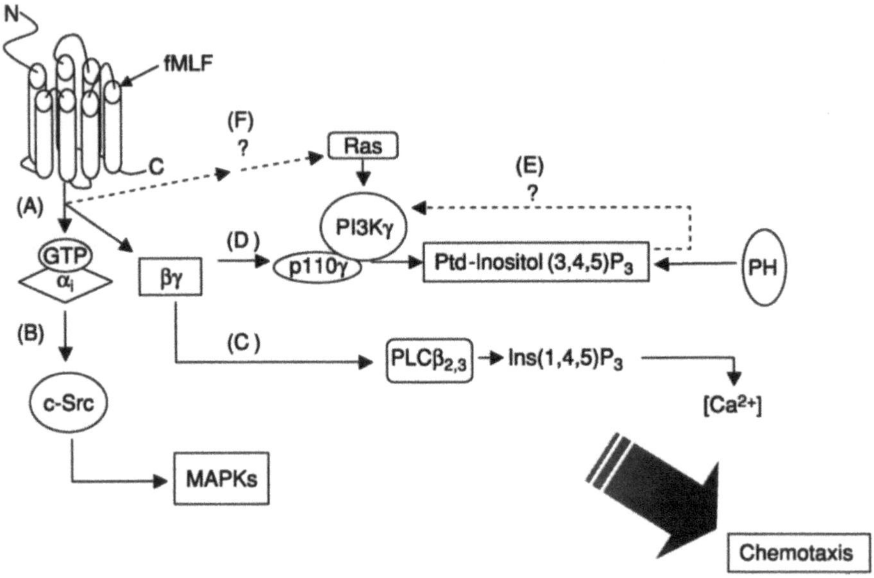

FIG. 3. Chemotaxis signaling events downstream of G protein activation. (A) Binding of a representative chemoattractant (fMLF) to its receptor triggers G protein activation and dissociation into the GTP-bound α_i subunit and the $\beta\gamma$ subunit complex. (B) The GTP-bound α_i subunit activates c-Src that in turn switches on MAP kinase signaling. (C) The $\beta\gamma$ subunit complex docks to and activates phospholipase Cβ that generates the soluble second messenger IP_3 and stimulates cytoplasmic calcium release. (D) The $\beta\gamma$ subunit complex also activates phosphotidylinositol-3-kinase-γ (PI3Kγ) which generates the membrane-bound lipid second messenger PIP_3. Multiple signaling proteins containing the conserved pleckstrin homology (PH) domain bind to PIP_3 and stimulate actin polymerization at the leading edge of the cell. (E) The pool of PIP_3 may amplify its own synthesis in a positive feedback loop that further activates PI3Kγ, thereby contributing to the high concentration of PIP_3 generated at the leading edge. (F) Receptor activation may also stimulate Ras activity by an unknown mechanism. Since Ras also activates PI3K, this represents another potential step leading to PIP_3 production (see text for additional discussion and references).

lipids in the plasma membrane (Stephens *et al.*, 1997; Stoyanov *et al.*, 1995). The P13Kγ-generated signaling lipids, including phosphatidylinositide-(3,4,5)-trisphosphate (PIP_3), serve as binding targets for signaling proteins containing specific plextrin homology (PH) and C2 domains, thereby recruiting proteins involved in actin polymerization to the leading edge of the chemotaxing cell (Fig. 3) (Firtel *et al.*, 2000). This recruitment brings additional effector molecules like protein kinase B/Akt and GTP exchange factors (GEFs) such as Rac and Cdc42 to the leading edge of the cell where they help polymerize actin filaments during pseudopod extension (Mackay *et al.*, 1998).

Much remains to be learned about the mechanisms by which PIP_3 accumulation is selectively amplified at the leading edge. A plausible sequence of events has been hypothesized. For example, it is proposed that the initial concentration of PIP_3 at the leading edge is small, but that it can act as a stimulant to Rho-GTPase activity and/or P13 kinase activity (Fig. 2). This positive feedback loop would build on itself to define the intracellular gradient of signaling molecules present in the polarized, chemotaxing cell (Wang *et al.*, 2002; Weiner *et al.*, 2002). If such feedback selectively activates P13Kγ at the leading edge, it could help account for the selective accumulation of PIP_3 at the leading edge. Moreover, it has been shown in *Dictyostelium* that the lipid phosphatase PTEN which hydrolyzes PIP_3 becomes depleted at the leading edge relative to its distribution at other regions of the plasma membrane, presumably so that it can selectively degrade PIP_3 at membrane locations other than the leading edge and maintain cell polarization (Funamoto *et al.*, 2002; Iijima *et al.*, 2002; Tamura *et al.*, 1998).

Recent studies of chemotaxis signaling in *Dictyostelium*, which shares many features with the mammalian chemotaxis signaling pathway (Chung *et al.*, 2002; Devreotes *et al.*, 2003; Merlot *et al.*, 2003) suggest that activation of P13Kγ may involve not only the G$\beta\gamma$ complex (Lopez-Ilasaca *et al.*, 1997) but also the small GTP binding protein Ras (Fig. 3) (Chung *et al.*, 2002). Both the mammalian and *Dictyostelium* P13K isoforms utilized in chemotaxis possess a well-characterized Ras binding domain (Pacold *et al.*, 2000). During chemotaxis signaling in *Dictyostelium*, GTP-bound Ras interaction with P13K is required for downstream signaling events like the activation of protein kinase B/Akt (Funamoto *et al.*, 2002; Lim *et al.*, 2001). This suggests that Ras binding is an upstream regulator of P13K activation in *Dictyostelium* chemotaxis. Further work is needed to determine the steps by which chemoattractant GPCRs activate Ras in *Dictyostelium*, as well as the relative contributions of G$\beta\gamma$ and Ras in regulating P13K both in *Dictyostelium* and mammalian chemotaxis pathways. One possible source of Ras activation is the known ability of certain GPCRs to transactivate tyrosine kinase receptors (Luttrell, 2002).

Increasingly, evidence suggests that multiple, overlapping signaling pathways are involved in the control of chemotaxis. Intracellular Ca^{2+} mobilization, as well as PIP_3 membrane targets, whose generation is regulated by $\beta\gamma$ subunits, plays a role in at least some chemoattractant signals (Li *et al.*, 2000). G protein $\beta\gamma$ subunits can directly activate phospholipase Cβ isoforms 2 and

3 (PLCβ2, 3) (Baggiolini *et al.*, 1997; Barr *et al.*, 2000; Sankaran *et al.*, 1998), whose activity generates inositol phosphate head groups that release Ca^{2+} signals from intracellular stores (Fig. 3). Alternatively, other second messengers like the CD38-synthesized cyclic-ADP ribose are known to open plasma membrane channels and thereby flood the cytoplasm with Ca^{2+}. CD38 knockout mice are more susceptible to bacterial infections, and their isolated neutrophils show impaired chemotaxis to fMLF, while chemotaxis to CXCL8 is not affected (Partida-Sanchez *et al.*, 2001).

III. G Protein-Coupled Receptors

Together, G protein-coupled receptors (GPCRs) form a large and diverse family of information processing receptors that regulate myriad sensory pathways involved in sight, smell, pain, neuroplasticity, and cell growth (Dryer *et al.*, 1999; Filipek *et al.*, 2003; Kotani *et al.*, 2001). Approximately 950 GPCRs are encoded in the human genome alone (Takeda *et al.*, 2002). All known GPCRs from a diverse array of species have been organized into eleven classes. The classes are compiled based on receptor sequence regardless of species of origin.

A current list can be found on the World Wide Web database, http://www.gpcr.org/7tm/, as described elsewhere (Horn *et al.*, 2003). Class A contains GPCRs that bind a variety of ligands from small peptides and lipids to nucleotides and represents some of the most thoroughly studied GPCRs. Among the members of this class are rhodopsin, the α and β adrenergic receptors, and the mammalian chemotaxis receptors for fMLF, C5a and chemokines. Classes B and C are comprised of hormone and pheromone receptors mostly from higher mammals but also include *Drosophila melanogaster* receptors as well. Classes D and E are dominated by yeast pheromone receptors and slime mold cyclic AMP receptors, respectively. Six additional classes have been designated and include the Frizzled/Smoothened class of Wnt ligand receptors, a class of nematode putative chemoreceptors and a class of vomeronasal mouse pheromone receptors, among others (Horn *et al.*, 2003). Large informatics efforts have been launched to analyze GPCR sequences in an attempt to define various levels of structural and functional relatedness (Attwood, 2001; Graul *et al.*, 2001; Hodges *et al.*, 2002; Horn *et al.*, 2003). From the standpoint of primary structure, even receptors from the same class often exhibit little overall sequence homology (Fig. 4); however, there are important common features of receptor architecture. On the other hand, functionally conserved receptors within the same class, like those involved in chemotaxis, can demonstrate a higher degree of primary sequence conservation.

Identity

	1.	2.	3.	4.	5.	6.	7.	8.	9.	10.	11.	12.	13.	14.	15.	16.	17.	18.	19.	20.	21.	22.	23.	24.	25.
1. FPR		32	25	28	25	25	24	30	29	28	31	28	25	26	26	24	26	24	24	24	28	25	28	24	19
2. C5aR	54		22	29	23	26	25	29	29	27	26	26	27	25	29	26	23	24	27	25	25	24	27	25	19
3. PAFR	48	46		23	21	26	24	25	26	23	24	25	23	26	24	25	25	26	25	24	22	23	24	19	16
4. LTB4R1	51	51	43		38	26	24	26	27	26	25	27	25	24	26	24	23	24	26	23	26	30	24	20	19
5. LTB4R2	45	42	39	54		22	24	25	25	29	22	25	23	23	22	22	20	24	24	22	24	32	22	20	19
6. CX3CR	51	48	50	48	42		34	32	32	32	32	30	41	42	40	41	39	34	32	39	32	29	32	21	19
7. XCR1	48	48	50	44	40	56		31	32	29	30	28	33	33	34	33	34	28	27	33	27	28	27	21	19
8. CXCR1	52	53	49	47	41	56	54		76	37	36	35	30	32	31	36	32	39	32	33	35	32	30	22	20
9. CXCR2	50	51	48	48	43	56	56	86		39	35	35	32	34	32	37	33	38	35	33	35	33	31	21	21
10. CXCR3	51	49	46	46	42	54	51	56	56		32	37	31	32	31	34	33	35	37	29	32	39	30	21	20
11. CXCR4	53	50	44	48	42	56	52	57	58	56		32	32	30	31	36	31	31	32	32	34	30	31	22	21
12. CXCR5	49	45	47	47	41	52	52	53	53	53	56		30	28	29	30	28	30	32	29	30	33	28	21	19
13. CCR1	50	52	50	49	42	66	56	58	59	57	55	54		55	63	47	56	33	35	40	35	33	32	22	18
14. CCR2	48	50	51	45	41	62	56	59	60	56	54	51	76		50	46	72	33	35	39	32	29	31	21	17
15. CCR3	51	52	50	47	41	63	56	56	57	54	55	53	81	73		45	51	33	34	38	33	31	30	22	18
16. CCR4	52	51	51	48	43	63	58	60	61	58	58	53	70	70	67		49	38	34	44	35	30	33	22	17
17. CCR5	50	52	51	47	42	62	57	59	61	56	56	54	76	83	74	72		33	34	41	36	31	31	21	18
18. CCR6	46	48	52	48	44	55	51	60	60	56	55	52	57	58	56	60	57		40	32	37	32	33	23	18
19. CCR7	47	48	48	49	44	56	51	56	59	58	57	55	57	55	55	60	58	66		30	40	37	37	22	19
20. CCR8	50	51	50	49	40	63	59	57	60	52	55	56	66	64	65	67	64	56	56		32	25	30	23	19
21. CCR9	49	52	49	50	43	56	50	57	59	52	57	54	58	59	57	58	63	59	62	58		32	34	21	20
22. CCR10	40	44	42	48	46	49	49	49	52	54	53	49	53	54	52	51	54	55	55	51	51		33	22	22
23. CCR11	48	50	52	47	43	54	52	57	57	52	51	48	54	56	53	56	57	55	58	56	60	53		21	20
24. rhodopsin	48	48	46	44	40	45	43	45	44	43	46	46	45	45	46	46	48	43	43	46	42	43	47		18
25. α_{1B} adren.	33	32	30	33	35	34	34	38	39	34	36	35	35	37	34	35	35	35	35	36	34	36	34	33	

Similarity

FIG. 4. Similarity/identity matrix generated for chemotaxis and non-chemotaxis GPCRs. Matrix was prepared for a representative set of 23 chemotaxis receptors, as well as for rhodopsin and the α_{1B}-adrenergic receptor, using the program MATGAT (Campanella *et al.*, 2003). The lower half of the matrix indicates overall sequence similarity between each pair of receptors, while the upper half indicates overall sequence identity.

IV. Conserved Elements of GPCR Structure

A. *Topology*

GPCRs are integral membrane proteins that, at least in the vast majority of cases, possess seven transmembrane α-helices (TM) alternating in their orientation as the protein sequence traverses the membrane (Fig. 5). The N-terminus is extracellular while the C-terminus is cytoplasmic, and the seven transmembrane helices are linked by three extra-cellular loops (ECL) and three intra-cellular loops (ICL) (Gether, 2000; Sakmar, 2002). These loops are important contact regions for ligands in the extracellular compartment and for heterotrimeric G proteins in the intracellular compartment (Kraft *et al.*, 2001; Miettinen *et al.*, 1999; Xie *et al.*, 1997). Additional

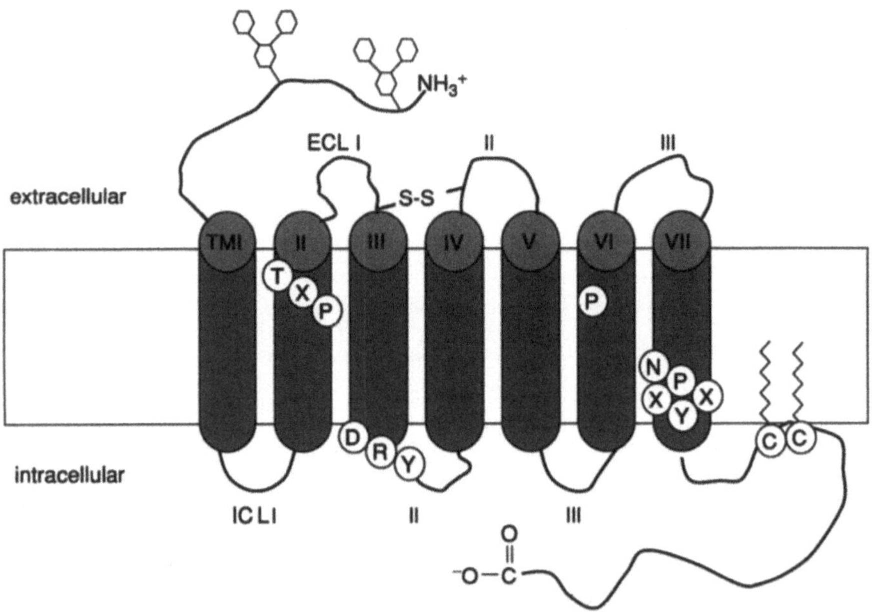

Fig. 5. Topology and common structural features of GPCRs from class A. The characteristic seven transmembrane domains, which are presumably α-helical, are shown as numbered cylinders TMI–VII. These are connected by intracellular loops ICLI–III and extracellular loops, ECLI–III. A conserved disulfide links ECLI and ECLII. The residues that define common sequence motifs, as discussed in the text, are denoted by white circles. Post-translational glycosylation and palmitoylation modifications are depicted at the N- and C-termini, respectively (see text for additional discussion and references).

ligand contacts can be found on the extracellular amino terminus while the intracellular carboxyl terminus is sometimes an important site of receptor regulation by kinases (Krupnick *et al.*, 1998).

To clarify the positions of specific residues in the standard TM domains, a uniform numbering scheme is helpful. The system used here to refer to residues in the TM domains consists of the residue number in the primary structure of the chosen receptor superscripted with the Ballasteros-Weinstein number based on the most conserved residue in the transmembrane helix (Ballesteros, 1995). For each helix, the most conserved residue is used as a common reference point, assigned the decimal value 0.50, that aligns the sequences of different receptors in the standard topology. For example, in the third transmembrane helix (TMIII) of CXCR1 (interleukin-8 receptor-α), the reference point is Arg135, designated as Arg135$^{3.50}$, where the superscript whole number identifies the helix and the decimal value indicates the position relative to the reference point. Residues N- or C-terminal to the reference position are given decimal values smaller or larger than 0.50, respectively. Thus, the adjacent residues are Asp134$^{3.49}$ and Tyr136$^{3.51}$ (Fig. 6). This system is applied only to the seven transmembrane helices, not to residues in the loops or tails due to the variable lengths of these domains.

In addition to conserved topological features, GPCRs often possess conserved sequence motifs as well. Sequence motifs and post-translational covalent modifications characteristic of class A GPCRs receptors, including chemotaxis receptors, are distributed among the various receptor domains. Some of these features are potentially important to the proper folding and delivery of the receptor to the plasma membrane while others are important for secondary structure or activation mechanism. The six conserved features discussed later are currently believed to play important roles in receptor structure, processing and/or signaling.

B. *Glycosylation Sites*

The N-terminal portion and ECLs of class A GPCRs possess 2–3 glycosylation sites indicated by the presence of a consensus motif for Asn-linked polysaccharides (Fig. 5). This post translational addition is important for the functional expression of certain chemokine receptors, like CXCR4, the platelet activating factor receptor (PAFR) and the receptor for fMLF. The addition of the polysaccharide appears to be important for protein trafficking to the membrane rather than an important aspect of ligand binding or receptor activation in each of these cases (Garcia Rodriguez *et al.*, 1995; Thordsen *et al.*, 2002; Wenzel-Seifert *et al.*, 2003).

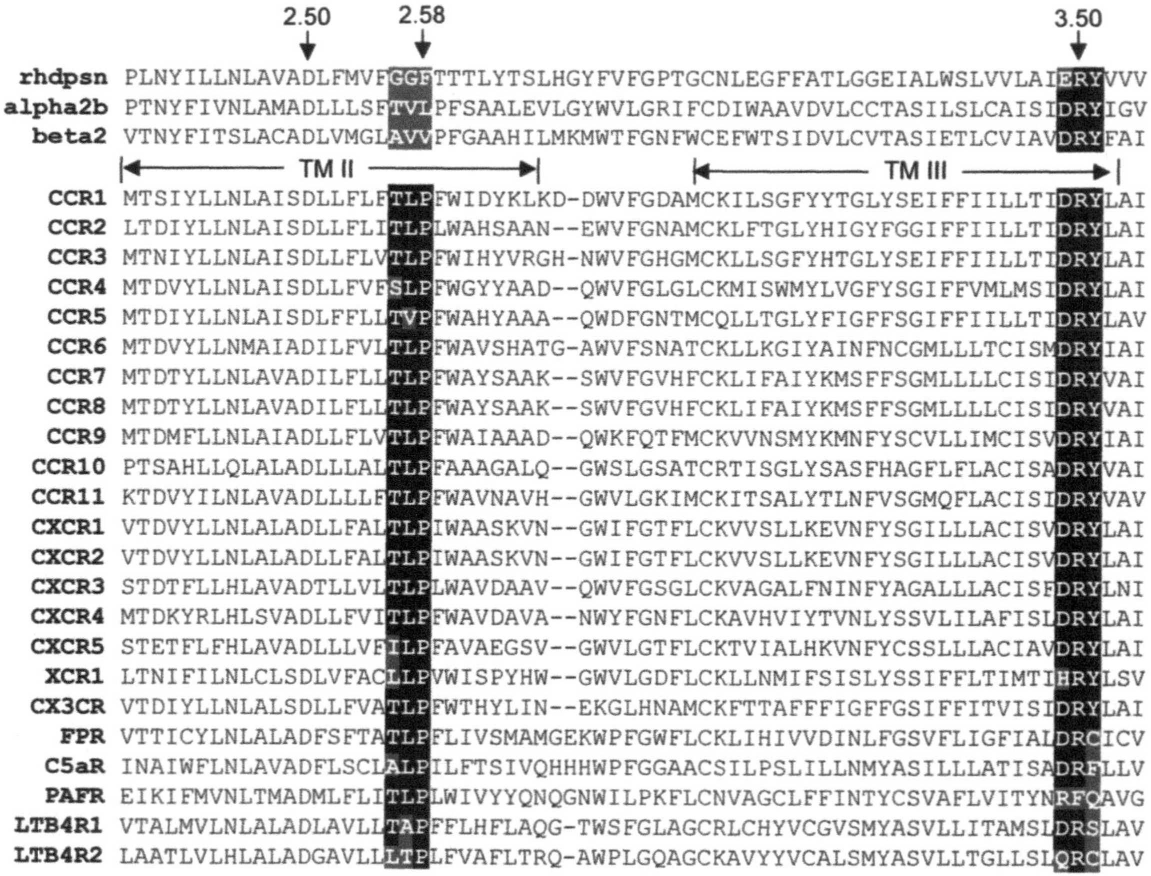
2.50
2.58
3.50
rhdpsn PLNYILLNLAVADLFMVFGGFTTTLYTSLHGYFVFGPTGCNLEGFFATLGGEIALWSLVVLAIERYVVV
alpha2b PTNYFIVNLAMADLLLSFTVLPFSAALEVLGYWVLGRIFCDIWAAVDVLCCTASILSLCAISIDRYIGV
beta2 VTNYFITSLACADLVMGLAVVPFGAAHILMKMWTFGNFWCEFWTSIDVLCVTASIETLCVIAVDRYFAI
TM II
TM III
CCR1 MTSIYLLNLAISDLLFLFTLPFWIDYKLKD-DWVFGDAMCKILSGFYYTGLYSEIFFIILLTIDRYLAI
CCR2 LTDIYLLNLAISDLLFLITLPLWAHSAAN--EWVFGNAMCKLFTGLYHIGYFGGIFFIILLTIDRYLAI
CCR3 MTNIYLLNLAISDLLFLVTLPFWIHYVRGH-NWVFGHGMCKLLSGFYHTGLYSEIFFIILLTIDRYLAI
CCR4 MTDVYLLNLAISDLLFVFSLPFWGYYAAD--QWVFGLGLCKMISWMYLVGFYSGIFFVMLMSIDRYLAI
CCR5 MTDIYLLNLAISDLFFLLTVPFWAHYAAA--QWDFGNTMCQLLTGLYFIGFFSGIFFIILLTIDRYLAV
CCR6 MTDVYLLNMAIADILFVLTLPFWAVSHATG-AWVFSNATCKLLKGIYAINFNCGMLLLTCISMDRYIAI
CCR7 MTDTYLLNLAVADILFLLTLPFWAYSAAK--SWVFGVHFCKLIFAIYKMSFFSGMLLLLCISIDRYVAI
CCR8 MTDTYLLNLAVADILFLLTLPFWAYSAAK--SWVFGVHFCKLIFAIYKMSFFSGMLLLLCISIDRYVAI
CCR9 MTDMFLLNLAIADLLFLVTLPFWAIAAAD--QWKFQTFMCKVVNSMYKMNFYSCVLLIMCISVDRYIAI
CCR10 PTSAHLLQLALADLLLALTLPFAAAGALQ--GWSLGSATCRTISGLYSASFHAGFLFLACISADRYVAI
CCR11 KTDVYILNLAVADLLLLFTLPFWAVNAVH--GWVLGKIMCKITSALYTLNFVSGMQFLACISIDRYVAV
CXCR1 VTDVYLLNLALADLLFALTLPIWAASKVN--GWIFGTFLCKVVSLLKEVNFYSGILLLACISVDRYLAI
CXCR2 VTDVYLLNLALADLLFALTLPIWAASKVN--GWIFGTFLCKVVSLLKEVNFYSGILLLACISVDRYLAI
CXCR3 STDTFLLHLAVADTLLVLTLPLWAVDAAV--QWVFGSGLCKVAGALFNINFYAGALLLACISFDRYLNI
CXCR4 MTDKYRLHLSVADLLFVITLPFWAVDAVA--NWYFGNFLCKAVHVIYTVNLYSSVLILAFISLDRYLAI
CXCR5 STETFLFHLAVADLLLVFILPFAVAEGSV--GWVLGTFLCKTVIALHKVNFYCSSLLLACIAVDRYLAI
XCR1 LTNIFILNLCLSDLVFACLLPVWISPYHW--GWVLGDFLCKLLNMIFSISLYSSIFFLTIMTIHRYLSV
CX3CR VTDIYLLNLALSDLLFVATLPFWTHYLIN--EKGLHNAMCKFTTAFFFIGFFGSIFFITVISIDRYLAI
FPR VTTICYLNLALADFSFTATLPFLIVSMAMGEKWPFGWFLCKLIHIVVDINLFGSVFLIGFIALDRCICV
C5aR INAIWFLNLAVADFLSCLALPILFTSIVQHHHWPFGGAACSILPSLILLNMYASILLLATISADRFLLV
PAFR EIKIFMVNLTMADMLFLITLPLWIVYYQNQGNWILPKFLCNVAGCLFFINTYCSVAFLVITYNRFQAVG
LTB4R1 VTALMVLNLALADLAVLLTAPFFLHFLAQG-TWSFGLAGCRLCHYVCGVSMYASVLLITAMSLDRSLAV
LTB4R2 LAATLVLHLALADGAVLLLTPLFVAFLTRQ-AWPLGQAGCKAVYYVCALSMYASVLLTGLLSLQRCLAV

C. Disulfide Bridge

A conserved disulfide bond is predicted to form between ECLI and ECLII in class A receptors (Fig. 5). While the existence of the bond has been tested in only a few cases, its importance in ligand binding is confirmed in CCR6 (Ai *et al.*, 2002) and in the β_2-adrenergic receptor (Dohlman *et al.*, 1990). Interestingly, in most chemokine receptors, two additional Cys residues are conserved in the N-terminus and ECLIII. It has been assumed that this Cys pair also forms a disulfide bond such that chemokine receptors possess two extracellular disulfides. However, for the CCR6 receptor it is known that expression and function is dependent only on the Cys residues in ECLI and ECLII and presumably the disulfide between them (Ai *et al.*, 2002). The Cys residues in the N-terminal tail and ECLIII of CCR6 are believed to exist as free sulfhydryl groups. More broadly, the tertiary structure provided by the disulfide linkage between ECLI and ECLII is assumed to be critical for appropriate receptor folding in all chemotaxis GPCRs.

D. E/DRY Motif

One of the more striking elements of sequence conservation is the Glu/Asp-Arg-Tyr sequence found at the cytoplasmic interface of TM III (Fig. 5). This so-called "$E/D^{3.49}$-$R^{3.50}$–$Y^{3.51}$" (E/DRY) motif is actually E-R-Y in rhodopsin and D-R-C in the receptor for fMLF (Fig. 6). Side chains of the E/DRY motif have been postulated to be involved in the structural dynamics of the receptor that lead to active conformations, as well as in G protein coupling (Cohen *et al.*, 1993; Miettinen *et al.*, 1999; Rasmussen *et al.*, 1999; Scheer *et al.*, 1997). In particular, these side chains are proposed to be involved in salt bridges and H-bonds that function as structural constraints maintaining the receptor in its inactive conformation. The location at the cytoplasmic-membrane interface of TMIII is consistent with the ability to impact signaling events that require contact with cytoplasmic G proteins.

FIG. 6. Partial sequence alignment of chemotaxis receptors with each other and with the best studied GPCRs. Shown are primary structures of helices TMII and TMIII from the three best studied GPCRs (rhodopsin, α_{1B}-adrenergic receptor, and β_2-adrenergic receptor) (top) as well as from 23 chemotaxis GPCRs (bottom). The alignment reveals two distinct conserved sequence elements: the TXP motif in TMII and the E/DRY motif in TMIII. The numbers indicate Ballesteros-Weinstein numeric reference points within each helix. Notably, the conserved proline at position 2.58 is not conserved among all class A GPCRs but is highly conserved among chemoattractant GPCRs (see text for additional discussion and references).

E. *TMVI Proline*

TMVI contains a conserved Pro residue near the extracellular end of TMVI at rhodopsin position $267^{6.50}$. The rhodopsin crystal structure indicates that the Pro residue introduces a kink in TMVI at this location. Biochemical characterization of a point mutation three positions away at $Cys285^{6.47}$ Thr in the β_2-adrenergic receptor, has suggested that this kink may function as a hinge during the activation of this receptor (Shi *et al.*, 2002). This mutation enhances the basal activity of the receptor in the absence of ligand, possibly by straightening the helix at the proline-induced kink (Shi *et al.*, 2002). Molecular modeling simulation of the dynamics of β_2-adrenergic receptor TMVI indicate that this hinge motion could be an important aspect of a structural change involving TMVI that has been proposed to occur during the activation of class A GPCRs. Since this Pro is conserved in chemotaxis GPCRs, such a hinge motion could be an important aspect of their activation as well.

F. *NPXXY Motif*

TMVII contains the conserved $Asn^{7.49}$-$Pro^{7.50}$-X-X-$Tyr^{7.53}$ (NPXXY) motif (Fig. 5). Its functional significance has been proposed based on preliminary studies of rhodopsin, β_2-adrenergic receptor and the chemotaxis receptors CCR5 and the receptor for fMLF. In rhodopsin, it has been suggested that this motif assists in maintaining the receptor in an inactive conformation. Specifically, in rhodopsin, $Tyr306^{7.53}$ on TMVII is argued to interact with Phe313 on helix VIII such that replacement of both residues by Cys followed by disulfide formation blocks receptor activation (Fritze *et al.*, 2003). However, these findings do not rule out the possibility that cysteine incorporation or disulfide formation blocks activation simply by perturbing receptor structure, thus additional studies are necessary to establish functional significance. In the β_2-adrenergic receptor (Barak *et al.*, 1994) the motif has been shown to be involved in sequestering the internalized receptor rather than in receptor signaling. Studies of the motif in two chemotaxis receptors, CCR5 and the receptor for fMLF, are described later.

G. *C-Terminal Palmitoylation*

Another conserved feature of class A GPCRs is a post-translational addition of two palmitate moieties covalently coupled via thioester bonds to the C-terminal tail at a Cys–Cys motif (Fig. 5) (Qanbar *et al.*, 2003). As a result the receptor C-terminus associates with the membrane creating, in effect, a fourth intracellular loop. The only chemotaxis GPCR shown thus far to be palmitoylated in this fashion is CCR5 (Kraft *et al.*, 2001). When the palmitoylation sites in the C-terminal tail of CCR5 are substituted with Ala, the receptor is not phosphorylated or internalized normally, suggesting that the palmitoylation sites are important for these regulatory events (Blanpain *et al.*, 2001).

These structural features help to define class A GPCRs and may be important components in the activation mechanism for individual receptors in this class. However, as the structural dynamics of class A receptors are beginning to be studied on a molecular level, important differences are emerging between the functional and structural roles of conserved sequence motifs in different class members. Thus, while the presence of conserved sequence motifs may in some cases help predict the mechanistic features of a novel receptor, careful experimental studies are still needed to test such predictions.

V. Rhodopsin Structure

Structural analysis of integral membrane proteins remains difficult (Hunte *et al.*, 2002). For many years the best GPCR structural models were based on micrographs and electron diffraction patterns obtained with bacteriorhodopsin and bovine rhodopsin obtained using cryo electron microscopy (Henderson *et al.*, 1990; Krebs *et al.*, 1998, 2003; Unger *et al.*, 1997). Currently, however, the high resolution X-ray crystal structure of bovine rhodopsin provides the best starting point for understanding GPCR structure (Palczewski *et al.*, 2000). This structure reveals the detailed packing arrangement of the seven transmembrane helices and the conformations of the external loops (Palczewski *et al.*, 2000). The structure captures rhodopsin in an inactive state and thus does not specifically reveal the mechanisms of activation and transmembrane signaling.

Since the activation mechanism likely involves displacements of transmembrane helices relative to one another, the helix contacts illuminated by the crystal structure provide a wealth of new information relevant to rhodopsin mechanism. For example, the tilt and central location of TMIII indicates that it can pack against and articulate with TMII and VI, while TMVI is significantly kinked by the presence of Pro267$^{6.50}$ near its extracellular end. These helical interfaces and proline-induced kinks comprise important aspects of current hypotheses of receptor activation (Hubbell *et al.*, 2003; Visiers *et al.*, 2002). Interestingly, TMs I, IV, and V also contain conserved prolines but are not kinked significantly. The extracellular loops and amino terminus were found to contain antiparallel β strands, with a strand provided by ECLII partially buried in the central pore where it is proposed to be part of the retinal binding pocket (Palczewski *et al.*, 2000; Sakmar, 2002). This structural feature effectively blocks access to the central pore and thus may not exist in receptors that reversibly bind ligands. The position of this EC loop emphasizes the possibility that rhodopsin is not necessarily the ideal structural and mechanistic model for ligand binding GPCRs including chemotaxis receptors. Finally, the rhodopsin structure reveals a short amphipathic helix, termed helix VIII, lying parallel to the plane of the membrane in the C-terminal tail distinct from TMVII. Potentially a site of intracellular effector molecule contact, the existence of helix VIII has yet to be confirmed in other GPCRs.

Numerous insightful reviews detail the use of the rhodopsin crystal structure as a model building template for other class A GPCRs (Archer *et al.*, 2003; Ballesteros *et al.*, 2001; Becker *et al.*, 2003; Filipek *et al.*, 2003). Modeling studies of GPCR structure have been advanced to such an extent that *ab initio* structure prediction methods using the minimum possible experimental structural information are underway (Vaidehi *et al.*, 2002).

VI. Proposed GPCR Activation Mechanisms

The three-dimensional description of rhodopsin provides a static view of intramolecular contacts that have been proposed to change during receptor activation. Extensive mutagenesis done on a variety of GPCRs has produced an enormous amount of biochemical data that has been instrumental in identifying residues and domains involved in receptor signaling. Key intramolecular movements hypothesized to be involved in receptor activation involve relative displacements of TMIII and TMVI. The rhodopsin crystal structure has confirmed previous descriptions of the location of TMIII with respect to the rest of the helix bundle. The central location of TMIII generates interfaces with TMII, IV, V, and VI, making TMIII a useful reference point for relative displacements of other helices.

A. *Relative Movements of Helices TMIII and TMVI*

Biochemical and biophysical characterization of rhodopsin and the adrenergic receptors have identified residues near the cytoplasmic ends of TMIII and TMVI as important constraints on receptor structure and function. In the rhodopsin structure, the conserved E/DRY motif found in TMIII is a central feature of the interface with TMVI. This structural motif in TMIII has been referred to as a "functional microdomain" because of its demonstrated functional significance in numerous GPCRs including the μ opiod receptor, 5-hydroxytryptamine$_{2A}$ receptor, m1 muscarinic receptor as well as rhodopsin and α_{1B} and β_2-adrenergic receptors (Ballesteros *et al.*, 2001; Hogger *et al.*, 1995; Huang *et al.*, 2002; Kim *et al.*, 1997; Shapiro *et al.*, 2002; Visiers *et al.*, 2002). The term functional microdomain has been used to refer to a domain that occupies a central role in the general dynamics of receptor activation that may be relevant to many structurally related molecules (Visiers *et al.*, 2002). However, it does not imply that the details of the structure or the conformational change are universally applicable.

In rhodopsin, substitution of $Glu134^{3.49}$ (the "E" in E/DRY) with Gln enhances the basal activity of retinal-free opsin suggesting that the mutation causes the receptor to adopt a conformation that resembles the light activated wild type protein (Cohen *et al.*, 1993). The acidic functional group of $Glu134^{3.49}$ is proposed to form a salt bridge with a primary amine on the adjacent $Arg135^{3.50}$ residue. In addition, $Arg135^{3.50}$ is proposed to form an ionic bond with $Glu247^{6.30}$ on TMVI. This network of weak interactions is proposed to constrain the receptor in the inactive state. The description of this "caged" $Arg135^{3.50}$ and its interactions with $Glu134^{3.49}$ and $Glu247^{6.30}$ have been confirmed by the crystal structure of rhodopsin in the inactive state and have been successfully modeled in other receptors (Palczewski *et al.*, 2000; Visiers *et al.*, 2002). By replacing the negatively charged $Glu134^{3.49}$ with the neutral Gln, the salt bridge network is disrupted, presumably allowing the ends of these two helices to separate (Kim *et al.*, 1997). The current interpretation is that upon activation the seven-helix bundle opens as the TMIII and TMVI domains are splayed apart as predicted by biophysical studies (see later), thereby exposing contact surfaces for intracellular molecules such as the heterotrimeric G proteins (Meng *et al.*, 2001).

The same network of salt bridges involving the E/DRY motif is believed to exist in α_{1B} and β_2 adrenergic receptors. Neutralization of the charge in the α_{1B}-adrenergic receptor at $Asp142^{3.49}$ by substitution with Ala results in a receptor with high basal activity, presumably as a function of disrupting the ionic interactions thus relieving some of the constraints at the interface of TMIII and TMVI (Greasley *et al.*, 2001; Scheer *et al.*, 1996). The β_2-adrenergic receptor behaves in a similar fashion when the charged residues of its E/DRY motif are neutralized by substitution with Asn or Ala (Rasmussen *et al.*, 1999). In the case of rhodopsin and the adrenergic receptors, it has been hypothesized that the carboxylic acid side chain at position 3.49 becomes protonated and therefore neutralized during normal ligand-mediated activation (Arnis *et al.*, 1994; Rasmussen *et al.*, 1999; Scheer *et al.*, 1997). An activation-triggered conformational change could break the weak salt bridges associated with the E/DRY motif, thereby allowing or driving relative movements of TMIII and TMVI.

The functional significance of the E/DRY motif is apparently not universal for class A GPCRs. Indeed, analogous residue substitutions in chemotaxis receptors that neutralize the charges of the E/DRY motif do not increase their basal activity. In particular, mutagenesis of the charged side chains at positions 3.49 and 3.50 in the E/DRY motif of CCR5, CCR3 and the receptor for fMLF caused loss of receptor function instead of increased basal activity (Auger *et al.*, 2002; Gosling *et al.*, 1997; Prossnitz *et al.*, 1995). Such results provide evidence that the microdomain is important for the native structure or mechanism of these receptors, but little mechanistic information can be deduced. Currently, the only chemotaxis receptor for which a strong case has been made for the involvement of TMIII and TMVI in the activation mechanism is the C5a receptor (see later) (Baranski *et al.*, 1999).

Biophysical approaches have been used to probe light-induced conformational changes in rhodopsin. Movement of TMVI upon receptor activation has been inferred from studies of rhodopsin using EPR (electron paramagnetic resonance) spectroscopy (Altenbach *et al.*, 1996; Farrens *et al.*, 1996; Hubbell *et al.*, 2000). Simultaneously introducing Cys residues on TMIII and TMVI followed by double labeling with a paramagnetic probe allows distance measurements to be made between these two helices. Upon light activation of rhodopsin, the calculated distance between spin labels on these two helices increases (Farrens *et al.*, 1996). It has been proposed that TMVI moves as a "rigid body" away from TMIII and that this tilting movement is allowed when the salt bridges involving the E/DRY motif are broken (Farrens *et al.*, 1996). A direct test of this hypothesis has been conducted by introducing disulfides between TMIII and TMVI in an attempt to prevent rhodopsin activation. In fact, when these two TM domains are covalently constrained by a disulfide, activation is not allowed (Cai *et al.*, 1999).

Tilting and rotation of TMVI has been detected in the β_2-adrenergic receptor as well (Ghanouni *et al.*, 2001). The attachment of fluorescent probes to native and introduced Cys residues in the β_2-adrenergic receptor on TMVI and TMIII allows direct detection of the immediate chemical environment of the fluorescent probe in the presence and absence of agonist (Gether *et al.*, 1997). This approach shows that both TMIII and TMVI undergo displacements upon ligand binding. The results are interpreted to indicate that the cytoplasmic ends of these two helices may move apart upon receptor activation. The movement of these helices is thought to be a result of removing the structural constraints endowed by the E/DRY motif.

Exactly how the cytoplasmic ends of TMIII and TMVI move away from one another remains difficult to pinpoint. As proposed by modeling studies of the β_2 adrenergic (Gether *et al.*, 1997; Shi *et al.*, 2002) and 5-hydroxytryptamine$_{2C}$ receptors (Visiers *et al.*, 2002), the highly conserved Pro residue in TMVI designated 6.50 that introduces a kink in TMVI in the rhodopsin structure (Palczewski *et al.*, 2000) could function as a hinge during receptor activation (see previously). Thus, instead of a TMVI rigid body movement as previously suggested (Farrens *et al.*, 1996), a hinge movement of the cytoplasmic end of TMVI is possible. Proline-induced kinks are proposed to function as hinges in other integral membrane proteins such as ion channels and gap junction proteins (Sansom *et al.*, 2000) and thus may represent a structural motif that plays an important role in transmembrane signaling in a wide variety of membrane proteins.

B. *Displacement of a TMII Hinge*

Another important motif described in the rhodopsin structure occurs in TMII, where the rhodopsin structure reveals a kink generated by consecutive Gly residues at Gly89$^{2.56}$-Gly90$^{2.57}$-Phe91$^{2.58}$, enabling the extracellular ends of TMII and TMI to contact one another. The kink may be functionally significant since a Gly90$^{2.57}$ Asp substitution constitutively activates chromophore-free opsin (Rao *et al.*, 1994). This mutation actually occurs naturally and manifests as night blindness. Replacement of Gly$^{2.57}$ with Asp presumably decreases flexibility at this helical turn and drives the receptor toward the active conformation. The Gly-Gly motif is not widely conserved among class A GPCRs, however, a model of chemokine receptor 5 (CCR5) predicts a similarly located and functionally important helical kink produced by residues Thr82$^{2.56}$-X83$^{2.57}$-Pro84$^{2.58}$ (TXP), a conserved motif found widely in the chemotaxis GPCRs of class A (Fig. 6) (Govaerts *et al.*, 2001).

The kinked end of TMII could be another example of a "functional microdomain" since multiple GPCRs within class A have been shown to have their activity regulated by structure changes in this domain. While the rhodopsin kink bends TMII towards TMI, the CCR5 model has TMII bent towards TMIII (Govaerts *et al.*, 2003). Substitution of Thr$^{2.56}$ in the putative kink region with Lys results in an increased basal activity for two chemokine receptors containing this motif, CCR2 and CCR5 (Arias *et al.*, 2003). According to molecular dynamics simulations, it is the presence of Thr82$^{2.56}$ two positions away from Pro84$^{2.58}$ that is important for defining the angle of the kink (Govaerts *et al.*, 2001). Modeling other side chains at that position results in a straightening of the helix suggesting that a hinge motion at this TMII Pro kink is a possible conformational change important for receptor activation (Govaerts *et al.*, 2001). There are many other chemotaxis receptors that contain this motif for which no data is available. Since the TXP motif is not found widely in other class A receptors, it appears that structural and mechanistic similarities may exist among chemotaxis receptors that are not found throughout class A (Fig. 6).

C. *Rearrangement of the TMVII-Helix VIII Interface*

TMVII contains a curious, but common motif near the cytoplasm-membrane interface. Molecular modeling of rhodopsin suggests that the hydrophobic side chain of Tyr306$^{7.53}$ in the Asn302$^{7.49}$–Pro303$^{7.50}$–X304$^{7.51}$–X305$^{7.52}$–Tyr306$^{7.53}$ (NPXXY) motif in TMVII makes contacts with the amphipathic helix VIII (Fritze *et al.*, 2003). Tyr306$^{7.53}$ on TMVII and Phe313 on helix VIII participate in a hydrophobic interaction proposed to constrain the inactive state of rhodopsin. When a rhodopsin double Cys mutant was engineered and putatively oxidized to link these two positions, light activation was blocked, suggesting that the activation mechanism could require a rearrangement of this interaction (Fritze *et al.*, 2003), or that the interaction plays an important role in native structure. Molecular dynamics simulations suggest that the disruption of the cytoplasmic ends of TMIII and TMVI could enable rearrangement of the TMVII-helix VIII interface involving NPXXY.

This motif has been shown to be important in the activation of the 5-hydroxytryptamine$_{2C}$ receptor as well (Prioleau *et al.*, 2002). The interaction of Tyr368$^{7.53}$ and Tyr375$^{7.60}$ has been inferred based on a double mutant receptor Tyr368$^{7.53}$ Phe/Tyr375$^{7.60}$ Ala that restores wild-type activity to the inactive Tyr368$^{7.53}$ Phe single mutant receptor. This result is in agreement with previous modeling studies suggesting that dynamics of NPXXY in TMVII are important in 5-hydroxytryptamine$_{2A}$ receptor activation (Konvicka *et al.*, 1998).

The NPXXY motif has been demonstrated to be important for internalization and down-regulation in the β_2-adrenergic receptor (Barak *et al.*, 1994, 1995). The only chemotaxis receptors examined with respect to the functional significance of the NPXXY motif are the receptor for fMLF (He *et al.*, 2001) and the chemokine receptor CCR5 (Aramori *et al.*, 1997). In the receptor for fMLF, Tyr301$^{7.53}$ has been shown to be important for both receptor internalization and MAP kinase activation since substitution with Phe impairs both of these functions (He *et al.*, 2001). In addition, mutation of Asn297$^{7.49}$ is not tolerated; thus, ligand affinity and downstream signaling are lost when the Asn297$^{7.49}$ Ala substitution is made (He *et al.*, 2001). Modeling studies of this domain in FPR have not been done. CCR5 signaling and receptor internalization is also dependent on Tyr297$^{7.53}$

such that the Tyr297[7.53] Ala mutant is unable to inhibit adenylyl cyclase in response to CCL4 (macrophage inflammatory protein-1β), a well characterized response to a known agonist (Aramori *et al.*, 1997). Interestingly, the ability of CCR5 to act as a co-receptor for HIV-1 is not impaired by the same substitution indicating that the receptor's ability to facilitate cell entry by HIV-1 is not dependent on its G protein coupling activity (Aramori *et al.*, 1997).

VII. Activation Mechanisms of Specific Chemotaxis GPCRs

Chemokines (abbreviated from chemoattractant cytokines) represent a robust group of at least 40 small secreted proteins, 8- to 14-kDa in size, that exhibit structural similarities and differences (Zlotnik *et al.*, 2000). Two classes predominate, C-X-C and C-C, that are distinguished by the number of residues between their two N-terminal cysteines. A new nomenclature for this large family of secreted proteins and their receptors is proposed that includes a class designation for both ligand and receptor (Murphy, 2002; Zlotnik *et al.*, 2000). The following sections discuss the best-studied chemokine receptors, CXCR1, CXCR2, CCR2, and CCR5, before turning to the non-chemokine receptors for C5a and formylated peptides. Although both chemokine and non-chemokine GPCRs involved in chemotaxis share limited sequence similarities, the chemokine receptors discussed herein are more homologous to each other than to the non-chemokine receptors. Figure 4 indicates that the sequence identity (i) between pairs of non-chemokine chemotaxis receptors ranges up to 38% but is typically 20–30%, (ii) between pairs of non-chemokine and chemokine receptors ranges up to 32% but is typically 20–30%, and (iii) between pairs of chemokine receptors ranges up to 76% but is typically 30–40%.

A. *CXCR1 and CXCR2*

Among the four receptors that are known to bind the CXC chemokine CXCL8, (CXC ligand-8, formerly interleukin-8) the receptors CXCR1 (CXC receptor-1) and CXCR2 (CXC receptor-2) are the most extensively characterized. These receptors have unique ligand binding specificities, however, both can couple to the same $G\alpha_i$ and $G\alpha_q$ GTP binding proteins and are present in neutrophils. CXCR1 is known to bind CXCL6 in addition to CXCL8, while CXCR2 is less selective, binding several CXC chemokines including, CXCL1, 2, 3, 5, 6, and 7 (Catusse *et al.*, 2003).

As for other chemokine receptors, the extracellular amino terminus occupies an important role in ligand recognition. When the amino terminal tails of CXCR1 and CXCR2 are switched, the ligand binding specificities are also switched (LaRosa *et al.*, 1992). In addition, the amino terminal portion of CXCR1 is an effective competitor of [^{125}I]CXCL8 binding site on HL-60 cells (Gayle *et al.*, 1993). Further, NMR structural analysis suggests that a short sequence from the N-terminus of CXCR1 (Pro21–Pro29) fits into a cleft on the ligand (Skelton *et al.*, 1999).

Specific residues important for ligand recognition in CXCR2 are known. When residues shared by CXCR1 and CXCR2 in the amino terminus and extracellular loop regions are replaced with alanine and ligand binding studies are performed, substitutions at seven specific positions perturb ligand affinity (Katancik *et al.*, 2000). Specifically, individual Ala substitutions at three positions specifically perturb affinity (greater than 2-fold change) for CXCL7 (Lys108, Lys110, Lys120). In contrast, CXCL8 binding affinity is significantly perturbed by five Ala substitutions (Glu7, Asp9, Glu12, Lys108, Lys120), while CXCL1 affinity is significantly perturbed by a subset of these substitutions (Glu7, Asp9, Glu12, Lys108). Clearly, the receptor-ligand interface is broadly distributed and different for each ligand (Katancik *et al.*, 2000). Depending on the ligand, residues on both the receptor N-terminus and the receptor loop ECLII can contribute to binding affinity.

The region of the receptor involved in the initial high affinity binding of ligand is apparently distinct from the structural elements involved in receptor activation (Ahuja *et al.*, 1996; Wu *et al.*, 1996b). Ligand binding and receptor signaling data obtained with chimeric receptors prepared with elements from CXCR1 and 2 show that signaling occurs even in the absence of a high-affinity binding site (Ahuja *et al.*, 1996). Specifically, when TMIV and ECLII from CXCR1 are used to replace the corresponding domains in CXCR2, the new chimera displays intracellular Ca^{2+} mobilization by CXCL8, CXCL7, and CXCL1 in the absence of detectable radioligand displacement. Moreover, examination of 20 Ala substitutions in native CXCR1 reveals 17 substitutions that reduce CXCL8 affinity but retain intracellular Ca^{2+} mobilization (Leong *et al.*, 1994). (As noted previously, intracellular Ca^{2+} mobilization occurs upon activation of many chemotaxis GPCRs and in those cases can be used as a reporter assay of receptor signaling.) At the opposite extreme, an antibody raised against CXCR1 residues not implicated in ligand binding allows CXCL8 binding but prevents receptor activation as measured by changes in intracellular calcium (Wu *et al.*, 1996b).

Together these results suggest a two-step activation model in which ligand first binds to the high affinity sequence on the receptor N-terminus thereby tethering it to the receptor. Subsequently, in an intramolecular docking event, the tethered ligand docks to a secondary site on the receptor that includes surface loop ECLII. It is this second docking event that is proposed to trigger receptor activation. Thus, the primary tethering step at the N-terminus provides strong ligand affinity and specificity, while the secondary intramolecular docking step is sufficient for activation of the transmembrane signal.

In CXCR1 the loop ECLII may play a direct role in the activation mechanism. Two residues on this loop not involved in ligand binding (Arg199, Arg203) have been replaced with His, together with two His residues introduced in a His-X_3-His configuration at the extracellular end of TMV to create an engineered Zn^{2+} binding site (Suetomi *et al.*, 2002). Functional analysis indicates that Zn^{2+} binds to these mutants and that the resulting modified receptor still binds CXCL8. However, the Zn^{2+} blocks calcium mobilization, presumably due to a linkage between ECLII and TMV generated by a strong coordination bond. These results are interpreted to suggest that constraint of relative ECLII-TMV movements by Zn^{2+} chelation prevents structural changes that are normally triggered by ligand binding. TMV dynamics have previously been implicated in the activation mechanism of the C5a receptor (DeMartino *et al.*, 1995). Given the often-repeated hypothesis that GPCRs have similar activation mechanisms involving displacements of TMIII and VI, relative displacements of ECLII and TMV during signaling may represent an alternative activation mechanism.

CXCR1 and CXCR2 possess the E/DRY and NPXXY functional microdomains discussed previously for rhodopsin and the adrenergic receptors, but the importance of these microdomains have not yet been examined. It is not known if chemokine receptors undergo conformational changes in these domains that resemble other class A receptors. Based on the complex relationship between ligand binding and receptor activation known to occur with these receptors, the mechanism of activation could be the same or significantly different from what is proposed for related GPCRs.

B. CCR2

Chemokine receptors for the ligand CCL2 (also termed macrophage chemotactic protein-1) come in two varieties that differ only in residues in the C-terminal tail, CCR2A and CCR2B. These receptors are post-transcriptional splice variants of the same gene (Charo *et al.*, 1994). There is little sequence homology between their C-terminal tails with CCR2A having fourteen more C-terminal residues than CCR2B. This primary sequence divergence explains, at least in part, their different signaling responses to specific ligands. Their signaling behavior appears to be cell type specific. When either variant is expressed in *Xenopus* oocytes, they exhibit nearly identical abilities to mobilize intracellular Ca^{2+} (Charo *et al.*, 1994). By contrast, Jurkat T cells transfected with CCR2A exhibit poor sensitivity to ligand CCL2 in chemotaxis and yield no detectable intracellular Ca^{2+} release when stimulated with this ligand, while CCR2B-transfected cells readily chemotax and mobilize Ca^{2+} in response to CCL2 (Sanders *et al.*, 2000). Such functional differences could arise from different post-translational modifications or cell trafficking arising from their contrasting C-terminal tails.

As for other chemokine receptors, a two-step activation model is proposed for CCR2B. In this model, the ligand is proposed to be tethered by the receptor N-terminus, followed by intramolecular docking to a secondary site that triggers receptor activation. When residues 2–36 of the amino terminus are fused to the transmembrane protein CD8 and expressed in human embryonic kidney cells, CCL2 binding affinity is the same as that observed for the wild type receptor (Monteclaro *et al.*, 1997), suggesting that the N-terminus alone is responsible for high affinity binding. Other studies have implicated ECLII as an additional contributor to high affinity ligand binding: the double mutation Asn104Asp/Glu105Asp in ECLII reduces ligand affinity by an order of magnitude even though the receptor retains the ability to mobilize Ca^{2+} (Han *et al.*, 1999). It could be argued, however, that the mutations in ECLII generate electrostatic or structural perturbations that are transmitted to the true high affinity site in the N-terminal region where they decrease the tethering affinity; thus, the contribution of ECLII remains unresolved. Overall, the primary determinants of ligand affinity appear to be located on the N-terminus that tethers the ligand to the receptor during the initial binding event. Subsequently, the secondary docking of the tethered ligand at ECLII is proposed to trigger receptor activation, and this secondary interaction may also enhance ligand affinity by a greater factor than observed in other members of the chemokine receptor family.

CCR2 binds a number of chemokines that are also agonists for other chemokine receptors. For example, CCL11 (eotaxin) binds both CCR5 and CCR2 (Ogilvie *et al.*, 2001). Interestingly, CCL11 activates CCR5-mediated Ca^{2+} mobilization, but is actually a CCR2 antagonist with respect to CCL2 stimulated chemotaxis (Ogilvie *et al.*, 2001). Since CCL11 can bind both chemokine receptors, it is likely that similarities exist between their N-terminal tethering sites, but the secondary ligand docking sites which bind the tethered ligand and trigger receptor activation are likely to be different.

The identification of point mutations that enhance receptor basal activity is a strategy used to better understand the differences between active and inactive receptor conformations. A substitution in TMII of CCR2, Thr94$^{2.56}$ Lys increases basal activity as measured by [^{35}S]GTPγS binding to cell membranes (Arias *et al.*, 2003). This position is part of the TXP motif conserved among many chemokine and small peptide receptors involved in chemotaxis, where "X" is a hydrophobic residue, usually Leu (Fig. 6). This motif is proposed to be involved in a Pro hinge movement in TMII of CCR5 (Govaerts *et al.*, 2001), although no direct evidence is yet available for this receptor.

The most interesting feature of the CCR2B signaling mechanism is a novel phosphorylation event implicated in receptor activation. Phosphorylation of Tyr139$^{3.51}$ located in the highly conserved E/DRY motif at the cytoplasmic end of TMIII has been found to be critical for CCR2B activation (Mellado *et al.*, 1998). Mutation of Tyr139$^{3.51}$ to Phe prevents this phosphorylation event such that no phospho-receptor is detected upon ligand stimulation, despite normal receptor expression and ligand binding profiles. Moreover, this mutant receptor fails to activate the Janus Kinase 2/STAT3 complex and to release cytoplasmic Ca^{2+}, providing strong evidence that receptor phosphorylation at Tyr139$^{3.51}$ is required for successful receptor activation (Mellado *et al.*, 1998; Vila-Coro *et al.*, 1999). The activation mechanism may involve disruption of the putative TMIII-TMVI interaction that the E/DRY motif is proposed to regulate, which could in turn enable displacement of TMVI away from TMIII.

Since phospho-activation represents a new type of activation mechanism, it is important to find out whether it can occur in other chemotaxis GPCRs possessing the E/DRY motif. Of the six chemotaxis receptors discussed in the present review, four (CXCR1, CXCR2, CCR2, and CCR5) possess the Tyr of the E/DRY motif and thus could, in principle, be activated by Tyr phosphorylation. By contrast, the receptors for fMLF and C5a lack the critical Tyr and could not be regulated the same way (Fig. 6). Both these receptors couple to Gα_i subunits and initiate chemotaxis despite their lack of a Tyr residue in the E/DRY motif. Thus, despite its importance in CCR2B phospho-activation, the significance and mechanism of the E/DRY motif may differ among chemotaxis receptors.

C. CCR5

The CC chemokine receptor 5 (CCR5) is of central interest not only because of its involvement in the cellular response to inflammation, but also because it is a key molecule utilized by Human Immunodeficiency Virus type 1 (HIV-1) to gain entry into immune cells (Dragic *et al.*, 2000). Having all the structural and G protein coupling attributes of other chemotaxis signaling GPCRs, this receptor also has a complicated pharmacology. There are at least eight known agonists (CCL2, CCL3, CCL4, CCL5, CCL8, CCL11, and CCL13) and one known antagonist (CCL7) for this heptahelical receptor (Blanpain *et al.*, 1999b). Perhaps this ligand promiscuity helps explain its use as a membrane docking site for the viral envelope glycoprotein gp 120 of HIV-1 and HIV-2 (both R5 strains of HIV) (Littman, 1998). This is a complex association first requiring a CD4-induced conformational change in gp 120 prior to docking CCR5 (Trkola *et al.*, 1996; Wu *et al.*, 1996a). The HIV-1 envelope acts as an agonist by mobilizing intracellular Ca^{2+} and initiating chemotaxis in CD4+ T-cells that are known to express CCR5, suggesting that the physiological docking mechanism results in CCR5 activation (Weissman *et al.*, 1997). However, such receptor activation may not be required for HIV entry into the cell since modified CCR5 receptors that fail to be activated by chemokine ligands can still be successfully utilized for HIV infection (Atchison *et al.*, 1996). Alternatively, it is possible that the docking of HIV protein to modified CCR5 receptors can trigger activation even though chemokine ligands do not; such ligand-specific activation has been observed for other CCR5 modifications (see later).

Chemokine docking to wild-type CCR5 and the subsequent receptor activation mechanism is proposed to follow similar principles outlined for other chemotaxis receptors, again involving an initial high-affinity tethering event followed by a low-affinity secondary docking and activation event (Blanpain *et al.*, 1999a, 2003; Navenot *et al.*, 2001). When CCR5 N-terminal residues Asp11 and Glu18 are changed to Ala, ligand affinity falls off dramatically for CCL3 (Macrophage inflammatory protein-1α), CCL4 and CCL5 (Blanpain *et al.*, 1999a; Navenot *et al.*, 2001). The Glu18Ala mutant exhibits detectable, though weak, ligand binding and retains the ability to signal in response to CCL3 and CCL5. However, cells expressing the Asp11Ala CCR5 mutant exhibit no detectable ligand binding and are unable to mobilize calcium in response to CCL3 or CCL5 (Navenot *et al.*, 2001). Thus, for this receptor, binding to the N-terminal tethering site may be absolutely required for secondary docking at the low affinity site and the associated receptor activation. Alternatively, the Asp11Ala mutant could require a higher ligand concentration for activation than can be obtained in practice, or Asp11 could be involved in a long-range conformational or electrostatic signal transmitted to the secondary docking site or the activation machinery. For a typical ligand such as CCL5 or CCL3, the core domain of the ligand binds to the amino terminus and extracellular loops of the receptor, while the amino terminal tail of the ligand contains the receptor activation domain (Blanpain *et al.*, 2003).

Given the wide promiscuity in ligand binding demonstrated by this receptor, residues implicated in binding and activation for one ligand may not be important for another. Indeed, CCL4 is capable of mobilizing intracellular calcium in cells expressing the Asp11Ala CCR5 mutant mentioned previously that is unresponsive to CCL3 and CCL5. This mutation does however increase the concentration of CCL4 needed for maximum response by more than order of magnitude (Blanpain *et al.*, 1999a).

Activation of CCR5 may involve the TXP motif. Molecular modeling of CCR5 places TXP near the extracellular end of TMII where it contacts TMIII (Govaerts *et al.*, 2001, 2003). This TMII-TMIII interface is proposed to involve residues Phe85$^{2.59}$ on TMII and Leu 104$^{3.28}$ on TMIII. The interaction of TMII and TMIII appears to be stabilized by Pro84$^{2.58}$ since the modeling suggests that Pro84$^{2.58}$ introduces a kink into TMII that is essential for the extracellular end of TMII to tilt toward its contact with TMIII in the inactive state. Evidence for a direct TMII-TMIII interaction is provided by double mutant CCR5 TMII Phe85$^{2.59}$ Leu/TMIII Leu104$^{3.28}$ Phe which restores ligand binding affinity for CCL3, CCL4, and CCL5 lost in the single mutant TMII Phe85$^{2.59}$ Leu, suggesting a successful rescue by side chain complementation at an interaction surface (Govaerts *et al.*, 2003). A mechanistic model proposes that the TMII kink is important to the activation mechanism of CCR5. Exactly how the kink is involved is not clear: simply straightening the helix by replacing Pro84$^{2.58}$ with an Ala is not sufficient to activate the receptor (Govaerts *et al.*, 2001). This model is significantly different than that proposed for rhodopsin, wherein the motif at approximately the same location in TMII is Gly89$^{2.56}$–Gly90$^{2.57}$. The crystal structure of rhodopsin does reveal a kinked conformation for TMII at this locus in the inactive state, however, the angle and orientation of the kink are significantly different from that modeled in CCR5. Interestingly, the Gly90$^{2.57}$ Asp substitution in rhodopsin is known to yield constitutive activity, highlighting this domain as an important part of the rhodopsin activation mechanism (Sieving *et al.*, 2001). Thus, both the Pro84$^{2.58}$ kink of CCR5 and the Gly89$^{2.56}$–Gly90$^{2.57}$ kink of rhodopsin may be involved in the activation mechanism, but they may differ significantly in their conformational dynamics.

Further support for a possible role of the TXP motif in CCR5 receptor activation is provided by the effects of mutagenesis at these positions on receptor function. Substitutions at Thr82$^{2.56}$ have little impact on ligand binding affinity but significantly weaken the ability of CCL3, CCL4, and CCL8 to activate the receptor as measured by intracellular calcium mobilization (Govaerts *et al.*, 2001). Moreover, the Pro84$^{2.58}$ Ala mutation reduces ligand affinity and blocks receptor activation by the same three chemokines (Govaerts *et al.*, 2001). These results illustrate the importance of Thr82$^{2.56}$ and Pro84$^{2.58}$ in receptor activation dynamics and demonstrate that ligand binding and receptor activation functions can be disconnected in mutants of CCR5. The most dramatic effects on ligand affinity and receptor activation are observed for the Pro84$^{2.58}$ Ala substitution, consistent with the idea that the putative Pro kink couples ligand recognition and receptor activation (Govaerts *et al.*, 2003). Interestingly, CCL5 appears to be more resilient to substitutions at Thr82$^{2.56}$ or Pro84$^{2.58}$, retaining nearly normal receptor affinity and activation, suggesting that this ligand uses a different activation mechanism or is able to correct the conformational perturbations triggered by TXP mutations.

Additional mutagenesis studies have begun to test other chemokine receptors for the functional significance of the TXP motif. Notably, Pro or Lys substitutions at the residue corresponding to CCR5 Thr82$^{2.56}$ yield constitutive receptor activation in CCR2 and CCR5 but not in CCR1, CCR2, CCR3, CCR4, CXCR2, or CXCR4 (Arias *et al.*, 2003). Overall, the limited success of constitutive activation by mutagenesis of the TXP motif indicates that some specialization of the motif or its local environment must exist for chemokine receptors. More broadly, the TXP motif is significantly more prevalent in chemotaxis GPCRs than other members of class A (Fig. 6), but the mechanistic importance of this motif may be limited to a smaller subset of the chemotaxis receptors that possess it.

Other models focus on the E/DRY motif as a central component of CCR5 activation. When the CCR5 DRY motif is changed to Gly-Gly-Ala, CCL4 binding affinity drops two orders of magnitude and there is no evidence of receptor-induced Ca^{2+} mobilization, inositol phosphate turnover, or chemotaxis (Gosling *et al.*, 1997). The effect of this mutation on ligand affinity suggests that this perturbation significantly alters the conformation of the extracellular regions involved in ligand docking, either by generating a large, widespread structural change, or by shifting the receptor toward the native off-state that possesses a lower ligand affinity, or by preventing G protein binding that also lowers ligand affinity. Thus it appears that the E/DRY motif is critical to the native structure, activation mechanism or G protein coupling of the CCR5 receptor.

It is unknown whether or not TMVI movements play a role in the CCR5 receptor activation mechanism. Overall, a molecular understanding of activation mechanism for CCR5 and other chemokine receptors requires further biophysical studies of transmembrane helix positions and displacements.

D. C5a Receptor

Before the characterization of chemokines the largest protein chemoattractant known was complement factor 5a. Complement protein 5 is proteolytically processed by the complement system during inflammation.

Proteolysis of this serum protein liberates the N-terminal 74 amino acids known as C5 anaphylatoxin (C5a). This small protein ligand plays broad cellular signaling roles that manifest in several leukocyte cell lineages where its receptor and compatible G protein subunits, $G\alpha_i$ and $G\alpha_{16}$ are expressed (Boulay *et al.*, 1991; Gerard *et al.*, 1991, 1994). The structure of C5a is known and is illustrated in Fig. 1C. (Zuiderweg *et al.*, 1989).

The receptor docking interface potentially involves three different surface loops of the C5a ligand (Huber-Lang *et al.*, 2003). A large ligand-receptor interface is implicated by a variety of experimental approaches. Several C-terminal peptide analogs of C5a are known and are useful in characterizing receptor binding and activation. C5a and a modified C-terminal peptide C-009 (designed to mimic residues 66–74 of C5a) are capable of displacing one another from the receptor. C5a and C-009 have IC_{50} values of 0.1 nM and 10 nM respectively when competition binding experiments are done with $[^{125}I]$C-009 (Siciliano *et al.*, 1994). The significantly lower affinity of C-009 indicates this peptide represents only part of the docking interface, although, remarkably, the peptide retains the ability to activate a proteolytically cleaved receptor missing its N-terminal tail (Siciliano *et al.*, 1994).

The C5a binding site at the amino terminus of the receptor is further localized to residues 2–22 (DeMartino *et al.*, 1994). When these residues are truncated, C5a affinity is reduced by as much as three orders of magnitude. Interestingly, another C-terminal peptide analog of C5a referred to as C-064, based on the penultimate 6 amino acids of the ligand but highly chemically modified, exhibits sub-nanomolar binding affinity for both the wild type and truncated receptor. In addition, this peptide ligand retains the ability to activate both full length and truncated receptors (DeMartino *et al.*, 1994). The affinity of this peptide ligand is about 10-fold lower than full length C5a, yet these results indicate that it is possible to design a small ligand that will mimic much of the behavior of a much larger protein ligand. Presumably the C-064 peptide binds exclusively to the secondary binding site and this interaction is sufficient to trigger receptor activation.

The results obtained with these peptides suggest that, like chemokine receptors, the C5a receptor may have a multi-step activation mechanism. The receptor N-terminus may serve as a high affinity tethering site, and following binding to this site the tethered ligand likely docks to a secondary site responsible for activation. The secondary activation site is found among, but does not necessarily include all of transmembrane domains III-VII (Siciliano *et al.*, 1994). The observation that C5a does not displace the C-terminal peptide analogs from the cleaved receptor suggests that ligand binding to the N-terminus of the receptor may generate a conformational change within the ligand or receptor that creates additional docking surfaces for ligand-receptor interactions. For example, initial binding to the receptor N-terminus could expose the C-terminal tail of C5a, which would then dock with high affinity to the activation site. Such a modified two-step mechanism would explain the ability of peptide ligands derived from the C-terminal tail to bind to the secondary site with high affinity and trigger receptor activation (Siciliano *et al.*, 1994).

Specific residues are implicated in the activation mechanism of the C5a receptor. A genetic screening procedure has been designed in yeast to identify residues in the TM domains of the C5a receptor that are critical for maintaining a functional receptor required for growth on media lacking histidine (Baranski *et al.*, 1999; Geva *et al.*, 2000). Comparison of growth in media containing or lacking ligand C5a allows identification of constitutively active receptors. These studies have generated a large number of interesting C5a receptor mutants, however, since multiple substitutions are typically found in each receptor isolate, the roles of specific side chains and positions are difficult to disentangle. This experimental strategy detects only receptors that gain function in the absence of ligand or retain normal function; nevertheless, several compelling insights are provided. Most relevant is the observation that five mutants with modifications in TMIII and sixteen with modifications in TMVI exhibit constitutive activity, suggesting that these TM domains are "hot spots" for receptor activation (Geva *et al.*, 2000). A relatively simple hypothesis states that it is the hydrophobic nature of the TMIII–TMVI interface that keeps the C5a receptor in its inactive state and that disruption of this interface by the introduction of polar side chains leads to receptor activation (Baranski *et al.*, 1999). This mechanism is similar to that proposed for rhodopsin except that the TMIII–TMVI interaction is proposed to be ionic rather than hydrophobic as for the C5a receptor (Cohen *et al.*, 1993; Greasley *et al.*, 2001; Palczewski *et al.*, 2000).

Individual position substitutions on TMIII at Ile124$^{3.40}$ and Leu127$^{3.43}$ give rise to receptors with constitutive activity (Baranski *et al.*, 1999). When changed to Asn and Gln respectively, the mutant receptors signal in the absence of C5a. Since these positions are functionally conserved as hydrophobic side chains in rhodopsin and the α_{1B} and β_2 adrenoceptors, it would be interesting if analogous substitutions induced the same phenotype. However, the same mutations have not yet been carried out in other receptors. Similarly, the Ile124$^{3.40}$ Asn/Leu127$^{3.43}$ Gln double mutant and Phe251$^{6.44}$ Ala mutants of the C5a receptor exhibit high basal G protein coupling activity. Notably, receptor endocytosis is observed to occur in a ligand independent manner in the Ile124$^{3.40}$ Asn/Leu127$^{3.43}$ Gln double mutant while the Phe251$^{6.44}$ Ala mutant did not internalize at all (Whistler *et al.*, 2002). Thus, it is proposed that each mutant yields a different receptor conformation that generates constitutive activity but has unique effects on other features of receptor regulation and processing, suggesting the possibility of multiple receptor conformations in the latter functions (Brady *et al.*, 2002).

It is not yet known whether other conserved functional motifs characteristic of class A GPCRs are involved in C5a receptor activation. The E/DRY motif, actually DRF in this receptor (Fig. 6), has not yet been investigated. Similarly, the conserved Pro of the TXP motif, ALP in this receptor (Fig. 6), could be involved in receptor activation but this possibility has not been examined. More generally, while some conserved Pro residues may play a role in activation, they have long been proposed to be important for normal receptor folding and membrane insertion (Kolakowski *et al.*, 1995).

E. *Formyl Peptide Receptor*

The source of the natural ligand for the human formyl peptide receptor (FPR) is unique among chemotaxis signaling GPCRs. Its primary ligand, formyl-Met-Leu-Phe (fMLF), typically arises from invading bacteria or damaged host cells since prokaryotic and mitochondrial proteins are typically N-formylated, while other eukaryotic proteins are not. This receptor and closely related family members are implicated in numerous pathophysiological situations from viral and bacterial infection to physically induced tissue damage (Le *et al.*, 2002). Three alleles of human FPR (FPR-26, FPR-98 and FPR-G6) are isolated and functionally characterized (Boulay *et al.*, 1991; Murphy *et al.*, 1993). They are members of a family of receptors that includes FPR like-1 and -2 (FPRL-1, -2) (Prossnitz *et al.*, 1997; Wenzel-Seifert *et al.*, 2003). The FPR alleles differ at three positions, $101^{3.28}$ (near the membrane extracellular interface in TMIII), 192 (in ECLII) and 346 (at the end of the C-terminal tail). Two of the three variant positions are $\text{Val}101^{3.28}$ in TMIII and Asn192 in ECLII in FPR26. These are changed to $\text{Leu}101^{3.28}$ in FPR-98 and Lys192 in FPR-G6. The other variant position, Glu346 in FPR-26 is an Ala in both FPR-98 and FPR-G6. Position 346 is just five residues from the C-terminal end of the molecule in a domain that is known to contain phosphorylation sites important in receptor downregulation (Prossnitz, 1997). FPR-26 is the most widely used allele when experimentally addressing fundamental questions about receptor function (Gripentrog *et al.*, 2000; Prossnitz *et al.*, 1999; Wenzel-Seifert *et al.*, 1998). Naturally occurring mutations that cause receptor dysfunction are implicated in chronic bacterial infection suggesting that FPR signaling is important for initiating a successful immune response (Jones *et al.*, 2003; Seifert *et al.*, 2001).

Because of the small size of formylated peptides, the two-step model involving ligand tethering followed by receptor activation described for other chemotaxis GPCRs is unlikely to occur in FPR and its isoforms. All three FPR alleles bind fMLF with similar nanomolar affinity, at 0.8–3.3 nM (Wenzel-Seifert *et al.*, 2003). Early work on the binding properties of formyl peptides indicates that the formyl group at the N-terminus of the ligand as well as the hydrophobic Phe and carboxyl terminus in the third position are important affinity and activation determinants (Freer *et al.*, 1980, 1982). More recent work has suggested that hydrophilic side chains like Asp and Glu in the third position are tolerated in chemotaxis and lysozyme release assays (Spisani *et al.*, 2003). However, these studies still emphasize that receptor binding affinity is dramatically reduced when the third position side chain is polar.

Chemical cross-linking of a photo-activated ligand has been used to identify receptor residues involved in ligand binding (Mills *et al.*, 1998). This method suggests that residues in the extracellular end of TMII, specifically, Val83$^{2.62}$-Arg84$^{2.63}$ -Lys85$^{2.64}$, lie in or near the ligand binding site. A testable hypothesis describing receptor side chains that interact with the ligand is proposed detailing a salt bridge between two side chains that may stabilize the receptor in an inactive state. This side chain interaction between TMII Lys85$^{2.64}$ and TMVII Asp284$^{7.36}$ is proposed to be broken by the insertion of the ligand between these two TM domains (Mills *et al.*, 1998). The prediction that side chain substitution could also break this salt bridge and trigger activation has not yet been tested, however. Another study proposes that Asp106$^{3.33}$ and Arg201$^{5.38}$ make direct contact with the formyl group of fMLF while Arg205$^{5.42}$ forms a salt bridge with its C-terminal carboxyl group (Mills *et al.*, 2000). Evidence supporting the involvement of the former residue is provided by the Asp106$^{3.33}$ Asn substitution, which decreases fMLF affinity and blocks receptor activation. Asp106$^{3.33}$ is on TMIII while Arg201$^{5.38}$ and Arg205$^{5.42}$ are on TMV. Comparing the locations of these residues with the corresponding positions in the rhodopsin structure indicates the potential for a salt bridge network between TMIII and TMV whose interface might be rearranged upon formyl peptide binding as part of receptor activation (Mills *et al.*, 2000). Exactly how this might happen remains unknown.

A large number of additional FPR-26 mutants have been described that perturb ligand binding or G protein coupling (Miettinen *et al.*, 1997, 1999; Mills *et al.*, 2000). Many of these mutants uncouple the ligand binding function of the receptor from its G protein coupling function. This behavior is often triggered by substitutions in the TMII or TMIII domain. For example, the Asp71$^{2.50}$ Ala substitution in TMII retains wild type fMLF binding affinity but is completely unable to couple to the G protein (Miettinen *et al.*, 1999).

Other FPR-26 mutants are known to partially block ligand-induced activation while retaining the ability to be phosphorylated and internalized by the cell (Prossnitz *et al.*, 1999). Ligand binding to the individual FPR-26 mutants Asp71$^{2.50}$ Ala in TMII and Arg123$^{3.50}$ Gly in TMIII is indistinguishable from wild type; however neither of these mutants exhibits ligand-induced Ca^{2+} mobilization (Prossnitz *et al.*, 1999). Interestingly, the Arg123$^{3.50}$ Gly mutant is phosphorylated and internalized following fMLF binding while these events are not observed for the Asp71$^{2.50}$ Ala mutant. Some receptor functions appear to be more tolerant of substitutions at Arg123$^{3.50}$ than Asp71$^{2.50}$, at least in FPR-26. It is suggested that these mutants structurally tune the receptor in different intermediate states of activation (Prossnitz *et al.*, 1995, 1999).

The Arg123$^{3.50}$ Ala substitution that blocks receptor signaling while retaining receptor phosphorylation and internalization is located in the highly conserved E/DRY motif, actually DRC in FPR-26. Other mutations within this motif have been specifically shown to block receptor signaling by uncoupling the receptor from its G protein. Specifically, Asp122$^{3.49}$ Ala and Arg123$^{3.50}$ Ala lose G protein coupling while retaining a nanomolar affinity for fMLP (Miettinen *et al.*, 1999). This result is in stark contrast to the constitutive G protein coupling activities of rhodopsin and the adrenergic receptors when the same substitutions are made in the E/DRY motif (Cohen *et al.*, 1993; Scheer *et al.*, 2000). In the case of FPR-26, it is believed that this motif interacts directly with the G protein and that removal of the charged side chains prevents important intermolecular contacts between the receptor and the G protein. It remains to be determined exactly what structural changes occur in FPR-26 when these mutants are expressed.

Mutants of the FPR-26 NPXXY motif are known to be important for normal ligand binding and signaling. Substitution of Asn297$^{7.49}$ with Ala resulted in loss of ligand affinity and signaling ability as measured by PLCβ and MAP kinase activation (He *et al.*, 2001). How this mutant perturbs the structure of the receptor in such a way that it loses signaling capacity is not yet known. Interestingly, FPR-26 retains signaling when the Tyr301$^{7.53}$ Phe substitution is introduced; however, this receptor is not internalized normally after ligand binding. Perhaps due to this less efficient internalization, cells transfected with this mutant are less sensitive than wild type receptor in a chemotaxis assay by an order of magnitude (He *et al.*, 2001). In rhodopsin, the corresponding residue, Tyr306$^{7.53}$, is hypothesized to mediate an interaction between TMVII and Phe313 on helix VIII (this cytoplasmic helix does not traverse the membrane) (Fritze *et al.*, 2003). Disruption of the TMVII-helix VIII interface in the 5-hydroxytryptamine$_{2C}$ receptor by a Tyr368$^{7.53}$ Asn substitution apparently releases a structural constraint and results in constitutively active G protein coupling (Konvicka *et al.*, 1998). While the available data for FPR-26 fail to show an increase in basal activity resulting from mutations in the NPXXY motif, it is clear that this motif is sensitive to side chain substitutions and more work is needed to determine if the motif plays an important role in the activation mechanism. Similarly, FPR-26 possesses the TXP motif on TMII implicated in the activation mechanism of CCR2 and CCR5 (Fig. 6), but the role of this motif in FPR-26 has not yet been investigated.

FPR-26 is described as having a high basal G protein coupling activity however, the basal activities of other FPR alleles are significantly lower (Wenzel-Seifert *et al.*, 2003). This contrast is surprising considering that only two residues differ between FPR-26 and FPR-28 or FPR-G6. Apparently, these subtle differences are enough to significantly modulate G protein coupling in the absence of ligand (Wenzel-Seifert *et al.*, 2003).

The basal activity of FPR-26 is sensitive to the presence of Na^+ ions (Seifert *et al.*, 2003; Wenzel-Seifert *et al.*, 1998). The Na^+ interaction with the receptor is not understood but may involve the highly conserved TMII Asp71$^{2.50}$ (Seifert *et al.*, 2001b). Since this residue is in the middle of TMII, the accessibility of a polar ion could only be made available by some kind of channel or pore, presumably the ligand-binding pocket. Interestingly, although the basal activities of the naturally occurring FPR alleles are measurably lower than that of FPR-26, the presence of Na^+ ions is able to reduce this the low level basal activity in the FPR-98 and FPR-G6 alleles (Wenzel-Seifert *et al.*, 2003). The basal activity of the C5a receptor is also known to be decreased as the concentration of Na^+ ions increases, while Na^+ effects are less dramatic in the platelet activating factor and leukotriene B_4 receptors (Seifert *et al.*, 2001b). No data is yet available to describe the impact of Na^+ on any of the chemokine receptors. Much remains to be learned about the mechanism of the observed Na^+ effects on the FPR and C5a receptors, and about the role of receptor basal activity in signaling.

VIII. Receptor Oligomerization

The classic model of GPCR signaling assumes that the mole ratio of ligand, receptor, and G protein is 1:1:1 in the functional ternary complex. Recently, however, increasing evidence indicates that many GPCRs function as oligomers, generally dimers. Dimerization is reported for rhodopsin (Fotiadis *et al.*, 2003; Liang *et al.*, 2003), β_2-adrenergic (Hebert *et al.*,

1996), D2 dopamine (Lee *et al.*, 2000, 2003) and the chemotaxis receptors CXCR2 (Trettel *et al.*, 2003), CXCR4 (Mellado *et al.*, 2001; Vila-Coro *et al.*, 1999), CCR2B (Rodriguez-Frade *et al.*, 1999), CCR5 (Issafras *et al.*, 2002) and C5a (Klco *et al.*, 2003). Moreover, dimer formation is correlated with receptor activation for CCR2B (Rodriguez-Frade *et al.*, 1999), CCR5 (Vila-Coro *et al.*, 2000), CXCR2 (Trettel *et al.*, 2003) and CXCR4 (Vila-Coro *et al.*, 1999). Notably, some chemokines are proposed to dimerize (Laurence *et al.*, 2000; Lowman *et al.*, 1997), and the resulting dimeric ligand could be essential for the ligand-induced dimerization proposed for specific chemokine receptors. Certain pairs of chemokine receptors can combine to form heterodimers, as reported for CCR2 and CCR5 in cells treated simultaneously with their ligands CCL2 and CCL5 (Mellado *et al.*, 2001). Currently, little evidence is available for the existence of receptor complexes larger than dimers (Issafras *et al.*, 2002).

Mouse rhodopsin is the first GPCR to be structurally characterized in its dimeric state in a native membrane. Atomic force microscopy yields images of the dimer in which the exposed faces of TMIV and V from one monomer are tentatively proposed to contact TMV and IV from the other monomer (Liang *et al.*, 2003), although higher resolution is needed to confirm this assignment. Such intermolecular contacts provide additional helix-helix interactions that may contribute to signal transduction (Fotiadis *et al.*, 2003). The role of dimerization in rhodopsin structure and function is not yet known.

The chemokine receptor CCR2B forms dimers during ligand-induced activation. Dimer formation and activation can also be triggered by bivalent antibodies against the CCR2B N-terminus (Rodriguez-Frade *et al.*, 1999). Moreover, co-transfection of CCR2B receptors with different epitope tags followed by ligand treatment, immunoprecipitation and western blotting yields a heterodimer containing both tags (Rodriguez-Frade *et al.*, 1999). Evidence suggesting that both subunits of the dimer may be required for signaling is provided by the Tyr139$^{3.51}$ Phe (the Y in DRY of CCR2B) substitution. When this substitution is introduced into one of the epitope tagged receptors it allows normal ligand-induced dimerization but acts as a dominant negative mutation, inhibiting signal transduction from receptors lacking the mutation as measured by Ca^{2+} mobilization (Rodriguez-Frade *et al.*, 1999). Thus, effective signaling appears to require not only dimer formation but also two functional receptors within the dimer. The CCR2B receptor forms a heterodimer with CCR5, yielding a highly sensitive signaling complex that can be activated by CCL5 and CCL2 at extremely low concentrations (Mellado *et al.*, 2001). These findings indicate that the signaling properties of heterodimers can differ substantially from the properties of their composite subunits, providing further strong evidence for functionally important interactions between two receptors within a dimer.

Although dimers of CCR5 are known from reports by several groups (Benkirane *et al.*, 1997; Issafras *et al.*, 2002; Vila-Coro *et al.*, 2000), the conditions that promote receptor association remain controversial. Some studies report that CCR5 is a constitutive dimer (Benkirane *et al.*, 1997) while others conclude that dimerization is induced by the ligand CCL5 or bivalent antibodies against the receptor N-terminus (Vila-Coro *et al.*, 2000). A recent study utilizing bioluminescence energy transfer (BRET) to detect proximity between donor and acceptor chimeric CCR5 molecules concludes that the receptor is self-associated, both in the endoplasmic reticulum and plasma membrane, suggesting that receptor oligomers may exist even prior to its arrival at the cell surface (Issafras *et al.*, 2002). Since the BRET signal does not increase with the addition of CCL5, no ligand-induced oligomerization event is detected by this approach. The BRET results do not allow the determination of the number of subunits in the oligomer; therefore, the observed self-association could represent receptor dimers or higher-order oligomers (Issafras *et al.*, 2002).

Certain chemokine receptors may form constitutive dimers. For example, CXCR2 is believed to exist as a dimer prior to ligand binding since some ($\sim$25%) though not all of the receptor population migrates as a stable dimer on SDS–PAGE (Trettel *et al.*, 2003). Moreover, when a truncated receptor missing only its N-terminal ligand binding tail is mixed with full length receptor, signaling is measurably impaired, suggesting that the active receptor is a constitutive dimer with two N-terminal tails that bind the dimeric form of the ligand CXCL8 (Trettel *et al.*, 2003). An alternative explanation is that the dimeric CXCL8 drives ligand-induced receptor dimerization, which would also require two receptor N-terminal tails. Thus, the available evidence does not resolve whether or not CXCR2 undergoes ligand-induced dimerization. For the related receptor CXCR4, ligand-induced dimerization is proposed to occur (Vila-Coro *et al.*, 1999). The CXCR4 receptor population is proposed to be monomeric until ligand CXCL12 (Stromal derived factor-1α) binds and drives dimer formation, an event that may play a central role in the activation mechanism (Vila-Coro *et al.*, 1999).

The chemotaxis receptors that do not bind chemokines also form dimers under certain conditions, but have not yet been found to undergo ligand-induced changes in oligomeric state. The C5a receptor is proposed to constitutively dimerize by association of respective TMIV domains, or possibly an interaction between TMI and TMII (Klco *et al.*, 2003). For this receptor, no evidence yet exists to support ligand-induced dimerization nor the formation of higher order oligomers (Floyd *et al.*, 2003). In contrast to other chemotaxis receptors, the formyl peptide receptor FPR-26 is proposed to be purely monomeric as deduced from its migration in SDS-PAGE. Surprisingly, the closely related receptors FPR-28 and FPR-G6, each differing from FPR-26 by only two residues, both yield SDS-resistant dimers (Wenzel-Seifert *et al.*, 2003). Such SDS-resistant dimer formation likely involves Ala346 since the Glu346Ala substitution in FPR-26 also produces an SDS-resistant dimer (Gripentrog *et al.*, 2003; Wenzel-Seifert *et al.*, 2003). The functional role of these dimers is not yet defined.

Overall, the functional significance of inter-receptor interactions within GPCR dimers remains unclear. Such interactions could be purely structural in certain cases. In other cases, allosteric interactions between the two subunits of a constitutive dimer or the more transient interactions formed during ligand-induced dimerization could play a central role in the mechanisms of receptor and G protein activation. GPCR oligomerization does not appear to be universal, since not all GPCRs can be shown to dimerize (Gripentrog *et al.*, 2003). Finally, many current studies use SDS-PAGE to analyze GPCR oligomers. While this approach is useful as an obvious first step its results may be irrelevant to native, membrane-bound receptors since studies of many transmembrane proteins have shown that detergent solubilization usually perturbs oligomeric state.

IX. Conclusions

Overall, the chemotaxis-initiating GPCRs characterized thus far exhibit at least two broad types of activation mechanism. The formyl-Met-Leu-Phe receptor exhibits single-step ligand docking and activation mechanism, wherein the small peptide ligand is believed to bind between the receptor transmembrane helices, thereby triggering rearrangements of these helices and transmembrane activation of the cytoplasmic G protein. The chemokine and C5a receptors, which are activated by macromolecular protein ligands, generally exhibit a more complex two-step mechanism. In the first step, ligand binds to a high affinity site on the receptor N-terminus and thereby is tethered. Subsequently, the tethered ligand docks to a second site formed by the extracellular loops and transmembrane helices, thereby triggering activation and the transmembrane signal. This second step is believed to resemble the single binding and activation step of receptors for small molecule ligands. The two-step mechanism generally requires the larger size of a protein ligand, which allows the ligand to bind simultaneously to the receptor N-terminus and the secondary site during receptor activation.

The structural changes that occur within chemotaxis GPCRs upon activation are not yet known. These ligand-induced structural changes may well differ from the light-induced structural changes proposed for rhodopsin. Moreover, even among the chemotaxis receptors themselves, the structural changes associated with activation likely differ due to specialization for different ligands, G proteins and functions. Evidence for such specialization is provided by the fact that chemotaxis receptors can exhibit different primary structures within the conserved sequence motifs believed to play roles in the activation mechanism.

Conserved sequence motifs characteristic of class A GPCRs have been investigated in a number of receptor systems, and in some cases their involvement in receptor activation is well documented. By contrast, studies of the significance and function of these conserved elements in chemotaxis GPCRs are still in the early stages. The most prominent motif uniquely conserved in chemotaxis receptors is the TXP motif located near the extracellular end of TMII. This motif is found in nearly all chemotaxis-related GPCRs, including both chemokine and non-chemokine receptors. Mutational analysis supports the involvement of the TXP motif in the activation of CCR2B and CCR5, but the motif has not yet been fully investigated in other chemotaxis GPCRs. In the case of CCR5, biochemical and molecular modeling studies suggest that the conserved Pro may induce a kink in TMII and that the kink may change conformation during activation. Although many class A receptors lack the TXP motif, rhodopsin contains a flexible Gly-Gly near the same position in TMII that may also undergo a conformational change during activation.

The E/DRY motif of TMIII is one of the more ubiquitous primary sequence elements found in class A GPCRs. Chemokine receptors contain this motif, however, it is noticeably absent or modified in a number of important chemotaxis GPCRs, including the receptors for C5a, formyl-Met-Leu-Phe and platelet activating factor. This motif is postulated to constrain rhodopsin and the adrenergic receptors in their inactive states via an extensive network of salt bridges and H-bonds evident in the rhodopsin crystal structure and modeled in other receptors. The motif may play a role in the ligand-induced displacement of TMIII and TMVI proposed by biophysical studies of rhodopsin and adrenergic receptors. While substitutions within the motif can generate constitutive activation of rhodopsin and α_{1B} and β_2-adrenergic receptor, no E/DRY motif substitutions have yet been found to generate constitutive activation of a chemotaxis receptor. However, in the case of chemokine receptors CCR2B and CXCR4, phosphorylation of the Tyr residue in the DRY motif is required for receptor activation, indicating that the motif plays a critical regulatory role in these and perhaps other chemokine receptors. This Tyr is absent in most non-chemokine chemotaxis receptors, indicating that its role is specialized rather than universal. Overall, the E/DRY motif is of central importance in the activation of at least a subset of chemotaxis receptors.

The importance of the TMIII-TMVI interface in receptor activation, as proposed for rhodopsin and adrenergic receptors, has not yet been extensively tested in chemotaxis GPCRs. Interestingly, however, genetic screening to identify activation "hot spots" in the C5a receptor yields a disproportionate number of mutations on TMIII and TMVI that lead to constitutive activation. Significantly, all hot spots were outside the E/DRY motif. Thus, while the significance of the localized E/DRY motif is not yet established in the C5a receptor, the TMIII–TMVI interface may well play an important role in the activation mechanism.

Recent evidence has suggested that many chemotaxis receptors, like other GPCRs, may form dimers. In some cases the dimer is proposed to be constitutive, as for CXCR2, C5aR, FPR-98, and FPR-G6. In other cases the dimer is proposed to be induced by ligand binding, as for CCR5 CCR2B, CXCR4, and possibly CCR5. Preliminary evidence suggests that one isoform of the receptor for formyl-Met-Leu-Phe may be purely monomeric, while two other very closely related isoforms form constitutive dimers. However, many studies use SDS–PAGE to detect dimer formation, and since membrane protein oligomeric structure can be significantly altered by detergent solubilization much work is still needed to probe the oligomeric structure of chemotaxis GPCRs in a native membrane system.

In summary, considerable diversity likely exists in the detailed ligand binding, transmembrane signaling, and G protein activation mechanisms of class A GPCRs from different functional families. Initial mechanistic studies of chemotaxis receptors also reveal considerable diversity even within this relatively small functional family. The initial ligand docking and activation steps of chemotaxis receptors can differ significantly, especially between receptors activated by macromolecular and small molecule ligands. Even the conserved sequence motifs found in the transmembrane helices of chemotaxis receptors deviate significantly in their primary structures, mechanisms and functional importance. Moreover, the role of receptor oligomers appears to vary among chemotaxis receptors. Further studies employing a wide range of experimental approaches are needed to reveal the fundamental secrets of signaling mechanisms in chemotaxis receptors and other GPCRs. In addition, further studies of GPCRs from different functional families are needed to resolve the universal and specialized mechanistic features of these important signaling machines.

References

Ahuja, S. K., Lee, J. C., and Murphy, P. M. (1996). CXC chemokines bind to unique sets of selectivity determinants that can function independently and are broadly distributed on multiple domains of human interleukin-8 receptor B. Determinants of high affinity binding and receptor activation are distinct. *J. Biol. Chem.* **271,** 225–232.

Ai, L. S., and Liao, F. (2002). Mutating the four extracellular cysteines in the chemokine receptor CCR6 reveals their differing roles in receptor trafficking, ligand binding, and signaling. *Biochem.* **41,** 8332–8341.

Altenbach, C., Yang, K., Farrens, D. L., Farahbakhsh, Z. T., Khorana, H. G., and Hubbell, W. L. (1996). Structural features and light-dependent changes in the cytoplasmic interhelical E-F loop region of rhodopsin: A site-directed spin-labeling study. *Biochem.* **35,** 12470–12478.

Amatruda, T. T., 3rd, Gerard, N. P., Gerard, C., and Simon, M. I. (1993). Specific interactions of chemoattractant factor receptors with G-proteins. *J. Biol. Chem.* **268,** 10139–10144.

Aramori, I., Ferguson, S. S., Bieniasz, P. D., Zhang, J., Cullen, B., and Cullen, M. G. (1997). Molecular mechanism of desensitization of the chemokine receptor CCR-5: Receptor signaling and internalization are dissociable from its role as an HIV-1 co-receptor. *EMBO J.* **16,** 4606–4616.

Archer, E., Maigret, B., Escrieut, C., Pradayrol, L., and Fourmy, D. (2003). Rhodopsin crystal: New template yielding realistic models of G-protein-coupled receptors? *Trends Pharmacol. Sci.* **24,** 36–40.

Arias, D. A., Navenot, J. M., Zhang, W. B., Broach, J., and Peiper, S. C. (2003). Constitutive Activation of CCR5 and CCR2 Induced by Conformational Changes in the Conserved TXP Motif in Transmembrane Helix 2. *J. Biol. Chem.* **278,** 36513–36521.

Arnis, S., Fahmy, K., Hofmann, K. P., and Sakmar, T. P. (1994). A conserved carboxylic acid group mediates light-dependent proton uptake and signaling by rhodopsin. *J. Biol. Chem.* **269,** 23879–23881.

Atchison, R. E., Gosling, J., Monteclaro, F. S., Franci, C., Digilio, L., Charo, I. F., and Goldsmith, M. A. (1996). Multiple extracellular elements of CCR5 and HIV-1 entry: Dissociation from response to chemokines. *Science* **274,** 1924–1926.

Attwood, T. K. (2001). A compendium of specific motifs for diagnosing GPCR subtypes. *Trends Pharmacol. Sci.* **22,** 162–165.

Auger, G. A., Pease, J. E., Shen, X., Xanthou, G., and Barker, M. D. (2002). Alanine scanning mutagenesis of CCR3 reveals that the three intracellular loops are essential for functional receptor expression. *Eur. J. Immunol.* **32,** 1052–1058.

Baggiolini, M. (1998). Chemokines and leukocyte traffic. *Nature* **392,** 565–568.

Baggiolini, M., Dewald, B., and Moser, B. (1997). Human chemokines: An update. *Annu. Rev. Immunol.* **15,** 675–705.

Baldwin, E. T., Weber, I. T., St Charles, R., Xuan, J. C., Appella, E., Yamada, M., Matsushima, K., Edwards, B. F., Clore, G. M., and Gronenborn, A. M. (1991). Crystal structure of interleukin 8: Symbiosis of NMR and crystallography. *Proc. Natl. Acad. Sci. USA* **88,** 502–506.

Ballesteros, J. A., Jensen, A. D., Liapakis, G., Rasmussen, S. G., Shi, L., Gether, U., and Javitch, J. A. (2001). Activation of the beta 2-adrenergic receptor involves disruption of an ionic lock between the cytoplasmic ends of transmembrane segments 3 and 6. *J. Biol. Chem.* **276,** 29171–29177.

Ballesteros, J. A., and Weinstein, H. (1995). Integrated methods for the construction of three-dimensional models of structure-function relations in G protein-coupled receptors. *Methods Neurosci.* **25,** 366–428.

Barak, L. S., Menard, L., Ferguson, S. S., Colapietro, A. M., and Caron, M. G. (1995). The conserved seven-transmembrane sequence NP(X)2,3Y of the G-protein-coupled receptor superfamily regulates multiple properties of the beta 2-adrenergic receptor. *Biochem.* **34,** 15407–15414.

Barak, L. S., Tiberi, M., Freedman, N. J., Kwatra, M. M., Lefkowitz, R. J., and Caron, M. G. (1994). A highly conserved tyrosine residue in G protein-coupled receptors is required for agonist-mediated beta 2-adrenergic receptor sequestration. *J. Biol. Chem.* **269,** 2790–2795.

Baranski, T. J., Herzmark, P., Lichtarge, O., Gerber, B. O., Trueheart, J., Meng, E. C., Iiri, T., Sheikh, S. P., and Bourne, H. R. (1999). C5a receptor activation. Genetic identification of critical residues in four transmembrane helices. *J. Biol. Chem.* **274,** 15757–15765.

Barr, A. J., Ali, H., Haribabu, B., Snyderman, R., and Smrcka, A. V. (2000). Identification of a region at the N-terminus of phospholipase C-beta 3 that interacts with G protein beta gamma subunits. *Biochem.* **39,** 1800–1806.

Becker, O. M., Shacham, S., Marantz, Y., and Noiman, S. (2003). Modeling the 3D structure of GPCRs: Advances and application to drug discovery. *Curr. Opin. Drug Discov. Devel.* **6,** 353–361.

Benkirane, M., Jin, D. Y., Chun, R. F., Koup, R. A., and Jeang, K. T. (1997). Mechanism of transdominant inhibition of CCR5-mediated HIV-1 infection by ccr5delta32. *J. Biol. Chem.* **272,** 30603–30606.

Blanpain, C., Doranz, B. J., Bondue, A., Govaerts, C., De Leener, A., Vassart, G., Doms, R. W., Proudfoot, A., and Parmentier, M. (2003). The core domain of chemokines binds CCR5 extracellular domains while their amino terminus interacts with the transmembrane helix bundle. *J. Biol. Chem.* **278,** 5179–5187.

Blanpain, C., Doranz, B. J., Vakili, J., Rucker, J., Govaerts, C., Baik, S. S., Lorthioir, O., Migeotte, I., Libert, F., Baleux, F., Vassart, G., Doms, R. W., and Parmentier, M. (1999a). Multiple charged and aromatic residues in CCR5 amino-terminal domain are involved in high affinity binding of both chemokines and HIV-1 Env protein. *J. Biol. Chem.* **274,** 34719–34727.

Blanpain, C., Migeotte, I., Lee, B., Vakili, J., Doranz, B. J., Govaerts, C., Vassart, G., Doms, R. W., and Parmentier, M. (1999b). CCR5 binds multiple CC-chemokines: MCP-3 acts as a natural antagonist. *Blood* **94,** 18900–18905.

Blanpain, C., Wittamer, V., Vanderwinden, J. M., Boom, A., Renneboog, B., Lee, B., Le Poul, E., El Asmar, L., Govaerts, C., Vassart, G., Doms, R. W., and Parmentier, M. (2001). Palmitoylation of CCR5 is critical for receptor trafficking and efficient activation of intracellular signaling pathways. *J. Biol. Chem.* **276,** 23795–23804.

Bokoch, G. M. (1995). Chemoattractant signaling and leukocyte activation. *Blood* **86,** 1649–1660.

Bokoch, G. M., Bickford, K., and Bohl, B. P. (1988). Subcellular localization and quantitation of the major neutrophil pertussis toxin substrate, Gn. *J. Cell Biol.* **106,** 1927–1936.

Bokoch, G. M., and Gilman, A. G. (1984). Inhibition of receptor-mediated release of arachidonic acid by pertussis toxin. *Cell* **39,** 301–308.

Boulay, F., Mery, L., Tardif, M., Brouchon, L., and Vignais, P. (1991). Expression cloning of a receptor for C5a anaphylatoxin on differentiated HL-60 cells. *Biochem.* **30,** 2993–2999.

Brady, A. E., and Limbird, L. E. (2002). G protein-coupled receptor interacting proteins: Emerging roles in localization and signal transduction. *Cell Signal.* **14,** 297–309.

Cai, K., Klein-Seetharaman, J., Hwa, J., Hubbell, W. L., and Khorana, H. G. (1999). Structure and function in rhodopsin: Effects of disulfide cross-links in the cytoplasmic face of rhodopsin on transducin activation and phosphorylation by rhodopsin kinase. *Biochem.* **38,** 12893–12898.

Campanella, J. J., Bitincka, L., and Smalley, J. (2003). MatGAT: An application that generates similarity/identity matrices using protein or DNA sequences. *BMC Bioinformatics* **4,** 29.

Catusse, J., Liotard, A., Loillier, B., Pruneau, D., and Paquet, J. L. (2003). Characterization of the molecular interactions of interleukin-8 (CXCL8), growth related oncogen alpha (CXCL1) and a non-peptide antagonist (SB 225002) with the human CXCR2. *Biochem. Pharmacol.* **65,** 813–821.

Charo, I. F., Myers, S. J., Herman, A., Franci, C., Connolly, A. J., and Coughlin, S. R. (1994). Molecular cloning and functional expression of two monocyte chemoattractant protein 1 receptors reveals alternative splicing of the carboxyl-terminal tails. *Proc. Natl. Acad. Sci. USA* **91,** 2752–2756.

Chung, C. Y., and Firtel, R. A. (2002). Signaling pathways at the leading edge of chemotaxing cells. *J. Muscle Res. Cell Motil.* **23,** 773–779.

Cohen, G. B., Yang, T., Robinson, P. R., and Oprian, D. D. (1993). Constitutive activation of opsin: Influence of charge at position 134 and size at position 296. *Biochem.* **32,** 6111–6115.

DeMartino, J. A., Konteatis, Z. D., Siciliano, S. J., Van Riper, G., Underwood, D. J., Fischer, P. A., and Springer, M. S. (1995). Arginine 206 of the C5a receptor is critical for ligand recognition and receptor activation by C-terminal hexapeptide analogs. *J. Biol. Chem.* **270,** 15966–15969.

DeMartino, J. A., Van Riper, G., Siciliano, S. J., Molineaux, C. J., Konteatis, Z. D., Rosen, H., and Springer, M. S. (1994). The amino terminus of the human C5a receptor is required for high affinity C5a binding and for receptor activation by C5a but not C5a analogs. *J. Biol. Chem.* **269,** 14446–14450.

Devreotes, P., and Janetopoulos, C. (2003). Eukaryotic chemotaxis: Distinctions between directional sensing and polarization. *J. Biol. Chem.* **278,** 20445–20448.

Dohlman, H. G., Caron, M. G., DeBlasi, A., Frielle, T., and Lefkowitz, R. J. (1990). Role of extracellular disulfide-bonded cysteines in the ligand binding function of the beta 2-adrenergic receptor. *Biochem.* **29,** 2335–2342.

Dragic, T., Trkola, A., Thompson, D. A., Cormier, E. G., Kajumo, F. A., Maxwell, E., Lin, S. W., Ying, W., Smith, S. O., Sakmar, T. P., and Moore, J. P. (2000). A binding pocket for a small molecule inhibitor of HIV-1 entry within the transmembrane helices of CCR5. *Proc. Natl. Acad. Sci. USA* **97,** 5639–5644.

Dryer, L., and Berghard, A. (1999). Odorant receptors: A plethora of G-protein–coupled receptors. *Trends Pharmacol. Sci.* **20,** 413–417.

Farrens, D. L., Altenbach, C., Yang, K., Hubbell, W. L., and Khorana, H. G. (1996). Requirement of rigid-body motion of transmembrane helices for light activation of rhodopsin. *Science* **274,** 768–770.

Feltner, D. E., Smith, R. H., and Marasco, W. A. (1986). Characterization of the plasma membrane bound GTPase from rabbit neutrophils. I. Evidence for an Ni-like protein coupled to the formyl peptide, C5a, and leukotriene B4 chemotaxis receptors. *J. Immunol.* **137,** 1961–1970.

Filipek, S., Stenkamp, R. E., Teller, D. C., and Palczewski, K. (2003). G protein-coupled receptor rhodopsin: A prospectus. *Annu. Rev. Physiol.* **65,** 851–879.

Firtel, R. A., and Chung, C. Y. (2000). The molecular genetics of chemotaxis: Sensing and responding to chemoattractant gradients. *Bioessays* **22,** 603–615.

Floyd, D. H., Geva, A., Bruinsma, S. P., Overton, M. C., Blumer, K. J., and Baranski, T. J. (2003). C5a receptor oligomerization. II. Fluorescence resonance energy transfer studies of a human G protein-coupled receptor expressed in yeast. *J. Biol. Chem.* **278,** 35354–35361.

Fotiadis, D., Liang, Y., Filipek, S., Saperstein, D. A., Engel, A., and Palczewski, K. (2003). Atomic-force microscopy: Rhodopsin dimers in native disc membranes. *Nature* **421,** 127–128.

Freer, R. J., Day, A. R., Muthukumaraswamy, N., Pinon, D., Wu, A., Showell, H. J., and Becker, E. L. (1982). Formyl peptide chemoattractants: A model of the receptor on rabbit neutrophils. *Biochem.* **21,** 257–263.

Freer, R. J., Day, A. R., Radding, J. A., Schiffmann, E., Aswanikumar, S., Showell, H. J., and Becker, E. L. (1980). Further studies on the structural requirements for synthetic peptide chemoattractants. *Biochem.* **19,** 2404–2410.

Fritze, O., Filipek, S., Kuksa, V., Palczewski, K., Hofmann, K. P., and Ernst, O. P. (2003). Role of the conserved NPxxY(x)5,6F motif in the rhodopsin ground state and during activation. *Proc. Natl. Acad. Sci. USA* **100,** 2290–2295.

Funamoto, S., Meili, R., Lee, S., Parry, L., and Firtel, R. A. (2002). Spatial and temporal regulation of 3-phosphoinositides by PI 3-kinase and PTEN mediates chemotaxis. *Cell* **109,** 611–623.

Gallin, J. I., and Snyderman, R. (1999). Inflammation: Basic principles and clinical correlates. Lippincott Williams & Wilkins, Philadelphia, PA.

Garcia Rodriguez, C., Cundell, D. R., Tuomanen, E. I., Kolakowski, L. F., Jr., Gerard, C., and Gerard, N. P. (1995). The role of N-glycosylation for functional expression of the human platelet-activating factor receptor. Glycosylation is required for efficient membrane trafficking. *J. Biol. Chem.* **270,** 25178–25184.

Gayle, R. B., 3rd, Sleath, P. R., Srinivason, S., Birks, C. W., Weerawarna, K. S., Cerretti, D. P., Kozlosky, C. J., Nelson, N., Vanden Bos, T., and Beckmann, M. P. (1993). Importance of the amino terminus of the interleukin-8 receptor in ligand interactions. *J. Biol. Chem.* **268,** 7283–7289.

Gerard, C., and Gerard, N. P. (1994). C5A anaphylatoxin and its seven transmembrane-segment receptor. *Annu. Rev. Immunol.* **12,** 775–808.

Gerard, N. P., and Gerard, C. (1991). The chemotactic receptor for human C5a anaphylatoxin. *Nature* **349,** 614–617.

Gether, U. (2000). Uncovering molecular mechanisms involved in activation of G protein-coupled receptors. *Endocr. Rev.* **21,** 90–113.

Gether, U., Lin, S., Ghanouni, P., Ballesteros, J. A., Weinstein, H., and Kobilka, B. K. (1997). Agonists induce conformational changes in transmembrane domains III and VI of the beta2 adrenoceptor. *EMBO J.* **16,** 6737–6747.

Geva, A., Lassere, T. B., Lichtarge, O., Pollitt, S. K., and Baranski, T. J. (2000). Genetic mapping of the human C5a receptor. Identification of transmembrane amino acids critical for receptor function. *J. Biol. Chem.* **275,** 35393–35401.

Ghanouni, P., Steenhuis, J. J., Farrens, D. L., and Kobilka, B. K. (2001). Agonist-induced conformational changes in the G-protein-coupling domain of the beta 2 adrenergic receptor. *Proc. Natl. Acad. Sci. USA* **98,** 5997–6002.

Goldman, D. W., Chang, F. H., Gifford, L. A., Goetzl, E. J., and Bourne, H. R. (1985). Pertussis toxin inhibition of chemotactic factor-induced calcium mobilization and function in human polymorphonuclear leukocytes. *J. Exp. Med.* **162,** 145–156.

Gosling, J., Monteclaro, F. S., Atchison, R. E., Arai, H., Tsou, C. L., Goldsmith, M. A., and Charo, I. F. (1997). Molecular uncoupling of C-C chemokine receptor 5-induced chemotaxis and signal transduction from HIV-1 coreceptor activity. *Proc. Natl. Acad. Sci. USA* **94,** 5061–5066.

Govaerts, C., Blanpain, C., Deupi, X., Ballet, S., Ballesteros, J. A., Wodak, S. J., Vassart, G., Pardo, L., and Parmentier, M. (2001). The TXP motif in the second transmembrane helix of CCR5. A structural determinant of chemokine-induced activation. *J. Biol. Chem.* **276,** 13217–13225.

Govaerts, C., Bondue, A., Springael, J. Y., Olivella, M., Deupi, X., Le Poul, E., Wodak, S. J., Parmentier, M., Pardo, L., and Blanpain, C. (2003). Activation of CCR5 by chemokines involves an aromatic cluster between transmembrane helices 2 and 3. *J. Biol. Chem.* **278,** 1892–1903.

Graul, R. C., and Sadee, W. (2001). Evolutionary relationships among G protein-coupled receptors using a clustered database approach. *AAPS PharmSci.* **3,** E12.

Greasley, P. J., Fanelli, F., Scheer, A., Abuin, L., Nenniger-Tosato, M., DeBenedetti, P. G., and Cotecchia, S. (2001). Mutational and computational analysis of the alpha(1b)-adrenergic receptor. Involvement of basic and hydrophobic residues in receptor activation and G protein coupling. *J. Biol. Chem.* **276,** 46485–46494.

Gripentrog, J. M., Jesaitis, A. J., and Miettinen, H. M. (2000). A single amino acid substitution (N297A) in the conserved NPXXY sequence of the human N-formyl peptide receptor results in inhibition of desensitization and endocytosis, and a dose-dependent shift in p42/44 mitogen-activated protein kinase activation and chemotaxis. *Biochem. J.* **352**(Pt 2), 399–407.

Gripentrog, J. M., Kantele, K. P., Jesaitis, A. J., and Miettinen, H. M. (2003). Experimental evidence for lack of homodimerization of the G protein-coupled human N-formyl peptide receptor. *J. Immunol.* **171,** 3187–3193.

Han, K. H., Green, S. R., Tangirala, R. K., Tanaka, S., and Quehenberger, O. (1999). Role of the first extracellular loop in the functional activation of CCR2. The first extracellular loop contains distinct domains necessary for both agonist binding and transmembrane signaling. *J. Biol. Chem.* **274,** 32055–32062.

He, R., Browning, D. D., and Ye, R. D. (2001). Differential roles of the NPXXY motif in formyl peptide receptor signaling. *J. Immunol.* **166,** 4099–4105.

Hebert, T. E., Moffett, S., Morello, J. P., Loisel, T. P., Bichet, D. G., Barret, C., and Bouvier, M. (1996). A peptide derived from a beta2-adrenergic receptor transmembrane domain inhibits both receptor dimerization and activation. *J. Biol. Chem.* **271,** 16384–16392.

Henderson, R., Baldwin, J. M., Ceska, T. A., Zemlin, F., Beckmann, E., and Downing, K. H. (1990). Model for the structure of bacteriorhodopsin based on high-resolution electron cryomicroscopy. *J. Mol. Biol.* **213,** 899–929.

Hodges, P. E., Carrico, P. M., Hogan, J. D., O'Neill, K. E., Owen, J. J., Mangan, M., Davis, B. P., Brooks, J. E., and Garrels, J. I. (2002). Annotating the human proteome: The Human Proteome Survey Database (HumanPSD) and an in-depth target database for G protein-coupled receptors (GPCR-PD) from Incyte Genomics. *Nucleic Acids Res.* **30,** 137–141.

Hogger, P., Shockley, M. S., Lameh, J., and Sadee, W. (1995). Activating and inactivating mutations in N- and C-terminal i3 loop junctions of muscarinic acetylcholine Hm1 receptors. *J. Biol. Chem.* **270,** 7405–7410.

Holmes, W. E., Lee, J., Kuang, W. J., Rice, G. C., and Wood, W. I. (1991). Structure and functional expression of a human interleukin-8 receptor. *Science* **253,** 1278–1280.

Horn, F., Bettler, E., Oliveira, L., Campagne, F., Cohen, F. E., and Vriend, G. (2003). GPCRDB information system for G protein-coupled receptors. *Nucleic Acids Res.* **31,** 294–297.
Huang, P., Visiers, I., Weinstein, H., and Liu-Chen, L. Y. (2002). The local environment at the cytoplasmic end of TM6 of the mu opioid receptor differs from those of rhodopsin and monoamine receptors: Introduction of an ionic lock between the cytoplasmic ends of helices 3 and 6 by a L6.30(275)E mutation inactivates the mu opioid receptor and reduces the constitutive activity of its T6.34(279)K mutant. *Biochem.* **41,** 11972–11980.
Hubbell, W. L., Altenbach, C., Hubbell, C. M., and Khorana, H. G. (2003). Rhodopsin structure, dynamics, and activation: A perspective from crystallography, site-directed spin labeling, sulfhydryl reactivity, and disulfide cross-linking. *Adv. Protein Chem.* **63,** 243–290.
Hubbell, W. L., Cafiso, D. S., and Altenbach, C. (2000). Identifying conformational changes with site-directed spin labeling. *Nat. Struct. Biol.* **7,** 735–739.
Huber-Lang, M. S., Sarma, J. V., McGuire, S. R., Lu, K. T., Padgaonkar, V. A., Younkin, E. M., Guo, R. F., Weber, C. H., Zuiderweg, E. R., Zetoune, F. S., and Ward, P. A. (2003). Structure-function relationships of human C5a and C5aR. *J. Immunol.* **170,** 6115–6124.
Hunte, C., and Michel, H. (2002). Crystallisation of membrane proteins mediated by antibody fragments. *Curr. Opin. Struct. Biol.* **12,** 503–508.
Iijima, M., and Devreotes, P. (2002). Tumor suppressor PTEN mediates sensing of chemoattractant gradients. *Cell* **109,** 599–610.
Issafras, H., Angers, S., Bulenger, S., Blanpain, C., Parmentier, M., Labbe-Jullie, C., Bouvier, M., and Marullo, S. (2002). Constitutive agonist-independent CCR5 oligomerization and antibody-mediated clustering occurring at physiological levels of receptors. *J. Biol. Chem.* **277,** 34666–34673.
Jones, B. E., Miettinen, H. M., Jesaitis, A. J., and Mills, J. S. (2003). Mutations of F110 and C126 of the formyl peptide receptor interfere with G-protein coupling and chemotaxis. *J. Periodontol.* **74,** 475–484.
Katancik, J. A., Sharma, A., and de Nardin, E. (2000). Interleukin 8, neutrophil-activating peptide-2 and GRO-alpha bind to and elicit cell activation via specific and different amino acid residues of CXCR2. *Cytokine* **12,** 1480–1488.
Kim, J. M., Altenbach, C., Thurmond, R. L., Khorana, H. G., and Hubbell, W. L. (1997). Structure and function in rhodopsin: Rhodopsin mutants with a neutral amino acid at E134 have a partially activated conformation in the dark state. *Proc. Natl. Acad. Sci. USA* **94,** 14273–14278.
Klco, J. M., Lassere, T. B., and Baranski, T. J. (2003). C5a receptor oligomerization. I. Disulfide trapping reveals oligomers and potential contact surfaces in a G protein-coupled receptor. *J. Biol. Chem.* **278,** 35345–35353.
Knall, C., Worthen, G. S., and Johnson, G. L. (1997). Interleukin 8-stimulated phosphatidylinositol-3-kinase activity regulates the migration of human neutrophils independent of extracellular signal-regulated kinase and p38 mitogen-activated protein kinases. *Proc. Natl. Acad. Sci. USA* **94,** 3052–3057.
Kolakowski, L. F., Jr., Lu, B., Gerard, C., and Gerard, N. P. (1995). Probing the "message:address" sites for chemoattractant binding to the C5a receptor. Mutagenesis of hydrophilic and proline residues within the transmembrane segments *J. Biol. Chem.* **270,** 18077–18082.
Konvicka, K., Guarnieri, F., Ballesteros, J. A., and Weinstein, H. (1998). A proposed structure for transmembrane segment 7 of G protein-coupled receptors incorporating an asn-Pro/Asp-Pro motif. *Biophys. J.* **75,** 601–611.

Koo, C., Lefkowitz, R. J., and Snyderman, R. (1983). Guanine nucleotides modulate the binding affinity of the oligopeptide chemoattractant receptor on human polymorphonuclear leukocytes. *J. Clin. Invest.* **72,** 748–753.

Kotani, M., Mollereau, C., Detheux, M., Le Poul, E., Brezillon, S., Vakili, J., Mazarguil, H., Vassart, G., Zajac, J. M., and Parmentier, M. (2001). Functional characterization of a human receptor for neuropeptide FF and related peptides. *Br. J. Pharmacol.* **133,** 138–144.

Kraft, K., Olbrich, H., Majoul, I., Mack, M., Proudfoot, A., and Oppermann, M. (2001). Characterization of sequence determinants within the carboxyl-terminal domain of chemokine receptor CCR5 that regulate signaling and receptor internalization. *J. Biol. Chem.* **276,** 34408–34418.

Krebs, A., Edwards, P. C., Villa, C., Li, J., and Schertler, G. F. (2003). The three-dimensional structure of bovine rhodopsin determined by electron cryomicroscopy. *J. Biol. Chem.* **278,** 50217–50225.

Krebs, A., Villa, C., Edwards, P. C., and Schertler, G. F. (1998). Characterisation of an improved two-dimensional p22121 crystal from bovine rhodopsin. *J. Mol. Biol.* **282,** 991–1003.

Krupnick, J. G., and Benovic, J. L. (1998). The role of receptor kinases and arrestins in G protein-coupled receptor regulation. *Annu. Rev. Pharmacol. Toxicol.* **38,** 289–319.

LaRosa, G. J., Thomas, K. M., Kaufmann, M. E., Mark, R., White, M., Taylor, L., Gray, G., Witt, D., and Navarro, J. (1992). Amino terminus of the interleukin-8 receptor is a major determinant of receptor subtype specificity. *J. Biol. Chem.* **267,** 25402–25406.

Laurence, J. S., Blanpain, C., Burgner, J. W., Parmentier, M., and LiWang, P. J. (2000). CC chemokine MIP-1 beta can function as a monomer and depends on Phe13 for receptor binding. *Biochem.* **39,** 3401–3409.

Le, Y., Murphy, P. M., and Wang, J. M. (2002). Formyl-peptide receptors revisited. *Trends Immunol.* **23,** 541–548.

Lee, S. P., O'Dowd, B. F., Ng, G. Y., Varghese, G., Akil, H., Mansour, A., Nguyen, T., and George, S. R. (2000). Inhibition of cell surface expression by mutant receptors demonstrates that D2 dopamine receptors exist as oligomers in the cell. *Mol. Pharmacol.* **58,** 120–128.

Lee, S. P., O'Dowd, B. F., Rajaram, R. D., Nguyen, T., and George, S. R. (2003). D2 dopamine receptor homodimerization is mediated by multiple sites of interaction, including an intermolecular interaction involving transmembrane domain 4. *Biochem.* **42,** 11023–11031.

Leong, S. R., Kabakoff, R. C., and Hebert, C. A. (1994). Complete mutagenesis of the extracellular domain of interleukin-8 (IL-8) type A receptor identifies charged residues mediating IL-8 binding and signal transduction. *J. Biol. Chem.* **269,** 19343–19348.

Li, Z., Jiang, H., Xie, W., Zhang, Z., Smrcka, A. V., and Wu, D. (2000). Roles of PLC-beta2 and -beta3 and PI3Kgamma in chemoattractant-mediated signal transduction. *Science* **287,** 1046–1049.

Liang, Y., Fotiadis, D., Filipek, S., Saperstein, D. A., Palczewski, K., and Engel, A. (2003). Organization of the G protein-coupled receptors rhodopsin and opsin in native membranes. *J. Biol. Chem.* **278,** 21655–21662.

Lim, C. J., Spiegelman, G. B., and Weeks, G. (2001). RasC is required for optimal activation of adenylyl cyclase and Akt/PKB during aggregation. *EMBO J.* **20,** 4490–4499.

Littman, D. R. (1998). Chemokine receptors: Keys to AIDS pathogenesis? *Cell* **93,** 677–680.
Locati, M., and Murphy, P. M. (1999). Chemokines and chemokine receptors: Biology and clinical relevance in inflammation and AIDS. *Annu. Rev. Med.* **50,** 425–440.
Lopez-Ilasaca, M., Crespo, P., Pellici, P. G., Gutkind, J. S., and Wetzker, R. (1997). Linkage of G protein-coupled receptors to the MAPK signaling pathway through PI 3-kinase gamma. *Science* **275,** 394–397.
Lowman, H. B., Fairbrother, W. J., Slagle, P. H., Kabakoff, R., Liu, J., Shire, S., and Hebert, C. A. (1997). Monomeric variants of IL-8: Effects of side chain substitutions and solution conditions upon dimer formation. *Protein Sci.* **6,** 598–608.
Luttrell, L. M. (2002). Activation and targeting of mitogen-activated protein kinases by G-protein-coupled receptors. *Can. J. Physiol. Pharmacol.* **80,** 375–382.
Ma, Y. C., Huang, J., Ali, S., Lowry, W., and Huang, X. Y. (2000). Src tyrosine kinase is a novel direct effector of G proteins. *Cell* **102,** 635–646.
Mackay, C. R. (2001). Chemokines: Immunology's high impact factors. *Nat. Immunol.* **2,** 95–101.
Mackay, D. J., and Hall, A. (1998). Rho GTPases. *J. Biol. Chem.* **273,** 20685–20688.
Maghazachi, A. A. (1999). Intracellular signalling pathways induced by chemokines in natural killer cells. *Cell. Signal.* **11,** 385–390.
Marasco, W. A., Phan, S. H., Krutzsch, H., Showell, H. J., Feltner, D. E., Nairn, R., Becker, E. L., and Ward, P. A. (1984). Purification and identification of formyl-methionyl-leucyl-phenylalanine as the major peptide neutrophil chemotactic factor produced by Escherichia coli. *J. Biol. Chem.* **259,** 5430–5439.
Mellado, M., Rodriguez-Frade, J. M., Aragay, A., del Real, G., Martin, A. M., Vila-Coro, A. J., Serrano, A., Mayor, F., Jr., and Martinez, A. C. (1998). The chemokine monocyte chemotactic protein 1 triggers Janus kinase 2 activation and tyrosine phosphorylation of the CCR2B receptor. *J. Immunol.* **161,** 805–813.
Mellado, M., Rodriguez-Frade, J. M., Vila-Coro, A. J., Fernandez, S., Martin de Ana, A., Jones, D. R., Toran, J. L., and Martinez, A. C. (2001). Chemokine receptor homo- or heterodimerization activates distinct signaling pathways. *EMBO J.* **20,** 2497–2507.
Meng, E. C., and Bourne, H. R. (2001). Receptor activation: What does the rhodopsin structure tell us? *Trends Pharmacol. Sci.* **22,** 587–593.
Merlot, S., and Firtel, R. A. (2003). Leading the way: Directional sensing through phosphatidylinositol 3-kinase and other signaling pathways. *J. Cell. Sci.* **116,** 3471–3478.
Miettinen, H. M., Gripentrog, J. M., Mason, M. M., and Jesaitis, A. J. (1999). Identification of putative sites of interaction between the human formyl peptide receptor and G protein. *J. Biol. Chem.* **274,** 27934–27942.
Miettinen, H. M., Mills, J. S., Gripentrog, J. M., Dratz, E. A., Granger, B. L., and Jesaitis, A. J. (1997). The ligand binding site of the formyl peptide receptor maps in the transmembrane region. *J. Immunol.* **159,** 4045–4054.
Mills, J. S., Miettinen, H. M., Barnidge, D., Vlases, M. J., Wimer-Mackin, S., Dratz, E. A., Sunner, J., and Jesaitis, A. J. (1998). Identification of a ligand binding site in the human neutrophil formyl peptide receptor using a site-specific fluorescent photo-affinity label and mass spectrometry. *J. Biol. Chem.* **273,** 10428–10435.
Mills, J. S., Miettinen, H. M., Cummings, D., and Jesaitis, A. J. (2000). Characterization of the binding site on the formyl peptide receptor using three receptor mutants and analogs of Met-Leu-Phe and Met-Met-Trp-Leu-Leu. *J. Biol. Chem.* **275,** 39012–39017.

Monteclaro, F. S., and Charo, I. F. (1997). The amino-terminal domain of CCR2 is both necessary and sufficient for high affinity binding of monocyte chemoattractant protein 1. Receptor activation by a pseudo-tethered ligand. *J. Biol. Chem.* **272,** 23186–23190.

Murphy, P. M. (1994). The molecular biology of leukocyte chemoattractant receptors. *Annu. Rev. Immunol.* **12,** 593–633.

Murphy, P. M. (2002). International Union of Pharmacology. XXX. Update on chemokine receptor nomenclature. *Pharmacol. Rev.* **54,** 227–229.

Murphy, P. M., and Tiffany, H. L. (1991). Cloning of complementary DNA encoding a functional human interleukin-8 receptor. *Science* **253,** 1280–1283.

Murphy, P. M., Tiffany, H. L., McDermott, D., and Ahuja, S. K. (1993). Sequence and organization of the human N-formyl peptide receptor-encoding gene. *Gene* **133,** 285–290.

Navenot, J. M., Wang, Z. X., Trent, J. O., Murray, J. L., Hu, Q. X., DeLeeuw, L., Moore, P. S., Chang, Y., and Peiper, S. C. (2001). Molecular anatomy of CCR5 engagement by physiologic and viral chemokines and HIV-1 envelope glycoproteins: Differences in primary structural requirements for RANTES, MIP-1 alpha, and vMIP-II binding. *J. Mol. Biol.* **313,** 1181–1193.

Neptune, E. R., and Bourne, H. R. (1997). Receptors induce chemotaxis by releasing the betagamma subunit of Gi, not by activating Gq or Gs. *Proc. Natl. Acad. Sci. USA* **94,** 14489–14494.

Nick, J. A., Avdi, N. J., Young, S. K., Knall, C., Gerwins, P., Johnson, G. L., and Worthen, G. S. (1997). Common and distinct intracellular signaling pathways in human neutrophils utilized by platelet activating factor and FMLP. *J. Clin. Invest.* **99,** 975–986.

Ogilvie, P., Bardi, G., Clark-Lewis, I., Baggiolini, M., and Uguccioni, M. (2001). Eotaxin is a natural antagonist for CCR2 and an agonist for CCR5. *Blood* **97,** 1920–1924.

Pacold, M. E., Suire, S., Perisic, O., Lara-Gonzalez, S., Davis, C. T., Walker, E. H., Hawkins, P. T., Stephens, L., Eccleston, J. F., and Williams, R. L. (2000). Crystal structure and functional analysis of Ras binding to its effector phosphoinositide 3-kinase gamma. *Cell* **103,** 931–943.

Palczewski, K., Kumasaka, T., Hori, T., Behnke, C. A., Motoshima, H., Fox, B. A., Le Trong, I., Teller, D. C., Okada, T., Stenkamp, R. E., Yamamoto, M., and Miyano, M. (2000). Crystal structure of rhodopsin: A G protein-coupled receptor. *Science* **289,** 739–745.

Parent, C. A., and Devreotes, P. N. (1999). A cell's sense of direction. *Science* **284,** 765–770.

Partida-Sanchez, S., Cockayne, D. A., Monard, S., Jacobson, E. L., Oppenheimer, N., Garvy, B., Kusser, K., Goodrich, S., Howard, M., Harmsen, A., Randall, T. D., and Lund, F. E. (2001). Cyclic ADP-ribose production by CD38 regulates intracellular calcium release, extracellular calcium influx and chemotaxis in neutrophils and is required for bacterial clearance *in vivo*. *Nat. Med.* **7,** 1209–1216.

Prioleau, C., Visiers, I., Ebersole, B. J., Weinstein, H., and Sealfon, S. C. (2002). Conserved helix 7 tyrosine acts as a multistate conformational switch in the 5HT2C receptor. Identification of a novel "locked-on" phenotype and double revertant mutations *J. Biol. Chem.* **277,** 36577–36584.

Prossnitz, E. R. (1997). Desensitization of N-formylpeptide receptor-mediated activation is dependent upon receptor phosphorylation. *J. Biol. Chem.* **272,** 15213–15219.

Prossnitz, E. R., Gilbert, T. L., Chiang, S., Campbell, J. J., Qin, S., Newman, W., Sklar, L. A., and Ye, R. D. (1999). Multiple activation steps of the N-formyl peptide receptor. *Biochem.* **38,** 2240–2247.

Prossnitz, E. R., Schreiber, R. E., Bokoch, G. M., and Ye, R. D. (1995). Binding of low affinity N-formyl peptide receptors to G protein. Characterization of a novel inactive receptor intermediate. *J. Biol. Chem.* **270,** 10686–10694.

Prossnitz, E. R., and Ye, R. D. (1997). The N-formyl peptide receptor: A model for the study of chemoattractant receptor structure and function. *Pharmacol. Ther.* **74,** 73–102.

Qanbar, R., and Bouvier, M. (2003). Role of palmitoylation/depalmitoylation reactions in G-protein-coupled receptor function. *Pharmacol. Ther.* **97,** 1–33.

Rao, V. R., Cohen, G. B., and Oprian, D. D. (1994). Rhodopsin mutation G90D and a molecular mechanism for congenital night blindness. *Nature* **367,** 639–642.

Rasmussen, S. G., Jensen, A. D., Liapakis, G., Ghanouni, P., Javitch, J. A., and Gether, U. (1999). Mutation of a highly conserved aspartic acid in the beta2 adrenergic receptor: Constitutive activation, structural instability, and conformational rearrangement of transmembrane segment 6. *Mol. Pharmacol.* **56,** 175–184.

Rediske, J. J., Quintavalla, J. C., Haston, W. O., Morrissey, M. M., and Seligman, B. (1992). Platelet activating factor stimulates intracellular calcium transients in human neutrophils: Involvement of endogenous 5-lipoxygenase products. *J. Leukoc. Biol.* **51,** 484–489.

Rodriguez-Frade, J. M., Vila-Coro, A. J., de Ana, A. M., Albar, J. P., Martinez, A. C., and Mellado, M. (1999). The chemokine monocyte chemoattractant protein-1 induces functional responses through dimerization of its receptor CCR2. *Proc. Natl. Acad. Sci. USA* **96,** 3628–3633.

Sakmar, T. P. (2002). Structure of rhodopsin and the superfamily of seven-helical receptors: The same and not the same. *Curr. Opin. Cell Biol.* **14,** 189–195.

Sanders, S. K., Crean, S. M., Boxer, P. A., Kellner, D., LaRosa, G. J., and Hunt, S. W., 3rd (2000). Functional differences between monocyte chemotactic protein-1 receptor A and monocyte chemotactic protein-1 receptor B expressed in a Jurkat T cell. *J. Immunol.* **165,** 4877–4883.

Sankaran, B., Osterhout, J., Wu, D., and Smrcka, A. V. (1998). Identification of a structural element in phospholipase C beta2 that interacts with G protein beta-gamma subunits. *J. Biol. Chem.* **273,** 7148–7154.

Sansom, M. S., and Weinstein, H. (2000). Hinges, swivels and switches: The role of prolines in signalling via transmembrane alpha-helices. *Trends Pharmacol. Sci.* **21,** 445–451.

Scheer, A., Costa, T., Fanelli, F., De Benedetti, P. G., Mhaouty-Kodja, S., Abuin, L., Nenniger-Tosato, M., and Cotecchia, S. (2000). Mutational analysis of the highly conserved arginine within the Glu/Asp-Arg-Tyr motif of the alpha(1b)-adrenergic receptor: Effects on receptor isomerization and activation. *Mol. Pharmacol.* **57,** 219–231.

Scheer, A., Fanelli, F., Costa, T., De Benedetti, P. G., and Cotecchia, S. (1996). Constitutively active mutants of the alpha 1B-adrenergic receptor: Role of highly conserved polar amino acids in receptor activation. *EMBO J.* **15,** 3566–3578.

Scheer, A., Fanelli, F., Costa, T., De Benedetti, P. G., and Cotecchia, S. (1997). The activation process of the alpha1B-adrenergic receptor: Potential role of protonation and hydrophobicity of a highly conserved aspartate. *Proc. Natl. Acad. Sci. USA* **94,** 808–813.

Seifert, R., and Wenzel-Seifert, K. (2001a). Defective Gi protein coupling in two formyl peptide receptor mutants associated with localized juvenile periodontitis. *J. Biol. Chem.* **276,** 42043–42049.

Seifert, R., and Wenzel-Seifert, K. (2001b). Unmasking different constitutive activity of four chemoattractant receptors using Na+ as universal stabilizer of the inactive (R) state. *Receptors Channels* **7,** 357–369.

Seifert, R., and Wenzel-Seifert, K. (2003). The human formyl peptide receptor as model system for constitutively active G-protein-coupled receptors. *Life Sci.* **73,** 2263–2280.

Shapiro, D. A., Kristiansen, K., Weiner, D. M., Kroeze, W. K., and Roth, B. L. (2002). Evidence for a model of agonist-induced activation of 5-hydroxytryptamine 2A serotonin receptors that involves the disruption of a strong ionic interaction between helices 3 and 6. *J. Biol. Chem.* **277,** 11441–11449.

Shi, L., Liapakis, G., Xu, R., Guarnieri, F., Ballesteros, J. A., and Javitch, J. A. (2002). Beta2 adrenergic receptor activation. Modulation of the proline kink in transmembrane 6 by a rotamer toggle switch. *J. Biol. Chem.* **277,** 40989–40996.

Siciliano, S. J., Rollins, T. E., DeMartino, J., Konteatis, Z., Malkowitz, L., Van Riper, G., Bondy, S., Rosen, H., and Springer, M. S. (1994). Two-site binding of C5a by its receptor: An alternative binding paradigm for G protein-coupled receptors. *Proc. Natl. Acad. Sci. USA* **91,** 1214–1218.

Sieving, P. A., Fowler, M. L., Bush, R. A., Machida, S., Calvert, P. D., Green, D. G., Makino, C. L., and McHenry, C. L. (2001). Constitutive "light" adaptation in rods from G90D rhodopsin: A mechanism for human congenital nightblindness without rod cell loss *J. Neurosci.* **21,** 5449–5460.

Skelton, N. J., Quan, C., Reilly, D., and Lowman, H. (1999). Structure of a CXC chemokine-receptor fragment in complex with interleukin-8. *Structure Fold Des.* **7,** 157–168.

Smith, C. D., Cox, C. C., and Snyderman, R. (1986). Receptor-coupled activation of phosphoinositide-specific phospholipase C by an N protein. *Science* **232,** 97–100.

Spisani, S., Turchetti, M., Varani, K., Falzarano, S., and Cavicchioni, G. (2003). Hydrophilic residues at position 3 highlight unforeseen features of the fMLP receptor pocket. *Eur. J. Pharmacol.* **469,** 13–19.

Sprang, S. R. (1997). G protein mechanisms: Insights from structural analysis. *Annu. Rev. Biochem.* **66,** 639–678.

Stephens, L. R., Eguinoa, A., Erdjument-Bromage, H., Lui, M., Cooke, F., Coadwell, J., Smrcka, A. S., Thelen, M., Cadwallader, K., Tempst, P., and Hawkins, P. T. (1997). The G beta gamma sensitivity of a PI3K is dependent upon a tightly associated adaptor, p101. *Cell* **89,** 105–114.

Stoyanov, B., Volinia, S., Hanck, T., Rubio, I., Loubtchenkov, M., Malek, D., Stoyanova, S., Vanhaesebroeck, B., Dhand, R., Nurnberg, B. *et al.* (1995). Cloning and characterization of a G protein-activated human phosphoinositide-3 kinase. *Science* **269,** 690–693.

Suetomi, K., Rojo, D., and Navarro, J. (2002). Identification of a signal transduction switch in the chemokine receptor CXCR1. *J. Biol. Chem.* **277,** 31563–31566.

Takeda, S., Kadowaki, S., Haga, T., Takaesu, H., and Mitaku, S. (2002). Identification of G protein-coupled receptor genes from the human genome sequence. *FEBS Lett.* **520,** 97–101.

Tamura, M., Gu, J., Matsumoto, K., Aota, S., Parsons, R., and Yamada, K. M. (1998). Inhibition of cell migration, spreading, and focal adhesions by tumor suppressor PTEN. *Science* **280,** 1614–1617.

Thelen, M. (2001). Dancing to the tune of chemokines. *Nat. Immunol.* **2,** 129–134.

Thordsen, I., Polzer, S., and Schreiber, M. (2002). Infection of cells expressing CXCR4 mutants lacking N-glycosylation at the N-terminal extracellular domain is enhanced for R5X4-dualtropic human immunodeficiency virus type-1. *BMC Infect. Dis.* **2,** 31.

Trettel, F., Di Bartolomeo, S., Lauro, C., Catalano, M., Ciotti, M. T., and Limatola, C. (2003). Ligand-independent CXCR2 dimerization. *J. Biol. Chem.* **278,** 40980–40988.

Trkola, A., Dragic, T., Arthos, J., Binley, J. M., Olson, W. C., Allaway, G. P., Cheng-Mayer, C., Robinson, J., Maddon, P. J., and Moore, J. P. (1996). CD4-dependent, antibody-sensitive interactions between HIV-1 and its co-receptor CCR-5. *Nature* **384,** 184–187.

Unger, V. M., Hargrave, P. A., Baldwin, J. M., and Schertler, G. F. (1997). Arrangement of rhodopsin transmembrane alpha-helices. *Nature* **389,** 203–206.

Vaidehi, N., Floriano, W. B., Trabanino, R., Hall, S. E., Freddolino, P., Choi, E. J., Zamanakos, G., and Goddard, W. A., 3rd (2002). Prediction of structure and function of G protein-coupled receptors. *Proc. Natl. Acad. Sci. USA* **99,** 12622–12627.

Verghese, M. W., Smith, C. D., and Snyderman, R. (1986). Role of guanine nucleotide regulatory protein in polyphosphoinositide degradation and activation of phagocytic leukocytes by chemoattractants. *J. Cell Biochem.* **32,** 59–69.

Vila-Coro, A. J., Mellado, M., Martin de Ana, A., Lucas, P., del Real, G., Martinez, A. C., and Rodriguez-Frade, J. M. (2000). HIV-1 infection through the CCR5 receptor is blocked by receptor dimerization. *Proc. Natl. Acad. Sci. USA* **97,** 3388–3393.

Vila-Coro, A. J., Rodriguez-Frade, J. M., Martin De Ana, A., Moreno-Ortiz, M. C., Martinez, A. C., and Mellado, M. (1999). The chemokine SDF-1alpha triggers CXCR4 receptor dimerization and activates the JAK/STAT pathway. *FASEB J.* **13,** 1699–1710.

Visiers, I., Ballesteros, J. A., and Weinstein, H. (2002). Three-dimensional representations of G protein-coupled receptor structures and mechanisms. *Methods Enzymol.* **343,** 329–371.

Wang, F., Herzmark, P., Weiner, O. D., Srinivasan, S., Servant, G., and Bourne, H. R. (2002). Lipid products of PI(3)Ks maintain persistent cell polarity and directed motility in neutrophils. *Nat. Cell Biol.* **4,** 513–518.

Weiner, O. D., Neilsen, P. O., Prestwich, G. D., Kirschner, M. W., Cantley, L. C., and Bourne, H. R. (2002). A PtdInsP(3)- and Rho GTPase-mediated positive feedback loop regulates neutrophil polarity. *Nat. Cell Biol.* **4,** 509–513.

Weissman, D., Rabin, R. L., Arthos, J., Rubbert, A., Dybul, M., Swofford, R., Venkatesan, S., Farber, J. M., and Fauci, A. S. (1997). Macrophage-tropic HIV and SIV envelope proteins induce a signal through the CCR5 chemokine receptor. *Nature* **389,** 981–985.

Wenzel-Seifert, K., and Seifert, R. (2003). Critical role of N-terminal N-glycosylation for proper folding of the human formyl peptide receptor. *Biochem. Biophys. Res. Commun.* **301,** 693–698.

Wenzel-Seifert, K., and Seifert, R. (2003). Functional differences between human formyl peptide receptor isoforms 26, 98, and G6. *Naunyn Schmiedebergs Arch. Pharmacol.* **367,** 509–515.

Wetsel, R. A. (1995). Structure, function and cellular expression of complement anaphylatoxin receptors. *Curr. Opin. Immunol.* **7,** 48–53.

Whistler, J. L., Gerber, B. O., Meng, E. C., Baranski, T. J., von Zastrow, M., and Bourne, H. R. (2002). Constitutive activation and endocytosis of the complement factor 5a receptor: Evidence for multiple activated conformations of a G protein-coupled receptor. *Traffic* **3,** 866–877.

Wilde, M. W., Carlson, K. E., Manning, D. R., and Zigmond, S. H. (1989). Chemoattractant-stimulated GTPase activity is decreased on membranes from polymorphonuclear leukocytes incubated in chemoattractant. *J. Biol. Chem.* **264,** 190–196.

Wilkinson, P. C., Komai-Koma, M., and Newman, I. (1997). Locomotion and chemotaxis of lymphocytes. *Autoimmunity* **26,** 55–72.

Wu, D., LaRosa, G. J., and Simon, M. I. (1993). G protein-coupled signal transduction pathways for interleukin-8. *Science* **261,** 101–103.

Wu, L., Gerard, N. P., Wyatt, R., Choe, H., Parolin, C., Ruffing, N., Borsetti, A., Cardoso, A. A., Desjardin, E., Newman, W., Gerard, C., and Sodroski, J. (1996a). CD4-induced interaction of primary HIV-1 gp120 glycoproteins with the chemokine receptor CCR-5. *Nature* **384,** 179–183.

Wu, L., Ruffing, N., Shi, X., Newman, W., Soler, D., Mackay, C. R., and Qin, S. (1996b). Discrete steps in binding and signaling of interleukin-8 with its receptor. *J. Biol. Chem.* **271,** 31202–31209.

Xie, W., Jiang, H., Wu, Y., and Wu, D. (1997). Two basic amino acids in the second inner loop of the interleukin-8 receptor are essential for Galpha16 coupling. *J. Biol. Chem.* **272,** 24948–24951.

Yoshimura, T., Matsushima, K., Tanaka, S., Robinson, E. A., Appella, E., Oppenheim, J. J., and Leonard, E. J. (1987). Purification of a human monocyte-derived neutrophil chemotactic factor that has peptide sequence similarity to other host defense cytokines. *Proc. Natl. Acad. Sci. USA* **84,** 9233–9237.

Zhang, X., Boyar, W., Toth, M. J., Wennogle, L., and Gonnella, N. C. (1997). Structural definition of the C5a C terminus by two-dimensional nuclear magnetic resonance spectroscopy. *Proteins* **28,** 261–267.

Zlotnik, A., and Yoshie, O. (2000). Chemokines: A new classification system and their role in immunity. *Immunity* **12,** 121–127.

Zuiderweg, E. R., Nettesheim, D. G., Mollison, K. W., and Carter, G. W. (1989). Tertiary structure of human complement component C5a in solution from nuclear magnetic resonance data. *Biochem.* **28,** 172–185.

AUTHOR INDEX

A

B

C

D

E

F

G

H

I

J

K

L

M

N

O

P

Q

R

S

T

U

V

W

X

Y

Z

SUBJECT INDEX

A

H

M

N

O

P

U

V

X

Y

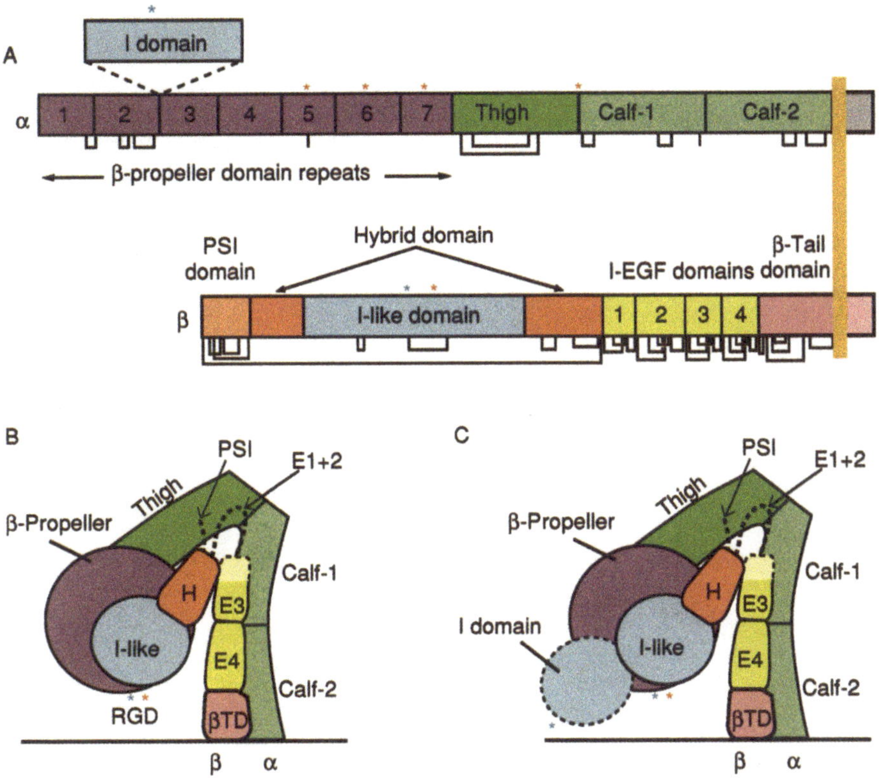

SPRINGER AND WANG, FIG. 2. Integrin architecture. (A) Organization of domains within the primary structure. Some α subunits contain an I domain inserted in the position denoted by the dotted lines. Cysteines and disulfide bonds are shown as lines below the stick figures. Red and blue asterisks denote Ca^{2+} and Mg^{2+} binding sites, respectively. (B) Arrangement of domains within the three-dimensional crystal structure of $\alpha_v\beta_3$ (Xiong *et al.*, 2001). Each domain is color coded as in A. (C) The structure in (B) with an I domain added.

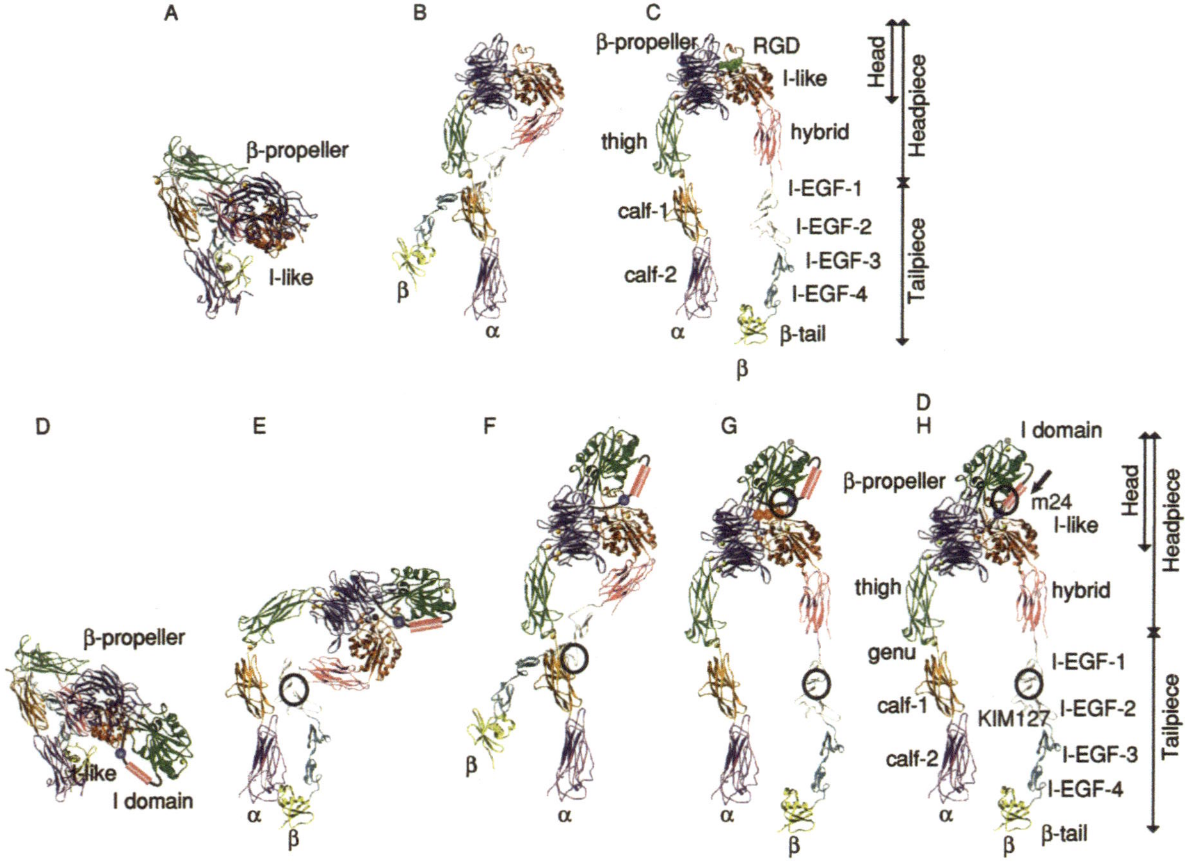
A
B
C
D
E
F
G
D
H
β-propeller
I-like
RGD
thigh
hybrid
Head
Headpiece
Tailpiece
calf-1
calf-2
I-EGF-1
I-EGF-2
I-EGF-3
I-EGF-4
β-tail
α
β
I domain
m24
genu
KIM127

SPRINGER AND WANG, FIG. 3. Conformational states for integrins. A–C. Model for $\alpha_v\beta_3$ integrin activation, with at least three conformations of the extracellular domain (Takagi *et al.*, 2002). (A) Bent, low affinity conformation. (B) Extended conformation with closed headpiece. (C) Extended conformation with open headpiece shown with bound RGD-mimetic peptide (green CPK). D–H. Model for $\alpha_L\beta_2$ integrin activation. (D) Bent conformation with low affinity. (E) and (F) $\alpha_L\beta_2$ with a closed headpiece and closed I domain in partially (E) or fully (F) extended states. (G) Extended conformation with open headpiece, and closed I domain, in the presence of α/β I-like allosteric antagonist, represented by three red spheres. (H) Extended conformation with open headpiece and open I domain. The models for all of the extracellular domains except for the I domain are based on conformational states of $\alpha_v\beta_3$ defined by negative stain electron microscopy (Takagi *et al.*, 2002), crystallography (Xiong *et al.*, 2002), NMR (Beglova *et al.*, 2002), and mapping of activation epitopes (Lu *et al.*, 2001a,c). The α_L I domain is a cartoon based on crystal structures (Shimaoka *et al.*, 2003b). The I domain is joined at the point of its insertion in the β-propeller domain but its orientation is arbitrary; it is shown at slightly larger scale for emphasis. The C-terminal I domain α-helix is represented by a red cylinder, and α_L Glu-310 in the linker as a blue sphere. The positions of activation epitopes m24 and KIM127 are circled only in the conformations in which they are thought to be exposed.

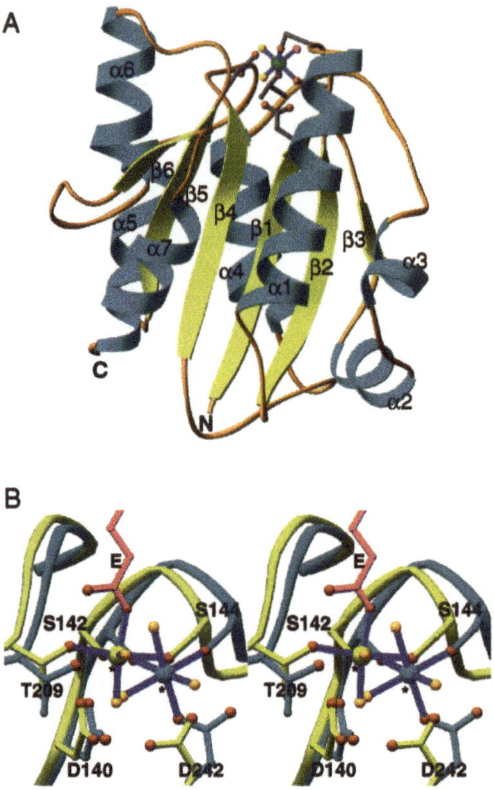

SPRINGER AND WANG, FIG. 4. α I domain structure and MIDAS conformational change. (A) Ribbon diagram of the α_M I domain in the open conformation (Lee *et al.*, 1995b). The β-strands (yellow), α-helices (cyan), and the N and C termini are labeled. The Mg ion is shown as a green sphere, and primary coordination bonds are blue. Side chains of residues that form primary or secondary coordinations to the metal ion are shown with grey bonds and carbon atoms and red oxygen atoms, and the oxygen of the ligand-mimetic Glu from another I domain is magenta. Coordinating water molecule oxygens are gold. Prepared with Ribbons (Carson, 1997). (B) Stereo view of alternative conformations of the α_M MIDAS. The backbone, coordinating side chain bonds, and metals (labeled with asterisks) are shown in yellow (open conformation) and cyan (closed conformation). The coordinating glutamate residue from the ligand-mimetic neighboring α_M I domain in α_M is in magenta. Primary coordination bonds to the metals are in blue. Oxygen atoms of the coordinating side chains and water molecules are red and gold, respectively. The 1IDO open α_M structure (Lee *et al.*, 1995a) was superimposed on the closed 1JLM α_M structure (Lee *et al.*, 1995b) using residues 132–141, 166–206, 211–241, 246–270, and 287–294.

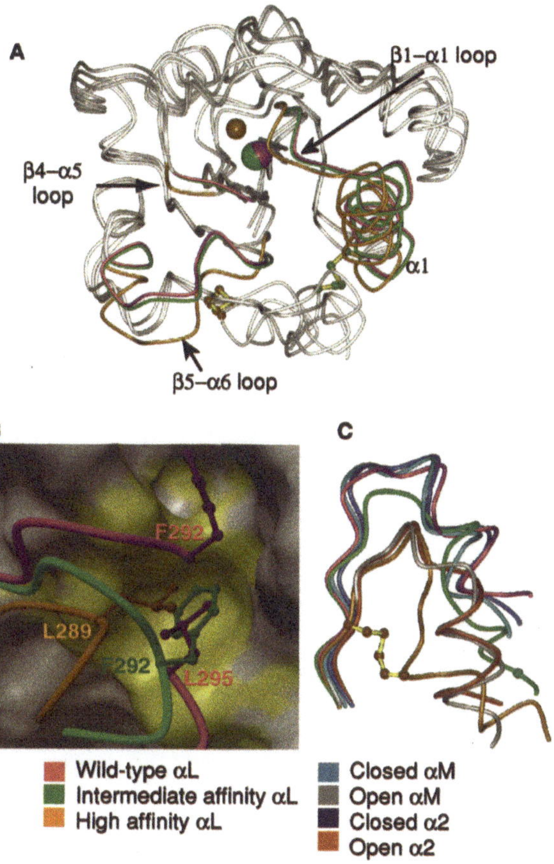

SPRINGER AND WANG, FIG. 5. Propagation of conformational change in the I domain. (A) The structures of the unliganded wild type, intermediate affinity, and high affinity αL I domains (Shimaoka *et al.*, 2003b). The three unliganded αL I domain backbones are shown superimposed and viewed centered on the MIDAS. Regions of the backbones that differ structurally are labeled and color keyed; other backbone regions are grey. The metal ions at the MIDAS and the atoms in the mutationally introduced, disulfide linked cysteine atoms are shown in the same colors as the backbone regions that differ; the cysteine sidechain bonds are yellow. The position of the missing metal ion in the high affinity structure is simulated with a smaller sphere. β6-α7 loop and α7 are shown in grey for clarity; differences in these regions are shown in (B) and (C). (B) The hydrophobic pocket that acts as a detent for the rachet-like movement of the β6-α7 loop. The backbone of the β6-α7 loop and the three residues that occupy the same hydrophobic pocket in the three different conformational states are color keyed. The pocket is shown as a GRASP van der Waals surface using the wild-type 1LFA structure with the residues from 287 to the C-terminus deleted. The upper hydrophobic pocket is also shown, which is occupied only in the closed conformation (by F292 which is shown in the wild type structure along with L295). For the closed conformation the β6-α7 mainchain trace is broken between F292 and L295 for clarity. On the otherwise grey GRASP surface, hydrophobic residues are colored yellow. (C) The C-terminal fragments encompassing the β6-α7 loop for the three unligated α_L conformations and open and closed α_2 and α_M I domain structures (see color keys). The sidechain bonds of Cys-287 and Cys-294 in the designed disulfide bridge in the high affinity mutant are shown in yellow; the Cα atom of Cys-299 in the designed disulfide of the intermediate mutant is shown as a green sphere.

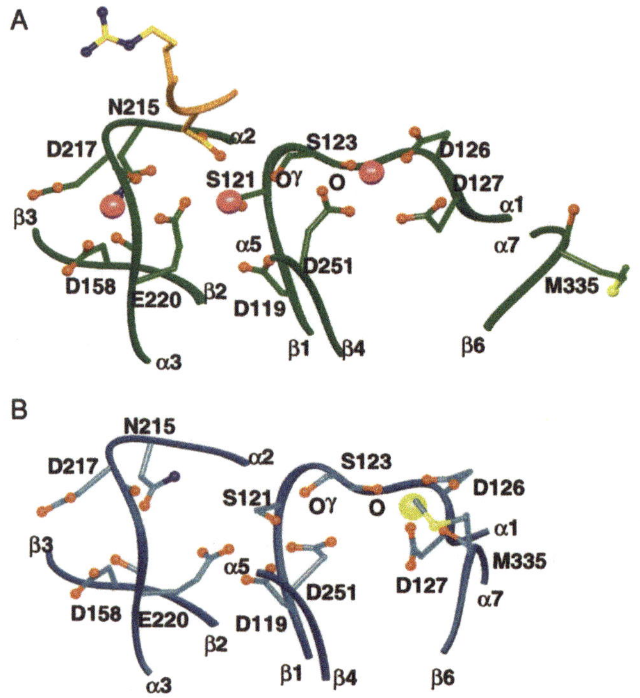

SPRINGER AND WANG, FIG. 7. The linear cluster of metal binding sites in the β I-like domain. (A) The liganded $\alpha_v\beta_3$ structure in Mn^{2+} with RGD (Xiong *et al.*, 2002) in gold. (B) The unliganded $\alpha_v\beta_3$ structure in Ca^{2+} (Xiong *et al.*, 2001). The structures were superimposed by using the I-like domain, so that equivalent positions in (A) and (B) are vertically aligned. The orientation is with the LIMBS, MIDAS, and ADMIDAS from left to right. Mn^{2+} and Ca^{2+} ions are large magenta and yellow spheres, respectively. All putative metal coordinating sidechains and backbone carbonyl groups are shown, with N, O, and S atoms in blue, red, and yellow, respectively. The carbonyl and sidechain oxygens of S123 are marked O and Oγ, respectively.

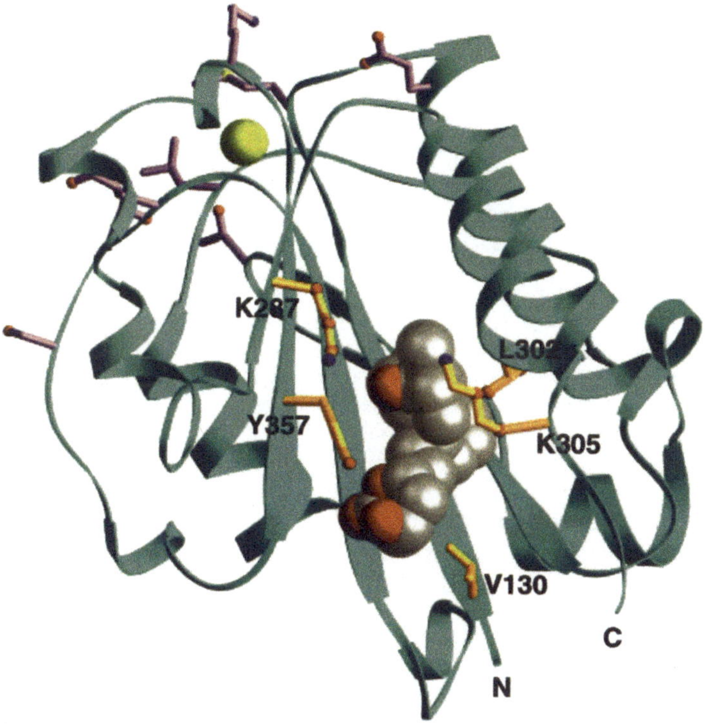

SPRINGER AND WANG, FIG. 8. Ribbon diagram of the α_L I domain in complex with an α I allosteric antagonist (Kallen *et al.*, 1999). The α I allosteric antagonist (shown by CPK with silver carbon atoms and red oxygen atoms) binds in the hydrophobic pocket underneath the C-terminal α-helix and stabilizes the I domain in the closed conformation. The side chains within the antagonist-binding pocket are shown with gold bonds and carbon atoms, red oxygen atoms, and blue nitrogen atoms. The residues critical for binding to ICAM-1 and ICAM-2 are shown with purple side chains and yellow sulfur, red oxygen, and blue nitrogen atoms. Note that these residues are located around the MIDAS, distal from the antagonist binding site.

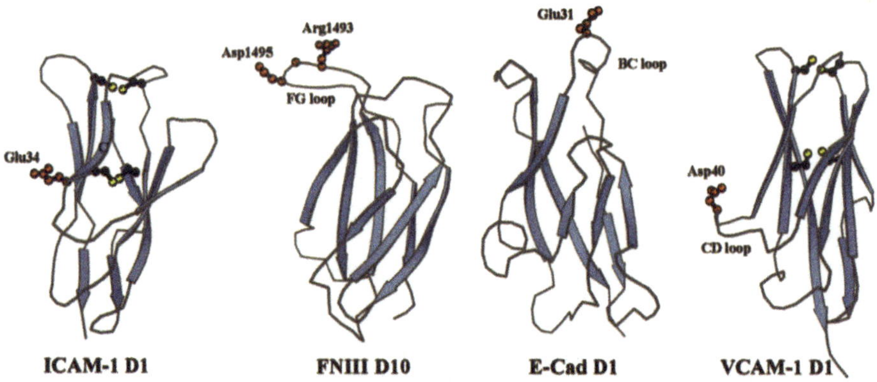

SPRINGER AND WANG, FIG. 9. Ribbon diagram of integrin-binding domain structures of four representative integrin ligands: ICAM-1 domain 1 (ICAM-1 D1), 10th fibronectin type III repeat (FNIII D10), E-cadherin domain 1 (E-Cad D1), and VCAM-1 domain 1 (VCAM-1 D1). In the figure, those key integrin-binding acidic residues are shown in red ball-and-stick model. They are located at the end of the C strand of IgSF domain (ICAM-1 D1), at the tip of the FG loop of FNIII domain (FNIII D10), at the tip of the BC loop of cadherin domain (E-Cad D1), and at the tip of the CD loop of IgSF domain (VCAM-1 D1), respectively. The figure demonstrates the diversity of integrin-binding modes to structurally very similar domains. The figure was prepared with MOLSCRIPT.

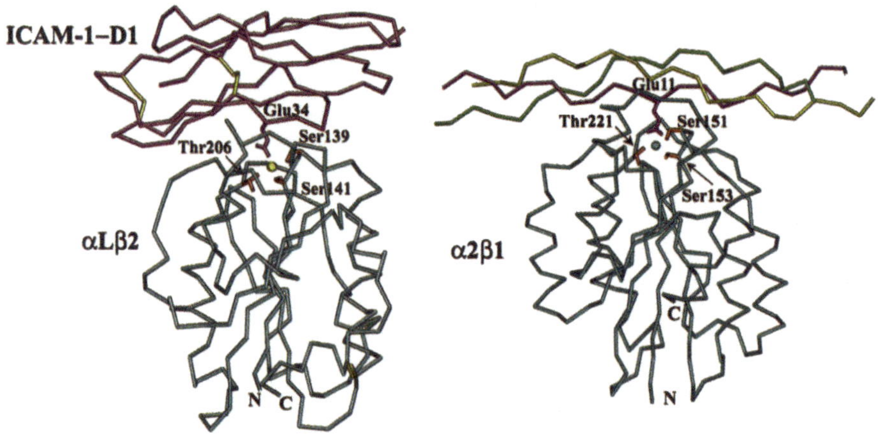

SPRINGER AND WANG, FIG. 10. Backbone diagram of two integrin/ligand complex structures. On the left panel is an intermediate affinity I domain from $\alpha_L\beta_2$ (in blue) in complex with ICAM-1 (in red). For clarity, only ICAM-1 D1 is shown. The sidechains of the key integrin-binding residue, Glu34, of ICAM-1 and MIDAS residues that directly coordinate the metal (in yellow) are illustrated. On the right panel is an I domain from $\alpha_2\beta_1$ (in blue) in complex with synthetic collagen-originated 21-mer triple-peptides (the major interacting middle strand is in red, and the other two strands are in yellow and green, respectively). Also, sidechains of the key integrin-binding residue, Glu11, of collagen and MIDAS residues of I domain are illustrated. The figure was prepared with SETOR.

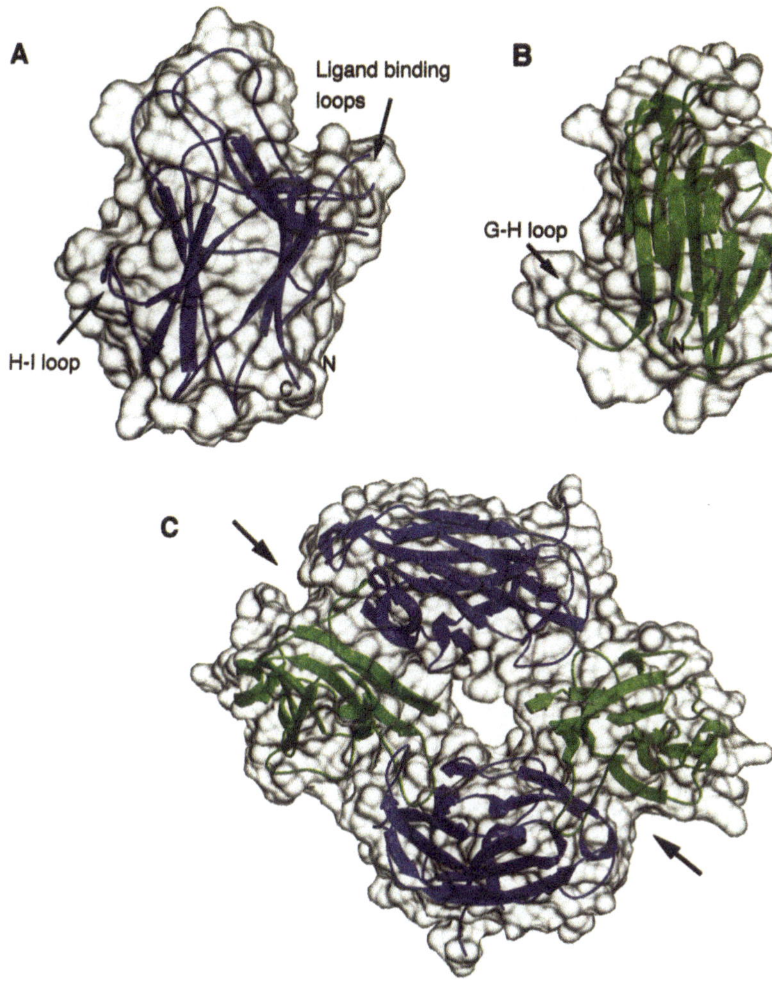

BARTON *ET AL.*, FIG. 2. Structures of the extracellular domains of Ephs and ephrins. The molecular surfaces (semi-transparent) are also indicated. (A) Structure of the ligand-binding domain of EphB2. The N- and C-termini of the molecule are labeled, as are the class-specificity loop (H-I) and the ligand-binding loops that are largely disordered in the absence of bound ephrin. (B) Structure of the extracellular receptor-binding domain of ephrin-B2. Indicated is the location of the receptor-binding G-H loop. (C) Structure of the EphB2/ephrin-B2 tetramer. Eph receptors are blue and ephrins are green. The high-affinity dimerization interfaces are indicated by arrows.

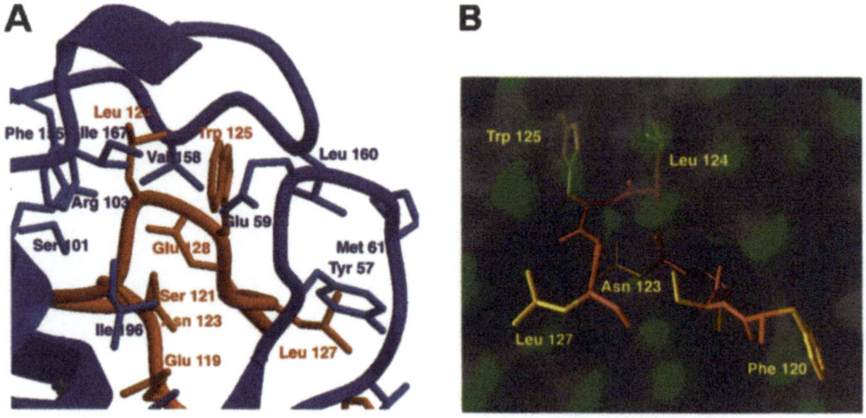

BARTON *ET AL.*, FIG. 3. Structure of the Eph/ephrin interface. (A) The channel on the surface of EphB2 viewed from the distal end of the ligand, showing the receptor (blue) and the incoming ephrin-B2 ligand (red). (B) Molecular surface representation of EphB2, viewed from position of the incoming ligand, showing the entrance of the hydrophobic channel with deeply inserted ephrin-B2 loop. The molecular surface of the receptor is color-coded according to curvature with concave regions in grey and convex – in green. Reproduced with permission from (Himanen *et al.*, 2001).

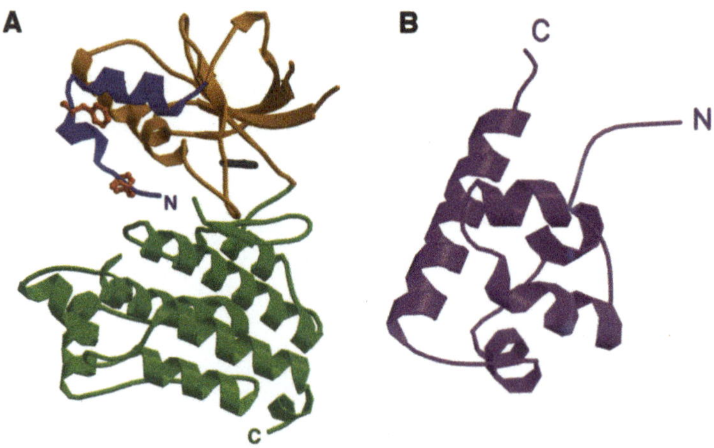

BARTON *ET AL.*, FIG. 5. Structures of the intracellular domains of Eph receptors. (A) Structure of the EphB2 kinase domain autoinhibited by the nonphosphorylated juxtamembrane region. The C-terminal kinase lobe is green and the N-terminal lobe is brown. The two regulatory tyrosine residues in the blue juxtamembrane region are shown in red. The ATP location in the active site is indicated in gray. (B) Structure of the monomeric sterile α motif (SAM) domain of EphA4. Reproduced with permission from (Himanen *et al.*, 2003b).

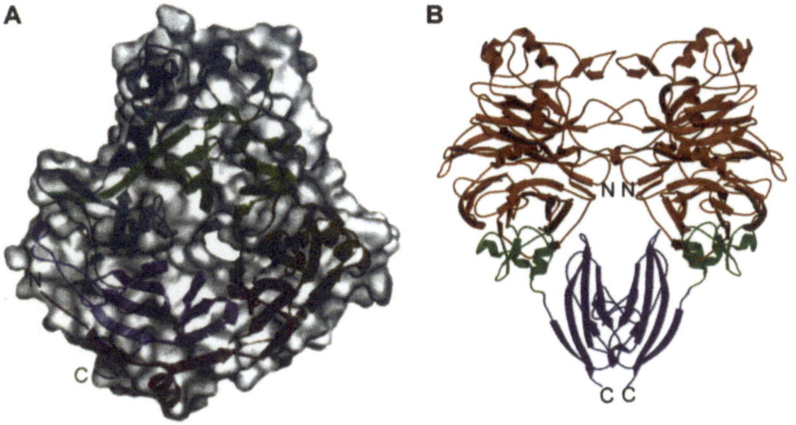

BARTON *ET AL.*, FIG. 8. Structures of Sema3A-65K and Sema4D. (A) The structure of Sema3A-65K viewed from the "top" face of the molecule. The molecular surface (semi-transparent) is also indicated. The individual Sema3A-65K pseudo-repeats corresponding to the individual β propeller blades are colored (from -N to -C terminus) in red (1), orange (2), yellow (3), green (4), cyan (5), blue (6) and magenta (7). (B) The structure of Sema4D homodimer. The sema domain is in red, the PSI – in green, and the Ig – in blue.

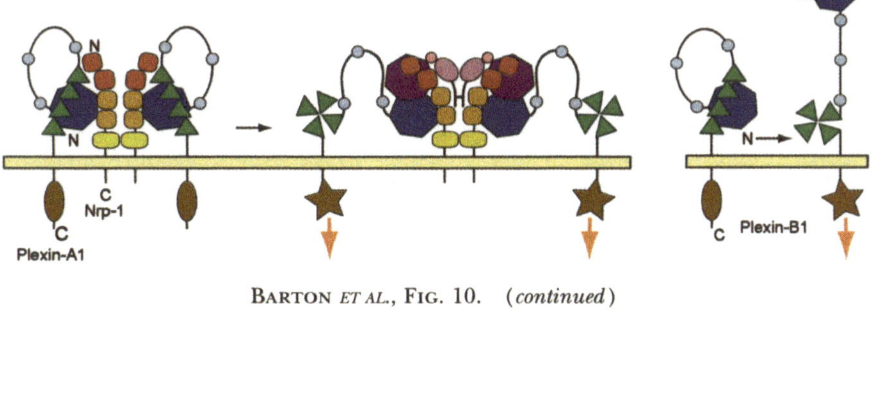

BARTON *ET AL.*, FIG. 10. (*continued*)

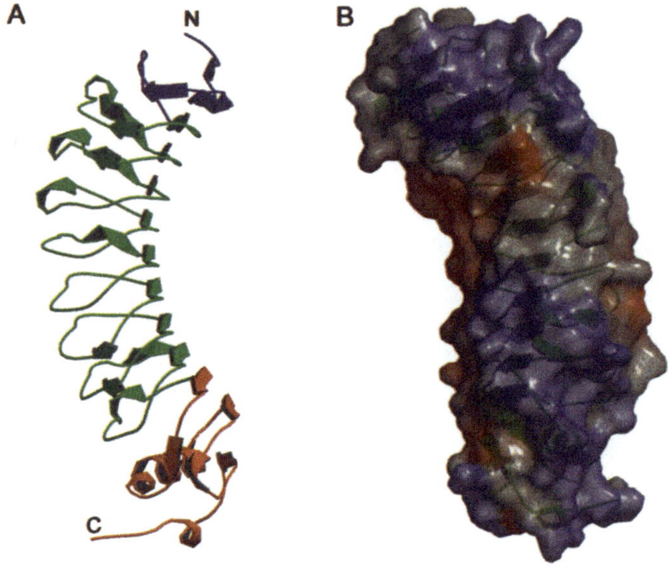

BARTON *ET AL.*, FIG. 12. (A) Side view of the NgR structure. The LRRNT domain is shown in blue, the LRRs in green, and the LRRCT domain in red. (B) Surface representation of the NgR structure showing the electrostatic potential in red and blue. Note the two large basic (blue) patches near the N- and C-termini and the diffuse acidic (red) region on the concave surface.

BARTON *ET AL.*, FIG. 10. Model for initiation of semaphorin signaling. (A) Initiation of Sema3A-mediated signaling via the Nrp-1/Plexin-A1 complex (see discussion in text). Sema3A is in magenta (Sema domain – heptagon; PSI domain – circle; Ig domain – oval); In plexin, the Sema domain is in dark blue, the IPT domains in green, and the PSI regions in light blue. The intracellular Sex-Plexin domain is in brown and the star corresponds to the activated form (the red arrow indicates signaling directing growth cone collapse). In Nrp-1, the A1 and A2 domains are in red, B1 and B2 in orange, and the MAM domain is in yellow. Sema3A binding results in a 2:2:2 ligand/receptor/co-receptor complex formation and the release of the plexin membrane-proximal extracellular region which in the absence of constrains adopts an active conformation (right panel). (B) Initiation of Sema4D signaling via the Plexin-B1 receptor. In this case no co-receptors are required. As with Sema3A, ligand binding disrupts the inhibitory semaphorin conformation resulting in activation of downstream signaling. Reproduced with permission from (Antipenko *et al.*, 2003).

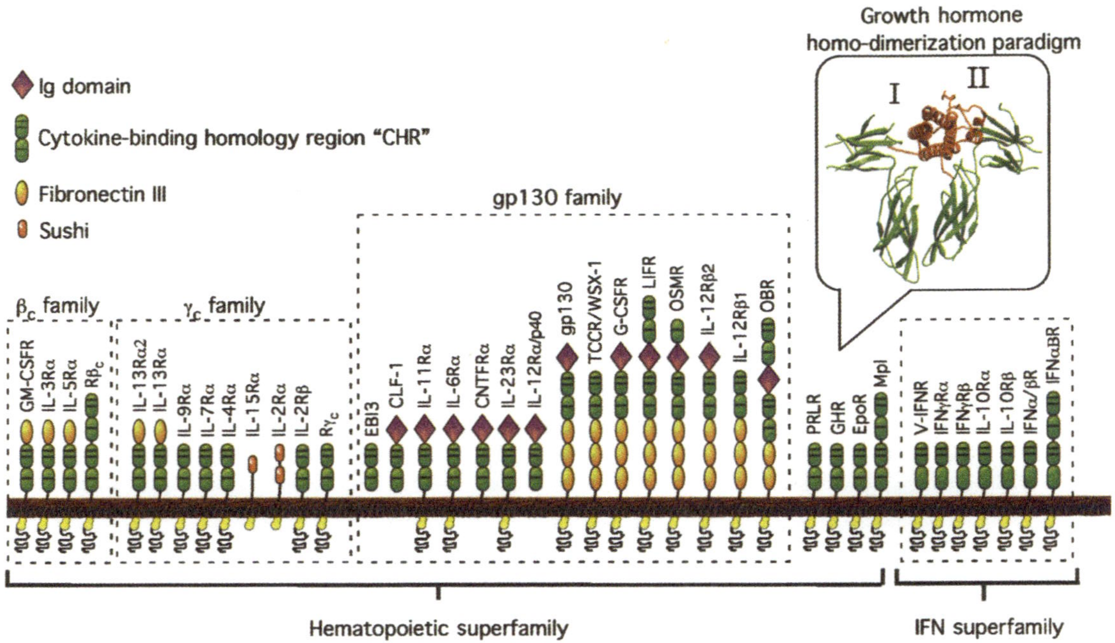

BOULANGER AND GARCIA, FIG. 1. Grouping of hematopoietic cytokine receptors by shared receptor usage. The majority of the hematopoietic cytokine receptors incorporate one of three shared signaling receptors, either the common beta (β_c) chain, the common gamma (γ_c) chain or gp130. The crystal structure of the growth hormone (red) complex (inset) was the first structure of a four helix bundle cytokine in complex with its receptor (green) (de Vos *et al.*, 1992) and established the paradigm of cytokine/receptor complex formation.

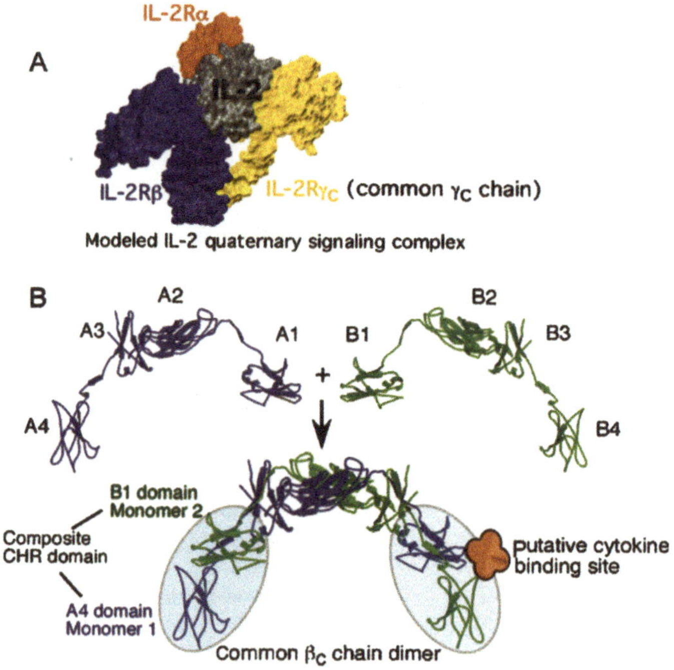

Boulanger and Garcia, Fig. 2. Common gamma (γ_c) and beta (β_c) receptors. (A) Modeled IL-2 ternary complex based on the growth hormone complex showing the partially modeled common gamma (γ_c) chain, (yellow) (Bamborough *et al.*, 1994). In this model the γ_c chain docks against the long axis of IL-2 and also forms contacts with the D2 domain of IL-2Rβ receptor in a similar fashion to the growth hormone structure (de Vos *et al.*, 1992). (B) Structure of the extracellular homodimer of the common β_c chain. (Carr *et al.*, 2001). Recent mutagenesis studies have identified the cytokine binding homology region (CHR) to be formed by the A4 domain of one monomer (blue) and the B1 domain of the second monomer (green) to form a composite epitope (shaded cyan region) (Murphy *et al.*, 2003). The cytokine ligand is drawn in red and docked onto the CHR domain.

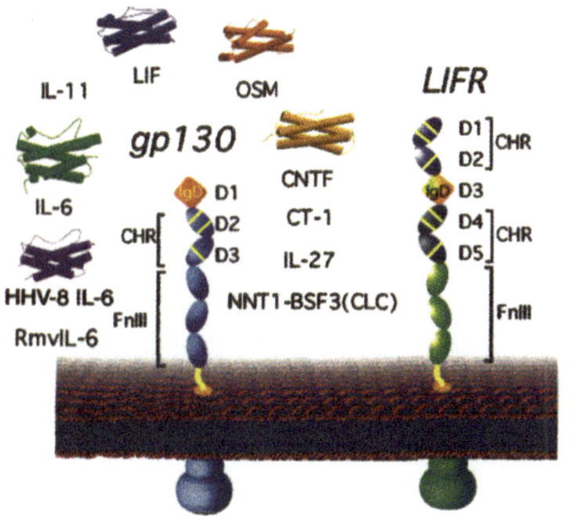

Cytokine	Site I	Site II	Site III
HHV-8 IL-6		gp130	gp130
Rm IL-6		gp130	gp130
IL-6	IL-6Rα	gp130	gp130
IL-11	IL-11Rα	gp130	gp130
LIF		gp130	LIFR
OSM		gp130	LIFR(*OSMR*)
CNTF	CNTFRα	gp130	LIFR
CT-1	CNTFRα	gp130	LIFR
NNT1-BSF3(CLC)	CNTFRα	gp130	LIFR
IL-27		TCCR	gp130

BOULANGER AND GARCIA, FIG. 3. The gp130 family of cytokines and receptors. Schematic representation of gp130 and leukemia inhibitory factor receptor (LIFR) oriented in a cell membrane. Of the four-helix bundle gp130 cytokines, structural information currently exists for human interleukin 6 (IL-6) (green) (Somers *et al.*, 1997), human herpes virus interleukin 6 (HHV-8 IL-6) (purple) (Chow *et al.*, 2001a), ciliary neurotrophic factor (CNTF) (orange) (McDonald *et al.*, 1995), leukemia inhibitory factor (LIF) (blue) (Robinson *et al.*, 1994), and oncostatin-M (OSM) (red) (Deller *et al.*, 2000). Lower panel is a detailed list of gp130 cytokines and the associated receptors incorporated into the final signaling complex.

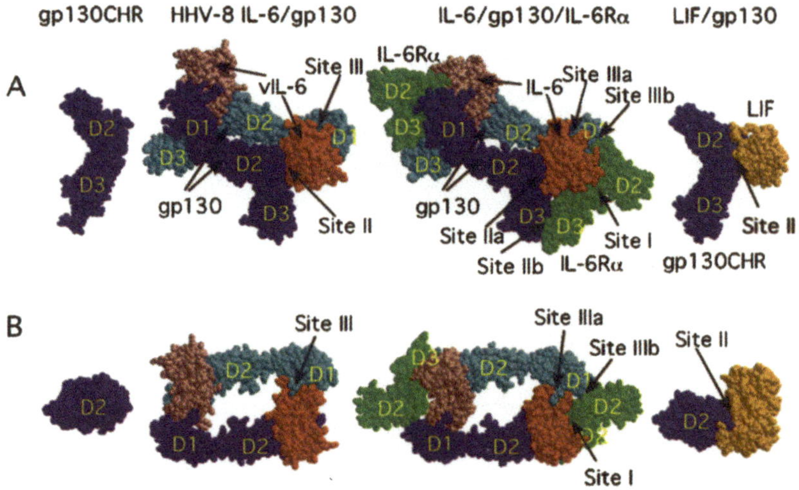

Boulanger and Garcia, Fig. 4. Current structural knowledge of gp130 cytokine/receptor complexes. Four crystal structures are shown in the same relative orientations based on the gp130 CHR, as seen from a tilted side view (A) and a view onto the top of the complexes (B). CPK models are shown for the crystal structures of gp130 (D2-D3–CHR domain) (Bravo *et al.*, 1998), HHV-8 IL-6/gp130(D1-D3) (Chow *et al.*, 2001a), IL-6/IL-6Rα/gp130(D1-D3) (Boulanger *et al.*, 2003b) and LIF/gp130(D2-D3–CHR domain) (Boulanger *et al.*, 2003a). In transitioning from the HHV-8 IL-6/gp130 complex to the IL-6/IL-6Rα/gp130 complex, the additional IL-6Rα docks onto the site I. CPK models were generated with MOLSCRIPT (Kraulis, 1991) and RASTER3D (Merritt and Murphy, 1994).

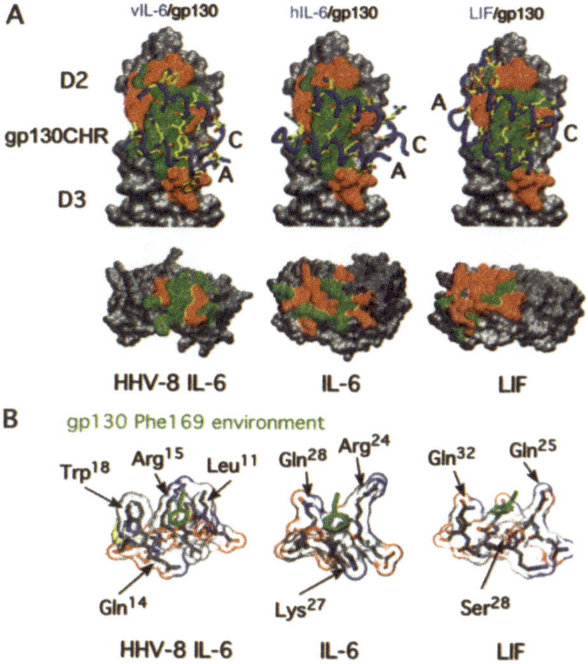

BOULANGER AND GARCIA, FIG. 5. Cross-reactivity of the cytokine-binding homology region (CHR) of gp130. (A) Upper panel: Contact surface of gp130 when bound to HHV-8 IL-6, IL-6 and LIF. Residues contributing to the buried surface area of the interface, as calculated using the Protein-Protein Interaction (PPI) server http://www.biochem.ucl.ac.uk/bsm/PP/server. Hydrophobic residues are colored as green surface and hydrophilic residues as red surface. Note the core hydrophobic region surrounded by discontinuous patches of polar residues. The A and C helices (blue) and the associated contact residues for each of the three cytokines are shown docked onto the surface of gp130. Lower panel: Surface representations showing the hydrophobic properties of the site II gp130 CHR binding epitope of HHV-8 IL-6, IL-6, and LIF. Note the significant hydrophilic contact surface of LIF relative to the primarily hydrophobic contact surface of HHV-8 IL6. (B) The packing of the conserved gp130 residue Phe^{169} (green) into surface pockets on HHV-8 IL-6, IL-6 and LIF. The binding affinity of the complexes appear to be consistent with the polarity of the Phe^{169} binding pocket (i.e., the more hydrophobic the pocket the higher the associated binding affinity) (HHV8 IL-6 > IL-6 > LIF). VMD (Humphrey *et al.*, 1996) was used to prepare MSMS surface representations (Sanner *et al.*, 1996).

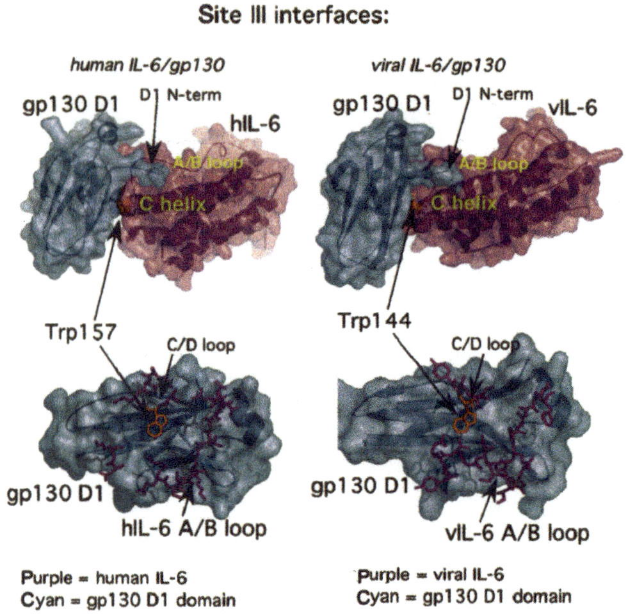

BOULANGER AND GARCIA, FIG. 6. Structural anatomy of the site III interface. The Ig domain of gp130 (blue) from the human and viral receptor complexes is shown in isolation with the respective cytokines (purple) at the site III interfaces. At the top is a "top view" of a transparent molecular surface with the ribbons of the structure within. In this view the convex tip of the cytokine is abutting into the bottom sheet of the gp130 Ig-domain. The penetrating N-terminal peptide from gp130 is also seen to insert into a groove on the cytokine surface. At the bottom of the figure, the view is directly into the cytokine binding sheet of the gp130 Ig-domain with the interacting loops of the cytokines shown as sticks, so that we are looking through the cytokine into the receptor. The centrally-disposed Trp residue at the heart of both site III interfaces is indicated in both views. Figure generated with PyMol (W. Delano).

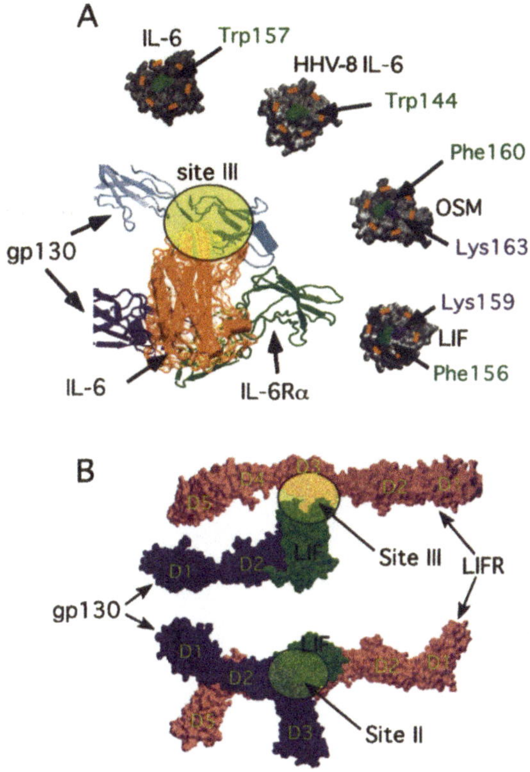

BOULANGER AND GARCIA, FIG. 7. The role of site III in the higher order assembly of gp130 signaling complexes. (A) Secondary structure depiction of one half of the hexameric IL-6/IL-6Rα/gp130 complex with the site III highlighted in yellow formed by the tip of the four helix bundle abutting into the D1 (IgD) domain of gp130 (Boulanger *et al.*, 2003b). Surface representations show the site III localization of the conserved tryptophan residue on IL-6 and HHV-8 IL-6 and the conserved phenylalanine/lysine pair on LIF and OSM. (B) Model of the extracellular signaling complex of LIF/gp130/LIFR where the site II of LIF (green) engages the CHR domain of gp130 (blue) (Boulanger *et al.*, 2003a) and the site III of LIF contacts the Ig domain of LIFR (purple).

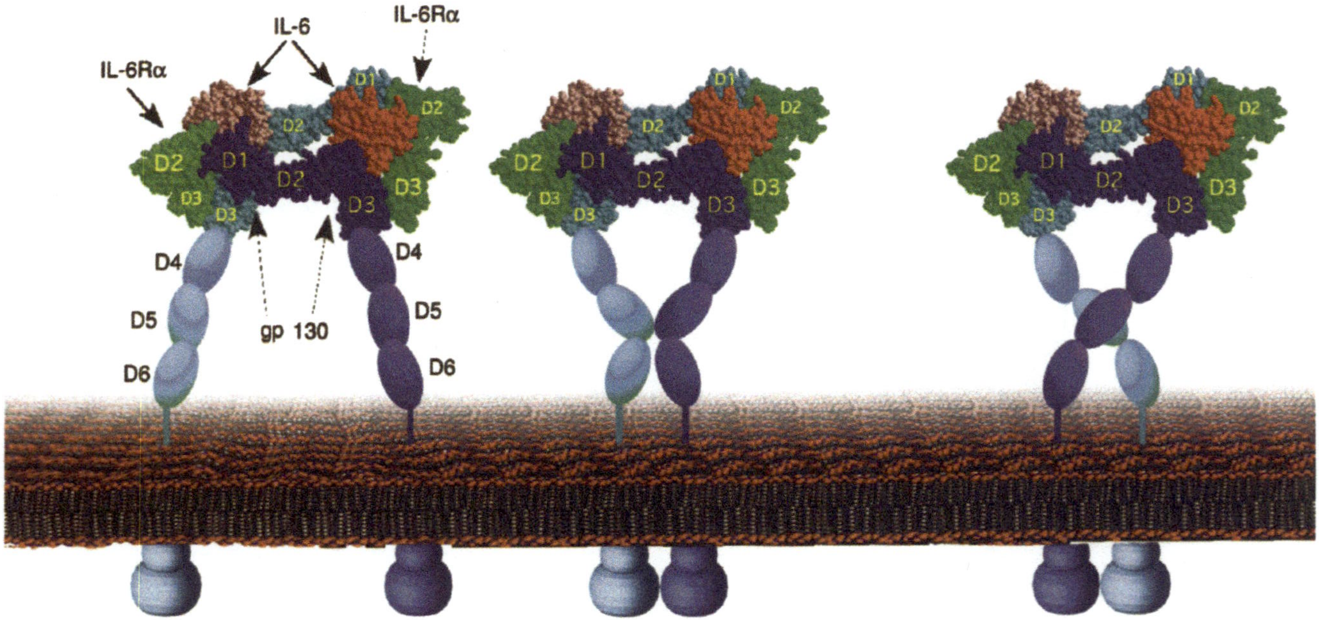

BOULANGER AND GARCIA, FIG. 8. Possible orientations of the "legs" of gp130. The CPK model of the IL-6/IL-6Rα/gp130 hexamer structure (Boulanger *et al.*, 2003b) is drawn along with schematic model of the three membrane proximal fibronectin repeats that connect the cap to the membrane. The two models to the right incorporate data suggesting an interaction between the gp130 legs (Boulanger *et al.*, 2003b), possibly the D5 domain (Kurth *et al.*, 2000). To date, no structural information exists describing the position and orientation of the three membrane proximal domains.

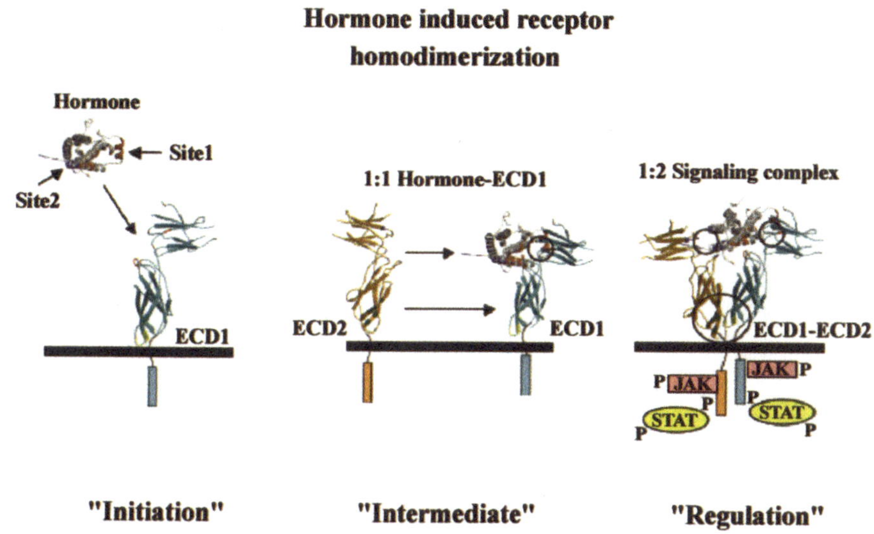

KOSSIAKOFF, FIG. 1. Mechanism of hormone-induced receptor homodimerization. The GH/PRL receptors are 3-domain single-pass receptors containing an extracellular domain (ECD), a transmembrane section of about 25 amino acids, and a cytoplasmic domain that forms the binding site for the tyrosine kinase activities. The "Initiation" step involves the hormone binding event to the ECD (ECD1). The segments of the molecules that are involved in the contact (Site1) and colored red (hormone) and yellow (ECD1). This first step forms a stable 1:1 "Intermediate" which then recruits a second receptor in the "Regulation" step through two sets of contacts (Site2); one to the hormone and the other through forming receptor "stem-stem" contacts (ECD1-ECD2). This step forms the stable homodimer, which organizes the cytoplasmic components to initiate binding and phosphorylation.

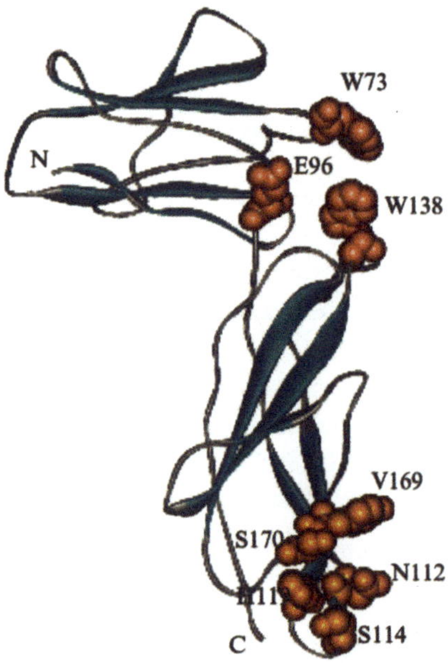

Kossiakoff, Fig. 2. Ribbon drawing of the hGH receptor ECD. The side chains of the ECD that are used by the receptor to bind at both Site1 and Site2 of the hormone are indicated. The side chains of the stem-stem contact are shown at the base of the C-terminal domain.

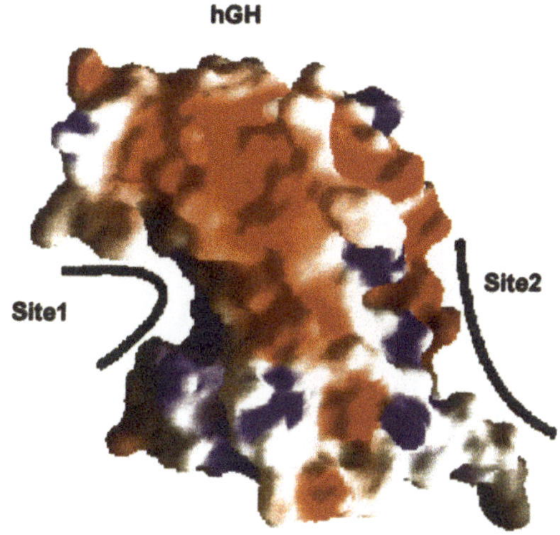

KOSSIAKOFF, FIG. 3. Molecular surface of hGH showing the different topographies of the Site1 and Site2 binding sites. In addition, the two sites possess quite different electrostatic properties (red negative charge, blue positive charge; image rendered between ± 10 kT).

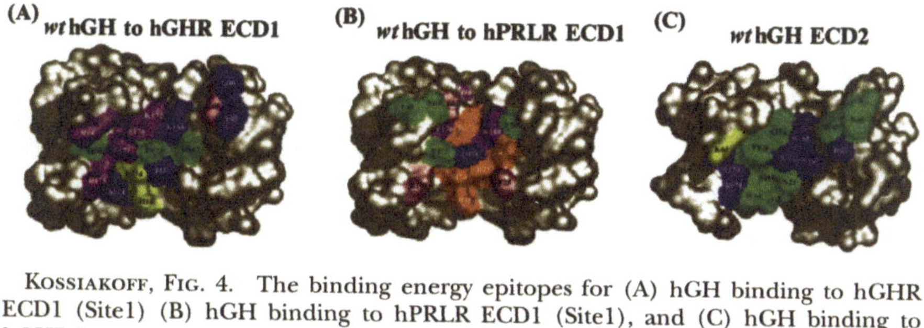

KOSSIAKOFF, FIG. 4. The binding energy epitopes for (A) hGH binding to hGHR ECD1 (Site1) (B) hGH binding to hPRLR ECD1 (Site1), and (C) hGH binding to hGHR ECD2 (Site2). Color code as decrease in binding based on Ala-scan mutagenesis: Red, no binding >100-fold; magenta, 15-80-fold; pink, 5-15-fold; green, 2-5-fold; blue, no change; yellow, >2-fold increase in binding.

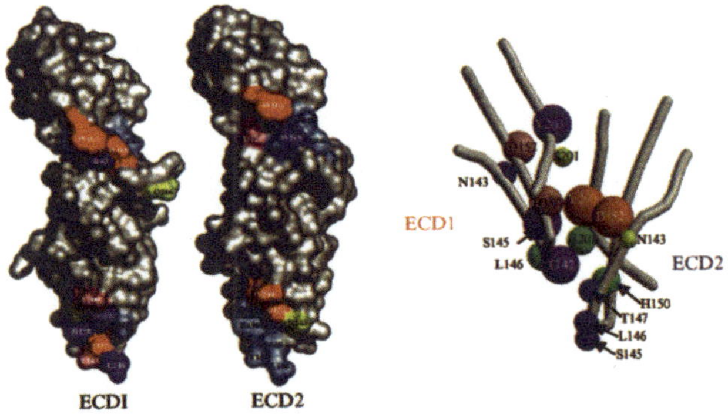

KOSSIAKOFF, FIG. 5. (Right) Structural relationship of residues in the stem-stem interface. Decreases in binding affinity due to individual wild-type to Ala substitutions are represented by color-coding and the size of the representative spheres. (Center and Left) Space filling representation of Ala-scan data for ECD1 and ECD2 binding to hGH Site1 and Site2 and the stem-stem interaction. The residues in the N-terminal domains of the ECDs (top patch) represent only subsets of the corresponding contact interfaces including most, but not all the energetically important positions. The residues in the N-terminal domains of ECD1 and ECD2 bind to Site1 and Site2 of hGH, respectively. Color code as decrease in binding based on Ala-scan mutagenesis: Red, >50-fold; magenta, 15-50-fold; blue, 3–15 fold; light blue, 0-2-fold; yellow, >2-fold increase in binding.

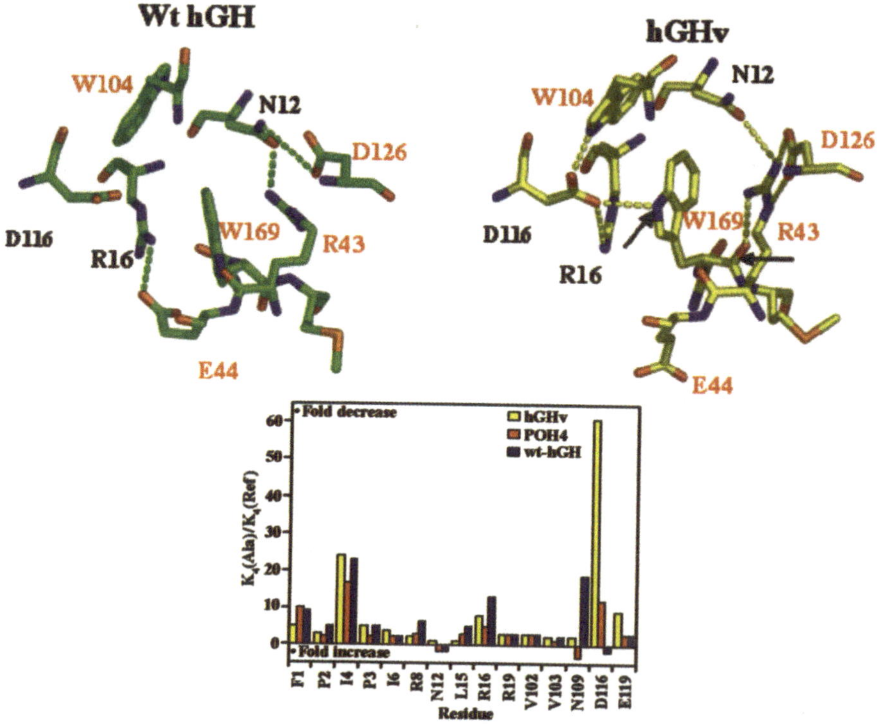

KOSSIAKOFF, FIG. 6. Site2 interfaces of hGHv and wt-hGH with their respective ECD2. (A) The *wt* hGH (green) and the hGHv (yellow) interface around D116. The hormone and receptor residues are labeled in black and red, respectively. The intermolecular H-bonds and salt bridges between the hormone and the receptor are depicted as broken lines. The arrows point to the peptide bond flip observed for $W169_{E2}$ between the two crystal structures. (B) Bar graphs of the fold decreases/increases for the Site2 Ala-scan of the R2 to all three hormones. The K_d is the reference Site2 dissociation constant K_d for each hormone. The asterisks indicate residues that when mutated to alanine results in no detectable Site2 binding affinity.

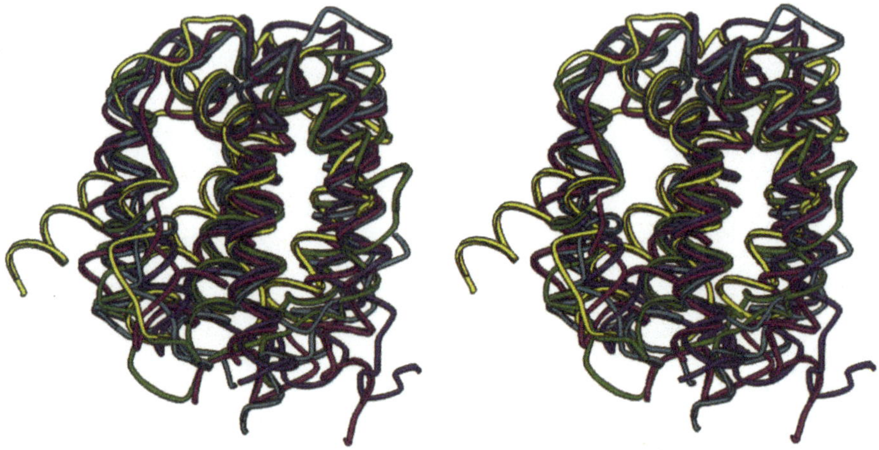

WALTER, FIG. 5. Superposition of the IL-10FM domains. The domains are oriented as shown in Fig. 2A. IFN-γ is colored yellow, IL-10 is blue, cmvIL-10 is magenta, IL-19 is cyan, and IL-22 is green.

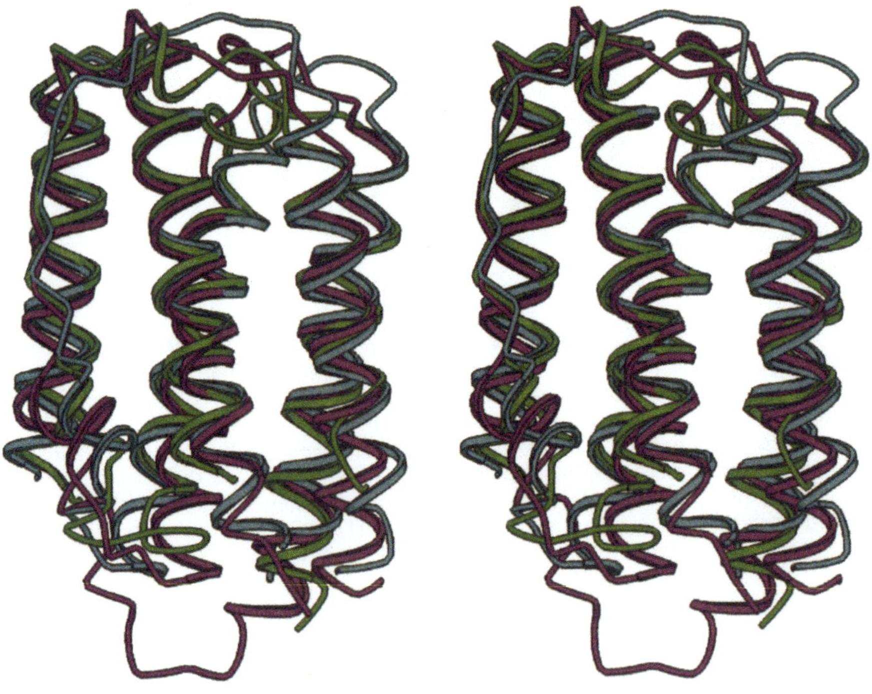

WALTER, FIG. 7. Stereoview comparison of the type I IFNs. IFN-α2b (green), human IFN-β (magenta), and ovine IFN-τ (cyan) are shown. The view is looking into the IFNAR2 binding site.

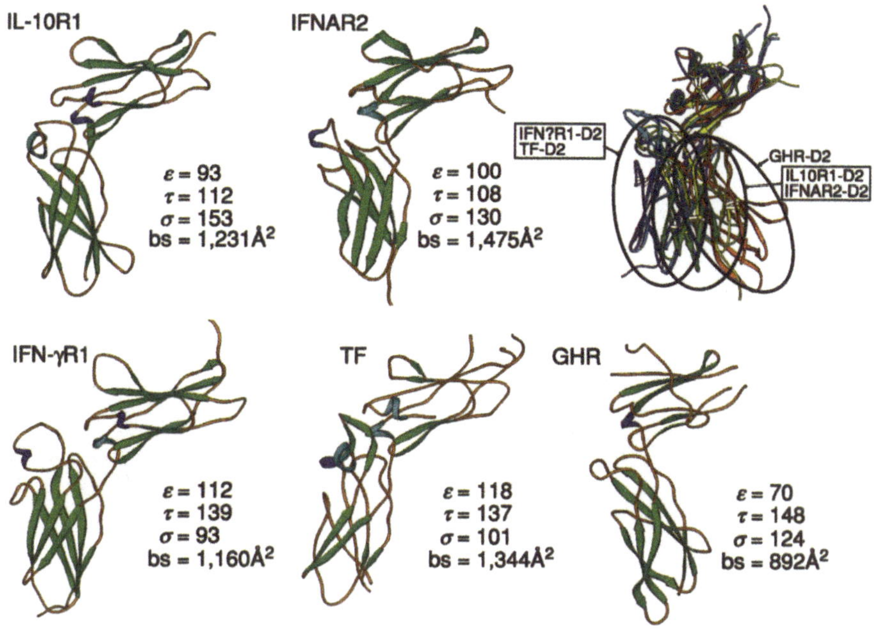

WALTER, FIG. 12. C2HR domain orientations. Each C2HR is labeled with its characteristic tilt (ε), roll (τ), and swivel (σ), parameters (Deivanayagam *et al.*, 2000) along with the amount of amino acid surface area buried in its domain interface (bs). The C2HRs adopt three distinct D1/D2 orientations. Similar orientations are observed for IL-10R1 and IFNAR2 (group 1), IFN-γR1 and TF (group 2), and GHR (group 3). The difference between each group is most obvious if shown by superimposing the D1 domains of the receptors.

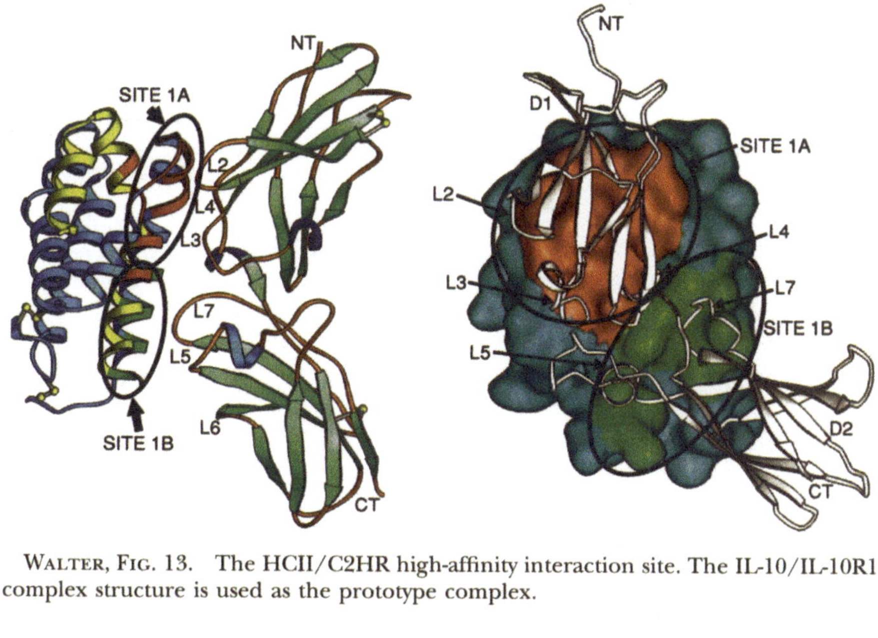

WALTER, FIG. 13. The HCII/C2HR high-affinity interaction site. The IL-10/IL-10R1 complex structure is used as the prototype complex.

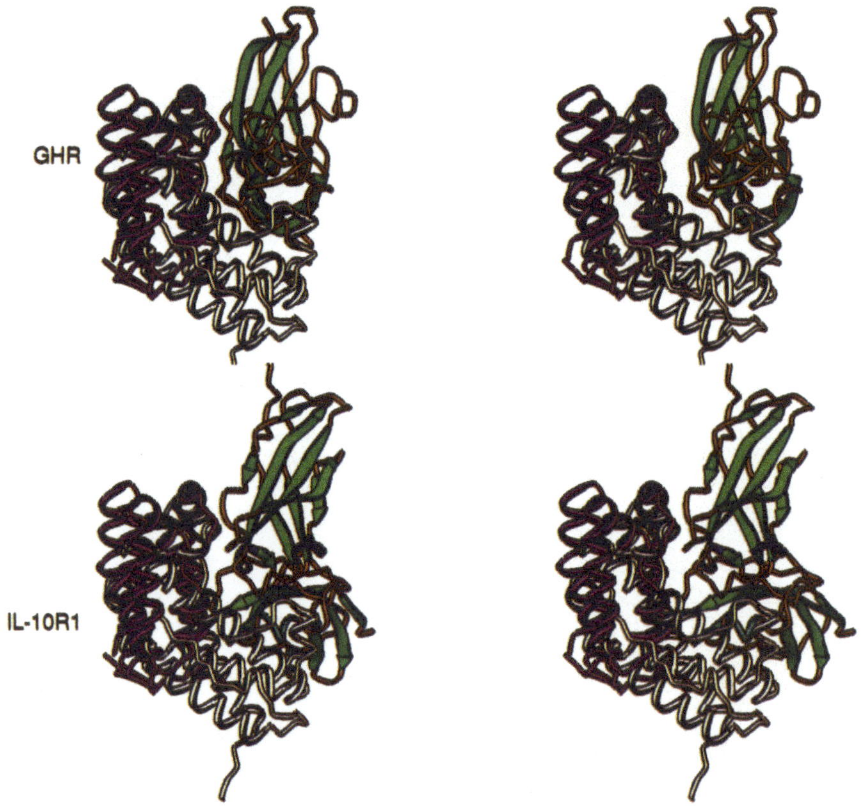

WALTER, FIG. 17. Stereoview comparison of the orientation of IL-10 and GH on their C2HRs. D1s of IL-10R1 and GHR were superimposed as in Fig. 10A. The resulting orientations of IL-10 (magenta) and GH (white) are shown with GHR (top stereoview) or IL-10R1 (bottom stereoview).

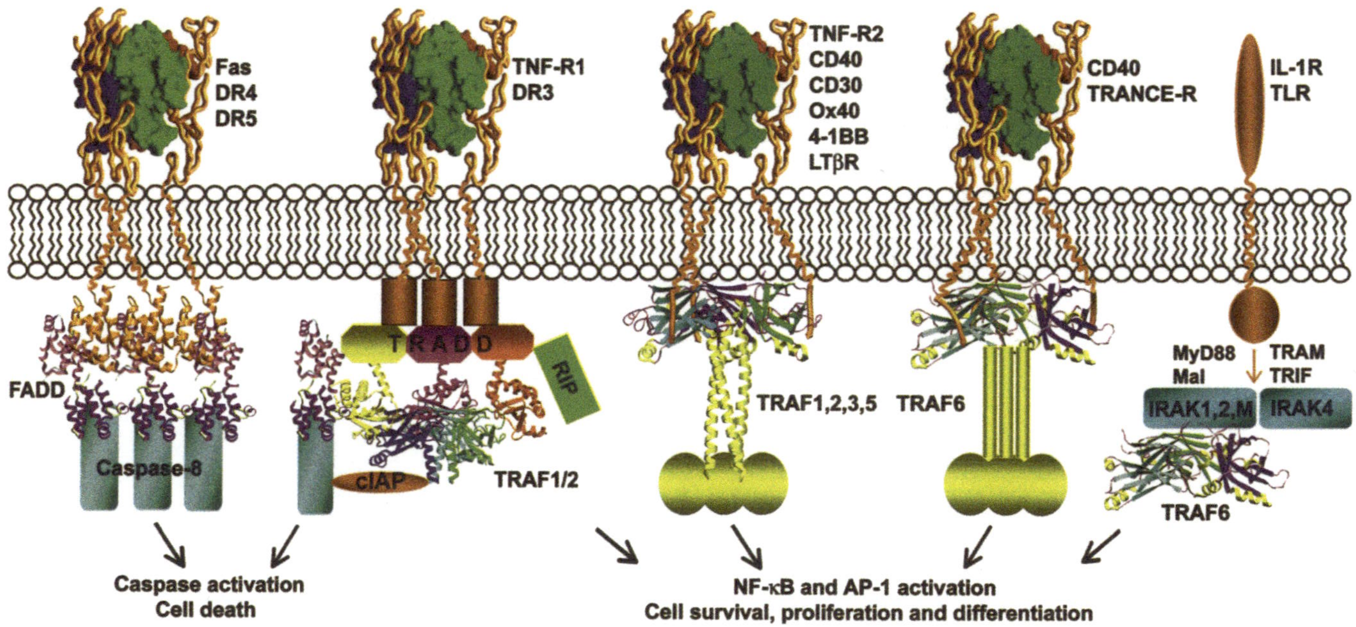

WU, FIG. 1. Intracellular signaling pathways for the TNFR superfamily and the IL-1R/TLR superfamily. Proteins with known structures are shown as ribbon drawings. Hypothetical transmembrane helices are built to connect extracellular rand intracellular domains of these receptors (shown in orange). The amino terminal domains of TRAFs are shown as yellow spheres.

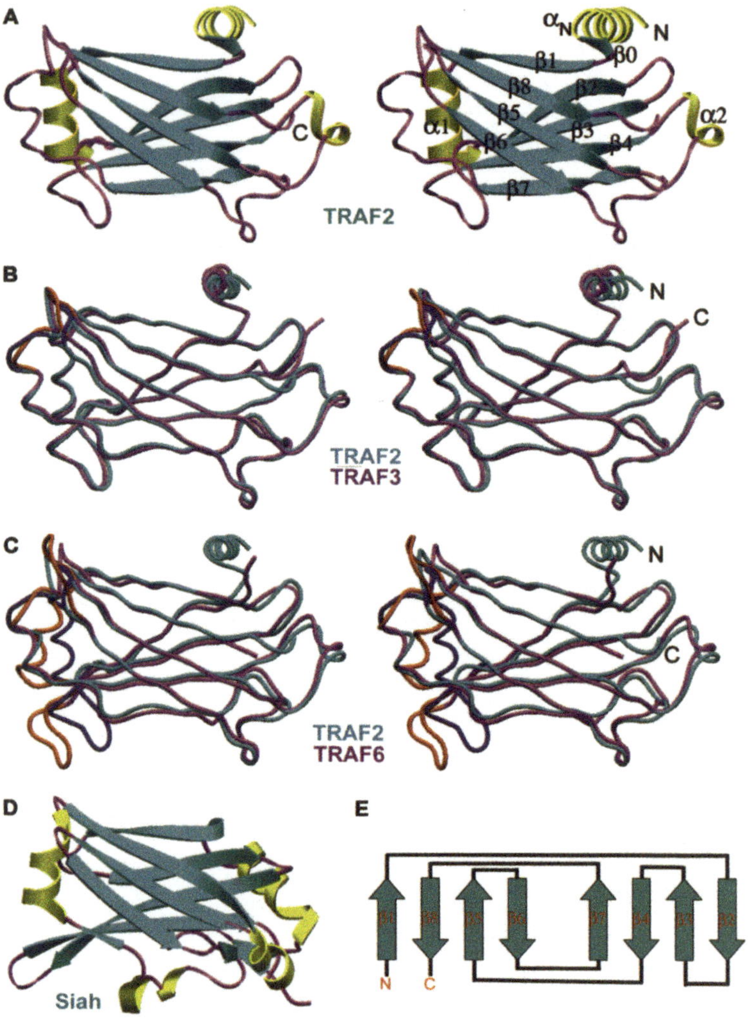

Wu, Fig. 2. TRAF domain structures. (A) Stereo drawing of the TRAF domain of TRAF2 with labeled secondary structures. (B) Superposition of the TRAF domain of TRAF2 (cyan) and TRAF3 (magenta). Regions with large differences between the two structures are shown in blue for TRAF2 and red for TRAF3. (C) Superposition of the TRAF domain of TRAF2 (cyan) and TRAF6 (magenta). Regions with large differences between the two structures are shown in blue for TRAF2 and red for TRAF6. (D) Ribbon drawing of Siah. (E) Topology of TRAF-C domains.

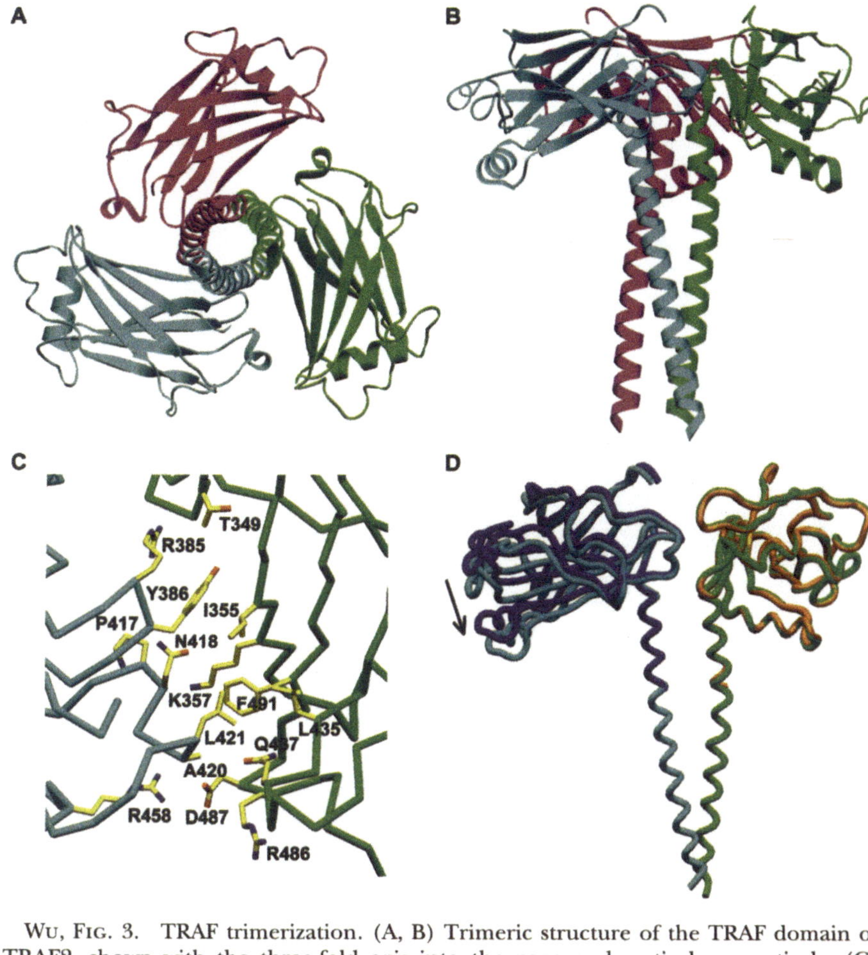

WU, FIG. 3. TRAF trimerization. (A, B) Trimeric structure of the TRAF domain of TRAF2, shown with the three-fold axis into the page and vertical, respectively. (C) Detailed interaction between the TRAF-C domains in the trimer. (D) Observed structural variation among TRAF domain trimers. When superimposed onto one of the protomers in the trimer, a neighboring protomer may exhibit a different tilt of the mushroom cap-shaped TRAF-C domain.

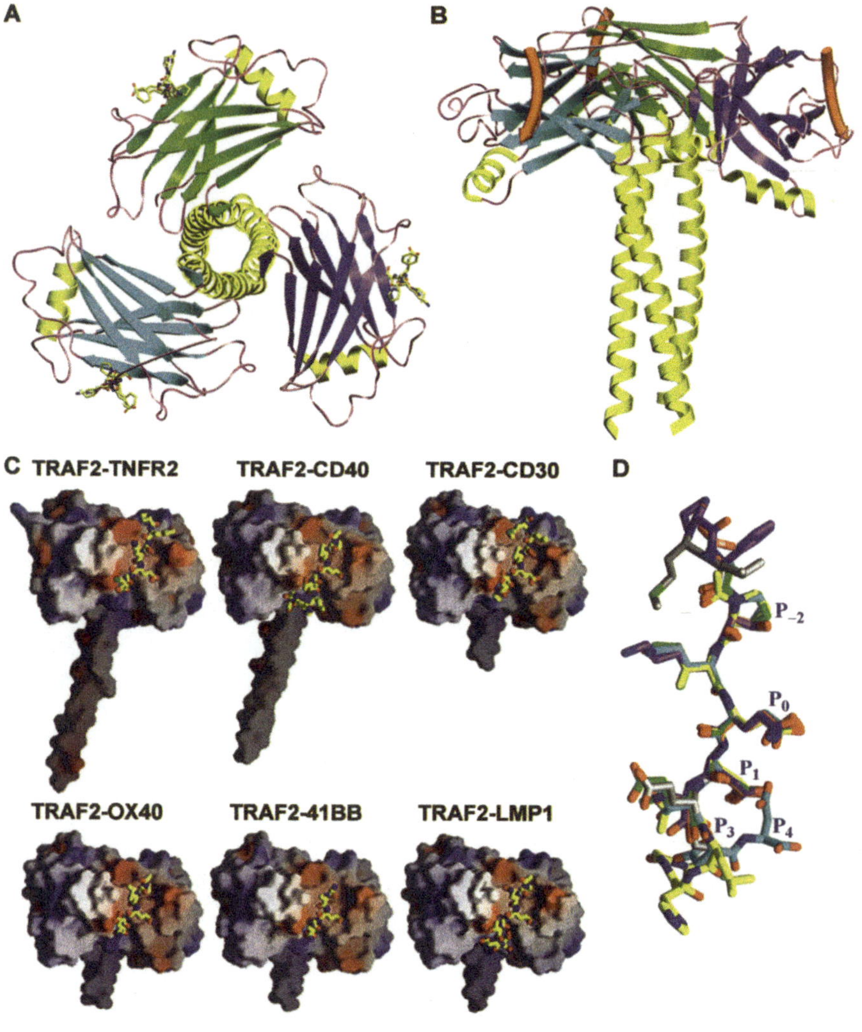

WU, FIG. 4. TRAF2-receptor interactions. (A) Ribbon diagram of a TRAF2-receptor complex, looking down the three-fold axis. The bound receptor chains are shown as stick models. (B) Ribbon diagram of a TRAF2-receptor complex with the threefold axis vertical. The bound receptor chains are shown as arrows. (C) Surface electrostatic representation of TRAF2-peptide complexes. (D) Superposition of bound receptor peptides, showing the structural conservation of the main chain conformations and the side chain conformations at P_2, P_0, and P_1 positions. Parts of this figure were modified from Ye *et al.* (1999).

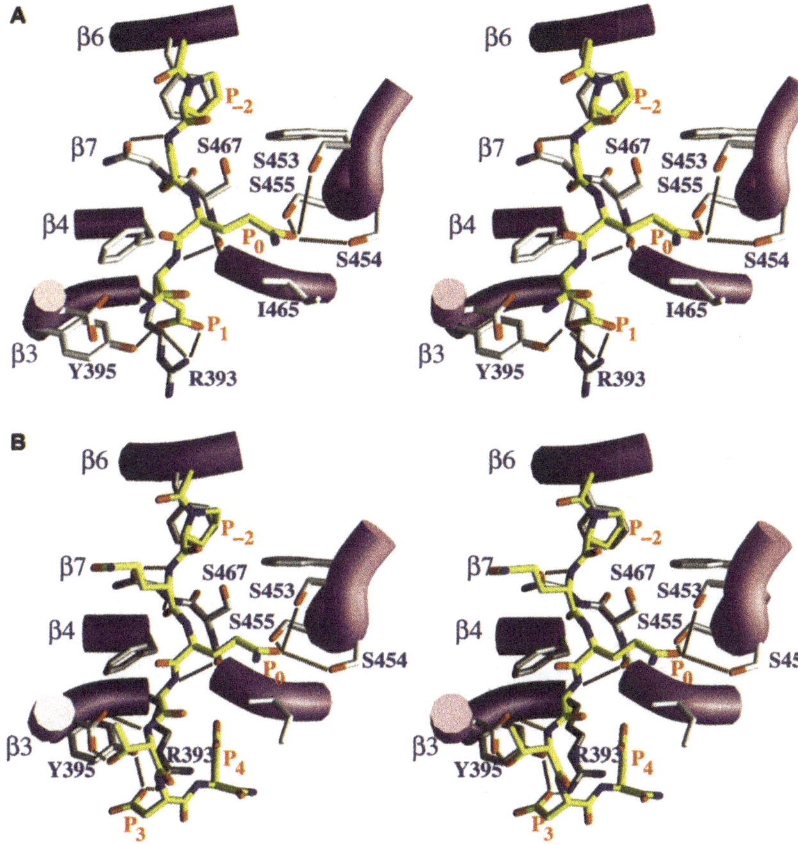

WU, FIG. 5. Detailed TRAF2-receptor interactions. (A) Interactions seen in the major TRAF2-binding motif. (B) Interactions seen in the minor TRAF2-binding motif. TRAF2 structures are shown as magenta worms and white stick models. The bound receptors are shown as yellow stick models. Modified from Ye *et al.* (1999).

	P_{-4}	P_{-3}	P_{-2}	P_{-1}	P_0	P_1	P_2	P_3	P_4	P_5	P_6
hTNF-R2 (422-432)	P	F	S	K	E	E	C	A	F	R	S
hCD40 (248-258)	A	A	P	V	Q	E	T	L	H	G	C
hCD30 (576-586)	M	L	S	V	E	E	E	G	K	E	D
hCD30 (559-569)	H	Y	P	E	Q	E	T	E	P	P	L
hCD27 (244-254)	T	I	P	I	Q	E	D	Y	R	K	P
hLTβR (386-410)	Y	P	I	P	E	E	G	D	P	G	P
hLTβR (400-410)	S	T	P	H	Q	E	D	G	K	A	W
hATAR (266-276)	T	V	A	V	E	E	T	I	P	S	F
hOx40 (260-270)	R	T	P	I	Q	E	E	Q	A	D	A
m4-1BB (230-240)	T	G	A	A	Q	E	E	D	A	C	S
m4-1BB (242-252)	R	C	P	Q	E	E	E	G	G	G	G
h4-1BB (232-242)	V	Q	T	T	Q	E	E	D	G	C	S
h4-1BB (244-254)	R	F	P	E	E	E	E	G	G	C	E
bLMP1 (204-214)	R	T	P	V	Q	E	S	G	Y	P	D
bLMP1 (219-229)	R	P	P	V	Q	E	T	G	G	G	G
bLMP1 (243-253)*	H	P	P	V	Q	E	T	G	G	G	G
bLMP1 (315-325)	H	P	P	V	Q	E	T	G	E	G	G
bLMP1 (359-369)	H	P	P	I	Q	E	T	G	N	G	G
hTANK (178-188)	S	V	P	I	Q	C	T	D	K	T	D
hLMP1 (202-212)	P	H	P	Q	Q	A	T	D	D	S	S
rLMP1 (315-325)#	P	Y	P	I	Q	A	T	D	G	G	N
rLMP1 (377-387)	P	H	P	I	Q	A	T	D	G	A	N
rLMP1 (425-435)	P	H	P	V	Q	A	S	D	G	G	D
Major Motif			p	x	Q/E	E					
Minor Motif			p	x	Q/E	x	x	D			

WU, FIG. 6. Sequence alignment of TRAF2 binding sequences, illustrating the two TRAF2-binding motifs. h: human; m: mouse; b: bovine; r: rat. Modified from Ye *et al.* (1999).

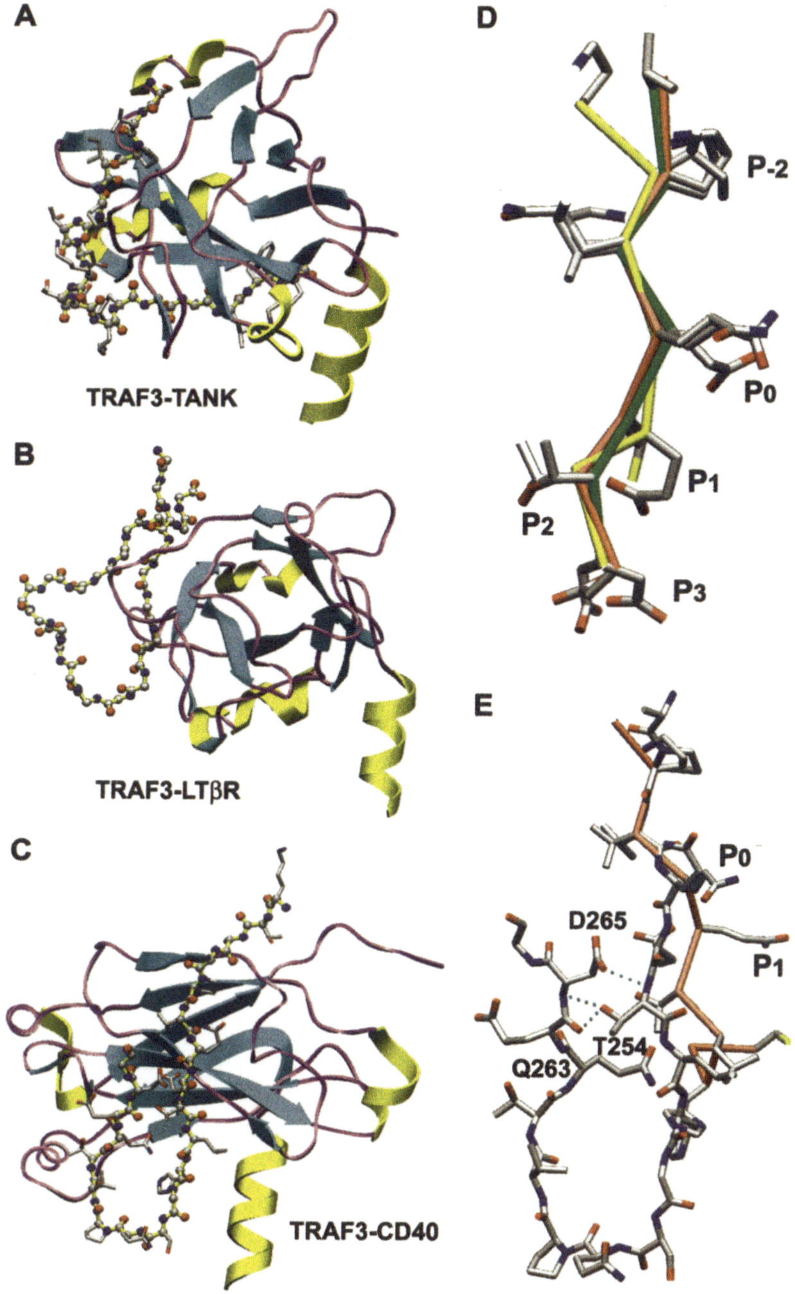

WU, FIG. 7. TRAF3-receptor interactions. (A, B, C) Ribbon diagrams of TRAF3-TANK, TRAF3-LtβR and TRAF3-CD40 complexes. (D) Superposition of the TANK peptide (green) and the LTβR peptide (yellow) with the LMP1 peptide (pink) in the TRAF2-LMP1 complex. Only the core residues are shown. (E) Details of the CD40 conformation in the TRAF3-CD40 complex. Shown in pink is the superimposed CD40 conformation seen in the TRAF2-CD40 complex.

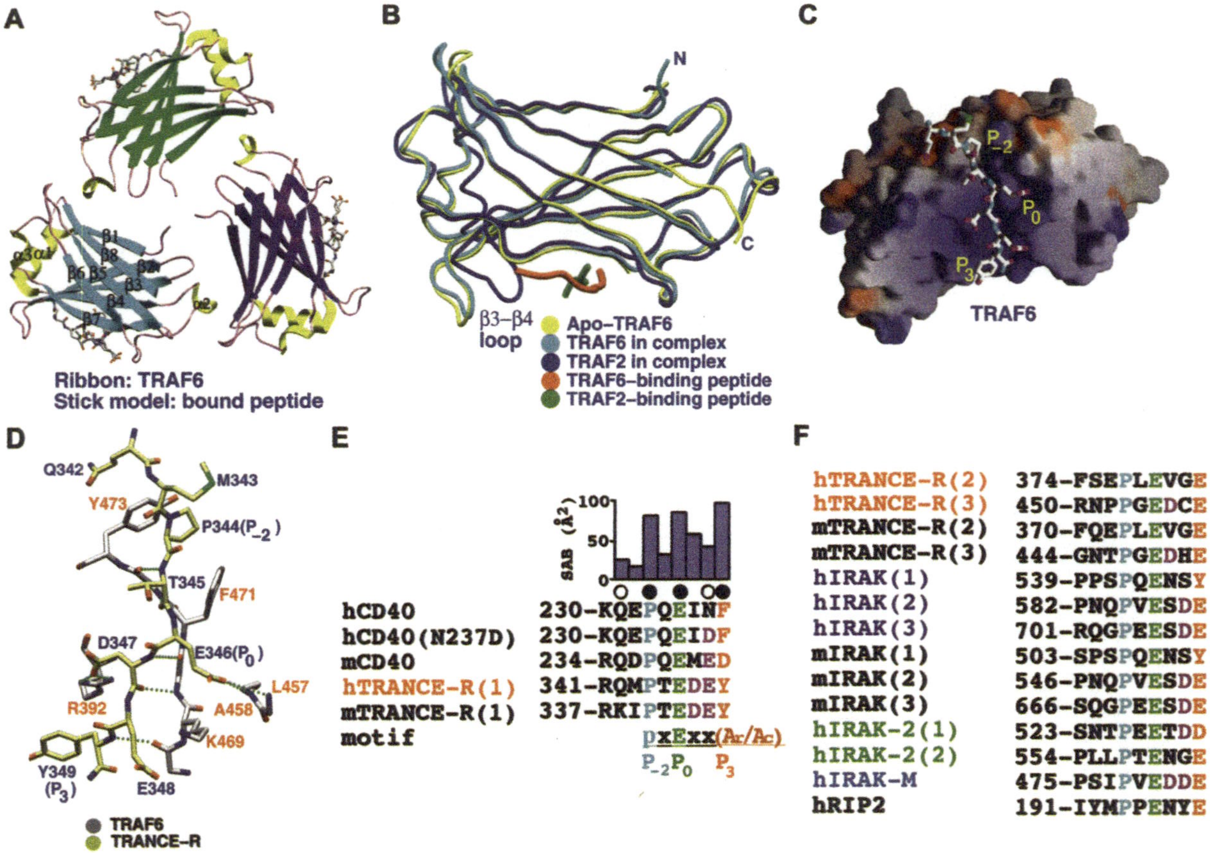

WU, FIG. 8. TRAF6-receptor interactions. (A) TRAF6 trimeric model. (B) Superposition of TRAF6 with TRAF2, with bound receptor peptides. (C) Electrostatic surface of TRAF6 in complex with the TRANCE-R peptide. (D) Detailed interactions between TRAF6 and TRANCE-R. (E) Sequence alignment of CD40 and TRANCE-R and the corresponding surface area burial (SAB) and the TRAF6-binding motif. (F) TRAF6-binding sequences identified based on the structurally defined motif. Modified from Ye *et al.* (2002).

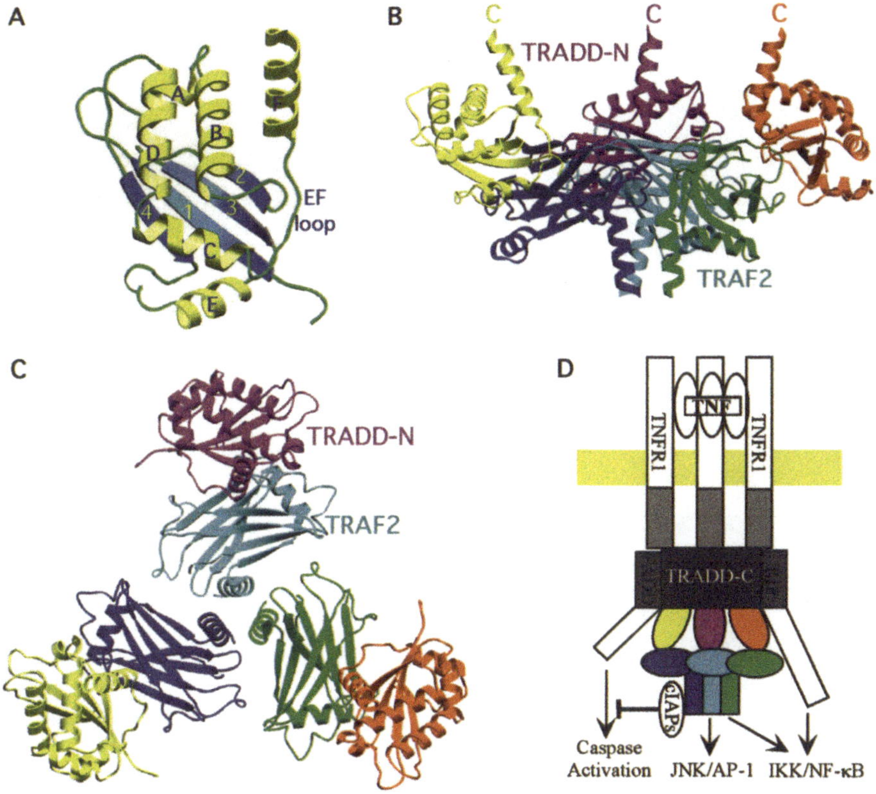

WU, FIG. 9. TRAF2-TRADD interaction. (A) Ribbon drawing of the TRADD-N domain. (B, C) Ribbon diagrams of the TRAF2-TRADD complex. (D) Schematic representation of the TNF-R1 signaling complex. Reproduced from Park *et al.* (2000).

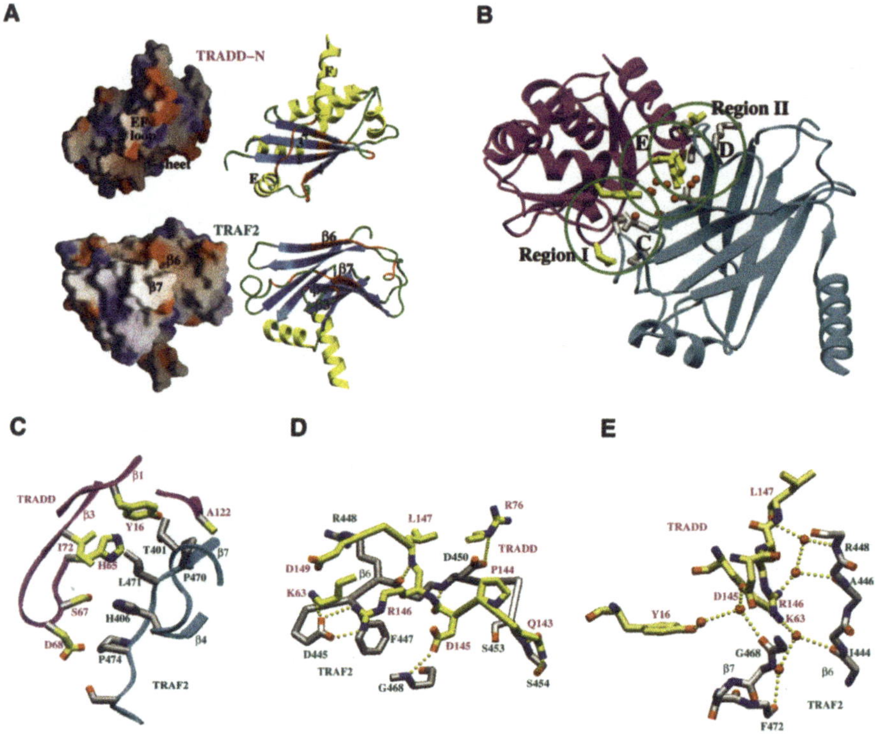

WU, FIG. 10. Detailed TRAF2-TRADD interaction. (A) Interaction surfaces and their locations on the individual structures (in red). (B) Molecular interactions at the two regions of the interactions. (C, D, E) Details of region I, region II and water-mediated interactions, respectively.

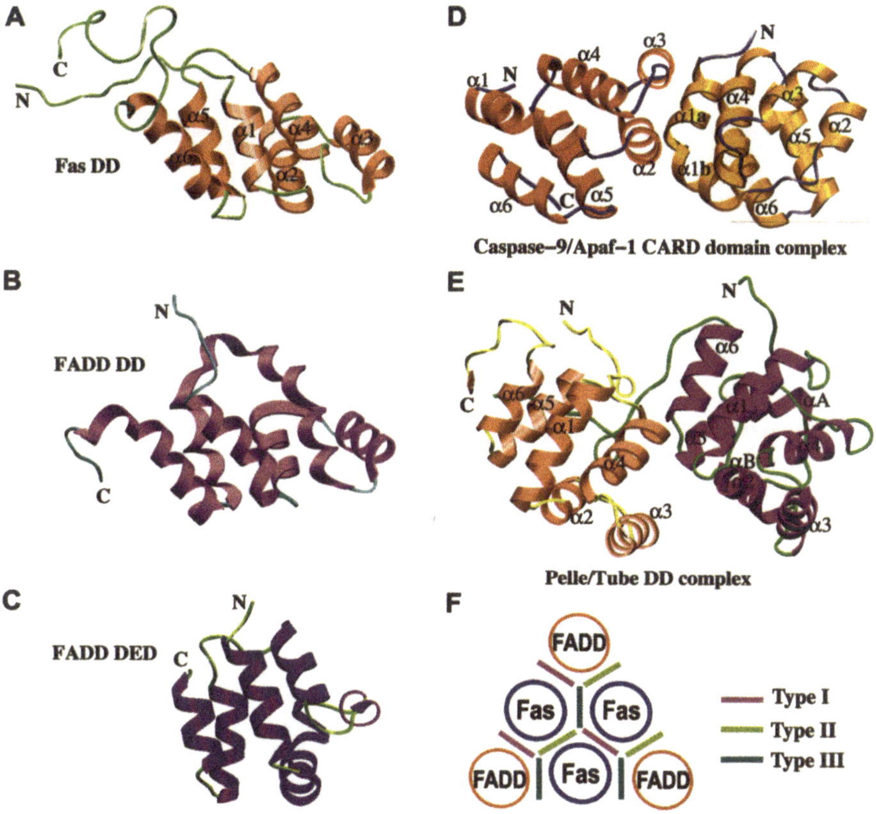

WU, FIG. 11. Structures of DDs and DEDs. (A) DD of Fas. (B) DD of FADD. (C) DED of FADD. (D) CARD domain complex of caspase-9 (red) and Apaf-1 (orange). (E) DD complex of Pelle (pink) and Tube (purple). (F) A hypothetical model of the oligomeric Fas-FADD complex, showing the potential involvement of three types of interfaces.

STRONG AND MCFARLAND, FIG. 1. Representative immunoreceptor–ligand complex structures. Views of a series of immunoreceptor–ligand complexes are shown, in a mix of ribbon and space-filling representations, highlighting the range of variation in these recognition events: $\alpha\beta$ TCR–classical pMHC class I (Garboczi *et al.*, 1996); KIR2DL–classical pMHC class I (Boyington *et al.*, 2000); Ly49A–classical pMHC class I (Tormo *et al.*, 1999); huNKG2D–MIC-A (Li *et al.*, 2001a); huNKG2D–ULBP3 (Radaev *et al.*, 2001); and muNKG2D–RAE-β (Li *et al.*, 2002). The views in the left-hand column are aligned looking down along the peptide-binding groove (when present). The views in the middle column are perpendicular to the dyad axis of the NKG2D receptor (a rotation of approximately 50° from the view in the left-hand column). The views in the right-hand column are oriented looking down onto the top (peptide-binding) surface of the platform domains of the MHC class I protein or homolog (a rotation of 90° from the view in the middle column). All the molecules in a column are aligned on the platform domain of the MHC class I protein or homolog. Domains are colored as indicated, with the receptors colored in blue and purple and the ligand proteins colored in yellow, orange, red and green; peptides, when present, are colored by atom-type (carbon: gray; oxygen: red; nitrogen: blue; and sulfur: yellow).

STRONG AND MCFARLAND, FIG. 2. Comparison of CTLD structures. Ribbon representations, colored by secondary structure (α-helices: yellow coils; β-strands: green arrows), are shown for a series of illustrative CTLD-containing immunoreceptors: Ly49A (Tormo *et al.*, 1999); CD69 (Llera *et al.*, 2001); CD94 (Boyington *et al.*, 1999); and huNKG2D (McFarland *et al.*, 2003). The structure of the archetype C-type lectin fold-containing protein, trimeric rat mannose binding protein (MBP), is shown at the top for comparison (Feinberg *et al.*, 2000). The structures are shown in two views, one perpendicular to the dyad axis of symmetry in the dimeric molecules (left) and one view from below, looking up onto the ligand binding sites (right; 90° from the view on the left for the dimeric molecules). The left-hand views are oriented so that the left CTLD or C-type lectin domains are superimposed, highlighting the differences in the fold of the short-form C-type lectins (MBP) and the CTLD-containing immunoreceptors, which are more homologous to long-form C-type lectins. N- and C-termini are labeled. A disaccharide ligand is shown (in a space-filling representation, colored by atom-type as in Fig. 1) for MBP; the NKG2D ligand-binding site is indicated with blue arcs. Bound ions are shown as spheres: calcium atoms (green) in the MBP structure; zinc atoms (gray) in CD69. The gray pointers mark the NKG2D stirrup loops, a structurally divergent feature of NKG2D that results in the distinctly saddle-shaped ligand-binding surface of this NKR.

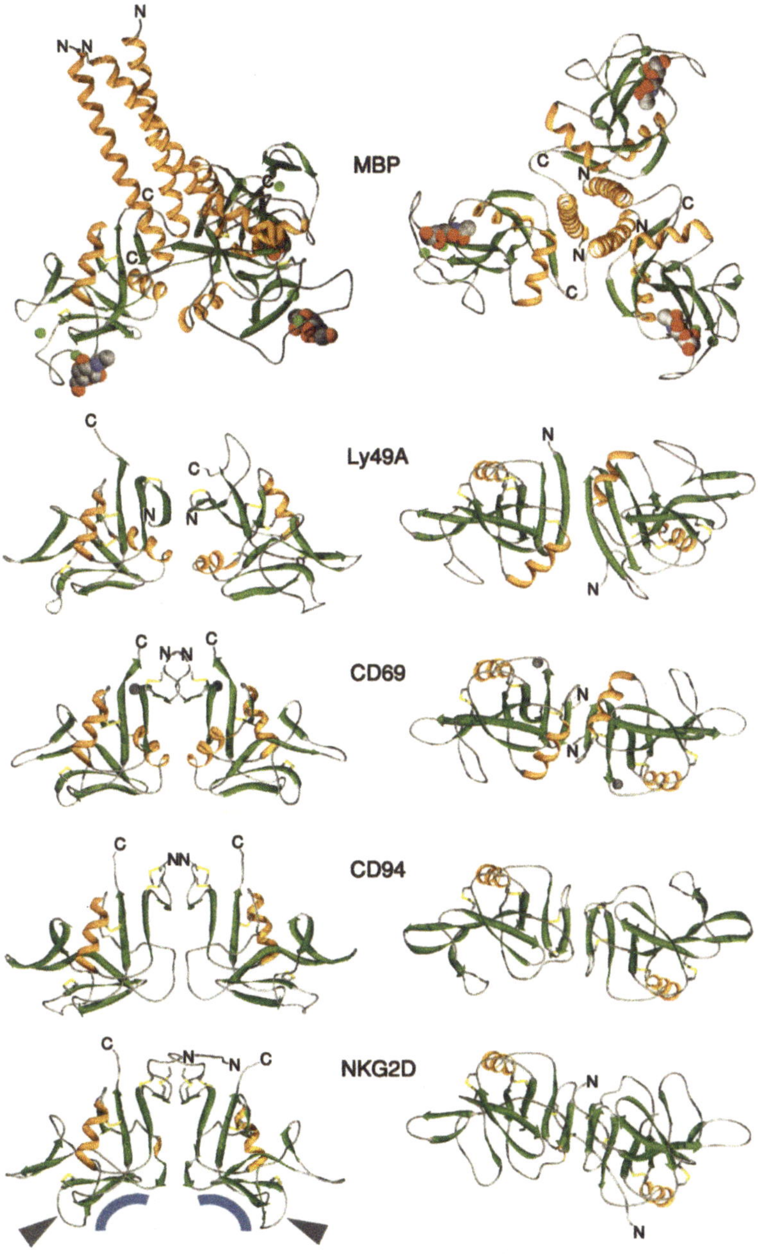

STRONG AND MCFARLAND, FIG. 2. (*continued*)

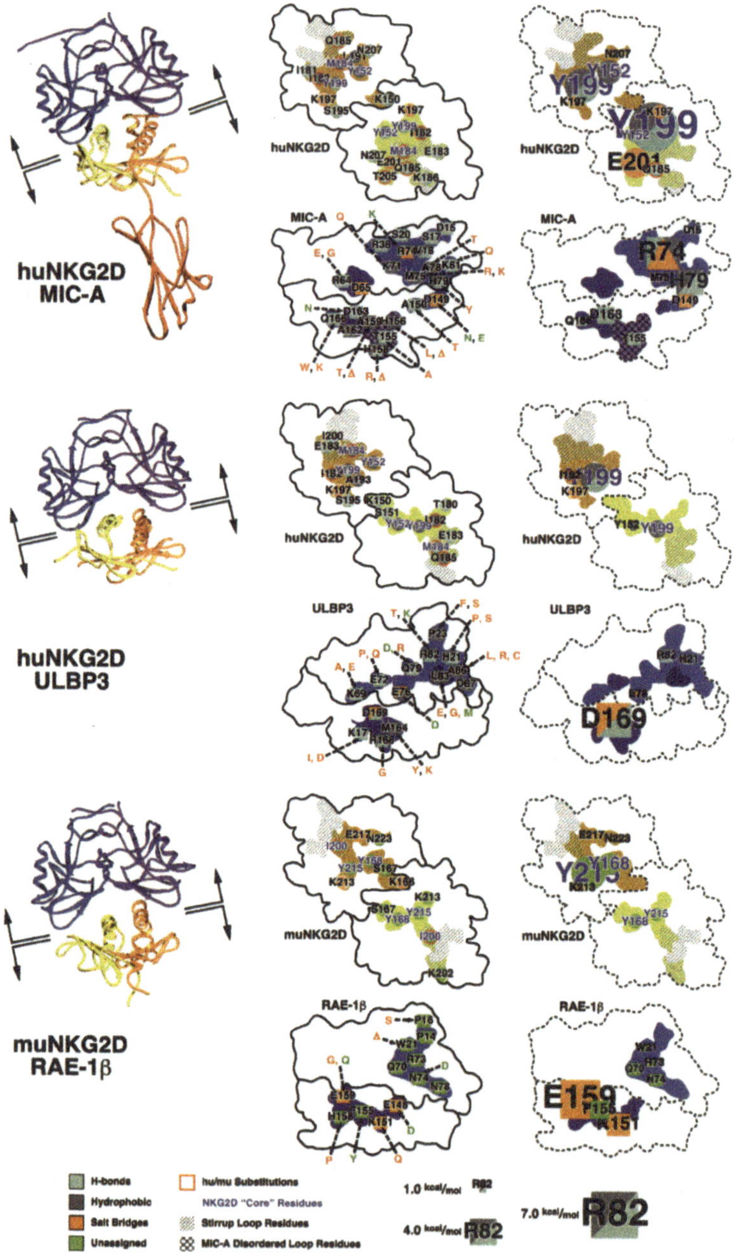

STRONG AND MCFARLAND, FIG. 7. (*continued*)

STRONG AND MCFARLAND, FIG. 7. Schematization of NKG2D–ligand interfaces. Residues making contacts in the NKG2D–ligand complexes are represented schematically on outlines of the separated proteins involved in the interaction: huNKG2D–MIC-A (top) (Li *et al.*, 2001a), huNKG2D–ULBP3 (middle) (Radaev *et al.*, 2001) and muNKG2D–RAE-1β (bottom) (Li *et al.*, 2002). On the left, the plane of the separation is indicated on ribbon representations of the complex structures, colored by domain as in Fig. 1. The interfaces are schematized by mapping the position of the contact residues (indicated by gnomons colored to reflect the nature of the bonds as indicated) onto outlines of the receptor (top) and the ligand (bottom). The overall footprint of one binding partner on the other is shown, colored by the domain making the footprint. Allelic and isoform substitutions affecting contact residues are indicated (conservative substitutions in green, non-conservative substitutions in red), as are portions of the receptors from stirrup loop residues (cross-hatched areas) and the portion of MIC-A from disordered loop residues (checkerboard area). The right-hand pair of interface schematics (dotted outlines) shows the same contact residue mapping, but with residue gnomons now scaled (as indicated) by calculated $\Delta\Delta G$ value (McFarland *et al.*, 2003); residues with $\Delta\Delta G$ values below the cut-off for the definition of binding "hotspots" (1 kcal/mole) are not shown. Gnomons with red outlines highlight NKG2D ligand contacting residues where sequence differences occur between murine and huNKG2D; NKG2D sequence positions that are ligand contacts in all six half-sites (NKG2D "core" residues) are indicated with blue labeling.

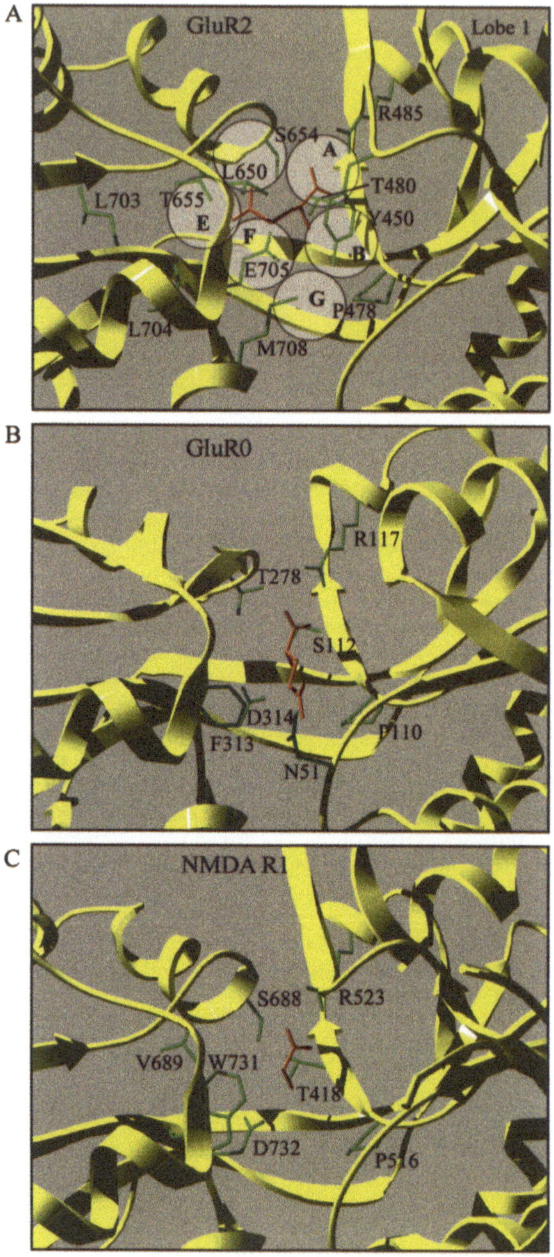

OSWALD, FIG. 2. The ligand binding sites of the S1S2 domain from (A) GluR2 (Armstrong and Gouaux, 2000), (B) GluR0 (Mayer *et al.*, 2001), and (C) NMDA R1 (Jin and Gouaux, 2003). The ligands (glutamate for GluR2 and GluR0 and glycine for NMDA R1) are shown in red. The relevant sidechains that make up the binding site are shown in green and are labeled. The subsites referred to in the text are shown as semitransparent circles.

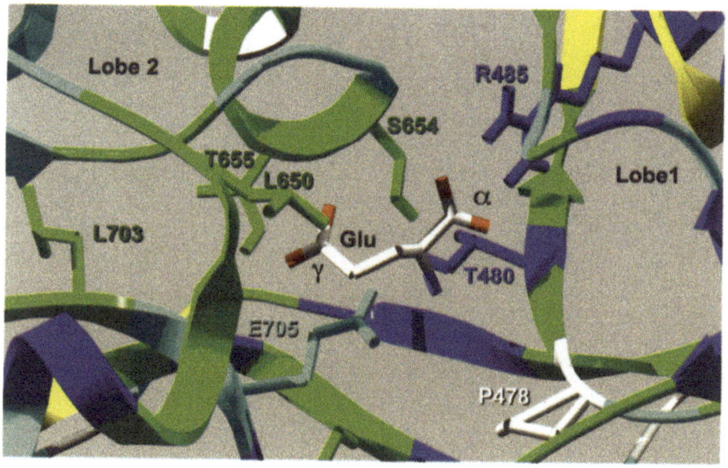

OSWALD, FIG. 3. The binding site of S1S2 GluR2 showing the amino acids important for interaction with glutamate with color coding according to the dynamics properties measured by NMR spectroscopy (McFeeters and Oswald, 2002). Blue and cyan represent residues that show only fast motions (ps-ns); whereas, residues colored yellow and green exhibit slower motions on the timescale of μs-ms.

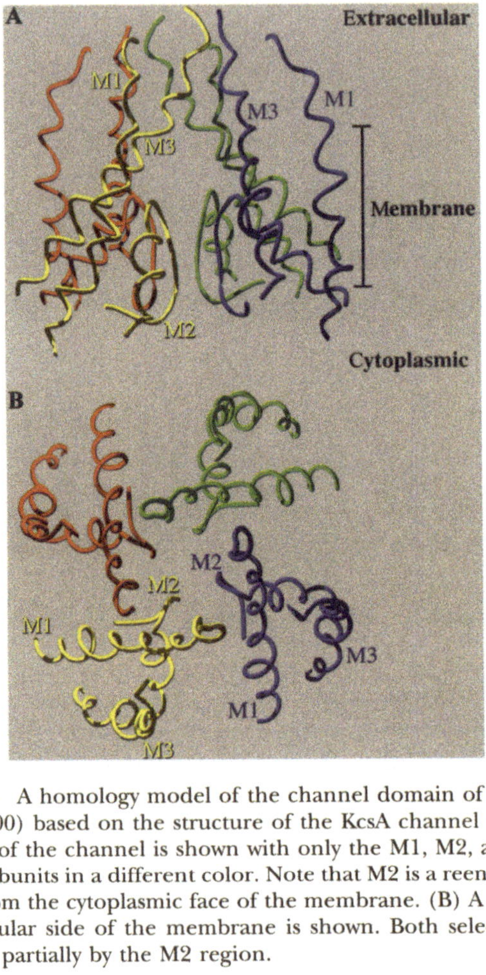

OSWALD, FIG. 4. A homology model of the channel domain of an NMDA receptor (Chohan *et al.*, 2000) based on the structure of the KcsA channel (Doyle *et al.*, 1998). (A) The side view of the channel is shown with only the M1, M2, and M3 regions and each of the four subunits in a different color. Note that M2 is a reentrant pore loop that enters the pore from the cytoplasmic face of the membrane. (B) A view of the channel from the extracellular side of the membrane is shown. Both selectivity and gating is controlled at least partially by the M2 region.

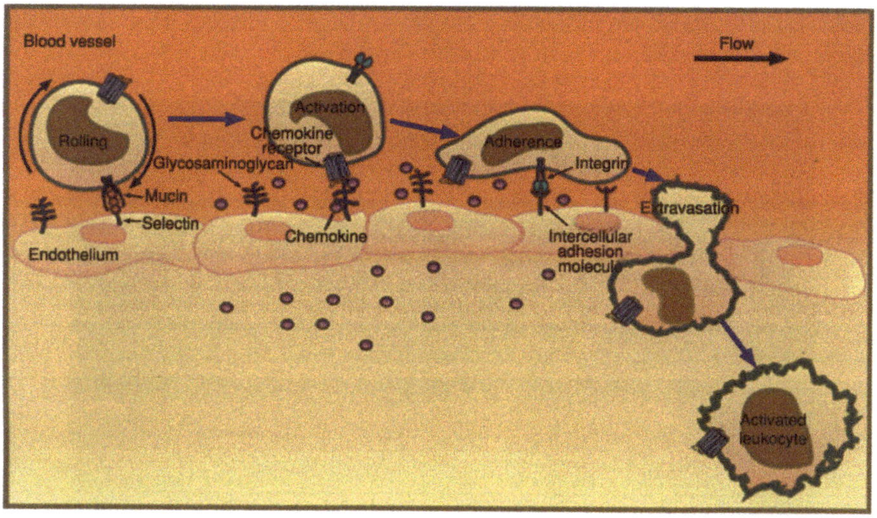

LAU *ET AL.*, FIG. 1. Illustration of the migration of a leukocyte from blood to extravascular tissue in response to chemokine production. Interaction of selectins on the endothelial surface with mucins on the leukocyte causes reversible tethering and a rolling behavior of the leukocyte. Chemokines are shown bound to the surface of the endothelial cell via interactions with glycosaminoglycans (GAGs); however, whether they simultaneously bind to both GAGs and chemokine receptors, as depicted in the figure, is unknown. In addition to GAGs, a promiscuous non-signaling chemokine binding protein (DARC) is thought to be involved in presentation of chemokines on endothelial cell surfaces. Tissue specific expression of these presentation molecules may play a role in the migratory patterns of receptor bearing cells. Subsequent binding of chemokines to chemokine receptors on leukocytes stimulates a wide variety of signaling pathways, the consequences of which include increased adhesiveness of the cells via activation of integrins, and cytoskeletal changes involved in cell movement. Migration toward subluminal chemokines occurs via one of two routes, either between the intercellular junction (shown here) or transcellularly through a pore in the endothelial cells.

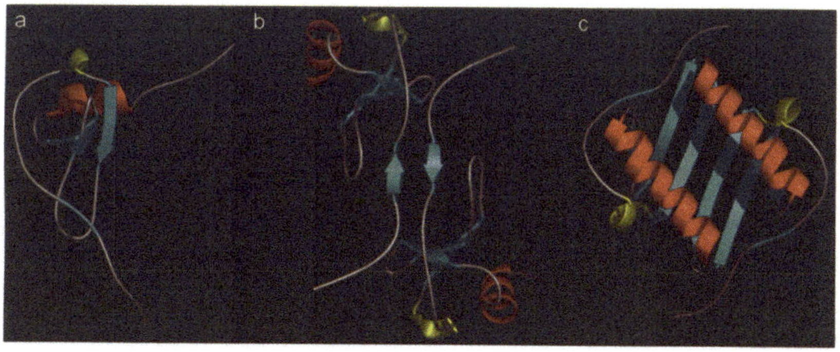

LAU *ET AL.*, FIG. 2. Ribbon diagram of (A) the MCP-1/CCL2 monomer, (B) the MCP-1/CCL2 dimer (from PDB code 1DOM), and (C) the IL-8/CXCL8 dimer (PDB code 1IL8). Figures were generated with Pymol (DeLano, 2002). The β-sheets are colored blue, the C-terminal helices are red, the 3_{10} helices are yellow, and loops, turns and the unstructured N-termini are white.

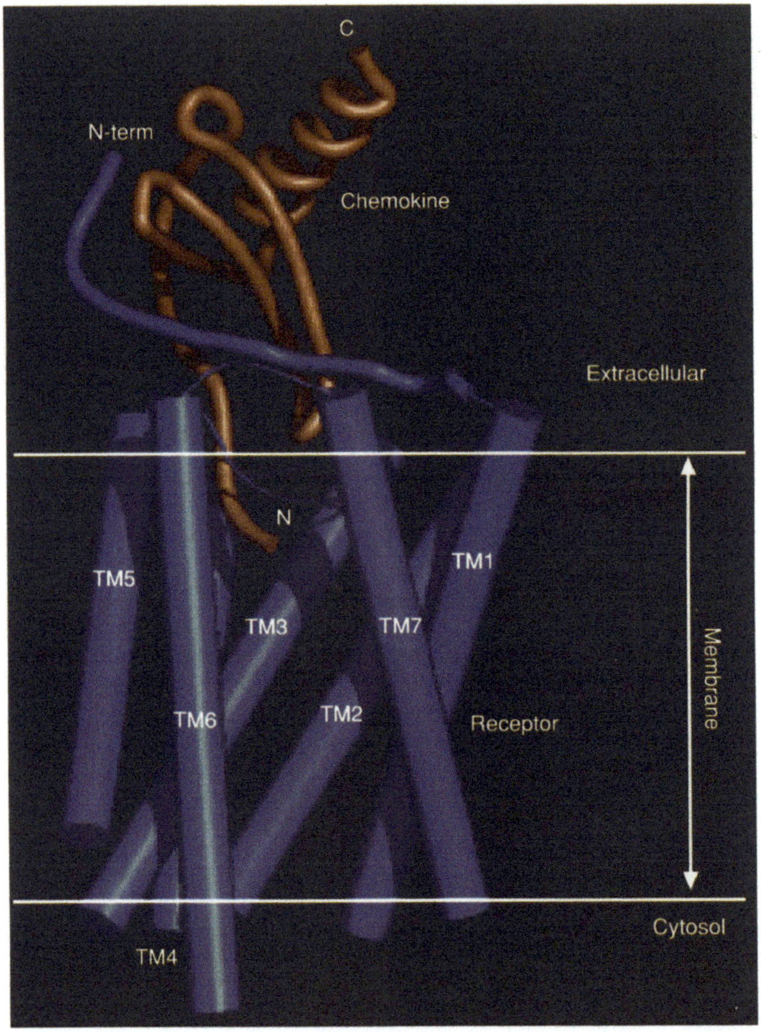

LAU *ET AL.*, FIG. 4. Schematic illustration of chemokine-receptor interactions. The transmembrane helices of the receptor, shown as blue tubes, were derived from the structure of rhodopsin (PDB code 1L9H). The chemokine (pink) and the N-terminal extracellular domain of the receptor were derived from the IL8-CXCR1 peptide complex shown in Fig. 5 (PDB code 1ILP). Except for the N-terminus, the loops of the receptor are not displayed. The figure illustrates the relative size of the receptor and ligand, and the interaction of the receptor N-terminus along one face of the chemokine. The interaction orients the N-terminal signal domain towards the receptor; as displayed here it is oriented towards the helical bundle, which may or may not be correct for some chemokines.

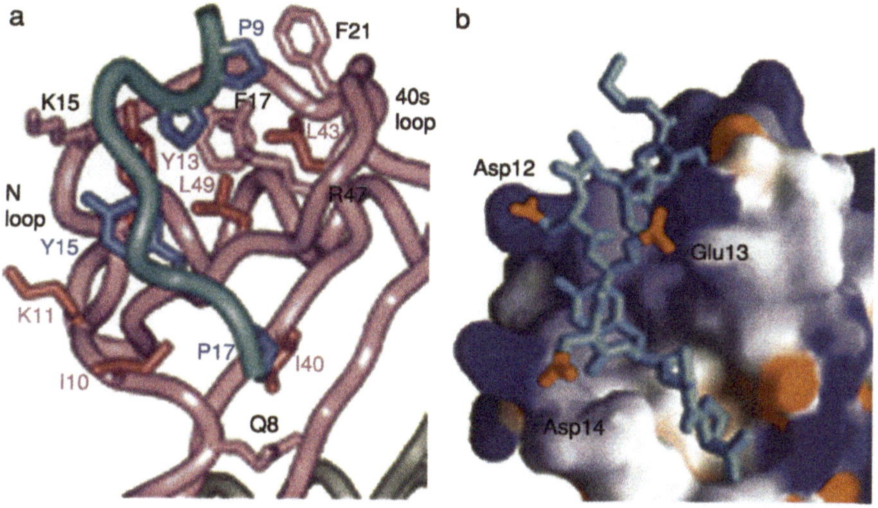

LAU *ET AL.*, FIG. 5. View of the structure of a modified peptide fragment from CXCR1 in complex with IL-8/CCL8. (A) The monomeric subunits of IL-8 are colored pink and grey, the peptide is cyan. The IL-8 residues surrounding the binding cleft and the peptide residues buried in the interface are shown as stick representations. (B) IL-8/CXCL8 is colored according to its electrostatic potential. The peptide is cyan with the terminal atoms of three acidic residues colored red. Note the interactions of the peptide within the hydrophobic pocket of IL-8 and those between the peptide acidic groups and the basic residues of IL-8/CXCL8. Whether or not Y15 is tyrosine sulfated is unclear, but Y15 is imbedded in a canonical tyrosine sulfation motif. The orientation is the same as in "a". Reprinted from *Structure* with copyright permission from Elsevier: "Skelton, N. J., Quan, C., Reilly, D., and Lowman, H. (1999). Structure of a CXC chemokine receptor fragment in complex with interleukin-8. *Structure* **7,** 157–168."

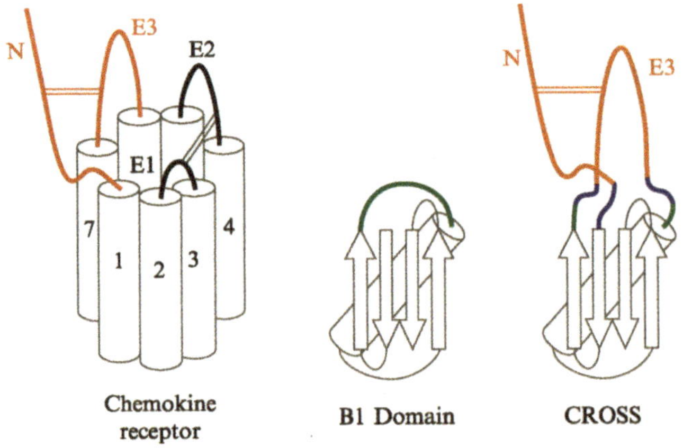

LAU *ET AL.*, FIG. 6. Schematic representation of the CROSS protein design. Left. Predicted topology of chemokine receptors. Membrane spanning helices are shown as cylinders, the extracellular loops are displayed as lines, and the disulfide bonds between the extracellular domains are double lines. Center. The B1 domain of protein G (PDB code 2GB1). β-strands are shown as arrows and the helix as a cylinder. The loop to be replaced by extracellular loop three (E3) of CCR3 is highlighted in green. Right. CROSS protein showing the chemokine receptor elements (red) displayed on the B1 scaffold. The N-terminus is attached via a $(Gly)_3$ linker (blue). The E3 loop from the chemokine receptor (red) is inserted between 2nd β-strand and the α-helix with a $(Gly)_2$ linker at each junction. Reprinted from *Protein Science* with copyright permission from Cold Spring Harbor Laboratory Press: "Datta, A., and Stone, M. (2003) Soluble mimics of a chemokine receptor: Chemokine binding by receptor elements juxtaposed on a soluble scaffold. *Protein Sci.* **12,** 2482–2491."

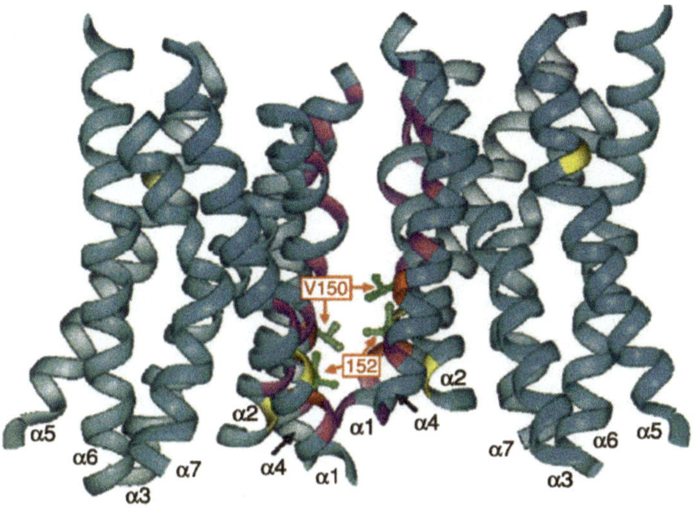

LAU *ET AL.*, FIG. 7. Structural model of CCR5 dimer association presented as a ribbon representation. TM1, TM2, and TM4 (labeled $\alpha 1$, $\alpha 2$ and $\alpha 4$) participate directly in the interaction surface. The positions of the mutated residues, V150 and 152, are indicated on each of the subunits. Reprinted from *Nature Immunology*, with copyright permission from the Nature Publishing Group: "Hernanz, P., *et al.* (2004). Identification of amino acid residues crucial for chemokine receptor dimerization. *Nature Immunology* **5,** 216–223."

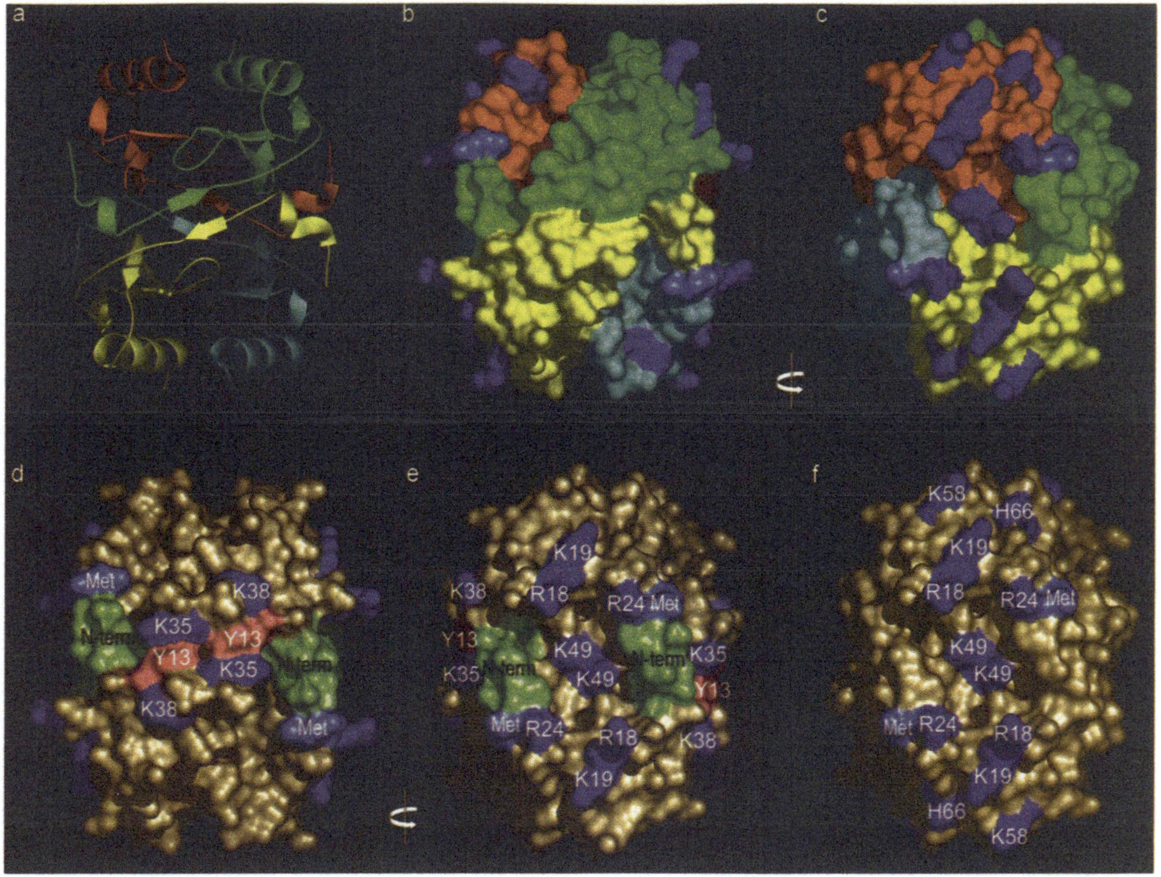
a
b
c
d
Met
K38
K35
N-term
Y13
Y13
K35
N-term
K38
Met
e
K19
K38
R18
R24 Met
Y13
K49
N-term
K35
K35
N-term
K49
Y13
Met
R24
R18
K38
K19
f
K58
H66
K19
R18
R24 Met
K49
K49
Met
R24
R18
K19
H66
K58

LAU *ET AL.*, FIG. 8. Structural views of the MCP-1/CCL2 tetramer (PDB code 1DOL) with each subunit color-coded red, green yellow, or blue, and functional epitopes highlighted as described: (A) Ribbon diagram showing the CC dimer subunits in front and back (green/yellow and red/blue). Green/red and yellow/blue subunits interact through a CXC-like motif at the top and bottom of the structure. (B) Surface topology model of the MCP-1/CCL2 tetramer in the same orientation as "a". GAG binding epitopes are colored in purple. (C) Same as "b", but rotated 90° to show the unique interface involving all four subunits. The GAG binding epitopes form a continuous basic strip when the tetramer is formed. (D) Same orientation as "a" and "b" but with the receptor (CCR2) binding epitopes highlighted. The N-terminal signal residues are more protected, and Y13, also required for chemotaxis, is largely buried, compared to the monomer structure of Figure 3a. (E) Same orientation as "c", but with the receptor binding epitopes highlighted along the interface involving all four subunits for comparison to the GAG binding site in "f". (F) Same orientation as "e", but with GAG binding residues highlighted. Note there is considerable overlap between the GAG and receptor binding sites. Thus in the context of a presentation function, oligomerization could enable simultaneous interaction with GAGs and the receptor. The burial of the signaling triggers could reflect an inhibitory function of GAG binding as has been observed *in vitro*, or protection from proteolysis.

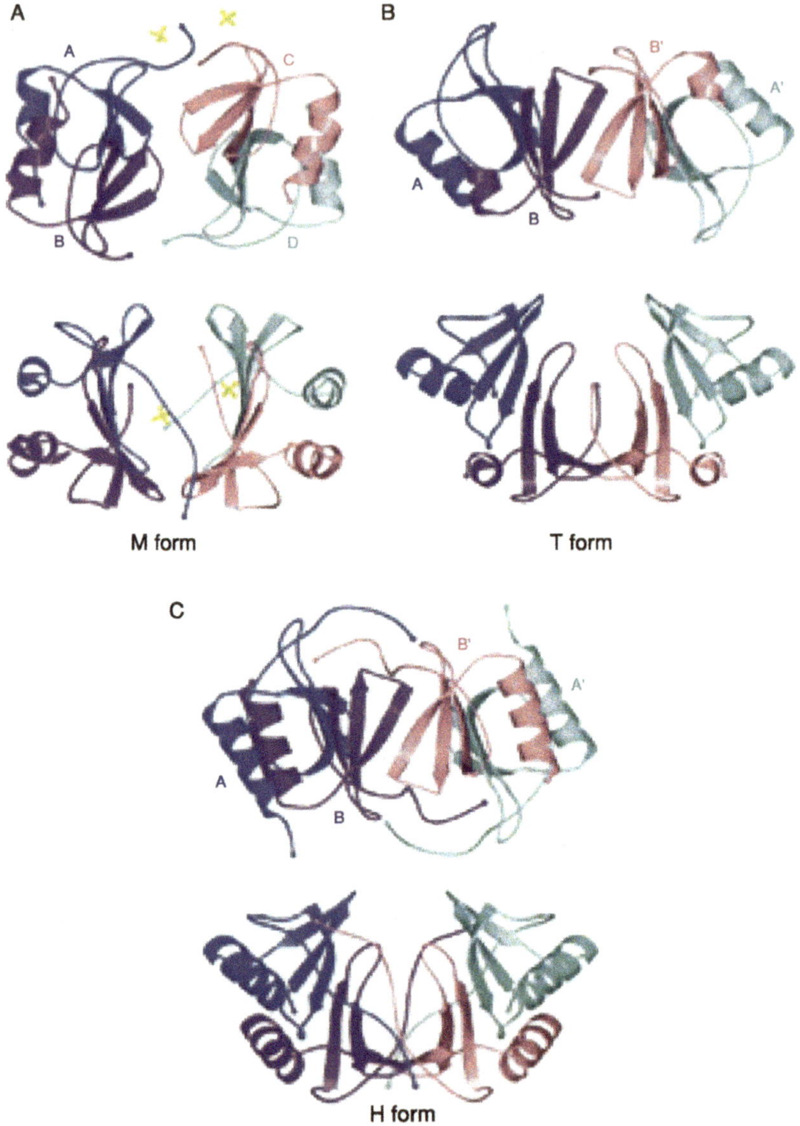

LAU *ET AL.*, FIG. 9. Ribbon structures of the three IP-10/CXCL10 tetramers. Two orthogonal views are shown for each. The M-form is similar to the PF4/CXCL8 and MCP-1/CCL2 structures that contain elements of CXC and CC dimers. The T and H forms contain a novel 12-stranded β-sheet. Reprinted from *Structure* with copyright permission from Elsevier. "Swaminathan, G. *et al.* (2003). Crystal structures of oligomeric forms of the IP-10/CXCL10 chemokine. *Structure* **11,** 521–532."

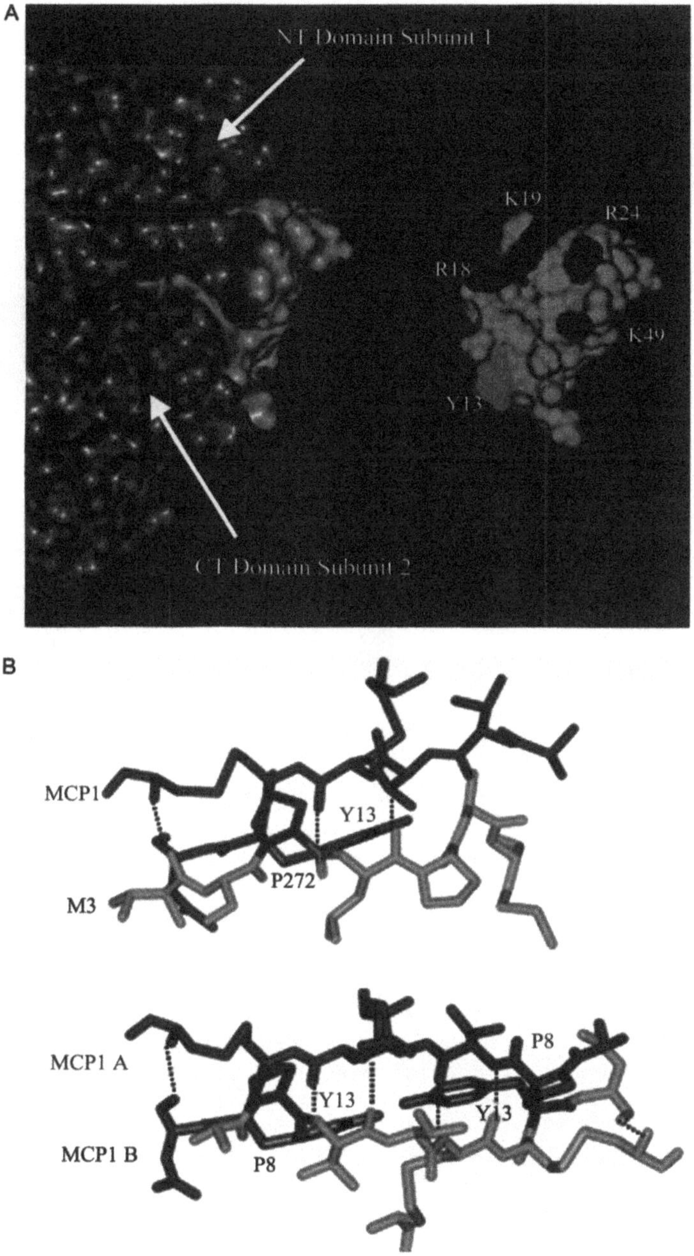

LAU *ET AL.*, FIG. 10. (*continued*)

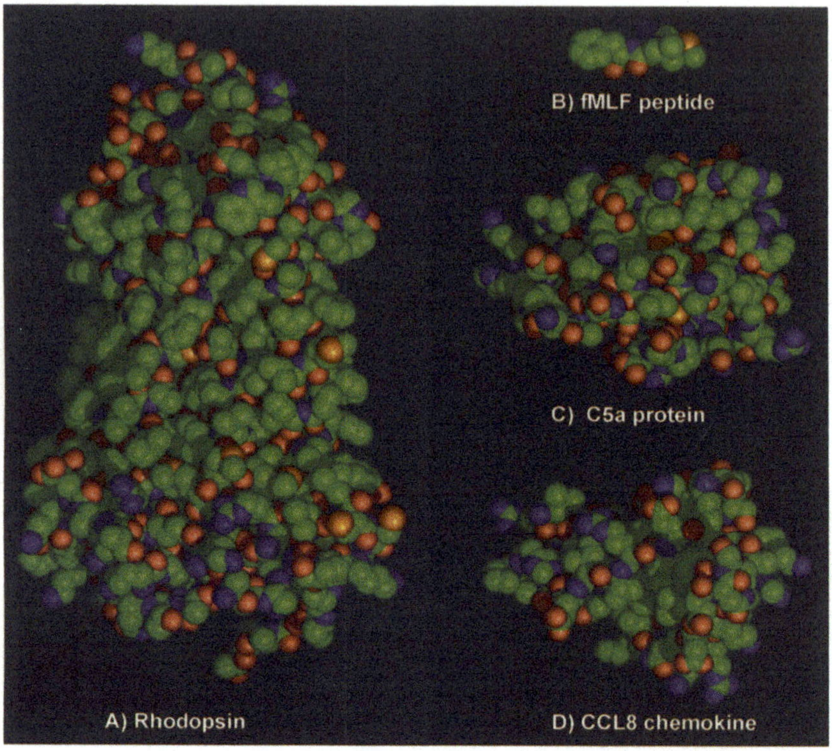

MILLER AND FALKE, FIG. 1. Space-filling models illustrating the relative sizes of a G protein-coupled receptor and three chemoattractant ligands. Shown are (A) rhodopsin, with its G protein-coupling surface oriented downward (Palczewski *et al.*, 2000), (B) the chemoattractant peptide formyl-Met-Leu-Phe, (C) the chemoattractant protein C5a (Zhang *et al.*, 1997), and (D) the chemoattractant chemokine CCL8 (or interleukin 8) (Baldwin *et al.*, 1991).

LAU *ET AL.*, FIG. 10. (A) Surface topology model of γ-herpesvirus M3 in complex with human MCP-1/CCL2. Only ~1/2 of the structure is displayed, focusing in on one of the chemokine binding clefts. M3 is colored purple, MCP-1/CCL2 is cyan with CCR2 receptor binding epitopes, defined by mutagenesis, highlighted in green (Y13) and blue (basic Lys of Arg residues). The N-terminal domain of M3 interacts with the basic patch at the top of MCP-1, as well as the adjacent hydrophobic region immediately below (see also Fig. 3a). The C-terminal domain of M3 pairs with the dimer interface of one subunit of MCP-1/CCL2, forming a β-strand similar to that observed in the MCP-1 homodimer. Note Pro272 of M3; it occupies a structurally homologous orientation as the Pro8 in the MCP-1 dimer that is required for homodimerization. (B) (bottom) Stick model of the dimer interface in MCP-1. One subunit is pink with P8 and Y13 sidechains highlighted in red. The other subunit is cyan with P8 and Y13 sidechains highlighted in dark blue. (top) M3:MCP-1 interactions in the C-terminal domain. MCP-1 is pink with Y13 highlighted in red. M3 is cyan with P272 highlighted in dark blue.